Spokes of the Wheel

Book 1:

The Science of Existence

Ishi Nobu

Spokes of the Wheel, Book 1: The Science of Existence

by Ishi Nobu

ISBN 978-1-948627-01-6

Produced in the United States of America.

2nd edition.

Table of Contents

❧ Prologue ❧

This book begins *Spokes of the Wheel. Spokes* explores existence: the wild cosmos, the untamed mind, and the social artifice which humanity has created.

Clarity: The Path Inside

Spokes Θ: *Unraveling Reality*

Spokes 1: *The Science of Existence*

Spokes 2: *The Web of Life*

Spokes 3: *The Elements of Evolution*

Spokes 4: *The Ecology of Humans*

Spokes 5: *The Echoes of the Mind*

Spokes 6: *The Fruits of Civilization*

Spokes 7: *The Pathos of Politics*

Spokes 8: *The Hub of Being*

Clarity explains reality and how to become enlightened. *Unraveling Reality* is an introduction to *Spokes*, touching upon major themes. *The Science of Existence* roams the universe and introduces the natural world. *The Web of Life* chronicles the wondrous diversity of life. *The Elements of Evolution* relates life's history and explains how organisms adapt. *The Ecology of Humans* assays the interfaces of the human body. *The Echoes of the Mind* pivots on how people feel, think, and behave. *The Fruits of Civilization* covers the consequences of human endeavor. *The Pathos of Politics* probes how polity has plagued humanity. *Spokes* culminates in *The Hub of Being*, an exposition toward realization.

❀ ❀ ❀

✄ Headings ✂

Spokes is heavily portioned, with symbolic conventions to delineate the hierarchy of material.

❀❀❀ delineates untitled segments, or a return to a subject after wandering off.

¤✧¤ demarcates subjects within a topical section.

Each chapter concludes with a synopsis. At the end of the book is a conclusion. In between are a lot of numbers that are often approximate; measurement is tricky.

For a fuller sense of terms that seem scantily defined in the text, please consult the glossary. There is a biographical section after the glossary.

¤✧¤

For more information about *Spokes of the Wheel*, visit: *ishinobu.com*. Research references for *Spokes 1* are at: *ishinobu.com/spokes-1/notes/*.

Real understanding of any scientific subject must include some knowledge of its historical growth; we cannot comprehend and accept modern concepts and theories without knowing something of their origins – of how we have got where we are. Neglect of this maxim can lead to that unfortunate state of mind which regard the science of the day as finality. ~ English cognitive scientist Colin Cherry

❧ The Universe ❧

One is nothing but an instrument on which the universe plays. ~ German composer Gustav Mahler

Earth is tucked into an infinitesimal spot on an arm of the Milky Way galaxy. The Milky Way is one of 2 trillion galaxies in the observable universe, which is spread out spherically, with a diameter of 90 billion light-years.[*]

To explain how our existence came to be, this instrument begins its play at a beginning....

❧ Beginning ❧

Universes are like petals of a flower: they unfold only when conditions are favorable.

3,800 years ago (YA), the Babylonians conceived a plurality of heavens and earths. 2,500 YA, Chinese philosopher Lao Tzu held that the universe originated out of nothingness.

The ancient Greeks thought existence eternal, comprising infinite space; so did a young Einstein. In the 6th century BCE, Anaximander of Miletus conceived the cosmos in a perpetual cycle of incarnation and reincarnation, powered by *apeiron*: an eternal coherence.

The primal essence of the existing objects is also the fact that when they perish, they return as dictated by necessity. ~ Anaximander

We do not know how or when the universe began. Astrophysicists have speculated by extrapolating backward, with guesswork about whether and how the dynamics of fundamental forces changed. Hence the following tale is one of speculative backfill.

The extant understanding of physics colors the picture: rendering a sketch rather than a portrait, as the yardstick for characterization of high-energy events on the periphery

---

[*] A light-year is how far light travels in a year at light-speed (as fast as light can travel). A light-year is ~9.461 trillion kilometers.

of quantum reality has smudges on the measurement lines; for good reason.

ഔ The Big Bang ക

What was God doing before *he made heaven and Earth? He was preparing hell... for those who pry too deep. ~ Latin theologian Augustine of Hippo (354–430)*

The prevailing cosmological model posits the universe explosively coming out of nowhere to create everywhere. The idea has been around at least since the 13th century.

⚸ History ଢ

Inspired by the recently rediscovered works of Aristotle, English scholastic philosopher, theologian, and scientist Robert Grosseteste wrote *De luce* ("The Metaphysics of Light") in 1225. In the book, he proposed that the universe expanded from a pinpoint of light. Grosseteste assumed that light and matter were somehow entangled.

In the 1920s, astronomers discovered that distant galaxies are moving away, indicating that space itself is expanding. This implied that, at some point in the past, the contents of the observable universe had been a hot, dense primordial fomentation.

All the matter in the universe was created in one big bang *at a particular time in the remote past. ~ Fred Hoyle in 1949 on a "hypothesis in conflict with the observational requirements"*

The term *Big Bang* was coined by English astronomer Fred Hoyle in a 1949 radio broadcast. Hoyle was no fan of the Big Bang. He instead favored the ancient Greek paradigm of a steady-state cosmos, where the universe eternally existed, but continuously accreted new matter as it expanded. That there was no evidence of this worried Hoyle not a whit.

German theoretical physicist Albert Einstein was disturbed by the prospect of the universe starting with an explosive singularity. By 1931 he had a model of a stable cosmos, but it held a fatal flaw: the universe had to be at least 10 billion years old. Einstein found that "unacceptable," as the cosmos could not possibly be that old.

Einstein abandoned his bias as new cosmological observations indicated the universe was not as static as he had hoped. Unconvinced, Hoyle and others took up the cause of steady-state a decade later.

The Big Bang theory was the 1931 brainchild of Monsignor Georges Lemaître, a Belgian Roman Catholic priest and astrophysicist.

> If the world has begun with a single quantum, the notions of space and time would altogether fail to have any meaning at the beginning; they would only begin to have a sensible meaning when the original quantum had been divided into a sufficient number of quanta. If this suggestion is correct, the beginning of the world happened a little before the beginning of space and time. ~ Georges Lemaître

According to Hoyle, the Big Bang imported religion into physics, by dint of it being proposed by a priest. The irony of that objection went unappreciated by Big Bang objectors. As it turned out, steady state adherents were the believers in a false religion.

Owing to pervasive noise, the Big Bang won out. The 1964 discovery of the cosmic microwave background (CMB) radiation by American astronomers Arno Penzias and Robert Wilson secured Big Bang as the most acceptable explanation of the origin and evolution of the universe. Lingering radiative scattershot suggested that, literally out of nowhere, a hellacious firecracker went off to start it all. This interpretation is wrong, as is the conventional construal of how and when the universe began.

৪ Story ৼ

> In the beginning there was nothing, which exploded. ~ English novelist Terry Pratchett

The conventional fiction is that 13.82 billion years ago (BYA), a quantum pinprick of infinite intensity exploded into existence. There were no separate physical forces as experienced today; no matter; just a singularity of energy: a universe's worth, taking no space to speak of, but pervading more dimensions than humans would ever perceive.

The "big bang" hypothesis of creation is more a confession of desperation and bewilderment than the outcome of logical argumentation rooted in the known (or even unknown!) laws of physics. ~ Belgian theoretical physicist Robert Brout *et al*

We do not know when the universe began, or how. The conventional 13.82 BYA origination date comes from the earliest light detected by a space telescope, in accordance with Robert Grosseteste's 1225 surmise.

The cosmic microwave background is the light that is the furthest away from us that we can see. ~ American astrophysicist Adam Riess

The bruited Big Bang was actually a quiet affair. No sound was made. But the misnomer does make for a catchy cosmological slogan.

Embedded within the embryonic universe were the ingredients for all that would ever be. The cosmos started as a viscous fluid of subatomic proto-matter. The earliest perturbations in energy density and flow rippled through time, defining the topology of the universe.

According to the Big Bang story, 1 trillionth (10^{-12}) of a second after the cosmos came into being, energy had spread sufficiently that electromagnetism and the weak nuclear force parted ways. Photons emerged. Other subatomic particles came into being.

By its spin, matter claimed the throne of everyday existence. Matter's mirror – antimatter – was largely shuttered from the observable 4 dimensions; exactly how remains unknown.

A universe was taking form. The cosmos was quark soup for a microsecond. Then came protons and neutrons, as quarks settled into being part and parcel of larger particles. This was followed by the formation of atoms. The time frame in which these events occurred remains highly speculative.

The standard hot cosmological model requires rather unnatural initial conditions at the big bang. One has to postulate that the universe has started in a homogeneous and isotropic state with tiny density fluctuations which are to evolve into galaxies. Homogeneity and isotropy must extend to scales far exceeding the causal horizon at the Planck time. In addition, the energy density of the universe must be tuned to be near the critical density with an incredible accuracy of $\sim 10^{-55}$. ~ Ukrainian American physicist Alexander Vilenkin

℘ Conceptual Expansion ♉

In 1912, American astronomer Vesto Slipher claimed that there were galaxies outside our own, having stars too far away to belong to the Milky Way. This radical notion was opposed by many in the astronomy establishment; a universe with more than 1 galaxy was ridiculous. In time, Slipher's conviction prevailed, as further work convinced astronomers and physicists that the universe was expanding, having grown far astray of being just the Milky Way.

American astronomer Edwin Hubble is often wrongly credited with the discovery of distant galaxies. Hubble was at Slipher's 1912 lecture and carried the conviction publicly.

Hubble's law, which characterizes the Doppler shift of receding galaxies, was based upon Slipher's data. The law was first derived by Georges Lemaître. Hubble confirmed the law, determined a more accurate measurement, and took credit.

Armed with an estimated size of the present universe, and a bogus guess about how long ago the cosmos debuted, cosmologists struggled to explain how the universe got from its supposed moment of inception to where it is today. There was an unfathomed gap between how small the universe supposedly started off as and how big it seems to be now; so the yawning gulf was filled with expansive stuffing.

℘ Cosmic Inflation ℞

It is said that there's no such thing as a free lunch. But the universe is the ultimate free lunch. ~ Alan Guth

According to a conjecture called *cosmic inflation*, 10^{-36} seconds after the Big Bang, the size of the universe mushroomed from next to nothing to the size of a dime: a 10^{78} size expansion or more in 3×10^{-36} seconds; faster than the speed of light. Apologists assert that cosmic inflation does not violate relativity theory because spacetime itself is expanding faster than the speed of light during that instant: absurd.

Cosmic inflation posits a supersizing of eye-watering magnitude in less than the blink of an eye, back when there were no eyes to blink. That's quite a magical moment.

After a miraculous instant of cosmic inflation, the universe proceeded to expand at a leisurely pace, as it does now.

¤ ✧ ¤

American cosmologist Alan Guth proposed cosmic inflation in 1980 to resolve issues with the way the universe is today: stable, flat, and livable; a probabilistically remote outcome given the projected conditions of the early universe.

Cosmic inflation was a hypothesis shoehorned to fit a few facts; but more a conceptual bandage to cover the gap between the universe's fabricated origination date and its current state. Unaddressed questions strike at its core plausibility. What caused inflation to start? How did it work? What caused it to stop? The balloon of cosmic inflation is deflated by answering none of those questions.

> The inflationary paradigm is fundamentally untestable, and hence scientifically meaningless. ~ American theoretical physicist Paul Steinhardt

To kick in, cosmic inflation required a special ingredient called *inflationary energy*, which combined with gravity to blow the universe up in a brief instant.

Cosmic inflators posit an imaginary particle – the *inflaton* – as invoking cosmic inflation. After pulling off its miraculous job, inflatons supposedly disappeared without a trace. Apologists wave this away by assuming that inflatons decayed into other particles as the universe matured, without proposing how this transition was achieved.

There is no evidence for inflatons, which were supposedly a *scalar field*. No fundamental scalar fields have been observed in Nature. Inflation advocates do not address this.

To work, inflationary energy had 2 stringent requirements. Both are unlikely if not impossible.

1st, inflationary energy had to have been incredibly dense, and its density constant, except for nominal random quantum fluctuations. Yet this assumption of steadiness in quantum effects is inherently shaky. Quantum effects dominated the energy flow at the supposed moment of cosmic inflation. There would have been nothing nominal about them.

Inflation was supposed to create a huge volume of space matching the observed large-scale features of our universe naturally. But unless the inflation energy curve had a very specific shape, the outcome would be "bad" – a huge volume with too high a density and the wrong distribution of galaxies. Given the range of possible values, bad inflation seems more likely. ~ Paul Steinhardt

2nd, inflationary energy requires that gravity worked in reverse: repelling rather than attracting. No physics theory supports this. Gravity as an expansive pressure by itself discredits the inflation speculation.

If gravity in reverse isn't hard enough to swallow, consider that, after an interval of just 3×10^{-33} seconds, inflation jerks to a halt, with the universe continuing to expand at a leisurely pace. Something had to have counteracted inflation, otherwise the universe would have been a quickly bursting bubble. But the inflation hypothesis has nothing to say of that. Instead, unless arbitrarily throttled, the math of inflation theory predicts that inflation never stops. Inflation is supposedly still sprouting an infinite multiverse by blowing up bits of spacetime.[*]

---

[*] This speculation about *parallel universes* emanates from the rancid math behind the cosmic inflation conjecture. The infinities that arise in quantum mechanics equations are sometimes similarly assuaged into a *many-worlds* interpretation. That idea originated with Erwin Schrödinger in a 1952 lecture, where he speculated about what his famous 1925 wave-particle duality equation might mean. The off-hand concept gained currency in the wake of the cosmic inflation.

The basic idea behind a many-worlds scenario is that otherwise useful equations which spout infinities mystically suggest the proliferation of empirical universes much like our own (or maybe not so much like our own). There are several variations on this theme, but all seem silly in the context in which they are presented: to address gaping, inexplicable mathematical holes.

Besides lacking evidence, cosmic inflation presumes continuous infinities at every scale of existence. Cosmic inflation is both a physical and mathematical absurdity.

> The part of the multiverse that we observe corresponds to a piece of just one such bubble. Scanning over all possible bubbles in the multiverse, everything that can physically happen does happen an infinite number of times. No experiment can rule out a theory that allows for all possible outcomes. Hence, the paradigm of inflation is unfalsifiable. ~ Paul Steinhardt

Another deflation for inflation is its requirement for density fluctuations in wavelengths at less than *Planck length*: the point at which space becomes so small as to be meaningless.

> The calculations are extrapolations into regions where we cannot trust them. ~ Canadian theoretical cosmologist and physicist Robert Brandengerger

Credence for cosmic inflation not only lacks evidence – the evidence indicates otherwise.

Cosmic microwave background (CMB) is a radiation pattern that originated ~378,000 years after the supposed Big Bang. Cosmic inflation supporters point to the incredible growth instant as inherently creating a near-homogenous CMB. This is incomprehensible, as whatever energetic irregularities existed prior to cosmic inflation should have been amplified by inflation, rather than smoothed.

Polarization patterns to the CMB were reported in 2014 which inflationists inscrutably interpreted as gravitational

---

Several prominent physicists now tout some multiverse rendition. Others dismiss the notion as purely metaphysical, for being beyond investigation.

Infinities in equations which supposedly represent reality simply mean that the mathematical expression, however proximately useful, is fundamentally amiss. But physicists are reluctant to part with an eminently handy model, especially when nothing better is at hand.

It seems likely that there are other universes, if only because ours exhibits aging. Our cosmos having an origination point suggests it came from some unknowable somewhere rather than out of nowhere. Existence as eternal seems a safe bet.

waves, based upon mathematical assumption and nothing else.

> We believe that gravitational waves could be the only way to introduce this B-mode pattern. ~ American cosmologist John Kovac

The work supporting this conclusion was shoddy: only a single frequency was measured, and that data was glommed onto an unreleased rough image of another survey. Multiple frequencies would be necessary to produce a credible assessment, and is the norm in such surveys.

Further, the report disregarded that cosmic dust could have mimicked the supposed signal. Yet that did not stop gullible partisans from overblown ecstasy.

> These results are a smoking gun for inflation. ~ Israeli American astrophysicist Avi Loeb, chair of the Harvard astronomy department, in 2014; jumping the gun to declare his cosmological religion vindicated

Careful review found the data worthless. Even if the data had been correct, the gravitational waves would have been opposite of what the cosmic inflation hypothesis predicts: getting weaker with scale, rather than stronger.

Einstein's general relativity theory predicted gravitational waves. In 2016, such waves were inferred in the wake of merging black holes. Their nature contradicts the inflation hypothesis. In short, cosmic inflation contradicts relativity theories which have been repeatedly confirmed.

The CMB is somewhat supportive of the Big Bang hypothesis, but utterly out of tune with cosmic inflation. Beyond altogether failing to account for cosmic inflation, the CMB damns the idea with its asymmetry.

> We live in a lopsided universe. ~ American science writer Ron Cowen

The temperature of CMB fluctuates more on one side of the sky, suggesting a curvature in space. This indicates that the universe, long presumed flat, is slightly curved – similar to a saddle. A curved universe knocks inflation out, as the asymmetry cannot be unaccounted for.

As with a sound wave, the CMB fluctuations can be analyzed by splitting them into their component harmonics – like a collection of pure tones of different frequencies or, more picturesquely, different instruments in an orchestra. Certain of those harmonics are playing more quietly than they should be.

In addition, the harmonics are aligned in strange ways – they are playing the wrong tune. These bum notes mean that the otherwise very successful standard model of cosmology is flawed. ~ astrophysicists Glenn Starkman & Dominik Schwarz

The 2012 discovery of the Higgs boson at ~125 gigaelectronvolts (GeV) puts another nail in cosmic inflation's coffin. The Higgs field supposedly gives matter its mass.

At the Higgs' inferred voltage, there simply would not be enough energy to inflate the universe as cosmic inflation claims. If somehow such energetic inflation was able to take place, cosmic stability via the Higgs field would have decayed, wiping out the universe.

What inflation predicted was actually the reverse of what we found. ~ Australian cosmologist David Parkinson

Besides disregarding that radiative energy does not begat gravity, the inflationary model does not take into account that any gravitational expansion would have distorted time as well as space. Gravity distorts *spacetime*, not just space.

Nor do inflationists consider that extra-dimensional (ED) dynamics may have been especially vigorous during early cosmic development, thus putting the 4D universe on a path that culminated in its current configuration. This is ironic, as the inflationary model assumes that the physics in the first few moments of the cosmos were much different than those that predominate now. If anything, the cosmic inflation conjecture is simplistic in supposing only 4 dimensions when general relativity showed there were more.

An abiding problem in cosmogony is that the earliest moments of the universe cannot be explained without an overarching physics' theory of everything. Cosmic inflation goes way beyond that. In requiring sudden spacetime disjuncture, cosmic inflation lacks any foundation in known physics.

So why the longevity of a such an absurdity? Astrophysicists like the inflation equations developed in the early 1980s because they correspond well with observations about the current cosmos and are relatively easy to work with. Despite the conundrums and contradictory evidence, the supernatural mechanism of cosmic inflation is generally accepted.

Cosmic inflation is called for solely because of the assumed Big Bang date. If instead the cosmos is much older, as it must be, no such mysticism need be conjured.

℘ Origination Date ℘

The idea of the big bang comes from a simple observed fact: galaxies in the universe are moving apart. If you play this trend back in time, galaxies (or their precursors) must have been all scrunched up 13.7 billion years ago. In fact, according to Einstein's general theory of relativity, they were scrunched into a single point of infinite density – the big bang singularity. But an infinite density is unrealistic: that relativity theory predicts it is a sign that the theory is incomplete. Without a singularity to demarcate the beginning of time, the history of the universe may extend further back. ~ German physicist Martin Bojowald

The existence of an initial singularity is disturbing: a singularity can be naturally considered as a source of lawlessness, because the spacetime description breaks down "there," and physical laws presuppose spacetime. ~ Brazilian astrophysicists Mario Novello & Perez Bergliaffa

There is compelling evidence for a cosmos exceeding the conventional longevity estimate of 13.82 billion years. There were thousands of galaxies that were ~3,000 light-years in diameter 13 BYA. Such extensive galactic formation cannot be accounted for within a billion years of when the universe supposedly started. Even more inexplicable are black holes.

Black holes at the centres of galaxies reach masses of over 10 billion times that of our Sun. Surprisingly, there were such massive black holes in the early Universe, just 800 million years after the Big Bang. How they grew to such mass so early after the Big Bang is a profound puzzle for physics. ~ Australian astronomer Christian Wolf *et al*

There is no astrophysical explanation for how unimaginably massive black holes could exist so quickly after the assumed Big Bang. There can be no better evidence that the bruited Big Bang is a bust, and that the standard cosmological model (ΛCDM) is a myth. The only profound puzzle is why astrophysicists stick with an obviously fictitious model when abundant evidence indicates its falsity.

From our perch in the cosmos, the farthest we can detect is 46.5 billion light-years away. As light speed delimits cosmological distance, the universe must be at least 46.5 billion years old. Our universe would only be that young if Earth were in the center of the universe, and if the cosmos did not extend beyond our detection. Both assumptions are unlikely. The cosmos has likely existed for over 100 billion years, and perhaps much longer: 500 billion years or more is entirely possible.

> All we can truly conclude is that the Universe is much larger than the volume we can directly observe. ~ NASA

Regardless of age, how our cosmos can come into being remains a central question. Cyclic cosmology is apt, and entirely fits the facts.

✺ Cyclic Cosmology ✺

> This world, which is the same for all, not one gods nor men has made. It always was and will be: an ever-living fire, with measures of it kindling, and measures going out. ~ Turkish Greek philosopher Heraclitus (535–475 BCE)

Like Anaximander, Heraclitus conceived Nature in an incessant, eternal cycle of creation. This idea has reappeared throughout history. Ancient Andeans believed that the cosmos periodically disintegrated and reconstituted.

> All things began in order, so shall they end, and so shall they begin again, according to the ordainer of order and the mystical mathematics of the city of heaven. ~ 17th century English author Thomas Browne

The *cyclic model* proposes that the existing universe is an expanding bounce from a previous cosmic contraction. Unlike the Big Bang of the standard cosmological model, the cyclic model accords well with known quantum effects and modern physics' models.

This is the latest stage in an eternal cycle of expansion, collapse and renewed expansion. ~ American theoretical cosmologist and physicist Michael Turner

Cyclic cosmology explains the relatively smooth universe, as the smoothing could have occurred during the preceding contraction, before the expansion began.

Space and time exist forever. The universe undergoes an endless sequence of cycles in which it contracts in a big crunch and re-emerges in an expanding big bang, with trillions of years of evolution in between. ~ Paul Steinhardt

Cyclic cosmology eliminates the need for cosmic inflation as an explanatory plug for the otherwise inexplicable. Einstein theorized a cyclic cosmology in 1930. Many other astrophysicists have developed their own models since.

Cyclic cosmology puts the beginning of this universe as just another cosmic bubble bursting into bloom, not the mythical origin point posited by ΛCDM, the standard cosmological model. The cyclic model supports the prospect that multiple universes exist (multiverse), and that existence is eternal.

A bounce takes place a short time before a would-be big bang. ~ American astrophysicists Lauris Baum & Paul Frampton

With this cosmos part of a multidimensional membrane, cruising an even higher-dimensional space, the universe's origin was an energetic intersection of membranes; something more than cosmic humdrum, but by no means the solitary singularity of a single Big Bang, with only this universe popping forth from literally nowhere.

Considering the inscrutability of a unique big bang, cyclic cosmology makes intuitive as well as factual sense. For one, cyclic cosmology accounts for cosmic microwave background radiation as a transference of energetic patterning from the universe's previous incarnation.

Central tenets of Hindu thought posit: 1) time as cyclical; 2) chaotic causality; and 3) interdependence between microcosmic and macrocosmic existence. Hinduism supposes that existence has neither a beginning nor an end.

Hindus have it that the life force of organisms is constantly recycled (*saaṅsāra*). Earlier acts have later influence (*karma*), often in subtle ways, and possibly across cosmic cycles of spacetime (*yugas*).

Under the Hindu conception: time, causality, and the microcosm of individuals and cosmic macrocosm are all linked in an interdependent mix. The cyclic model of cosmogony supports the tenet of cyclical time, as does Einsteinian relativity.

Objection to the cyclic model is based upon classical physics laws of thermodynamics, which blithely assume that the universe is only 4 dimensions, and a closed energy system. Abundant evidence and modern physics instruct otherwise, rendering the objections archaic. In contrast to the irregularities and improbabilities of a singularity followed by faster-than-light inflation, cyclic cosmology seems sensible.

> The reason why the universe is eternal is that it does not live for itself; it gives life to others as it transforms. ~ Lao Tzu

Explaining cosmic creation is just the beginning of contention in storytelling about how the universe operates. The Big Bang hypothesis fails to address the critical question of how everything can emerge from nothing. Cyclic cosmology neatly answers the question by putting it off: this universe is simply a single incarnation of many, in an eternal cycle of creation.

ഔ The Standard Model ങ

The conventional cosmology story comes courtesy of the Lambda cold dark matter (ΛCDM) model, which emerged in the late 1990s. Further exploration of ΛCDM, beyond its 13.82 BYA Big Bang and mythical cosmic inflation, reveals further inanities which discredit it.

The canard of cosmic inflation arose from thinking that the universe began with the 1st discernible light. Instead, the early universe was nearly pitch black, filled with a miasma of light elements: hydrogen, helium, and lithium, until stars formed and baked heavier elements, creating a faintly luminous byproduct.

Not until stars formed were there loci of light. Dispel the notion of the universe starting like the torching of a firecracker and there is no need for the physics-defying folly of cosmic inflation.

ဆ Dark Matter ၷ

In 1932, Dutch astronomer Jan Oort could not account for the orbital velocity of stars in the Milky Way. Much matter seemed missing.

In 1933, Swiss astronomer Fritz Zwicky had the same problem figuring galactic speeds: they were moving so fast that the galaxies shouldn't hold together. He thought that there must be invisible matter giving them extra gravity. He termed it *dark matter*.

Confirming observations followed, all viewed using the same model. Surveys to figure the large-scale structure of the universe created more credence for dark matter.

ΛCDM is simple but strange. It implies that almost all matter is unseen, and most energy undetected. Beyond the pale lies most of the mass holding the universe together, and the energy propelling the cosmos outward: dark matter and dark energy, respectively.

ΛCDM's lame accounting of cosmic matter for requisite gravitational oomph is astonishing. Under ΛCDM, materiality is mostly mirage.

> Our inventory of stuff that makes up our universe amounts to a humbling 5%. ~ American astrophysicist Paul Hamilton *et al*, in the context of the ΛCDM model

The supposed dark matter does not form atoms. Dark matter has been instead surmised to emanate from some massive, slow-moving exotic particle that accretes into clumps that attract ordinary matter, thereby acting as gravitational seeds for galactic formation and growth. Extensive search has found no such particle. In 2016, after a 20-month search, the most sensitive detector ever created found nothing to indicate that dark matter exists.

> We've been looking where our best guess told us to look for dark matter all these years, and we're starting to wonder if we

maybe guessed wrong. ~ American theoretical astrophysicist
Dan Hooper

   The problem of dark matter, raised decades ago by the dynamical studies of clusters of galaxies and by the flat rotation
curves of galaxies, is still resisting to explanations. ~ Swiss theoretical astrophysicist André Maeder

Dark matter has a historical analogy in cosmic aether,
which was also presumed but could not be found. Dismissing
cosmic aether turned a page in the history of physics theorization, just a disabusing dark matter will do so for astrophysics.

The very idea of dark matter being necessary for requisite
cosmic mass is absurd, especially in the high proportion of
dark matter estimated. Matter always emanates from quantization of localized fields. If stars and galaxies would fly
apart absent dark matter, so too should ordinary matter, as
it would lack sufficient cohesiveness.[*]

The astrophysicists formulating a need for dark matter
via ΛCDM improperly treated the basic relationship between
matter, mass, and gravity.[†] The claim of non-detectable dark
matter itself disproves ΛCDM.

ଽ Dark Energy ଓ

A 1998 astrological survey indicated that the universe
was expanding at an accelerating rate. The idea of *dark energy* arose to explain it. In 2016, additional data on stellar
luminosity discredited dark energy as causing accelerating
expansion. Dark energy was simply a case of insufficient data
viewed through the wrong lens.

   The apparent manifestation of dark energy is a consequence
   of analyzing the data in an oversimplified theoretical model –
   one that was in fact constructed in the 1930s, long before there
   was any real data. A more sophisticated theoretical framework,

---

[*] A similar dilemma arises at the quantum scale, where virtual particles are hypothesized as necessary to provide subatomic particles with enough mass for existence to occur. Both the standard
   astrophysics and quantum physics models fail for bad math.
[†] ΛCDM is incongruous with quantum physics' Standard Model.

accounting for the observation that the universe is not exactly homogeneous and that its matter content may not behave as an ideal gas – two key assumptions of standard cosmology – may well be able to account for all observations without requiring dark energy. Indeed, vacuum energy is something of which we have absolutely no understanding in fundamental theory. ~ Indian theoretical physicist Subir Sarkar

The Milky Way has far fewer neighbors than theory suggests that it should. The galaxy is in an abyss about 2 billion light-years wide.

With the Milky Way in a void, the apparent rate at which the universe is expanding depends upon how it is measured. Measurements based upon the cosmic microwave background (CMB) radiation suggest a slower expansion rate than measurements of nearby supernovas. The actual expansion rate is probably even slower than the CMB rate.

If you don't account for the void effects, you could mistake this relationship to indicate there is too much expansion. ~ American astrophysicist Benjamin Hoscheit

Dark energy can also be discounted by facile assumptions about the structure of the universe which conjure it.

Einstein's equations of general relativity that describe the expansion of the universe are so complex mathematically that, for a hundred years, no solutions accounting for the effect of cosmic structures have been found. Coarse approximations to Einstein's equations may introduce serious side-effects, such as the need for dark energy in models designed to fit observational data. ~ Hungarian astrophysicist László Dobos in 2017

ஐ Light Matter's Modest Missing Half இ

It is always dark. Light only hides the darkness. ~ Daniel McKiernan

ΛCDM hypothesizes that a lot of light matter is too shy to show itself. At least half of the matter in the universe is missing.

The word *missing* is a colloquial way of putting it. ~ American astrophysicist Neta Bahcall

Baryonic matter forms the ordinary atoms and ions that comprise stars, planets, dust, and gas. Galaxies tally but 10% of cosmic baryonic mass. Another 10% is swaddled in space as warm gas. Some 30% coagulates in cold blobs of space gas. The other 50% is presumed to be a thin paste of hot plasma between stars. The conjecture comes as a best guess of hiding in not-so-plain sight, as astronomers have yet to figure a way to spot the hot gruel.

ℬ Too Bright ℛ

Something is very wrong. ~ American astronomer Juna Kollmeier on ΛCDM

A recent census of celestial objects that produce high-energy ultraviolet light created a cosmic accounting conundrum. According to ΛCDM, the universe is far brighter than it should be, based upon the number of light-emitting objects identified.

There is 5 times as much light-emitting ionized gas than ultraviolet sources could produce in the modern, nearby universe. Strangely, for the early, more distant universe, UV sources and ionized gas match up. Hydrogen, which makes up the vast bulk of cosmic gas, may be misunderstood.

ℬ Scale-Invariant Cosmology ℛ

It appears as one of the fundamental principles of Nature that the equations expressing basic laws should be invariant under the widest possible group of transformations. ~ English theoretical physicist Paul Dirac

Quantum physics explains the emergence of matter from coherent, localized energy fields, emanating out of a ground state (vacuum) which itself seethes with tremendous energy. Quantization of energy is the fundamental mechanism by which existence manifests. Therefore, materiality must have a spatial metric in its origination.

Vacuum at the quantum level is not scale-invariant, since some units of mass, length and time can be defined on the basis of the Planck constant. However, large-scale empty space differs by an enormous factor (10^{39}) from the quantum scales. Empty space at large scales could have scale invariance, since by definition there is nothing to define a scale. ~ André Maeder

Actuality emerges at the quantum scale, with physics which are unique to the subatomic realm. There is no reason to think that empty ambient space is dependent upon scale, certainly not like quantum mechanics are in relation to the ground state. In other words, there appears to be a fundamental discontinuity between quantum quantization and the ambient world, which is ruled by electrodynamics. At the stellar scale, gravity becomes more pronounced in its effects.

Galileo discovered that the laws of physics have no scale (Galilean invariance). Einstein's general relativity theory revealed matter warping the very spacetime fabric that defines existence. The relation between mass and gravity were considered constant, thereby scale-invariant.

The equations which describe electrodynamics, in absence of charges and currents, are scale-invariant; so too cosmic interplay via gravity.

Empty space at large scales is scale-invariant. ~ André Maeder

A cosmological model assuming scale invariance corresponds with all astronomical observations, and all macroscopic physics theories. Dark matter and dark energy disappear with a scale-invariant cosmological model: Nature goes about its business without need to invoke a mystical glue to hold the universe together, or phantom energy to nudge comportment with actuality.

Cosmologists are often wrong, but never in doubt. ~ Azerbaijanian physicist Lev Landau

Though ΛCDM is widely accepted, the model is clearly invalid. Its failure came from built-in assumptions that astronomers relied upon to interpret their spotty data. From these unexamined axioms sprang dark matter, dark energy, and missing light matter.

The universe is nearly flat as a sheet. For that to be true, there must be a critical level of mass/energy density which has not been explained. The seemingly stable universe is an inscrutable fact. Despite all the stargazing and hypothesizing, we know little about cosmic construction.

Scale-invariant cosmology is just one example of the maxim that strangeness in science owes to having the wrong worldview. With the right model, a rough sketch of existence may be made. That Nature is so intricate that it defies comprehension is simply an endorsement for awe.

> Nature is not embarrassed by difficulties of analysis. ~ French engineer and physicist Augustin Fresnel

ಐ Cosmic Matter ಚ

From the onset of the cosmos, energy expanded and dissipated. Once matter formed, the early universe was a hot, dense plasma of emerging photons, electrons, and protons.

Depending upon the account you cotton to, it took anywhere from 10 seconds to 10's of billions of years for the cosmos to cool enough for atoms to form: protons captivating electrons via the music of emerging electromagnetism.

Neutrons weigh 1.00137841917 that of protons; exactly the ratio needed for *nucleosynthesis*: the creation of atomic nuclei in stars. Further, the electrical charge of electrons neatly balances that of protons. Without these precise balances, there would be no matter in the universe.

At 3,000 Kelvin (K), electrons slowed enough to be snared by the gravitational force of atomic nuclei and set up housekeeping as atoms. Only the lightest elements – hydrogen and helium – spontaneously arose as primordial gases.

The universe was still mostly dark, though scattered with matter, and seething with energy. The slightest variations in gravitational densities acted as seeds for the distribution of what would become stars and galaxies.

> Reionization is one of the major milestones in the universe's history. ~ American astronomer Brant Robertson

A peculiar transition *may* have happened 13.6–12.8 BYA: *reionization*. Something stripped the electrons off atoms.

Radiation bursts from star formation in the 1st generations of galaxies may have caused reionization, though how it came about remains mysterious. Whether reionization even occurred is less than certain. The cosmic particle soup had thinned enough prior to reionization that photons could travel freely, turning most the universe's matter into the glowing ionized plasma that abides to this day.

¤ ✧ ¤

From the appearance of primordial gas clouds it took hundreds of million years for the nascent stars to accrete. The 1st stars brought light and warmth into the cosmos, as well as melting hydrogen and helium together, forging the heavier elements, including carbon, nitrogen, and oxygen. After a few hundred million years of stars making matter, all the natural elements had emerged.

As they were the first to sup on primordial matter, the earliest stars were monstrously large. Some were the size and luminosity of 100 million suns.

The distortion of gravity was slow to make its cosmic presence felt; only with the advent of the first stars did gravity emerge in any significant way. As gravity came into play, swirls of matter coalesced into nebulas, forming galaxies ~13.2 billion years ago.

Gravity is not the only force binding cosmic matter. Magnetism was instrumental in shaping accretion disks that became stars and black holes, and in its flux begetting glue to nascent galaxies.

Metals – such as life-essential iron – are rather evenly spread throughout the cosmos. This is the legacy of an energetic episode for matter creation, when exploding stars and black holes at the hearts of young galaxies were especially vigorous.

ൈ Dust ൙

The primordial cosmos consisted of hydrogen and helium, with slight traces of lithium. The heavier elements – needed for planets and everything else made of matter as we think of it – are manufactured via supernova explosions. Such star dust was sucked into new stars as they formed. Hence, ever-heavier elements were made.

The first supernova stars had already lived their lives and were gone 13.62 billion years ago.* By 13.22 BYA, the relentless progression of dust trade-up was well underway.

Even now, most of the ordinary matter in the cosmos is hydrogen and helium. The most abundant molecule in the universe – H_2 – primarily forms on the surfaces of dust grains. Heavier elements make up only 1% of galactic mass. Half of heavy matter is bound in dust grains which are blown into existence by the aftermath of a supernova; thus, from dust to dust.

Dust largely defines the interstellar medium. In absorbing ultraviolet radiation from stars, dust emits electrons that are the main heat source of interstellar gas. This bit of warmth from dust helps molecules survive the harshness of deep space. Hence, cold, diffuse clouds of molecular hydrogen course the vastness of space. These clouds may be as frigid as 7 K and as diffuse as 300 light-years.

Over billions of years, the persistent gentle nudge of gravity corrals the molecules in interstellar space closer to each other. They warm as they snuggle. The excitation hastens further condensation. Stars form in these regions. Dust is the subtle conductor of star formation.

Absorbing radiation imparts momentum to dust grains, driving them away from newly formed stars, or even an entire galaxy. Such winds transfer vast amounts of matter between galaxies. Large galaxies, such as the Milky Way, may have amassed half their matter from neighboring star clusters up to a million light-years away. Thus, dust plays an essential role in the evolution of galaxies.

Dust dies by shock: destroyed by shock waves that emanate from the remnants of a supernova. The shock waves are partly comprised of high-speed dust grains, traveling in excess of 1,000 km second. These fast-moving grains are also

---

* There are considerable discrepancies in chronologies among the various accounts of cosmic evolution, all of which are highly speculative. On the previous page, the first stars formed 13.72–13.27 billion years ago; now, a mere page later, supernova stars had already blown their lights out 13.62 BYA.

subject to shocks as they come to rest in the interstellar medium. When it comes to cosmic dust, what comes around goes around.

ഇ Galaxies ര

> The Force binds the galaxy together. ~ Obi-Wan Kenobi in the movie *Star Wars* (1977)

A *galaxy* is a cluster of star systems and stellar remnants, swirling in an interstellar mixture of gas, dust, and massive matter.

> The structures in our present universe are the outcome of more than 10 billion years of evolution. Slight irregularities imprinted at very early eras led to increasing contrasts in the density from place to place, until overdense regions evolved into bound structures. ~ English astrophysicist Martin Rees

We do not know when galaxies emerged. The 1st galaxies we are aware of coalesced by 13.57 BYA.[*] There were already mature galaxies with billions of stars 12.3 BYA.

> The number of galaxies is much bigger than anyone would have guessed. And the real number could be even higher. ~ English astrophysicist Christopher Conselice

There are now ~4 trillion (4,000,000,000,000) galaxies in the universe; roughly half light and half dark. Each light galaxy may contain millions or even billions of stars. Almost all visible star systems have planets.

Within a few billion years of galactic formation, there were 10 times as many galaxies as there are today. Cosmic evolution reduced the number of galaxies through extensive merging.

To this day, mysterious filaments of galactic attraction thread the universe in an invisible gravitational web. Galaxies form along these filaments, with massive black holes as their hearts. The gravitational influence of filaments entices

---

[*] That galaxies existed only 250 million years after the supposed Big Bang strongly indicates that the standard cosmological model is bogus, as there is no astrophysical explanation for such rapid galactic evolution.

molecular hydrogen gas to coalesce, and so permits star formation. Galaxies run on gas.

Acting as a black backdrop to the glittering cosmopolitan cosmos, dark galaxies are conjectured to be as plentiful as the light variety. Some are utterly devoid of stars; others have a relative few. Black holes and gravitational filaments corral dense gas globules that shed no light. Little is known about the ecologies of galaxies, light or dark.

✠ Galaxy Classification ℞

> Equipped with his 5 senses, man explores the universe around him and calls the adventure *science*. ~ Edwin Hubble

Astronomers have traditionally classified galaxies by how they look. Edwin Hubble developed his galaxy morphology in 1926. The *Hubble sequence*, still used today, has 3 galaxy classes: ellipticals, spirals, and lenticulars.

Elliptical galaxies appear as featureless ellipses of light.

Spiral galaxies have a central concentration of stars – a galactic bulge – with arms forming a spiral structure. How fat the bulge is depends upon how rapidly the galaxy is spinning. 70% of the galaxies near the Milky Way are spiral.

Lenticulars have a bright central bulge, surrounded by a thinner disk, but without the spiraling effect.

Hubble also defined 2 classes of irregular galaxies outside the Hubble sequence: one of star clusters without a central bulge, and another smoother configuration with an asymmetric appearance.

New galactic types are still being found. 300 million light-years away is a galaxy with the same mass as the Milky Way, but with only 1% of the star shine.

> That's just something we never knew could happen. ~ Dutch astronomer Pieter van Dokkum in 2016

A galaxy's *visual morphology* reflects the dynamics of its formation and evolution. Galaxies form a spiral pattern out from a central core via *density waves*: oscillations in the galactic gravitational field that sashays stars back and forth. Galactic structures are frequently shaped by tidal interactions with other galaxies.

Before describing the importance of black holes and quasars in galactic formation and dynamics, a brief digression into the history of astrophysics.

In the late 18th century, English natural philosopher and geologist John Michell and French mathematician and astronomer Pierre-Simon Laplace contemplated the prospect of an object with gravity too strong for light to escape: a *black hole.*

German physicist Karl Schwarzschild mathematically conjectured simple black holes in 1915, the same year Einstein introduced general relativity. The *Schwarzschild radius* is the size of the event horizon for a simplified abstraction of black holes: massive, non-rotating, and spherically symmetric objects. A black hole's *event horizon* defines the rim of no return for matter/energy, as the gravitational pull approaches infinity.

Einstein was pleasantly surprised to learn of Schwarzschild's exact solutions for general relativity's field equations, as he could only produce an approximate solution.

Whereas Einstein had used a rectangular coordinate system to approximate the gravitational field near the black hole mathematical construct, Schwarzschild developed a polar, spherical coordinate system, which afforded more elegant mathematical expression.*

Einstein considered black holes purely a mathematical construct. He did not think that black holes could actually form. Following Einstein's lead, mainstream physicists disregarded all results to the contrary, though a minority maintained that black holes were possible. It was not until the close of the 1960s that consensus conviction turned toward acceptance that black holes existed.

---

* Schwarzschild's triumphal equations were created while he was in the German army during World War I. Schwarzschild was suffering from a painful autoimmune disease (pemphigus) which he developed while at the Russian front, yet he managed to write 3 outstanding physics papers in 1915: 2 on relativity theory, and 1 on quantum theory. Schwarzschild died the following year.

Via quantum mechanics, not general relativity, English theoretical physicist Stephen Hawking predicted in 1974 that black holes must emit radiation (*Hawking radiation*), though at a temperature inversely proportional to the mass of the black hole.

There is no escape from a black hole in classical theory, but quantum theory enables energy and information to escape. ~ Stephen Hawking

30 years later, Hawking had convinced himself that black holes do not exist. Hawking's repudiation stemmed from a paradoxical conundrum.

The *equivalence principle* of relativity assumes that the laws of physics are identical everywhere. Someone falling into a black hole would feel the same as if floating free in space, at least until ripped apart by gravitational intensity.

At the quantum level, approaching a singularity would be very energetic, with excited particles bustling about. Someone entering the event horizon would be fried to a crisp by sizzling subatomics.

Such a quantum firewall poses serious problems. It violates the relativity axiom of equivalence. And it breaks the mathematical symmetry of quantum theory.

So, Hawking proposed that a black hole isn't really a black hole. Instead, it is a cosmological shredder, which merely mangles matter before releasing it; an utterly unsupported hypothesis.

A different resolution of the paradox is proposed, namely that gravitational collapse produces apparent horizons but no event horizons behind which information is lost. ~ Stephen Hawking, concerned about the cosmic integrity of information

By trading an event horizon for one which is only apparent, Hawking's proposal attempts to leave both relativity and quantum theories intact. Instead, it denies what a black hole must be.

✄ Black-Body Radiation ✄

A *black body* is an idealized object that absorbs all incident electromagnetic radiation. The only consummate black bodies are black holes.

Black bodies at a uniform temperature emit an electromagnetic signature termed *black-body radiation*. The radiation from a black body depends only on the body's temperature.

In 1879 Jožef Stefan mathematically stated the law pertaining to radiant energy after considering the mathematical relation between temperature and radiation in black bodies. With this equation, Stefan was able to make the first approximate estimate of the temperature of the Sun's surface.

At the atomic and molecular level, radiation typically exerts a positive pressure. As an energy source, radiation temporarily charges particles. This is known as the *Stark shift*.

The Stark shift of black-body radiation is roughly proportional to the 4th power of the black body's temperature. The hotter the body, the higher the shift.

Stark shifts induced by black-body radiation can combine, creating an attractive force that overwhelms the repulsive radiation pressure. Despite outgoing radiative energy flow, a hot black body can attract nearby neutral molecular matter rather than repel it. This attraction happens because radiated atomic matter can be drawn to a higher radiation intensity; in this instance, a hot black body.

Up to a point, the hotter the body, the stronger the attraction. But, above a few thousand degrees Kelvin, attraction turns to repulsion. The attractive force of black-body radiation rapidly decays with distance: to the 3rd power. The force of transition is stronger for small bodies.

The attractive black-body force of a black hole the size of a dust grain at 100K is much stronger than its puny gravitational tug. In contrast, a large, hot black hole relies upon its gravitational pull. By virtue of their tiny black bodies, minute black holes take advantage of their black-body allure to get a head start on growth.

♌ Black Holes ♐

> The black holes of Nature are the most perfect macroscopic objects there are in the universe: the only elements in their construction are our concepts of space and time. ~ Indian astrophysicist Subramanyan Chandrasekhar

A *black hole* is a cosmic singularity of no return once within, past the *event horizon* that defines entrance into a black hole.

> The event horizon is not a physical barrier. ~ Scottish astrophysicist Paul McNamara

Most matter drawn into a black hole is spun off before it reaches the event horizon. This expelled energy flow is termed a *quasar* when the emitted radiation is luminous (light and infrared warmth being the only spectral range of radiation celebrated by humanity).

A black hole grows as matter is absorbed. However logical that seems, how it happens is not known. The mystery emanates from the nature of the singularity. A black hole is literally a perfectly spherical hole in the universe – spacetime simply ceases to exist.

It is bizarre that these infinite voids provide for the gyre of existence by anchoring galaxies and driving their dynamics. As Lao Tzu stated: what is not makes what is useful.

¤ ✧ ¤

> Black holes and their host galaxies coevolve, with the feedback from the black hole inducing star formation. ~ Israeli astronomer Benny Trakhtenbrot *et al*

Small density fluctuations in the early universe led to perturbations that sprouted in the fertile ground of gravity. These grew to the point that they disconnected from the global expansion of the universe. Such centers became self-gravitating; forming halos within which gas condensed to form stars, black holes, and galaxies.

Black holes have peppered the cosmos since its salad days. They were abundant in the early universe; swapping the beginning of something for a gaping maw of nothing. These seeds of nothingness grew by consuming whatever fell into their path.

One black hole has a mass 12 billion times that of the Sun, accreting surrounding substance at the maximum rate afforded by the laws of physics. This black hole had an awesome girth 12.92 BYA, only 900 million years after the supposed Big Bang. Other contemporaneous monstrous black holes have been found. Astrophysicists have no explanation for how such massive black holes were possible so early in the cosmos' supposed history.

> We expected as we looked further back into time that the black holes would be smaller and smaller because they hadn't had as much time to grow. ~ American astrophysicist Rob Simcoe

The existence of massive black holes just hundreds of millions of years after the Big Bang strongly indicates that the conventional dating of this universe's birth is wrong. The cosmos must be much older.

Black holes were the nursery in which early galaxies grew up; both a great attractor and generator of galaxy-making material. Black holes were more massive relative to their respective galaxies when the universe was young.

Besides the intense implosion of a massive gas cloud, a black hole can form after a supernova explosion, with the remnants collapsing, forming a forceful gravitational pull that sucks in surrounding mass in an ongoing accretion process.

Black holes are everywhere and come in all sizes. Some are swollen to 50 billion times the mass of the Sun.

A massive black hole is a gyre, gaining girth and power while emitting energetic streams that may stretch for millions of light-years. A star coming close to a supermassive black hole may be ripped apart by the hole's tidal pull, with stellar debris spun off as a quasar.

There is an ancient quasar at the edge of the observable universe that appears to be 12.9 billion years old, powered by a black hole of 2 billion solar masses. The quasar emits 60 trillion times the light of the Sun.

This enhancement of star formation by outflows would have been even more important in a younger universe, where dense clumps of gas were much more common. ~ Australian astrophysicist Stanislav Shabala

Black holes typically account for 0.1% of a galaxy's mass, but one has been observed that is a whopping 14% of galactic girth.

A spinning black hole draws matter that rotates around it. On the way to a black hole, incoming material picks up its pace. In the competition between speed and gravity, speed wins. Over 99% of the matter drawn to black holes is ejected.

Feasting makes a black hole faster. The larger the black hole, the quicker its spin. A supermassive black hole at the center of a nearby galaxy has been clocked at 1.08 billion kilometers per hour: close to the speed of light. Black holes continuously spew cosmic rays, the most energetic radiation in the universe.

Galaxies and black holes have grown in tandem throughout cosmic history. There is a correlation between the mass of a galaxy's central black hole and the velocity of stars in its galaxy. All the galaxies near the Milky Way have about 700 times more mass in their stellar bulges than deposited in their black holes, irrespective of the galaxy's size. This appears to be an evolved situation.

The Milky Way grew by capturing dwarf galaxies which had originally formed from black holes. The galactic merger process can result in a dwarf black hole recoiling rather than merging with the massive black hole at the heart of the Milky Way. More often, galaxies are joined as their black holes interact.

When black holes encounter one another, they dance together for a while in a close embrace. The footfalls of black hole ballet are *gravitational waves* that ripple spacetime itself. These waves carry for untold light-years, creating a web of galactic interactions. Eventually, the black holes merge into one – their mutual gravitational attraction irresistible.

Black holes are not confined to being center stage in the galaxy. Galaxies may have millions of black holes roaming about, each with the mass of anywhere from 1,000 to 100,000 suns, swallowing anything in their path, shaping galactic dynamics. There an estimated 400 million black holes in the Milky Way galaxy.

While black holes often engender galactic formation, they can also slowly suffocate galaxies. The spin of a black hole determines what role it plays in the galaxy about it.

The lives of galaxies and their supermassive black holes are inextricably intertwined. ~ American astrophysicists Timothy Heckman & Guinevere Kauffmann

As matter is sucked toward a black hole, a disk of infalling gas and dust forms around the rim. On the journey inwards, owing to quantum effects, incoming debris emits large amounts of X-ray and ultraviolet radiation; radiation so strong that it diverts part of the inflow. This causes strong outflowing winds, with velocities up to several hundreds of kilometers per second.

Outflowing jets from massive black holes engender star formation by plowing through galactic gas, thereby creating hot gas filaments, the raw material from which stars form. A black hole's energetic jet causes a supersonic shock wave on a gas cloud in its path, heating and compressing the gas.

The shock wave ionizes the gas cloud: stripping electrons from the gaseous atoms. After the shock wave subsides, the ions recombine, emitting radiation, which takes energy out of the cloud. This cooling effect causes the gas cloud to contract further. When the knot of gas reaches a critical density it collapses to form a star.

☿ Quasars ♐

Quasars are immensely bright. From the central point in a galaxy, they emit as much energy as thousands of giant galaxies from a region as tiny as the solar system. ~ American astrophysicist Robert Antonucci

Quasars are the shiny companions to black holes. Powered by regurgitation from supermassive black holes, quasars appear as stunningly bright, distant stars. More than 200,000 quasars have been spotted. Quasars are powered by both the spin toward the black hole and the rotation of the black hole itself.

A quasar's brightness corresponds to how much matter a black hole is consuming. When black hole intake slackens, the light goes out and black hole output is downgraded in human esteem. A typical quasar only lasts a few hundred million years.

Quasars are not always solely outflow. 1 out of 10,000 quasars feed the black hole from which they are formed. How that happens is not understood.

Quasars appear linked in a cosmic web of filaments and clumps in the vast voids where galaxies are scarce. The dynamics of large quasar group formation remain a mystery.

Some of the quasars' rotation axes are aligned with each other, despite the fact that these quasars are separated by billions of light-years. ~ Belgian cosmologist Damien Hutse-mékers

While the cause of disparate quasar alignment is uncertain, it is a clear indication of synchronism on a vast cosmic scale.

The alignments hint that there is a missing ingredient in our current models of the cosmos. ~ Belgian astrophysicist Dominique Sluse

ϑ Galaxy Dynamics ଞ

Galaxy formation history may be telling us something about the places in the universe where life can form. ~ Swedish astrophysicist Kambiz Fathi

At different scales, accretion disks of dust and gas are the cradles of galaxies, stars, and planets. Accretion disks are common because the coalescing force of gravitation is offset by angular momentum, forming a disk.

The growth of galaxies in the early universe was vigorous. Within just 3 billion years of galactic formation, there were already thousands of galaxy clusters.

The nascent universe was much smaller. Star density was 10 times greater than today. Each galaxy cluster contained hundreds of thousands of galaxies. Some were massive galaxies, with several hundred billion stars, formed by collisions of smaller galaxies.

Most of the earliest galaxies were elliptical, having many stars, but insufficient dust and gas to fuel organic expansion. The most massive galaxies are giant ellipticals.

The mass of a galaxy directly relates to the mass of its central black hole. Mass determines how fast a galaxy spins. Spin slows as a galaxy grows.

> Regardless of whether a galaxy is very big or very small, if you could sit on the extreme edge of its disk as it spins, it would take you about a billion years to go all the way round. ~ American astrophysicist Gerhardt Meurer

Just 3 billion years after galaxies got going there were already spent elliptical galaxies: no longer forming new stars. In contrast, spiral galaxies, like the Milky Way, contain much material for star formation.

Young galaxies furiously create stars. The more gas a galaxy has, the more sparkling stars are ignited. A galaxy's magnetic field nudges huge clouds of gas and dust into pregnant concentrations that give birth to stars.

> Through self-excitation, a magnetic field is created from virtually nothing, whereby the complex movement of the conductive plasma serves as an energy source. ~ German physicist Frank Stefani

Massive cosmic magnetic fields pervade the universe and persist for billions of years. Small-scale fluctuations of astrophysical plasma create large-scale, persistent magnetic fields which shape the material dynamics of galaxies. Like rivers of energy, plasmas flow in a certain direction. There are also plasmatic counter-streams.

Like water, plasmas have abiding internal structures which have been observed during star formation and star death. Supernovas are tremendous plasma producers.

Plasmas also pulse on a galactic scale. The coherent self-organizing of plasma among seeming chaos produces the energetic seeds from which galaxies and star systems are born.

¤ ✧ ¤

A massive black hole forms and builds a galaxy from its quasar emissions. The mass of a black hole in a galaxy's center typically ranges between a million and a billion times that of the Sun.

Galactic formation dynamics act as a thermal gyre: pulling in dense, cold gas, and ejecting hot gas back into intergalactic space. A galaxy ends up with a fraction of the raw material it processes.

Galaxies grow from the inside out. Most galaxies have a bulge at their center, as does the Milky Way.

Black hole growth and star formation typically go together. If the nearby environment of a black hole is gas poor, gas accretion is slow. Radiation emission is correspondingly low.

Black holes at the heart of a galaxy not only spin, they also move across their host galaxy, altering galactic dynamics. The speed at which a black hole spins distinctively affects the spacetime around it: another factor in the gyre of a black hole.

Galactic gyres follow fluid dynamics, with viscosity something of a mystery. The level and nature of turbulence determine what stays and what flies away. In a galactic butterfly effect, small disturbances can affect stabilities and mass transfers at a much larger scale.

The structures and sizes of galaxies vary. Galaxy range from dwarfs of 10 million (10^7) stars to giants with a hundred trillion (10^{14}) stellar lights.

Galaxies typically spread from 1,000 to 100,000 parsecs in diameter, separated by millions of parsecs (*megaparsecs*) of intergalactic space. The space between galaxies is a tenuous gas, with less than 1 atom per cubic meter.

A *parsec* is an astronomical length unit: about 3.26 light-years, just under 31 trillion (3.1 x 10^{13}) kilometers (km). A *light-year* is ~9.461 trillion km: how far light can travel in a vacuum in 1 Julian year (365.25 days).

¤ ✧ ¤

Big galaxies are crashing into other big galaxies to make even bigger galaxies. ~ American astrophysicist Adam Bolton

Galaxies are attracted to each other under the influence of their gravity. Galaxies may collide, merge, or pass through each other. Large galaxies grow by absorbing smaller ones.

Even with no major collisions, the interstellar medium of gas and dust interact, triggering bursts of star formation. Collisions gas up galaxies, further triggering star birth bursts.

Stellar collisions can severely distort the galaxies involved, forming oddly shaped galactic artifacts, such as tail-like structures. Stellar orbits about a galaxy can be thrown off course.

Relatively passive pass-throughs between galaxies can leave lasting connections. Tendrils of cold hydrogen gas can be pulled from one galaxy toward another, creating a tenuous bridge between the two.

☿ The Milky Way ♌

The Milky Way formed 13.2 BYA. Our galaxy is now a stellar spiral disk with 4 major arms and 2 dozen smaller ones. The Milky Way's present shape is a product of its evolution, which continues. The Milky Way is a gyre of streaming debris, replenished by gas cloud encounters, and consuming satellite galaxies which are drawn to it. So far, the Milky Way has devoured 15 other galaxies.

The Milky Way is at least 200,000 light-years in diameter, with over 400 billion stars, and at least 640 billion planets, cumulatively weighing in at 1.54 trillion suns.

Like the planet Tatooine in the Star Wars movies, many millions of Milky Way planets orbit 2 stars.

> We used to think that the Earth might be unique in our galaxy. But now it seems that there are literally billions of planets with masses similar to Earth orbiting stars in the Milky Way. ~ German astronomer Daniel Kubas

The Milky Way spins at 250 kilometers per second. 1 revolution takes 240 million years.

At the center of the Milky Way lies a ponderous, barely spinning black hole, equivalent to 4 million solar masses. The black hole continues to accrete matter and energy: constantly snacking on hot gas. This central black hole is surrounded by a well-ordered magnetic field that regulates the flow of material into it.

The Milky Way and its neighbor, the *Andromeda* galaxy, are reckoned to be halfway through their life cycle, with some 4 billion years left. That estimate does not account for a continuing influx of new matter that keeps galaxies dynamic.

The Milky Way and Andromeda are encircled by a ring of 12 large galaxies ~24 million light-years (MLY) across. All these galaxies lie on a sheet that is 34 MLY across, but only 1.5 MLY thick.

Somewhat analogous to the atmosphere of a planet, galaxies such as the Milky Way have halo clouds of hydrogen gas and other incidental matter. These clouds are not evenly distributed, but cluster as residue from star formation.

Massive stars age quickly. Within a few million years, their stellar wind sheds matter, as a prelude to exploding as supernovae, spraying their contents into a cloud which forms galactic halos.

Cloud matter gets recycled back into the galaxy, seeding it with the fuel to trigger another burst of star system births. Galactic fuel clouds can also drift in from other galaxies along a gravitational filament.

These influxes are simply a continuation of the dynamics by which the Milky Way came to be. While matter accretion is one route to growing a galaxy, about 25% of the star clusters in the Milky Way immigrated from other galaxies.

Many of the stars in the halo that surrounds the Milky Way travel in groups. These small star clusters spend most of their time outside the disk-like structure that gives the Milky Way its name. They are the remnants of small galaxies that were cannibalized by the Milky Way.

A star cluster carries from 100,000 to a million stars. The Milky Way has swallowed up to 6 dwarf galaxies on its journey so far. There are hundreds, if not thousands, of small satellite galaxies swarming around the Milky Way.

The dwarf galaxy *Sagittarius* has smacked into the Milky Way several times. In an early encounter, 80% to 90% of its mass was stripped from Sagittarius. This set off a cascade of instabilities that resulted in the formation of the spiral arms in the Milky Way, along with ringlike structures on the galaxy's outskirts. Sagittarius strikes again 10 million years from now, fated to slap the southern face of the Milky Way disk.

Milky Way vitality was flagging until 5 BYA, when its star population suddenly burgeoned. Half of all the Milky Way's stars were produced during this period.

Like most galaxies, the Milky Way has had a calamitous evolution. 100 MYA, the Milky Way was banged by a smaller galaxy (not Sagittarius). Like a gong, the Milky Way reverberates from that encounter, and will continue to do so for the next 100 million years.

The Milky Way's spinning galactic disk is warped, bowing to immense gravitational dynamics. Gravitational effects shape the structure of the cosmos.

2 satellite galaxies – the *Magellanic Clouds* – orbit the Milky Way. Their gravitational tug distorts the galactic disk.

2.5 billion years from now, one of those satellite galaxies will collide with the Milky Way, infusing the Milky Way with more stellar material and altering its shape.

For all the drama that has beset the galaxy in its evolution, the major arms of the Milky Way are uncommonly symmetrical. The Milky Way may be a rare beauty in spiral galaxies.

¤ ✧ ¤

To put it mildly, the Milky Way is on the move. Our home galaxy is coursing through the cosmos at 2.15 million kilometers per hour.*

The Shapely Supercluster is a galactic concentration 650 million light-years away that exerts a powerful pull on the Milky Way. That is not the whole story of why the Milky Way is bustling at such ferocious speed.

Behind the Milky Way, on the far side of the constellation Lacerta (the lizard), is a relative void. This vast patch of nada has a striking dearth of galaxies compared to the rest of the cosmic neighborhood. But somehow it is exerting a repulsive force, pushing the Milky Way on its way.

---

* Please fasten your seat-belt and observe the "no smoking" sign.

The Shapley attractor is really pulling, but then almost 180°
in the other direction is a region devoid of galaxies, and this
region is repelling us. So, we have a pull from one side and a
push from the other. It's a story of love and hate, attraction and
repulsion. ~ Israeli cosmologist Yehuda Hoffman

⚨ Galactic Webs ⚮

The cosmic web formed very early in the history of the universe, starting with small initial fluctuations in the primordial
universe. ~ American astrophysicist Behnam Darvish

Galaxies are not isolated. They are instead interactively
distributed via a cosmic web of gravitational filaments.

The filaments are like bridges connecting the denser regions
in the cosmic web. ~ Behnam Darvish

Where intergalactic gravitational filaments meet are
dense galactic clusters of galaxies, which began as modest
fluctuations away from homogeneity. Galaxy distribution ultimately reflects subtle variations in the early universe.

These galactic filaments are themselves dynamic gyres,
growing as tendrils, sprouting new galaxies in a variety of
formations and with different growth patterns. By this, galaxies are organized in a hierarchy of associations.

Filaments engender interaction between galaxies,
thereby enhancing star formation. This dynamic began early
and continues today.

Galaxies flow in currents, swirl in eddies and collect in pools.
~ German cosmologist Noam Libeskind & Canadian astronomer Brent Tully

Clusters of galaxies form superclusters comprising tens
of thousands of individual galaxies. Superclusters fit into galactic sheets and filaments that fly through the immense
voids that comprise 90% of the volume of the universe.

Galactic superclusters are the largest known arrangements in the universe. Even larger structures are suspected.

The Milky Way lies within the Laniakea supercluster,
which encompasses 100,000 galaxies stretched out over 160
megaparsecs (520 million light-years). Laniakea weighs

roughly 10^{17} (a hundred quadrillion) solar masses; 100,000 times that of the Milky Way.

Laniakea is an elaborately organized gyre. Within it, galaxies flow inwards toward a gravitational valley called the *Great Attractor*. *Laniakea* is Hawaiian for "immeasurable heaven"; an oddly inapt name in that the supercluster has an approximate measure and is a relatively small part of a much larger universe.

℘ Large Quasar Groups ℞

As quasars are the bright half of galactic formation (black holes being the shadowy opposite), astronomers refer to an oversized galactic cluster as a *large quasar group*. Owing to the dynamics of black hole affiliation, quasars tend to clump together.

ෂ Uniformity ඥ

Everyone knows that the universe is inhomogeneous. To idealize such a complex structure with a homogeneous solution is a bold idealization. ~ German cosmologist Thomas Buchert

Einstein's theory of general relativity underpins modern cosmology. Matter curves spacetime around it by the force of gravity.

The spatial aspect of gravity is apparent with our feet on the ground and things not flying about willy-nilly. That gravity loosens its tug with distance celebrates celestial orbits without stretching the imagination.

It is harder to envision that clocks are inherently inconstant depending upon the gravitational environment: that time is zippy in galactic cores and grows sluggish in the voids between galactic clusters. But it is a proven fact.

When Prussian astronomer Nicolaus Copernicus laid out the *Copernican principle* in the 16th century, it scandalously cast Earth as *not* the center of the universe. Modern cosmology extended this into the *cosmological principle*: that Earth is nowhere special at all, nor is anywhere else.

ᛤ Cosmological Principle ᛤ

The premise of the cosmological principle is that the distribution of matter in the universe is homogeneous and isotropic when viewed on a large-enough scale. Homogeneity implies that the cosmos is much the same everywhere. Isotropy presupposes that the universe looks roughly the same from any viewpoint.

Both assumptions are false. The cosmos today is made of clusters of galaxies, strung along filaments of matter distributed around the boundaries of massive bulbous voids, which account for 10% of matter density, but cover over 60% of cosmic volume.

The cosmological principle supposes that the voids and galaxy clusters average out into uniformity. If the cosmological principle holds, no galactic structure should exceed 370 megaparsecs. The Milky Way galaxy is 0.1 Mpc. Typical galactic clusters are 2–3 Mpc. Large quasar groups are typically 200 Mpc or more across.

The cosmological principle has been unhinged by several observations. A large quasar group has been found that is 1,240 Mpc (4 billion light-years). One spherical void in the cosmos is 1.8 billion light-years in diameter. A cosmic construction 5.6 billion light-years across has been found: far larger than the theoretical limit that could support the cosmological principle.

> This structure contradicts the current models of the universe. We don't understand at all how it came to exist. ~ Hungarian astronomer Lajos Balazs in 2015

Finally, the Hercules-Corona Borealis Great Wall is a multiple galactic superstructure that is 3,000 Mpc (10 billion light-years) across.

There is no evidence of cosmic homogeneity; to the contrary. The very idea of universal uniformity is an affront to Nature's demonstrated fondness for diversity at every scale.

Cosmic microwave background (CMB) radiation has been expanding and cooling with the universe since it formed. The CMB is the only indicator of directional continuity. It belies isotropy, as the cosmos looks decidedly hotter in the one direction than the other from where we sit. This dipole anisotropy is conventionally explained by Earth's movement through space via the Doppler shift, which explains relative motion. Approaching objects appear warmer, receding objects cooler.

Take account of all the celestial motions through the lens of Doppler shift, and the hot and cold patches of the CMB supposedly melt away for the most part. But Earth's place within the galaxy skews the view, as the Milky Way heads toward a greater density of matter while moving away from a void. This gives the impression of slowing cosmic expansion ahead, while the void behind produces the opposite effect. These polar discrepancies account for nearly all the dipole anisotropy, and so color our view of the entire universe. Earth *is* in a special cosmic position, as is most everywhere else.

Via gravity, matter causes spacetime to warp, creating a closed or open curvature to the cosmos, depending upon total mass. How much matter is in the universe determines its fate. Too much matter and spacetime will eventually ball up into a crunch. Too little and spacetime curves outwards, creating an open geometry of eternal cosmic expansion.

Based upon appearances coupled to the cosmological principle, the universe was long presumed *flat*: a sweet-spot matter density that avoided much curvature. Einstein's equations predicted a slightly open universe, with eternal expansion gradually slowed by gravity. Cosmological observations tentatively confirm this.

General relativity conjoins space and time. If space expands at different rates in different places based upon gobs of matter, so too time flows at different rates depending upon matter density. Relativity theory argues that the cosmological principle is unsound.

Without the cosmological principle, the age of the universe is complete conjecture. Further, the large-scale size and shape of the cosmos becomes unknowable without a point of view that cannot be had.

Until recently no one doubted that the universe was homogeneous, so everyone just used the simplest models. Once you say that the universe can be very different in other parts of space, then you open up a can of worms. It would just be incredibly more complex to do cosmology. ~ American physicist Paul Halpern

The foundation of modern cosmology is flawed. At the most basic level, human comprehension of the physical universe is slight.

❧ Stars ☙

You must have chaos within you to give birth to a dancing star. ~ German philosopher Friedrich Nietzsche

A stellar mass forms when gravity squeezes a dense portion of a molecular cloud into a ball of plasma, consuming and releasing heat.

When temperatures approach 10^7 K, atomic collisions become so energetic that they spark nuclear fusion. Hydrogen molecules lock into an irreversible embrace, becoming helium while releasing prodigious energy in an exothermic reaction. A galaxy gains a shiner.

When a star is born, it can have a mass 0.1 to 100 times that of the Sun. This property controls a star's influence on its environment, its lifetime and even its ability to host habitable planets. ~ English science writer Nate Bastian

As a clump of gas tries to collapse under gravity, it hots up. The heat creates radiation pressure that opposes gravity. Unless a star can shed some of this heat, collapse stalls.

Silently, one by one, in the infinite meadows of heaven, blossomed the lovely stars, the forget-me-nots of the angels. ~ American poet Henry Wadsworth Longfellow

The first stars were hydrogen gas ablaze, rather terrible at shedding heat. These protostars accumulated hydrogen fuel, but the high pressure prevented them from forming a dense core. This left them unable to collapse into fusion reaction, which would drive much of the surrounding gas back out into space.

Instead, early stars gorged on gas until they had built a massive, diffuse core. The first stars could have been a million times as massive as the Sun.

The above scenario is one of many that may have transpired. Feedback loops that act on hydrogen gas as it collapses could have fragmented collapsing clouds, creating stars just a few tens that of solar masses.

While gigantic stars would have lived fast and died young, smaller stars churned through their nuclear fuel more slowly. Regardless of size, the earliest stars ended their existence in fiery supernovae before collapsing into black holes. Supernova explosions seeded the interstellar medium with an initial inventory of heavier elements, including oxygen, carbon, and silicon, while leaving behind diminished, dense neutron stars.

Hundreds of millions of supernovas have come and gone in this greedy cycle of gorge and regurgitate, with the residue as starter material for further cosmic construction. Star dust is the material medium of a maturing universe.

The tug of gravity and nudge of coherence gamed the cosmos into galaxies. A homogenous beginning begat a heterogeneous universe from the earliest blips in being. Galaxies evolved in a vast variety of formations.

Likewise, stars shine with a wide range of expanse, from 10% that of the Sun to at least 100 times more massive. A star's size depends upon how much fodder is found in the vicinity.

Star formation is an accretion ballet based upon fluid dynamics. The planetary ecosystem that evolves nearby has much to do with the feeding of a growing star, and how cosmic building material is distributed.

¤ ✧ ¤

A faint star in the constellation Leo (The Lion) has been shining for over 13 billion years. Based upon its scarcity of source material, the star's longevity defies comprehension.

A widely accepted theory predicts that stars like this, with low mass and extremely low quantities of metals, shouldn't exist because the clouds of material from which they formed could never have condensed. ~ French cosmologist Elisabetta Caffau

The way in which stars form depends on their gestation habitat. Star origination is but an episode in galactic evolution which resembles an organic process in its interrelated growth and decay over an astronomical time frame. The local dynamics of star-making depend upon the galactic ecosystem.

Stars do not typically form in isolation. They are instead born in batches, cradled together within a cloud of dusty gas. These clouds are surrounded by a fog of hydrogen that interacts with the clouds. Depending upon galactic dynamics, hydrogen fog may decrease cloud pressure, suppressing star formation instead of fanning its flames.

Star systems come in myriad forms. There can be single stars, binary stars, triple stars, even quintuple star systems. ~ American astronomer Lewis Roberts

Planets can form within all the different types of star systems, in a vast diversity of sizes and orbits. Multiple star systems produce an abundance of massive planets.

Star formation peaked when the universe was a few billion years old. It has declined steeply ever since, as the supply of molecular hydrogen gas that fuels new stars has dwindled.

70% of the original gas is locked up in white dwarfs, neutron stars, and planets. 66% of the rest is spread thin in the intergalactic medium. Only a small portion is shed by stars at various stages of their lives, or recycled wholesale by supernovae.

The Milky Way suddenly stopped birthing stars after it formed its thick, saucer-like disc ~9 billion years ago. The galaxy resumed forming stars after this sudden die-off, but at a much slower rate.

Star formation boils down to a battle between gravity and other things, like turbulence. ~ American astronomer Katherine Alatalo

The Milky Way's bulging disc and bar-like concentration of stars at its center stir galactic gas, injecting energy that keeps gas from coalescing, thus arresting star formation. The mass of stars in the Milky Way's central bulge is ~20 billion times the mass of the Sun.

Half the stars in the Milky Way have a companion, traveling as a binary system. Infant triplets are not unusual.

Partnered stars are often torn apart by a collision on the galactic dance floor sometime during their lives, often in their infancy. By this, the population of binaries is diminished before the stars spread out into the wider galaxy.

⚸ Fusion ♌

Fusion is chemical fury. Even the lightest element, hydrogen (H), does not easily give up its molecular independence (as H_2). H_2 is formed by sharing electrons; easily done because solitary electrons like the company of another.

Atomic protons are anti-social: they naturally repel each other. It takes enormous energy to cajole their union and bang on fusion.

Once stellar fusion starts it sustains itself. The feast isn't over until the food runs out.

Stars prefer light eating: a steady diet of hydrogen is imminently combustible. At the extreme pressures during star formation, and as an active star, hydrogen exists in a unique phase: as a honeycombed 3D matrix interlaced with free-floating molecules. Some of the hydrogen slurry fuses into helium because of the intense gravity inside a star.

As a star evolves, the number of atoms in its core decreases, but its luminosity increases. The Sun has gained 30% in brightness in the past 4.5 billion years.

Hydrogen consumption typically lasts for billions of years, with helium accumulating along the way. This is a star's stable period. Then the contraction that birthed the star pauses.

Stars are insatiable consumers. A star has to do what a star has to do – stay hot. Once the hydrogen stock is depleted, contraction commences again. The star heats up more.

At about 10^8 K, helium burns. Helium atoms do not normally bond, but within stars their electrons interactively dance in a way to form an attraction, thanks to hellish temperatures and ferocious magnetic fields.

The pressure in a stellar core is relentless. Heavier elements form, including carbon (^{12}C) and 2 isotopes of oxygen (^{16}O & ^{8}O).

Helium doesn't have the kick of hydrogen. Stars eat through their helium supply within a few hundred million years.

Some stars burn out after consuming their helium, making molten masses of carbon and oxygen; leaving themselves a legacy as a *white dwarf*. A white dwarf may be the size of Earth, or as massive as the Sun.

The *carbon–nitrogen–oxygen (CNO) cycle* is 1 of the 2 sets of fusion reactions by which stars combust; converting hydrogen to helium as a pathway to even heavier elements. The *proton–proton chain reaction* is the other star combustion process. The proton–proton chain reaction predominates in smaller stars, while CNO feeds the fire in stars more massive than 1.3 times the mass of the Sun.

Small stars in binary systems dance to a different dynamic. A white dwarf may pull in its partner, thereby gathering enough energy to collapse before erupting in a stellar explosion.

Bulkier stars burn on: the ones that are at least 8 times more massive than the Sun. Increasingly heavier elements collect through fusion, up to the point of iron. Iron is the final hurrah in a normal star's natural life.

♋ Massive Stars ♌

The most massive stars reach the highest core temperatures because they can release the most gravitational potential energy. ~ American astronomer Jennifer Johnson

The heavyweight stars that drive galactic evolution are truly massive and bright.

These stars are absolute behemoths. They have 15 or more times the mass of our Sun and can be up to a million times brighter. These stars are so hot that they shine with a brilliant blue-white light. ~ Dutch astronomer Hughes Sana

75% of these high-mass stars exist in pairs. Vampirism is common: the smaller star sucks matter off the surface of its larger neighbor. 1/3rd of these stars eventually merge.

A supersized star's afterlife is spectacular. Stars at least 12 times the size of the Sun burn through an increasingly heavy chemical cocktail, including such spicy elements as silicon, sulfur, and scandium, to a manganese tang, right down to the iron core, all in about a day.

Suddenly lacking the energy to maintain their full volume, burned-out stars implode under their own immense gravity, collapsing thousands of kilometers in mere seconds. In their cores, they even crush protons and electrons into neutron mush. Then these stars fiercely explode.

Neutrinos are the engine that drives the exploding star. ~ American physicist John Cherry

Neutrinos are wispy subatomic particles that interact gravitationally among themselves and other particles but are otherwise stand-offish. During star implosion, massive numbers of neutrinos are zipping all about: colliding, changing flavors (type), and in doing so driving energetic interactions and the transformations of other particles. The least substantial subatomic particle is most impressive when other matter is most pressed.

A supernova stretches millions of kilometers, briefly shining brighter than a billion stars. The show may last but a month; but what a show. Gazillions of intensely excited particles fuse, creating a blizzard of element alphabet soup.

Such supernovae typically leave behind charred remains: a neutron star or even a black hole. Stars that are sized 20–25 suns don't explode. Instead, they implode. The crushing cascade of density bottles up the energy into a black hole.

⚡ Neutron & Quark Stars ⚡

A *neutron star* is a stellar husk packed with neutrons. Neutron stars are as compressed as normal matter can be.

In their extant state, these stars can collapse no further because of *quantum degeneracy pressure*: a property of the Pauli exclusion principle, which states that no 2 fermions (subatomic particles of matter) can occupy the same place at the same time.

Owing to furious quantum mechanics, neutron stars stay hot. Their spinning, at up to 1,000 revolutions per second, emits intense electromagnetic radiation that is received as directional pulses. Such a star is termed a *pulsar*: a portmanteau of *pulsating star*.

Pulsars directionally pulse radiation at regular intervals, so precise that some rival atomic clocks for their accuracy in timekeeping. Sometimes these pulses become temporally distorted – what has been termed *starquakes*. The cause is uncertain.

> When a topological defect passes through a pulsar, its mass, radius, and internal structure may be altered, resulting in a pulsar 'quake'. ~ Australian physicist Victor Flambaum

A *neutron* is made up of 3 subatomic quarks: 2 down and 1 up. While protons and neutrons are the core of normal matter, they do not occupy the lowest net energy possible. Under intense gravity, a greater degree of subatomic stability exists.

As it rotates, a neutron star is constantly shedding magnetic field energy. In cosmological time, the star's spin slows. Centrifugal forces that kept the utmost gravity at bay weaken, yielding further squish.

As a spinning neutron star slows down, deep within, under intensifying gravity, neutron quarks shirk to an even lower energy level than normal matter. Neutrons convert to *hyperons*: a soup of up, down, and strange quarks.

The initial nucleation of strange quark matter runs amok once hyperons start to form. Core density increases, melting the star from the inside out. Quarks are liberated from their normally bound state.

A neutron star evolves into a *quark star*. The seed of strange quark matter spreads until it reaches the iron-rich crust. It then separates from the crust, collapsing into even greater density.

An intense shock wave is generated when the collapse concludes. A spectacular explosion ejects the crust and leftover neutrons in a *quark nova*.

Quark nova bits slam into earlier supernova remnants, causing another outburst of light, as happened when the star first exploded. The 2nd blast can occur anywhere from seconds to years after the original supernova. Such double explosions, in rapid succession, have been observed in multiple instances.

More conservative thinkers are just not open to the idea that free quarks exist in neutron stars. ~ American astrophysicist Fridolin Weber

⚕ Hypernovae ☡

Hypernovae, which are supernovae at least 140–200 solar masses, are even more explosive than their more petite sisters – leaving absolutely nothing behind as core material. Part of hypernova explosive power stems from production of matter-antimatter particle pairs, which are particularly antagonistic toward each other in such a high-energy setting.

These hypernovae are rare, but particularly potent in seeding the next generation of cosmic matter consumers and doing so with the most energetic explosion possible.

If matter becomes hot enough it can emit photons so energetic that they can collide and convert into other particles, notably pair production of an electron and a positron, the electron's antiparticle. Pair production results in matter at much lower pressure. This deadweight intensifies the collapse of a hypernova, causing a runaway reaction that results in an energy release that exceeds the star's entire gravitational energy. The inevitable explosion obliterates the hypernova, leaving behind only an expanding cloud of the elemental debris synthesized from the terminal fury.

හ Planets infinity

Most stars have planetary systems, probably like our own. ~ American astronomer Debra Fischer

A *planet* is a major celestial body, massive enough to be rounded by its gravity, but not so sizable as to catch fire in thermonuclear fusion.

The term *planet* is ancient, tied to mythology as firmly as to astronomy. Early cultures often considered planets themselves divine, or at least the emissaries of deities.

Even after astronomy supplanted astrology, solar planets were named after mythological beings, as were many moons. Then, with the 1851 discovery of 2 moons orbiting Uranus, English astronomer William Lassell started a tradition of naming Uranus' satellites for characters in the works of William Shakespeare and Alexander Pope.

¤ ✧ ¤

Planets are a byproduct of star formation. They emerge not long after a star gets its start. As a young star gathers sustenance from surrounding clouds of gas and dust, the incoming material forms a flat, spinning disk around the aspiring star.

Planets start as small clumps within a stellar disk, managing through gravitational attraction to put on weight. As a planet pulls in more material, it leaves a wake in its trail, creating a gap in the stellar disk of dust.

Despite planets creating gaps, a star still gets fed. Streamers of gas from the outer portion of a disk are pulled in by the planets, adding to their girth. Alas for the nascent planets, much of the pulled plumes pass the planet that enticed them. In overshooting, the filaments head inward, further feeding the star.

As the universe aged, an unfathomable number of star systems formed. While stars now number 100–300 billion trillion, the number of planets within and without star systems are many times that.

The growing embryos of large planets suck in hydrogen and helium from the disk encircling an infant star. These become gas giants, such as Jupiter in our solar system.

Planets that form too close to a star risk being eaten. As a star grows, it may gobble a gas giant in close orbit.

Smaller planets accumulate debris and form a rocky core. Cosmologists do not know how large a rocky planet may be before its girth lets it become a gas giant. A rocky planet has been found that is 17 times Earth's mass; far outside the hoary 10-times rule of rocky cores turning into gas Goliaths. This illustrates how little is known of planet formation dynamics.

Solar systems with rocky planets first formed after some unknown threshold was crossed in accumulating heavy elements in natal interstellar clouds. From this horizon arose a vast variety of planetary *metallicity*: elements exclusive of hydrogen and helium.

> Unlike the gas giants, the occurrence of smaller planets is not strongly dependent on stars with a high content of heavy elements. Planets that are up to 4 times the size of Earth can form around very different stars; also stars that are poorer in heavy elements. ~ Dutch astrophysicist Lars Buchhave

22–40% of the planets around low-mass stars like the Sun are in a habitable zone. Even those that are not, such as Mercury, may have regions livable to extremophilic microbes. Further, as Europa and Titan evidence, planetary moons may offer a home to organisms. In the face of enormous cosmic diversity, the universe seems to favor life.

♂ Starless Planets ♀

The Milky Way is home to 200 billion free-floating planets that do not orbit a star. Many were tossed from a star system: flung for having the wrong trajectory to fit in.

Planets can come together on their own, rather than synthesizing from the leftovers of a star on the make. Small round clouds shear from the dusty pillars of gas sculpted by young stars. These balls of debris are pushed from the center of a nebula by radiative pressure from the hot stars there. So sped on their way, such clouds develop the spin to coalesce into planetary free floaters that never catch fire, nor become attached to a star system. Their hardscrabble upbringing means that freeborn planets invariably are compact, with dense cores.

Brown dwarfs, sometimes called failed stars, are bodies with a mass between that of planets and stars. Akin in outcast status, they are big brothers to the rogue rocks that roam the cosmos without a star to call their own.

ঙ The Solar System ଔ

It's time for the human race to enter the solar system.
~ American politician Dan Quayle

Our solar system got its start less than 5 billion years ago. A shock wave from a supernova explosion created a debris cloud that collapsed to form the Sun and its satellites. The cloud also cradled the birth of hundreds of thousands of other star systems. By that time, among the trillions of star systems throughout the universe, many millions had already matured to have planets teeming with life.

Until the mid-1990s, cosmologists embraced the *core-accretion theory*: a standard model which explained star system formation using a few basic principles of physics and chemistry. The theory accounted for every major feature of the solar system, yet it turned out to be inapt.

From the mid-1990s, exoplanets in other star systems kept being found that were inexplicable, such as gas giants the size of Jupiter in close, tight orbits around their stars. The accumulation of such discoveries trashed the core-accretion theory. Space oddities continue to be found, leaving no coherent account of how planetary systems come into being.

One feature found in many systems is planetary migration. All sorts of planets grow to full size in the middle to outer part of a solar disc before moving inwards. This can sometimes cause other planets to shift their orbits outwards. Such shenanigans occurred during the development of our solar system.

Our star system itself is coursing the cosmos at 20 kilometers per second. The solar system makes a rotation around the Milky Way galactic disk every 230 million years.

ঞ History ৯

Our solar system closely resembles other observable planetary systems within our galaxy. ~ Dutch astronomer Martin Bizzarro

For millennia, most humans thought themselves at the center of the universe: the Earth stationary, while celestial bodies moved through the sky. Ancient Greek astronomer Aristarchus of Samos first speculated that the Earth orbited the Sun.

As Christianity became the dominant European religion, church authorities estimated Earth's age by counting the number of generations since Adam made his appearance in the biblical book of Genesis. The answer: Earth got its start between 4000–7000 BCE.

Then there was the issue of Earth being the center of the universe. The Catholic Church had no doubt of it (faith and doubt being antithetical). Looking into the heavens with a more open mind led to a different conclusion.

1,800 years after Aristarchus, Copernicus developed his heliocentric system in 1513, with the Earth revolving about the Sun. Given what was *not* known at the time, the Copernican notion seemed ridiculous.

The size of the Earth was known. Nothing could explain the power it would take to make such a massive orb move.

Conversely, the motion of celestial bodies was easily explained. They were stirred by swirling aether, a substance not found on Earth. No less an authority than Aristotle had stated such in the 4th century BCE, and Aristotle was esteemed by the Church at this time.

What Copernicus proposed had profound implications for the size of the cosmos, and even of individual stars.

Stars look to have fixed widths. Both ancient Egyptian astronomer Ptolemy and Danish astronomer Tycho Brahe had measured them with their naked eyes.

Knowing nothing about optics or the nature of light, stars under the Copernican conception would be absurdly enormous, and the girth of the universe unimaginably ample.

Early supporters of Copernicus felt compelled to invoke God in his defense.

> Grant the vastness of the Universe and the sizes of the stars to be as great as you like – these will still bear no proportion to the infinite Creator. ~ German mathematician and astronomer Christoph Rothmann in the late 16th century

The day soon came when God and Copernicus were not so aligned in the sights of Catholic authorities. Italian astronomer-mathematician Galileo Galilei was convicted of being "vehemently suspect of heresy" by the Catholic Church in 1633 for buying into Copernicus' heliocentricity, forced to recant, and spent the rest of his life under house arrest. Compared with others, he got off lightly.

The Catholic Church amassed quite a track record of solemn folly when it came to science. Into the 17th century, Christendom regarded fossils as images of God's creation, put on Earth for man's admiration: God the decorator; nice touch.

It was not until the 17th century that the mathematics of the planets orbiting the Sun were worked out. Isaac Newton mathematically described gravity from an everyday point of view. A couple of centuries later, Einstein suggested a radical view of gravity: as an entropic distortion rather than a fundamental force. Einstein's gravitational theory changed astrophysics to a degree greater than Einstein himself was able to accept.

♒ Emanuel Swedenborg ♒

The inner self is as distinct from the outer self as heaven is from Earth. ~ Emanuel Swedenborg

In 1734, Swedish scientist, theologian, and Christian mystic Emanuel Swedenborg developed the *nebular hypothesis*: that the solar system formed by swirling accretions of matter; a surmise explaining star system formation which has stood as essentially correct for 3 centuries.

Swedenborg also had prescient concepts concerning the cerebral cortex, the nervous system, and the functions of the pituitary gland. Swedenborg was a century ahead of others in anticipating nerve cells.

Swedenborg regarded eating meat as "something profane."

There are at least 3 reputable incidents of Swedenborg's psychic power.

In 1758, the Queen of Sweden asked him to tell her something about her deceased brother. The next day he whispered in her ear a fact that turned the Queen pale. She explained that what Swedenborg had told her was something only she and her brother knew about.

In 1759, Swedenborg made an accurate prediction of fire threatening his home when he was hundreds of kilometers away at the time.

The 3rd incident was Swedenborg telling a woman who had lost an important document where it was located; information known only to someone recently deceased.

Swedenborg wrote about the relation between the finite and the infinite, and how the soul and body interconnect. Swedenborg believed that higher knowledge was received wisdom.

❀❀❀

The self-congratulatory religious belief that humans are the most important life in the universe, ever at the center of things, is not easily forsaken. Early cosmologists convinced themselves that Earth lay near the heart of the galaxy.

The 1st map of the Milky Way, compiled by German-born English astronomer William Herschel and his sister Caroline in 1785, showed the solar system in the middle of a starry puddle. Instead, the Sun stirs 27,400 light-years from the galactic center, on a spiral arm termed the Orion–Cygnus Arm, sauntering around the Milky Way. As to our place in the cosmos, we have no idea where the center of the universe is.

꧁ The Sun ꧂

We need not hesitate to admit that the Sun is richly stored with inhabitants. ~ William Herschel in 1818

4.56 BYA, accretion of a hydrogen-laden protostellar core, covered in a cloud of dust, hit a critical point. Gravity was working its magic: compressing accreted matter, generating heat. The molecular cloud collapsed into a condensed ball; as perfect a sphere as Nature can make. Along with 1,000+ siblings, the Sun emerged from a stellar nursery that had formed 30 million years earlier.

This cluster of stars was birthed further out in the galaxy. The Sun astronomically inched up the galactic arm, fortuitously never suffering the casualty of collision with another star.

The Milky Way's frenzy of birth stars peaked 10 billion years ago. That makes the Sun a late bloomer, galactically speaking.

50 of the Sun's siblings should still be in the neighborhood of 300 light-years. 400 in the family are likely within 3,000 light-years.

By the time the Sun was visible, the dust had practically cleared, leaving a ball of hydrogen (74.9%), helium (23.8%), and various heavier elements (1.3%). About 12 million years after formation, the Sun began to radiate. It's been hotting up ever since.

Most stars like the Sun rotate relatively quickly in their infancy, spinning once every few (Earth) days before slowing down as they age. The early Sun was a slow rotator: completing a rotation just once every 9 days. This slow rotation was fortuitous for engendering life in the solar system, as rapidly rotating stars fling more plasma into space, battering planets in their orbit.

¤ ✧ ¤

The Sun and its atmosphere are layered. From the inside out, the solar interior comprises: a core, a radiative zone, and

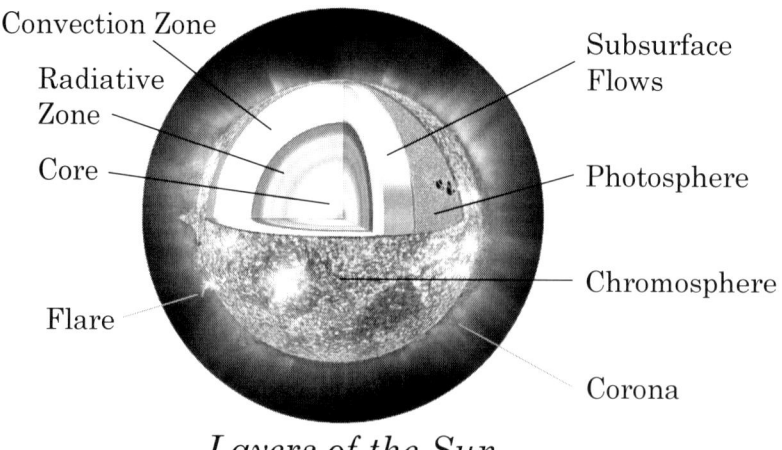

Layers of the Sun

a convective zone. The solar atmosphere consists of the photosphere, chromosphere, a narrow transition region, and the corona. Beyond is the solar wind.

The *core* extends a quarter of the way to the Sun's surface. Though only ~2% of the Sun's volume, the core is 15 times the density of lead, and has nearly 50% of the Sun's mass.

In the core, which is at 16 million K, fusion reactions consume hydrogen to form helium and send energy to the surface, which blazes away at 6000 K.

When matter burns, it is a chemical reaction that merely rearranges the electrons of the elements involved. One billionth of matter's mass is lost. In contrast, fusion changes the nature of the inflicted elements. Fusion is a billion times more efficient than burning in energy release. The Sun is not burning, but it is converting ~4.1 billion tonnes of matter into energy every second.

The *radiative zone* extends from the core to 70% of the way to the surface (*chromosphere*); making up 32% of solar volume and 48% of the Sun's mass. Light created in the core scatters throughout the radiative zone. It may take a million years for a photon to make its way out.

The core of the Sun produces gamma rays. As that energy radiates outward, it is absorbed and re-emitted. By the time the energy reaches the surface and heads into space, emissions are primarily visible light and infrared radiation (heat).

In the *convection zone*, heat generated at the core moves to the surface by overturning convective motions, which take cellular form. Though the convection zone accounts for 66% of the Sun's volume, it is only 2% of solar mass.

The roiling convection cells which dominate this zone come in various sizes and shapes. Banana-shaped cells transport angular momentum toward the Sun's equator, providing a critical assist in maintaining differential rotation in convection zones between equatorial and polar regions. This differential affects the Sun's magnetic field flux.

The *photosphere* is the lowest layer of the Sun's atmosphere, emitting the light we see. The photosphere is 500 km thick. Most of the light comes from the lowest third.

The *chromosphere* is entirely made up of spiky structures called *spicules*, which are typically 1,000 km across, and up to 10,000 km high. Spicules last about 15 minutes.

Outside the chromosphere is the *corona*, a wispy crown of blazing heat: 1 million K. The corona is 300 times as hot as the Sun's surface, owing to 10-million-K streams that shoot up from the surface.

Corona

High-speed plasma jets and magnetic tornadoes propagate from the chromosphere to the corona, distributing heat and depositing their own energy. While much of the gaseous heat streams outward, some falls back onto the Sun's surface as coronal plasma rain.

The mechanism behind all coronal irruptions is *magnetic reconnection*: the rapid shifting of magnetic topography in the Sun's plasma, with magnetic energy converted to heat, kinetic energy, and particle acceleration.

> Magnetic reconnection is the dominant mechanism by which solar wind energy enters Earth's magnetosphere. This energy is subsequently dissipated by geomagnetic substorms and aurorae. ~ American astrophysicist Roy Torbert *et al*

The strength of the Sun's magnetic field is typically only twice that of Earth's. But the Sun's magnetic field frequently becomes highly concentrated, spiking to 3,000 times normal. Twists and kinks in the magnetic field develop because of discrepancies in the Sun's spinning, both from the inside out, and between the equator and poles.

The Sun spins about every 25 days, at an angle of 7.25° to the planetary disk around it. Because of its tilt, a good gander at the solar poles eludes Earthlings.

Earth spins once on its axis every 24 hours (a day), and travels around the Sun every 365 days (a year).* In contrast, the Sun's schedule is convoluted. While a day at the equator lasts 25 days, polar regions are pokier, taking a few days longer to complete a rotation.

The layers of the Sun invoke a complex convection dynamic that modulates energy production and distribution. The Sun's core rotates nearly 4 times faster than its visible outer shell (photosphere); and there is even a slight difference in rotation between the top and bottom layers of the photosphere.

The interior of the Sun has complicated rotation because of turbulence, but the outer layer of the Sun isn't turbulent. The photosphere is stable, so it's surprising that there would be any gradient in its rotation. ~ American astronomer Jeff Kuhn

Photons created in the Sun's core ricochet around on their way out, gaining a bit of momentum from each atom they bounce off. When a photon finally leaves the photosphere, it carries its acquired momentum with it.

We tend to think that the Sun is yellow. That impression owes to atmospheric filtering which reduces higher energy blue light. In space sunlight looks white.

As photons radiate from the Sun at different angles, the gas of the photosphere gets a backward push from the photonic brawl to get out. The shove associated with each departing photon is negligible, but the barrage adds up, and accounts for the rotational discrepancies between photosphere layers.

The uneven spin creates distortion in the Sun's magnetic field, with knock-on effects. The Sun's magnetic field gets wound up and then releases pent-up energy as a *solar flare* or as a more furious expulsion from the corona (*coronal mass ejection*).

---

* Earth orbits the Sun every 365.242199 days, coursing at 108,000 km/h, covering 940 million kilometers each orbit.

As the equator spins, it drags the magnetic field that connects the Sun's poles. This creates a solar magnetic cycle.
This cycle, from magnetic twist to energy discharge, runs
11.07 Earth years; corresponding precisely with alignment of
the Sun, Venus, Earth, and Jupiter – a resonance of celestial
bodies.

> Planetary tidal forces act as minute, external pace setters. The
> impulse for this oscillation requires almost no energy. If you
> only just give a swing small pushes, it will swing higher with
> time. ~ Frank Stefani

Corresponding to the magnetic cycle presides a cyclic
prevalence of sunspots. Their greatest appearance is at the
solar maximum, during which the corona sloughs off the
magnetic fingerprint of the previous cycle by sweeping it to
the poles.

The fluxing solar magnetic field reverses direction around
the time of the solar maximum. When this does not happen,
the Sun is heading for a *solar minimum* (of sunspots). While
eruptions still occur, spots do not form on the surface because
their magnetic power is not strong enough to overcome the
convective mixing of gas at the surface, of which the jet
stream is part. Previous spots disappear.

The Sun's magnetic turbulence drives various energy
flows. Incited by magnetism, tornadoes 1,500 km wide rise
from the surface into the corona. Each lasts 10–15 minutes.
There are 11,000 tornadoes raging on the Sun at any time.

Sunspots, solar flares, and the solar wind are dynamically
interwoven with the Sun's magnetic fluctuations. The Sun's
magnetic field extends into space, rendering an *interplanetary magnetic field*.

Sunspots are eruptions onto the surface, from perturbations in deeper magnetic currents, which themselves interrelate to the convective mixing on the surface. Sunspots create
local drops in temperature, rendering the surface gas darker;
hence the spots. Sunspots are indicative of solar flares and
coronal mass ejections. These outbursts disrupt satellite and
radio communications on Earth.

There is an east-west jet stream near the Sun's surface, influenced by convection currents below. Changes in the flow of the jet stream presage changes in the pattern of sunspots by 4 years.

Energetic outflows from the corona power the *solar wind*, which blows a plasma of charged particles into the surrounding interstellar medium, creating a bubble-like *heliosphere*. As an interplanetary plasma medium inflated by the solar wind, the heliosphere defines the domain of the Sun. It creates an interstellar interaction with the dust and gas clouds that it moves through.

The Sun moves at 83,500 kilometers per hour.[*] The heliosphere carries the Sun's magnetic field, which is overridden by planets that have their own magnetosphere, such Earth and Jupiter. The galaxy's own magnetic field sculpts the heliosphere's boundary with interstellar space.

¤ ✧ ¤

The Sun has consumed about half of its hydrogen, leaving another 5 billion years of peaceful production. After that, the Sun will become a *red giant*: outer layers expanding as the hydrogen fuel in the core is consumed, causing the core to contract and heat up.

Following its red giant phase, the Sun will throw off its outer layers, creating a *planetary nebula*: a recycling of material to the interstellar medium, providing fodder for further galactic evolution. The remaining core will slowly cool and fade, over billions of years, transforming the now-bright beacon into a diminished white dwarf.

A day without sunshine is like, you know, night. ~ American comedian Steve Martin

---

[*] The heliosphere is not moving fast enough to create a shock wave, as long thought.

ಞ   The Planets   ಛ

> Since nothing prevents the Earth from moving, I suggest that we should now consider also whether several motions suit it, so that it can be regarded as one of the planets. For it is not the center of all the revolutions. ~ Nicolaus Copernicus

The Sun captured 99.8% of local system mass.* Jupiter grabbed most of the leftover matter, leaving Saturn as an also-ran. Jupiter's early formation is the reason that the inner solar system has such puny planets.

The early formation of the solar system was a carnival of debris collisions. The basic idea that tiny grains stick together into a gathering ball and swoop up gas conceals many levels of intricacy.

A chaotic interplay among various mechanisms occurs during star system development. A divergent diversity of outcomes is possible. The formation of the solar system is physics applied to happenstance.

Though shifts in the solar system were affected by the sweeping motions of both gas giants, Jupiter's movements were decisive in the solar system's early evolution, owing to its girth: 2.5 times the mass of all other planets combined.

4 rocky bits stabilized into the inner planets, all terrestrial: Mercury, Venus, Earth, and Mars.

In geological time, planets form fast. Jupiter gained its great girth within a very few million years of the solar system starting up. Transforming nebular dust into a rocky planet takes only tens of millions of years.

Planetary core formation begins within a million years of the first solids beginning to condense. Once a few hundred kilometers in radius, a planet on the make is large enough to retain heat.

Thermodynamics spur transformation. Chemical components differentiate by weight. Molten metals sink to form a core. Lighter liquefied silicate rises to form a crust.

The inner 4 have rocky outer shells and metallic cores. The family resemblance ends there.

---

* About 1 million Earths could fit inside the Sun.

Earth and Venus are roughly the same mass, size, and composition, but Earth is swathed in a life-sustaining atmosphere, while the Venusian atmosphere is acid-laced, crushingly dense, and hot enough to melt lead. Venus sports a single, immobile shell of rock. Earth is encased in tectonic plates topped by cruising continental crusts. Earth is oceanic. There is no sign that Venus ever hosted an ocean.

Earth's churning iron core generates a magnetic field. Earth has a large moon that sways its tides, and it rotates 365 times per orbit. In stark contrast, Venus is moonless and bereft of a magnetic field. Venus rotates, but backward, and less than once per Venusian year, which is 224.7 Earth days.

The more diminutive pairing of Mercury and Mars is another study in contrast. Mars, with 11% of the mass of Earth, has twice the mass of Mercury, but its core-generated magnetic field sputtered out early on.

While smaller, Mercury is denser than Mars. Minuscule Mercury is still spinning a magnetic field, albeit a weak one, centered far from the center of the planet.

Of the first 4 planets in the solar system, only Mercury lacks an atmosphere. Mercury is so close to the Sun that it is constantly scoured by the solar wind.

Mercury's huge metallic core is cooling. It has shrunk 11 km since its formation. Part of this owes to Mercury's thermal character. With no atmosphere to retain heat, Mercury's surface ranges from 100 K (–173 °C) during the night to 700 K (427 °C) during the day at the sunny equator; the greatest variation among solar planets.

A migrant that made its way in from further out in the solar system, Mercury is the most deeply cratered of the solar planets.

With a rotational axis perpendicular to its orbital plane, Mercury's poles never tip to the Sun. Hence, many of its polar craters never see the Sun. Deep within they preserve a trillion tonnes of water ice.

Closer to the rim of craters, where the ice warms to a watery sheen, organic matter has been discovered. While "organic" should not be confused with "biological," the possibility of mercurial microorganisms exists.

Mars formed rather quickly, within 2 to 4 million years. This hasty accretion created a planetary structure with a less intense convection dynamic, making for modest magnetic mojo.

Mars should be 1.5 to 2 times the mass of Earth. Mars is instead only 10%. The gas left over from Jupiter's formation meddled with the rocks forming Mars, making them fall apart rather than clump together.

¤ ✧ ¤

Jupiter began forming from an ice asteroid 4.5 BYA, 4 times further from the Sun than it is now. Over 700,000 years, Jupiter carved a spiraling path to its current orbit.

4.4 BYA, the newborn giants – Jupiter and Saturn – orbited in a tight circle, their orbits influencing each other. Beyond them were Neptune and Uranus, with Uranus on the outside.

As Jupiter and Saturn sauntered into place 4.1 BYA, their gravitational tug flung asteroid fragments willy-nilly. This high-speed slam fest lasted hundreds of millions of years, all the while altering the participants' chemistry via the altercations.

During this time, Saturn swung into an orbital period twice that of Jupiter. This further scattered cosmic debris about, bombarding Earth and the other inner planets.

Saturn's shift drove Uranus and Neptune outwards, into the comet belt, causing them to fling these cold bits all over, including hurtling more meteorites toward the inner planets as they sweep clear their orbit. In their swirl they switched places, with Neptune now further out.

Once Jupiter and Saturn settled in, a raft of rocks, numbering in the millions, were held between the tug of the Sun and Jupiter, and so formed an asteroid belt between Mars and Jupiter.

The gravitational perturbations from Jupiter kept the asteroids and debris from accreting into a planet. The extra orbital energy from Jupiter's gravity instead caused collisions that shattered the protoplanets.

From the debris, Saturn spun its stunning signature ring system, including remnants from a vanished moon. Jupiter too has a ring, but quite faint.

During formation, the 2 gas giants gobbled much material that would have otherwise made moons. Only the late starters survived to spin about as satellites.

Jupiter captured 63 sizable moons compared to Saturn's 62.* Jupiter has 4 large satellites, all discovered by Galileo. Saturn has only 1 big moon: Titan.

♎ Titan ♎

Titan is bloody complicated. ~ American astrophysicist Ralph Lorenz

Titan is an exceptional moon. Spawned from giant asteroid impacts, Titan is the only satellite known to have a dense atmosphere, with a layered temperature profile like Earth, though much colder: 93 K (–180 °C).

Solar irradiation produces polymer aerosols, which give Titan its famed orange glow. Titan's atmosphere is laden with organic molecules, formed by sunlight striking the atmospheric methane.

Titan has weather and seasons. The southern hemisphere has lingering clouds during the summer.

Methane plays the role that water does on Earth. Titan has a methane cycle like Earth's water cycle. Hydrocarbon gases condense and fall as methane rain. Titan's surface sports methane seas, lakes, and networks of rivers.

As Titan revolves around Saturn every 16 days, it has tides that are pulled by proximity to Saturn.

Titan does have water, but underground: with a thick ice layer near the surface, and a salty, ammonia-laden watery ocean underneath, heated from below by the core.

Titan has a rock-iron core. Titan does have mantle dynamics, though in fits and starts, unlike Earth's continual movements.

---

* In 2018 Jupiter's moon total was upped to 79. The new outer moons were at most only a few kilometers in diameter.

Titan lacks Earth's plate *tectonics*: the movement of large crustal plates which bump and grind to produce terrestrial effects – a lithosphere in motion. Still, Titan has a geographically active surface. Methane-laden lavas flow from cold volcanoes, replenishing the atmospheric methane that is constantly decimated by solar ultraviolet rays. The volcanic eruptions also carry iced ammonia to the surface, which may mix with the methane and nitrogen to create a prebiotic brew. From that life could emerge. Preliminary evidence hints that methane-munching bacteria may reside on Titan's surface.

> Methane has the disadvantage that it is nonpolar, and hence a poor solvent for the polar compounds necessary for the complex interactions required for life. ~ English ecologist Andrew Clarke

♎ Europa ♎

Europa is the smallest of the 4 Galilean satellites of Jupiter, the 6th closest around the gas giant. Slightly smaller than the Moon, Europa is the 6th largest moon in the solar system.

Europa is a silicate rock with an iron core. Its tenuous atmosphere is primarily oxygen.

Situated past the planetary snow line, Europa's surface is H_2O ice, 15–25 km thick. Europa's face is one of the smoothest in the solar system, albeit pockmarked in patterned scratches.

Europa's surface

Underneath the ice is a dark, global saltwater ocean, 160 km deep. Turbulence in Europa's subsurface sea, inspired by Jupiter's gravitational tugs, causes chaotic cracks on its surface, prompting water plumes that rise 20 times the height of Mt. Everest.

Europa has subduction-driven plate tectonics like Earth, though on Europa, it is an icy shell that submerges into a warmer mantle.

Warmed by internal heat from the core, which creates global convection currents, there may be microbial life in the stormy subsurface sea of Europa.

Saturn is 60% the size of Jupiter, but less than a 3rd as massive, making it the least dense planet in the solar system. Saturn is the only planet less dense than water.

Saturn's famous rings are only 100 million years old or less and are now slowly fading. The rings, which are a composite of rocky bits and ice, are about halfway through their life. 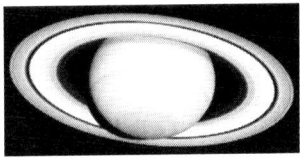 The rings are held in place by magnetic field lines.

> We just missed out on seeing giant ring systems of Jupiter, Uranus, and Neptune, which have only thin ringlets today.
> ~ English astronomer James O'Donoghue

Gas giants are mostly hydrogen and helium. Jupiter's mass creates enormous gravitational pressure that squeezes most of its hydrogen into a metallic fluid that conducts electricity. Compression radically alters electron orbitals in atoms. Chemistry is a different beast under high pressure.

Jupiter has a core of iron, rock, and ice which weighs 10 times as much as Earth. Because of the intense pressure (40 million Earth atmospheres), Jupiter's core temperature is 16,000 K; hotter than the Sun's surface. The convection at the core boundary may be so extreme as to cause the core to slowly dissolve near its boundary, eroding the core into the fluid hydrogen and helium that surrounds it.

Past Saturn are 2 more planets: Uranus and Neptune. Uranus is the lightest of the outer planets. Uranus is slightly larger than Neptune. Both planets are about 4 times the size of Earth. Both have rocky cores.

Neptune's density, and the relative ease of aggregating matter in the inner system, indicates that Neptune formed closer to the Sun before migrating outward.

While the other planets orbit the Sun axially standing up, Uranus orbits the Sun on its side (a 97° axial tilt). Uranus has a much colder core than the other gas giants, and so radiates very little heat into space.

All the planetary orbits have small deviations, which cannot be accounted for by Newtonian physics. Einstein's general relativity theory adequately explains these peculiarities.

Mercury has an especially eccentric orbit: neither circular nor elliptical, but instead rosette-like, its perihelion (closest point to the Sun) precessing (gyral rotation) at more than 43 arc seconds per century. Yet Mercury has the smallest axial tilt (2.11°).

The 4 outermost planets have magnetic fields, though each is at a different tilt to the axis of rotation. Saturn's magnetic field is perfectly aligned with its rotation axis. Jupiter is slightly tilted. Neptune tilts 47°. Uranus is askew a whopping 60°.

Roaming outside the planets – past Neptune – are smaller chunks, still held in sway by the Sun: the Kuiper belt. A Kuiper belt resident, Pluto, was counted as the 9th planet upon its discovery in 1930. Pluto has at least 3 moons.

In 2006, Pluto's mass was unchanged, but it diminished in stature: no longer regarded as a planet proper; instead, merely a "dwarf planet." Pluto was downgraded for untidiness: not having cleared other objects from its orbit.* However rotund a rock one may be, having an icy composition does not win friends.

2 planetoids have been found outside the Kuiper belt. There are likely many more, but observation from Earth is problematic, as objects so far out are faint. The outer solar system is dark.

All told, there are 8 major planets: the iron-core inner 4 (Mercury, Venus, Earth, Mars) and 4 outer gas giants (Jupiter, Saturn, Uranus, and Neptune), along with 5 dwarf planets (Pluto, Ceres, Eris, Haumea and Makemake). Riding planetary shotgun in the system are 162 major satellites.

The planets have a roughly circular orbit, but comets and other cosmic bits have much more elliptical orbits. Some travel as far as the Oort cloud.

---

* Historically, determining a planet had been based upon how the celestial body formed. The criterion of orbit-clearing for defining a planet had previously been mentioned in only 1 publication, dated 1802, and its reasoning since disproven: by that criterion, there are no planets. The downgrading of Pluto demonstrated core incompetence in the International Astronomical Union.

The Oort cloud is teeming with a trillion icy objects. It is nearly a light-year from the Sun; almost a quarter the distance to Proxima Centauri, the closest star to the Sun. The Oort cloud was formed from gyral scatterings created by the orbital wanderings of the 4 gas giants.

Earth

Those who dwell among the beauties and mysteries of the Earth are never alone or weary of life. ~ American marine biologist Rachel Carson

Life may have found a precarious hold on a few orbs in the solar system. Venus might have once supported life. Mars almost certainly did and may still.

Mars once had regular wet seasons, rivers, and lakes. Even now there are discharges of methane from the surface of Mars into the atmosphere. This may be from *serpentinization*: a geological process of rock oxidation and hydrolysis via heat and water. Or it may be microbial methanogens.

Europa is another candidate for life. Even nominally hellish Mercury may harbor microbes, nestled in lakes which never see the Sun.

Earth was something special. Its ability to hold an atmosphere made a difference.

4.55 BYA, Earth came into being by accretion: currents of particles swirling around the Sun collided and coalesced, as with other planets. Molten iron sank to the center, forming the planetary core.

Earth's continued formation was by violence for at least 800 million years, pilloried by 20^{18} tonnes of cosmic debris. Enormous impacts occurred as recently as 1.8 BYA.

Impacting meteorites were stirred into Earth's mantle by massive convection processes. The vast bulk of the planet's precious metals, including gold, came from space after its formation.

The bombardment continues to this day, but it has been reduced to a fine drizzle. Over 3,600 tonnes of extraterrestrial dust a year – 9 tonnes a day – settle on Earth's surface.

By 4.4 BYA a crust had formed. A 100 million years later vast oceans covered the surface. Life made its debut on Earth ~4.1 BYA.

ॐ Dechlorination ॐ

Chlorine is an extremely reactive element – a blatant oxidizing agent that readily strips electrons from those elements it deals with. Though chlorine is an essential dietary element in minute quantities, it is not biologically friendly.

The composition of ancient meteorites indicates that Earth should have 10 times the chlorine that it does. Mars has more than twice the chlorine of Earth despite having suffered much less cosmic assault.

The 4 halogen elements, including chlorine, do not readily dissolve in metals; nor do they often combine with other elements to form rock minerals. Hence, chlorine is concentrated on the surface. Much of Earth's chlorine that is not in the ocean lies in salt deposits and brines.

The relentless bombardment of early Earth engendered life later by scouring much of the chlorine off the planet. If not, the world's oceans would have been too salty for complex life to evolve.

Chlorine-rich seas would have reduced precipitation. With less rain, there would have been less erosion, and fewer nutrients washing into the ocean to foster life.

ॐ The Moon ॐ

> When a finger points to the Moon, the imbecile looks at the finger. ~ Chinese proverb

~4.51 BYA, much of Earth's iron had sunk towards the core when a planetesimal the size of Mars – *Theia* – smacked the Earth at an oblique angle. Theia was the goddess who gave birth to Selene, the Moon.

Like a caroming billiard ball, Theia rebounded, but was captured within Earth's gravitational pull. Theia took with it a divot from Earth: adding a clump to what would become the Moon.

On its bounce back into space 4.5 BYA, the Moon was, for a while, but 20,000 kilometers from Earth: exerting tremendous pull that buckled Earth's crust with each lunar rotation. Earth spun on its axis in a 5-hour day. The solar cycle was the same, but there were over 1,750 days per year. Meanwhile, the *lunar cycle* – from one full moon to the next – was a mere 1.5 days.

Early on, the Earth and Moon loomed large in each other's skies. Because the Earth and Moon were tidally locked from the beginning, the still-hot Earth radiated its heat on the near side of the Moon. While the far side cooled, the Earth-facing side remained molten. This temperature gradient crucially affected crustal formation on the Moon.

The 2 sides of the Moon are strikingly different. The near side is low and flat, rich in rare earth elements. The far side is mountainous and heavily cratered. The evolution of the Moon accounts for the bifurcation.

> The thermal gradient created by Earthshine produced the chemical gradient responsible for the crust thickness dichotomy that defines the lunar highlands. ~ American astronomer Jason Wright *et al*

Early meteoroid impacts on the Moon's near side punched through the crust, releasing vast lakes of basaltic lava that crafted the large, dark plains which form the Moon's visage. These basins were dubbed *maria* – Latin for "seas" – by early astronomers who mistook them for actual oceans. Their darkness owes to heavy iron concentration, which is less reflective than surrounding crust. Maria cover 16% of the lunar surface, almost all on the visible side of the Moon.

Meanwhile, bolides that struck on the far side hit crust too thick to puncture, making their mark with craters and highlands, but scant maria.

The Moon is now a bit squashed, with an equatorial bulge. Tidal heating during Moon formation thinned the polar crust while thickening crust in regions in line with the Earth. That, along with rotational forces, left a lemon-shaped bulge on the side facing the Earth, and a counterpart distention on the opposite side.

Whereas the combined mass of the outer planets' satellites is less than 0.1% of their parents, the Moon is ~1% of Earth's mass. Even more important, the Moon contributes 80% of the angular momentum of the Earth–Moon system. For the outer planets, this figure is less than 1%. In sum, the Moon's effect on Earth is much greater than other moons in the solar system.

The Moon-forming event birthed Earth's seasons, facilitating the equator-to-pole heat conduit that rendered the planet more habitable.

The impact from the emergent Moon blew away much of Earth's early atmosphere, which had been captured by gravity from the solar nebula before the nebular cloud dissipated.

Gases were released from Earth's mantle into the atmosphere. Nitrogen, which is relatively unreactive, outgassed early, and remains the predominant atmospheric gas (78%).

Earth's atmosphere had enough carbon dioxide to attenuate solar infrared radiation. This helped stabilize Earth's surface temperature.

As the Moon receded from the Earth, and its orbit stabilized, Earth's day length increased slightly, and the amplitude of Earth's ocean tides lessened. Ocean tides are produced by a combination of the Sun and the Moon's gravity, along with Earth's rotation, creating bulges of water on opposite sides of the planet.

Collisions like the pairing of the Moon and the Earth are cosmically common. 8% of Earth-sized planets may have captured a moon. These typically occur as a star system is forming, as happened here.

The gravitational pull of a big nearby satellite keeps a planet from tilting too much on its axis, and so helps stabilize planetary rotation. The tidal effects from a moon's orbit may also be beneficial, as they are on Earth. A moon may encourage the prospects of a planet birthing life and sustaining it.

The Moon does more than sway ocean tides. It also tugs on Earth's crust, triggering massive earthquakes along fragile fault lines, especially during new and full moons, when the Earth, Moon and Sun are aligned. Thus, the Moon has contributed to the Earth's geophysical evolution.

The gravitational field of the Moon is the most varied in the solar system. Gravity anomalies, termed *mascons*, come from matter compression – often caused by meteorite impacts, but sometimes due to dense basaltic lavas.

The Moon was long thought dry, but water molecules are widely distributed over the lunar surface, as well as locked up in icy crust enclaves within, bound to phosphates in volcanic rocks. The Moon's mantle has much water locked within, as Earth does.

The Moon would long ago have become a cold rock if not bound to the Earth. Surrounding the metal core of the Moon is a slightly viscous deep mantle, kept warm by Earth's gravity. Tidal heating occurs via viscous dissipation.

Early in the Moon's history, its rotation slowed; becoming locked into synchronous rotation by frictional gravitational forces, caused by tidal effects from Earth (*tidal periodicity*). This tidal interaction pulls the Moon slightly along its orbit, causing it to move further away from Earth 3.8 cm per year. The Moon is now 384,400 km away from Earth.

The lunar cycle – from one new moon to the next – is 29.53 days. The Moon constantly shows the same face because of its synchronous rotation about the Earth. It takes precisely as long for the Moon to orbit the Earth as it does to revolve.

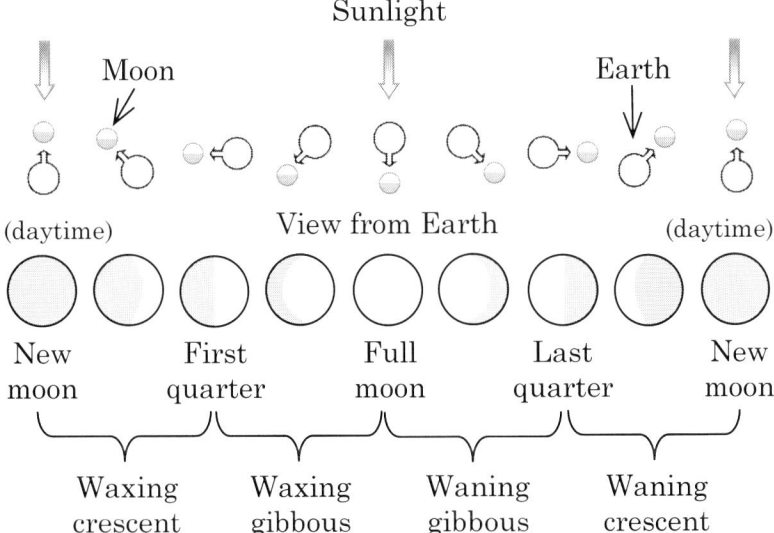

3,475 km in diameter, the Moon is the 5th largest satellite in the solar system, but the largest relative to its parent planet. Earth's diameter is 12,756 km; only 3.67 times that of the Moon.

In the last billion years, the Moon has shrunk by 200 meters in diameter. Why the Moon is a prune is not yet known.

The Moon has been cooling since its fiery birth but may not be dead yet. Explosive releases of underground gas have occurred within the past 10 million years.

The Moon dramatically shaped the Earth on its way to being a lasting rhythmic influence: the *lunar cycle* to which much life, especially nocturnal creatures, respond.

⚘ Life Under the Moon ⚘

Ecologists have long viewed the darkness of a moonless night as a protective blanket for nocturnal prey species. Moonlight alters predator-prey relations in more complex ways than previously thought. ~ American biologist Laura Prugh

At full moon the Earth stands between the Moon and the Sun. The view of the Moon is like a brightly lit coin in the sky.

In the following nights, as the Moon circles back toward the Sun, that coin slowly shrinks. Yet the sky seems darker than just dwindling light would allot; and it is. The Moon rises 50 minutes later each evening, carving a channel of darkness between the Sun dropping below the horizon and the Moon appearing.

Predators ply that channel. In doing so, the early waning Moon instilled innate fears in potential prey, where darkness spells danger.

During the full moon and days thereafter, few nocturnal reef fish are to be found, as they are more easily spotted by those that would make a meal of them. In contrast, the dark nights around the new moon cue fish that swimming about is safer.

Rabbits stay close to their burrows during the full moon and the days that follow. The darkness of the new moon lets them travel long, exposed distances.

Conversely, cheetahs and wild dogs in Africa have more active nights once the lunar cycle has waxed past half-full. Illuminating the hunt raises the odds of a kill.

Eagle owls and other avian predators take advantage of the Moon toward fullness to vigorously hunt and seek out new territory.

Lunar favor depends upon an animal's senses. Nearly half of all mammals are nocturnal, experiencing lunar cycles with light levels that change 3 orders of magnitude every month.

Animals active at night are adapted to the lifestyle. Moonlight benefits visually oriented prey. The prospects for lurking predators are lessened under the Moon's glow.

Many bat species become less active as the moon waxes full. Nocturnal insect prey have a better chance of spotting a threat, and echolocation yields no edge for luminosity.

Many marine organisms move up and down in the sea depending on the level of moonlight, to keep their light level constant.

The Moon is an environmental cue to many species, providing coordination with an animal's innate circadian rhythm. Coral synchronously spawn on full moon nights, their clocks aligned by *cryptochrome*: a protein sensitive to blue light. Galápagos marine iguanas travel for hours to arrive at the shoreline in time to graze on algae at low tide.

Gardening folklore suggests that planting crops according to the phases of the Moon yields a better harvest. The gravitational tug of the Moon does affect plants slightly. Plants can feel the Moon via the water that runs through them, most sensitively in the *pulvinus*: the joint where leaf meets stem.

ℬ Final Formation ℛ

4 BYA, the Sun brightened to 70% of its current light level, while the intense solar ultraviolet output dropped dramatically: by more than 30 times.

By absorbing more of the Sun's energy, Earth failed to ice over when the Sun was dimmer. Earth's surface was darker. The continents were much smaller, so the oceans, which are typically much darker than land masses, absorbed more heat.

Earth's early atmosphere was a brew of greenhouse gases that helped stabilize global temperature. Carbon dioxide (CO_2) and methane (CH_4) prevented the planet from freezing and triggered synthesis of a rich variety of organic molecules via ultraviolet radiation in the upper atmosphere.

Bombardment from space continued after Earth was moonstruck, cratering both Moon and Earth. The celestial siege of Earth eased somewhat after practically sterilizing the planet's surface, but bringing water, hydrogen, nitrogen, and a wealth of minerals and organic compounds that would transform the planet.

Jupiter was instrumental in both seeding Earth and in sweeping up errant projectiles, some of which formed the array of moons and asteroids coming under Jupiter's sway.

> Discount the "Jupiter as shield" concept. Jupiter was responsible for the vast majority of the encounters that "kicked" outer planet material into the terrestrial planet region, delivering the volatile-laden material required for the formation of life. Saturn assisted in the process far more than has previously been acknowledged. ~ American planetary physicist Kevin Grazier

To this day, Jupiter is Janus-faced toward Earth. While it does vacuum some debris, it also sometimes hurls objects Earth's way.

In 1770, a large comet whizzed by, missing Earth by a mere million miles. The comet had come into the outer solar system 3 years earlier, its path determinedly far from Earth. But the comet passed close to Jupiter, which diverted it to a new course: a cosmic whisker away from collision with the blue planet that Jupiter only sometimes protects.

The comet made 2 passes around the Sun before heading out. In 1779, the comet again passed close to Jupiter, which summarily slung it out of the solar system.

Though still subject to upheaval, Earth's crust was complete within 100 million years after its birth: a solid but deformable shell.

Nearly half of the crust's mass is made up of oxygen and over 27% silicon. Both are major components of rocks. Metals used in manufacture, such as iron and aluminum, comprise ~18% of the crust.

The lithosphere sorted itself into continents above sea level, resulting in land surface. Volcanism accomplished this. Frequent eruptions subducted hot surface materials, eventuating in a cool, thick crust by altering convection dynamics between Earth's crust and mantle.

> The mantle's viscosity is extremely dependent on its temperature. ~ Australian geophysicist Craig O'Neill *et al*

An abrupt transition to tectonics began ~3 BYA, once the lithosphere had sufficiently cooled. Before that, the upper mantle was too hot to convey rock without melting it.

Despite voluminous bombardment, early Earth mineral variety was quite limited. Of the 4,500 chemical species on Earth today, up to 2/3rds are attributable to biological activity. The earliest life engendered mineral evolution.

Late arrivals from space added to land mass. Meteorite impacts shifted mantle convection patterns, triggering plumes that heated the crust from below. Continents evolved.

In chewing rocks for sustenance, the earliest microbes were instrumental in creating the continents. Their waste products – sedimentation – acted as a viscous lubricant for tectonic plate subduction, thereby facilitating the rise of vast land masses. Without the lubricating sediment, Earth might have been a water world, dotted with small volcanic islands.

Earth's oldest rocks were volcanic artifacts (*igneous*). As the surface cooled, torrential storms ensued, begetting erosion. From surface debris emerged the 2nd great family of rocks (*sedimentary*).

The heat and crushing turmoil of tectonics led to melting and recrystallization of older rocks, producing a 3rd rock family (*metamorphic*). From these mountains were made.

The atmosphere in the late Hadean eon comprised gases released by volcanic activity, primarily large volumes of carbon gases (CO, CO_2, CH_4) which helped keep the surface warm.

Water vapor was in the air, as oceans had already formed; but free oxygen of any form (O_2, O_3 (ozone)) was entirely absent.

⚡ Tilt & Spin ⚡

Thanks to being whacked by its soon-to-be Moon, Earth's rotational axis tilts at 23.4°. This tilt brought seasonal variations.

The obliquity of Earth – the orientation of its spin axis to solar orbital plane – has changed over time. Even minor changes in obliquity cause major climatic shifts. The tilt of the Earth's axis as it spins gives rise to the seasons. The aspects of periodicity in Earth's spatial and orbital changes are known as *Milankovitch cycles.*

Earth's shape and spin result in a difference in gravitational pull between the poles and the equator, with equatorial objects lighter by 0.6%. Even at the same latitude, gravity varies from place to place because of several factors, including the bulge about the Earth, elevation, such as mountain ranges, and the moon's gravitational influence.

Earth's rotation has a gravitational effect upon itself, causing the diameter at the equator to be 27 km greater than its diameter through the poles.

Earth spins about its axis at a rate of 0.5 km per second. Its revolution about the Sun moves at 30 km/sec.

⚡ Water Supply ⚡

The solar system's water supply was inherited as ice from interstellar space. The water included prebiotic matter that would later integrate into life on Earth. Similar ices are likely to be found around other protoplanetary disks. There's a surprising amount of cosmic water.

> Water is pervasive throughout the universe, even at the very earliest times. ~ American astrophysicist Matt Bradford

The solar system has a planetary *snow line*: the zone beyond which ice could have condensed on emergent planets.

Although water covers 70.9% of Earth's surface, it accounts for far less than 1% of the planet's mass. Uranus and Neptune, which formed well past the snow line, are loaded with tens of percents of water by mass.

During Earth's birth, the inner solar system was hot enough to melt lead. The inner planets – those as far out as

Mars – would have been born dry, had they started out where they are now; which was certainly not the case for Mercury. If Earth did not start out as a hot dry rock, then it either moved into its current orbit after formation or made much of its water itself.

Earth looks to have been born wet, not dry. The chemical signature for water found deep within the mantle suggests that much of the planet's water was primordial. Water was generated within the mantle by combining fluid hydrogen and the silica in quartz, both of which would have been abundant in Earth's early mantle. Earth's crust is now 59% silica.

Fluid hydrogen and silica form water at 1700 K and pressure 20,000 times that of Earth's atmosphere. These requirements were easily met in the planet's mantle.

The swirling in the solar system that caused the planets to coalesce from bits of cosmic dust dragged an emerging Jupiter about before it settled into its current orbit. Earth's primordial water supply suggests that Earth, like Jupiter, came toward the Sun during its formation, likely in tow of Jupiter (and Saturn).

A *chondrite* is a stony meteorite, formed by accretion from dust and small grains. Many chondrites in the early solar system acquired a coating of ice.

Jupiter's promenade dragged chondrites into collision with Earth, seeding it and the Moon with water. Hence, the bombardment of Earth – that largely started and abated by Jupiter's planetary evolution – supplemented Earth's water supply. The leftovers which did not rain down form the asteroid belt that ranges between the orbits of Mars and Jupiter, where the solar system's snow line is situated.

As Earth cooled, a crust formed, as well as an atmosphere bearing water; bringing rain, and in time, oceans. This allowed surface temperatures to drop to less than 102 °C as early as 4.4 BYA.

Earth may have sported its first ocean at this time, when the planet was but 150 million years old. The Sun's evaporative blaze was 30% less than now.

3 billion years ago, the ocean was 67 °C. By 1.5 BYA, the ocean had cooled to 27 °C; a warm soup supporting life.

Sponged up in the Earth's interior, 410–600 km down, is at least 25 times the water in the oceans. If not for this lubrication, there would be no plate tectonics and no continents. Without continents, there would be no transport of life-sustaining nutrients from rivers into the oceans.

ဆို Structure ဂ္ဂ

Earth is a thermodynamic engine powered by its own finite internal reserves of heat that are gradually brought to the surface and radiated to space. ~ American geophysicists Don Anderson & Scott King

The Earth is a geologic onion of layers, alternately described by chemical or rheological properties. *Rheology* is the study of matter flow.

There are at least 2 main core layers, multiple layers of mantle, with a transition zone between the upper and lower mantle, and a 2-part outer layer, capped by a crust.

Owing to gravitational force, the planet gets denser further down. The center of the Earth is 6,370 km below sea level.

♂ Core ☊

Earth's cores are nested layers. The outer core is a molten metal sea, floating over a solid inner core that is roughly the size of the Moon, and is almost entirely iron.

The viscous outer core is mostly iron, with some nickel. 10% of outer core material comprises light alloys, made of silicon, oxygen, sulfur, carbon, and other elements.

Radioactive material within the core decays, giving off heat. Earth's core has cooled by some 1000 degrees (K) since its formation. Such cooling is necessary to sustain the planet's geomagnetic field.

At the heart of the inner core lies the innermost core: a solid ball of pure iron 600 km in diameter. Everything else got squeezed out of the innermost core.

The innermost core may be as hot as 6,923 K; yet the otherwise liquid iron is frozen because of the extreme pressure.

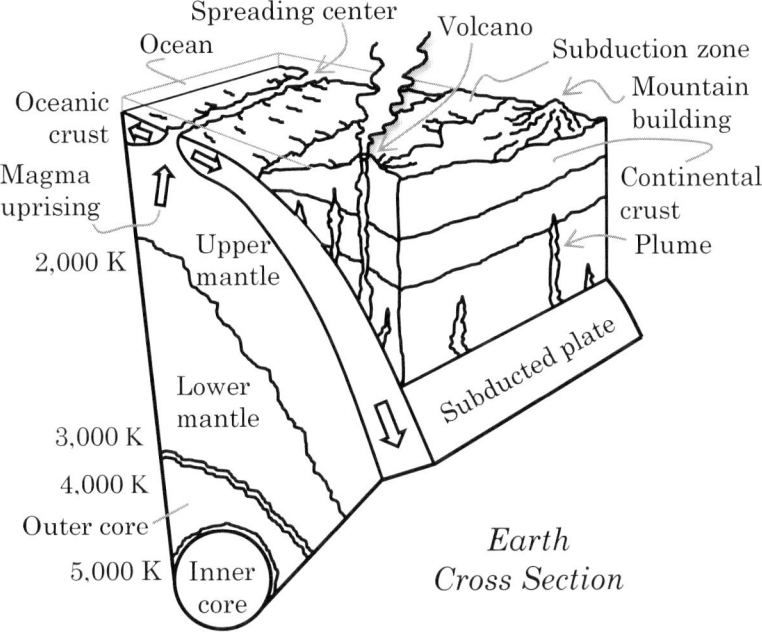

*Earth
Cross Section*

Earth was initially a growing ball of molten rock. As the orb grew, the heavy metals in the rocks descended into the planet's interior. The frictional heat from sinking made for a relatively chilly metallic center surrounded by hot viscous rock.

Rising pressure inside the planet caused the cool core to condense. The innermost core solidified ~100,000 years after Earth's accretion began.

Altogether, Earth's cores comprise a turbulent engine: generating over 15 terawatts of heat energy at the core-mantle boundary. All this is fueled by energy left over from the cosmic collisions that formed the planet.

♉ Earth's Magnetic Field ♋

Earth's inner core rotates eastward, slightly faster than the rest of the planet. Outside the spinning inner core, flows of electrically conductive liquid-metals near the boundary of the outer core and mantle fashion fluctuations that create massive, shifting electromagnetic currents.

This geodynamic generates Earth's magnetic field, which first developed 4.2 BYA. The timing of the core condensing and rotation was critical to propagating a strong magnetic field. The magnetic field extends for several thousand kilometers outside of Earth, creating a protective blanket that deflects much of the solar wind.

Geophysicists long had a hard time explaining how Earth's magnetic bodyguard came on duty so early in the planet's history. The answer lies in the dynamics of hot metal under pressure.

Thermal energy nominally transfers freely from atom to atom via conduction. The atoms are unmoved. But when the heat flow exceeds what a material can handle through conduction, atoms become restless. Convection emerges.

In metals such as iron, free-moving electrons ferry electromagnetic charge as well as heat. How readily they do so depend upon how much resistance they encounter.

Earth's early core would have been more conductive and less convective if not for the tremendous pressure involved. Containment built both heat and resistance.

Pressure in the core squeezes the iron and nickel to more than 1.6 times its normal density. The electrons within are especially excited.

Above 1,970 K, thermally energized electrons more than scatter off vibrating atoms: they increasingly collide with each other. This electron–electron bashing drives electromagnetic generation.

Resistivity doubles while thermal conductivity drops. Pressure amplifies the electromagnetic effect of convection. Hence, the early emergence of a powerful magnetic field from the core.

Though predominantly iron, Earth's core is almost 20% nickel, which plays a crucial role in generating the magnetic field.

Under pressure, nickel behaves differently from iron. At high pressure, the electrons in nickel tend to scatter much more than the electrons in iron. As a consequence, the thermal conductivity of nickel and, thus, the thermal conductivity of Earth's core, is much lower than it would be in a core consisting only of iron.
~ Italian physicist Alessandro Toschi

> If Earth's core consisted only of iron, the free electrons in the iron could handle the heat transport by themselves, without the need for any convection currents. Then, Earth would not have a magnetic field at all. ~ Austrian physicist Karsten Held

By sheltering the planet from high-energy solar radiation and wind, the magnetic field helped preserve early Earth's oceans from evaporation and provided some protection for nascent life. By lessening ionization from the solar wind, the magnetic field also kept atmospheric nitrogen from escaping.

Other rocky worlds in the solar system have not been so fortunate. Lacking a magnetic shield, the Sun has stripped away their atmospheres.

⌇ Van Allen Belts ⌇

From 1,000 km above Earth's surface stretches layered belts of highly charged particles. Via the first artificial satellites, the belts were first noticed in 1958, and named after their discoverer: American astrophysicist James Van Allen.

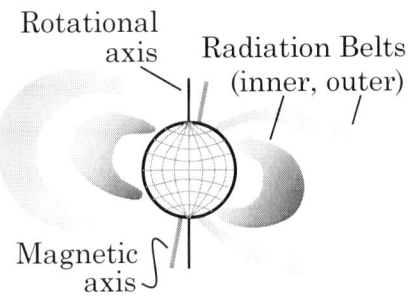

Van Allen Belts

This plasma zone arises from the planet's magnetic field, which holds the belts in place.

The Van Allen belts are 2 pronounced concentric doughnut-shaped rings that strip atoms of their electrons and accelerate the subatomic particles to near lightspeed. No real gap exists between the 2 zones; simply gradations in radiation intensities.

As a particle approaches a magnetic pole, the increased field strength bounces it back to the other pole. Hence the belts are most intense over the equator, and effectively absent above the poles. Over time, particles collide with atoms in the atmosphere and are knocked out of the belt.

The inner belt (1,000–6,000 km) largely comprises protons, energized to 30 million electron volts (MeV). Many of the protons are produced by decay of neutrons, which wither from the intense radiation.

The outer belt (13,000+ km), fed from particles both atmospheric and solar in origin, has lower-energy protons. The most energetic particles in the outer belt are electrons, reaching several hundred MeV.

There is also a 3rd belt between the inner and outer belts, which modulates the activity of the outer belt.

Particles in the belts stream in spiral paths along the force lines of Earth's magnetic field. Synchronicity in the frequency of electromagnetic waves and electrons in transit keeps the belts enlivened.

At 11,600 km altitude, there is an extremely sharp boundary at the inner edge of the outer belt that acts as a shield, blocking ultrarelativistic electrons from whizzing closer to Earth's atmosphere. This boundary is a mystery, as Nature typically abhors strong gradients.

> It's almost like these electrons are running into a glass wall in space. The invisible shield blocking these electrons is an extremely puzzling phenomenon. ~ American astrophysicist Daniel Baker

Solar flares disrupt the belts, which in turn invokes auroras and magnetic storms. Even during less turbulent times, the Van Allen belts endanger man-made satellites with their intense, fluctuating radiation.

☄ Polarity Reversal ☄

> Sometimes you won't have a flip for about 40 million years; other times there will be 10 flips in 1 million years. On average, the duration between two flips is a few hundred thousand years. The last flip was around 780,000 years ago, so we are actually overdue for a flip. ~ Chinese geophysicist Huapei Wang

Earth's magnetic field sporadically reverses polarity. The duration of continued polarity (a *chron*) varies by tens of millions of years, with an average of 450,000 years.

Field reversal typically takes 4,000 years, though it may occur in as little as a decade. Changing continental configurations via tectonics may trigger geomagnetic field reversal.

The Earth's magnetic field is currently weakening 5% each decade. Magnetic north is moving toward Siberia. Current field intensity is twice the historical average, so polarity reversal is not likely for many millennia.

♂ Mantle ♑

2,900 kilometers thick, the *mantle* is the thickest layer, comprising 68% of Earth's mass and 84% of its volume. The mantle is dense and rigid at depth.

The predominant mineral in the lower mantle is *bridgmanite* ($(Mg,Fe)SiO_3$): a ferromagnesian silicate mineral with different phases at different depths, with more or less iron. 38% of the Earth's volume is bridgmanite.

Closer to the surface, the mantle becomes increasingly viscous. Toward its upper boundary, there is a sharp increase in energy (seismic) waves.

The mantle is mostly iron-magnesium silicate rock but mixed with many other minerals. Although the mantle is of solid stuff, the high temperatures within render the silicate sufficiently ductile to deform and flow, albeit on a geologic time scale, where a millennium is a New York minute.

The upper and lower mantle are separated by a 250 km *transition zone* (410–660 km down), where mineralogy modes change. This transition zone has a major role in Earth's geodynamics, particularly influencing mantle convection: slowing slab subduction and plume ascent.

Water is introduced into the mantle where oceanic plates spread apart and new ocean bedrock forms. This water may make its way to great depths, where its pressurized presence fuels partial melting of mantle rocks, creating magmas laden with water. Thusly water recycles through the mantle.

Water ordinarily has an ordered, polymerized structure. Pressure and heat within the mantle create supercritical conditions which disorganize water molecules. The flowing hydrogen-bond network that exists in surface water is literally crushed. Supercritical water turns into an aggressive solvent. This property of water is crucial in promoting the geodynamics of Earth's crust and mantle.

♂ Crust ☿

Earth's relatively rigid outer shell, the *lithosphere*, is made up of the crust and uppermost mantle layer. Oceanic crust is thinner (5–10 km) but denser than continental crust (20–90 km; average 35 km). Currently, 1/3rd of the crust is continental, 2/3rds oceanic.

Crust composition differs. Oceanic crust is rich in iron and magnesium. Continental crust, derived from oceanic crust over eons, and fed by volcanoes, has more granite.

The continental crust contains the oldest rocks: up to 4 billion years. Nevertheless, continental crust changes constantly, due to erosion, sedimentation, volcanic activity, and tectonics.

Oceanic crust is constantly recycled by tectonic plate subduction and regenerated by magma plumes rising from the bottom of the mantle. No location in the oceanic crust is older than 200 million years.

Mantle plumes are *not* rapidly rising jets of magma. Instead, they are broad upwellings – thousands of kilometers across – of magma and hot rock.

> Surface plates, their motions, and their return to the mantle via subduction control mantle dynamics and heterogeneity. Ultimately, it is the cooling of the Earth, modulated by internal heating, that provides the energy for convection. ~ Don Anderson & American marine geologist James Natland

Little violent mixing of materials occurs deep within Earth. Subducted volcanic rock may travel through the mantle and resurface, largely intact, after billions of years. The deep mantle is a graveyard of ancient tectonic slabs.

From planetary wear and tear over the course of billions of years, rocks and minerals ground to a powder on the surface, mixing with moisture and microbes to become dirt. Life was well on its way by the time dirt was young.

❧ Day Length ❧

300 million years ago a day was 21 hours, while a year was 450 days. Since then the Earth's rotation has slowed, lengthening the day.

The fluid outer core and solid mantle create an angular momentum that largely determines the rate of Earth's rotation. There is a persistent wobble every 5.9 years that stutters day length. Other jitters happen when a patch of the molten outer core temporarily gets stuck to the mantle, ratcheting angular velocity. This also affects the Earth's magnetic field.

The oceans, land, and atmosphere also have some significance in rotational fluctuations. For example, the force of the wind against mountain ranges can change the length of a day by a millisecond or so over the course of a year.

❧ The End ❧

The universe is flat with only a 0.4% margin of error. ~ NASA

The cosmos seems to be currently swelling at a comfortable clip. Whether the expansion will go on endlessly, or cyclically reverse course and contract into a singularity, depends upon the shape of the universe, and the nature of energetic forces that govern cosmologically.

3 topological alternatives exist: open, flat, or closed. If the cosmos is open or flat, the fan out will continue (open) or eventually halt (flat).

We live in an accelerating universe now and so, as time goes on, the density of galaxies is going to thin out. ~ English astronomer Tony Hewish

In the eventuality termed the *Big Freeze*, cosmic energy is spread so thin that entropy succeeds in turning the universe into an enormous popsicle. Star formation has already started to fizzle.

Stars are formed in galaxies. There was a peak in the rate at which galaxies formed, and that time has passed. ~ English astronomer Alan Heavens

The universe is curling up on the sofa and becoming a couch potato. ~ German astronomer Jochen Liske

Under the *Big Rip* scenario, cosmic expansion gets the upper hand. All matter and light energy will be ripped apart.

Under the rip scenario, you get this wild expansion that essentially rips spacetime apart. The universe would vanish. ~ English Mexican cosmologist Carlos Frenk

If the universe is a single shot of existence in a closed container, a *Big Crunch* will be the symmetrical opposite of the Big Bang. The universe collapses into a dimensionless singularity.

The latest research shows that the universe's expansion is accelerating, so there is no reason to expect a collapse from cosmological observations. Thus, it will probably not be a Big Crunch that causes the universe to collapse. ~ Danish physicists Frederik Colding & Jens Krog

Cyclic cosmology posits a *Big Bounce*. Universes go through endless incarnations. Cosmic contraction follows the current expansion if there is a phase transition that causes matter to take on a bit more mass; a result of fundamental bosonic forces, especially the Higgs field, changing the key of their tune.

The phase transition will start somewhere in the universe and spread from there. Maybe the collapse has already started somewhere in the universe and right now it is eating its way into the rest of the universe. ~ Jens Krog

While the shape of the universe and semblance of forces applied appears to be comprehensible, this facile understanding is an illusion of our quite limited perspective, and of feeble mental capabilities overcome by overconfidence.

≈ Synopsis ≈

There is a theory which states that if ever anyone discovers exactly what the Universe is for and why it is here, it will instantly disappear and be replaced by something even more bizarre and inexplicable. There is another theory which states that this has already happened. ~ English writer Douglas Adams

Cosmic Considerations

➢ The mainstream conjecture for the origin of this cosmos is a *Big Bang* appearance of something from nothing, where everything that was ever to be in this universe came packaged in an infinitesimal singularity that exploded in a violent blossoming; no telling where it came from.

➢ The explanation for how the cosmos got from nada to now involves filling in big blanks between what can be observed and what happened from the supposed onset of the universe. As most of the universe is beyond view and measure, astrophysics is necessarily guesswork, and much of the guessing has been laughably bad. The prevailing cosmological model is wrong on all key aspects, including the age of the universe, cosmic inflation, the cosmological principle, dark matter, and dark energy.

➢ The foolishness of conventional cosmogony begins by assuming that this universe started with our detection of the earliest light. There is no reason to think that the cosmos began with stars lighting up. Instead, this universe is probably hundreds of billions years old, with a dark beginning and a long, shadowy weaving of energy/matter patterns before turning the stellar lights on.

➢ The *cyclic model* posits that the existing universe is but a bounce back from a previous incarnation. The cyclic model implies a *multiverse*: universes coming into being on a vast canvas of endless time. The cyclic model accords well with the known facts.

How the universe ends – whether by expansion into oblivion or a cosmic crunch – does not impinge on the potential validity of a multiverse or continuing cosmic incarnations.

➤ The *dynamics of the cosmos unfolding* are a symphony of interdimensional complexity and interdependence, myriad in manifestations. The forming of star systems and galaxies are exemplary of the interwoven intricacies and varieties that pervade every facet of Nature.

➤ There are at least 10^{21} stars in the *cosmic firmament*; the majority having planets orbiting them.

> It is indeed a feeble light that reaches us from the starry sky. But what would human thought have achieved if we could not see the stars? ~ Jean Perrin

➤ The *Sun*, as with all blazing stars, is an intricate gyre of energy, manufacturing the materials that in the next stellar incarnation create planets and embody life.

> We are stardust, billion-year-old carbon. ~ Canadian musician Joni Mitchell in the song "Woodstock" (1970)

Earth

➤ Earth's fulsome formation was funded by *bombardment* that yielded the Moon, metals, water, and an organic chemistry set; all the essential elements for engendering life. With its richness and variety, Earth is a natural paradise.

➤ *Earth's structure* comprises layers that dynamically interrelate. The movements of tectonic plates are influenced by mantle heat from below, and water flow that is interconnected to the oceans. The Earth's protective magnetic field, generated by the core, is affected by movements in the mantle and crust.

✑ Physics ✑

Not only is the Universe stranger than we think, it is stranger than we can think. ~ Werner Heisenberg

While *physics* is derived from the Greek for "knowledge of Nature," its locus is the study of moving matter. A central concern of physics is *energy*: what it takes to get matter to *work*. *Work* is the product of a force applied to matter. Work refers to a transfer of energy, often to matter. Work can be said to be energy in transit.

To understand motion is to understand Nature. ~ Italian polymath Leonardo da Vinci

German physician, chemist, and physicist Robert Mayer enunciated in 1841 one of the original statements on the conservation of energy: that "energy can be neither created nor destroyed." The implications of this statement came to be interpreted too literally by those insufficiently meticulous about their philosophy of physics.

Energy is usually presented in the following way: "energy can neither be created nor destroyed but only transformed." If energy cannot be destroyed, it must be an existing thing. If its form changes, it must be something real as well. Thus, that statement can easily lead to the concept of energy as something material. Robert Mayer did not find, however, anything like a substance but rather a methodology for dealing with phenomena. ~ Portuguese physicist Ricardo Lopes Coelho

Energy is nothing but a convenient concept: a way of characterizing observed changes in matter. Energy is merely a quantitative property – a measure of what it takes to put matter in motion. *Energy does not exist.*

The subtle and intellectually difficult concept of *energy*, with its associated mathematical splendor, permits the integration and use of mechanics. ~ American biomechanist Steven Vogel

Changing motion – the velocity of matter accelerating or decelerating – concerns physics greatly. The relation between energy and speed is *quadratic*: double the energy is required to incrementally change speed.

The energy associated with motion is *kinetic energy*. Gustave-Gaspard Coriolis developed the modern concept of kinetic energy in 1829.

Temperature is an approximate measure of the kinetic energy of molecules. More formally, the kinetic energy of an object is measured by its movement. The measure of kinetic energy (E) depends upon the mass (m) of a moving object and its velocity (v): $E = \frac{1}{2}mv^2$.

Energy related to position is *potential energy*: energy stored that may be released. Scottish mechanical engineer William Rankine coined the term *potential energy* in 1853. A ball lifted into the air has potential energy, because, if released, gravity would work on the ball to have it drop to the ground – in the process, turning potential energy into kinetic energy.

An atom has potential energy which is released if the atom's nucleus starts to decay, thus radiating energy; or if the electrons of an atom are diverted to work elsewhere: somewhere other than clouding the home nucleus with a wondrous whirl.

Matter has mass. Mass is not weight, though it is commonly expressed as such. Instead, *mass* is a measure of matter's *inertia*: indisposition to a change of motion, regardless of whether the object is moving or at rest.

Mass is measured in kilograms (kg). *Weight* is a force, measured in newtons (N).

Energy does *not* exist in a myriad of forms: mechanical, thermal, chemical, radiant, atomic, and quantum. Theoretically, different forms of energy are interconvertible. Actuality renders this energetic fluidity a polite fiction seemingly full of exceptions. As constrained as matter in its specificity, energy works its magic in distinct domains. The appearances of energy are always statements made on matter: either retarding or accelerating transitions which, ultimately, are nothing more than expectations and their confounding.

> It is wrong to think that the task of physics is to find out how Nature is. Physics concerns what we can say about Nature.
> ~ Danish physicist Niels Bohr

❧ History ☙

According to one mode of expression, the question: "What are the laws of Nature?" may be stated thus: What are the fewest and simplest assumptions, which being granted, the whole existing order of Nature would result? ~ English philosopher John Stuart Mill in 1843

Historically, physics long fell under the appellation of *natural philosophy*. An increasing emphasis on empiricism and mathematical description in the 17th century turned natural philosophy into *natural science*, though that term was applied in hindsight.

Galileo, perhaps more than any other single person, was responsible for the birth of modern science. ~ Stephen Hawking

Besides his championing heliocentrism when it was controversial, Galileo furthered *kinematics* (classical physics' theory of motion), and materials science, particularly the ability of materials to withstand stress without failure; in other words, the energetic integrity of matter.

♒ Isaac Newton ♒

A man may imagine things that are false, but he can only understand things that are true, for if the things be false, the apprehension of them is not understanding. ~ Isaac Newton

In 1687, English physicist and alchemist Isaac Newton published *Mathematical Principles of Natural Philosophy* (often referred to simply as *Principia*), creating the mathematical edifice of *classical mechanics* with a model of universal gravitation and *3 laws of*  *motion*: 1) a body has constant velocity unless acted upon by an external force; 2) acceleration is proportional to force and inversely proportional to mass; and 3) the mutual forces of action and reaction between 2 bodies are equal, opposite, and collinear (straight, not funky).

The famous book of Mathematical Principles of Natural Philosophy marked the epoch of a great revolution in physics. The method followed by its illustrious author Sir Newton spread the

light of mathematics on a science which up to then had remained in the darkness of conjectures and hypotheses.
~ French polymath Alexis Clairaut in 1747

Newton's conceptual world was based upon absolute
space and time, which were taken to be independent foundations of reality.

Absolute *space*, in its own nature, without regard to anything
external, remains similar and immovable.

Absolute, true, and mathematical *time*, for itself, and from its
own nature flows equably without regard to anything external,
and by another name is called duration. ~ Isaac Newton

The esteemed image of Newton is of a rational practi-
tioner of pure reason. Far from it. Newton believed in an al-
mighty God. Newton was convinced that *The Bible* contained
secrets in the form of numerological codes.

Newton was obsessed with alchemy, writing over 1 million words on the subject. He spent untold hours trying to
replicate alchemical recipes. Instead of the first king of reason, Newton was the last of the magicians.

Newton spent half his life muddling with alchemy, looking
for the philosopher's stone. That was the pebble by the seashore
he really wanted to find. ~ American writer Fritz Leiber

In 1797, English physicist Benjamin Thompson showed
that a seemingly infinite amount of heat could be generated
from a finite amount of material. This demonstration of kinetics was instrumental in establishing modern thermodynamics; though, ironically, Thompson's finding of infinite
energy was ignored.

৪১ Energy ୧୧

Aristotle used the word *energy* (*energeia*) in the 4th century BCE. Energeia was a qualitative concept which included
motion of all kinds, including pleasure and happiness. This
vibrant quality would become treated qualitatively as physics evolved into a mathematical discipline.

In 1676 German philosopher and mathematician Gottfried Leibniz began to develop the idea that a system had a

vis viva: a "living force". At the time *vis viva* seemed opposed to the theory of conservation of momentum advocated by Isaac Newton and René Descrates. In the 1730s French physicist, mathematician, and natural philosopher Émilie du Châtelet understood that Leibniz was referring to conservation of kinetic energy, which is distinct from conservation of momentum. The prior opposition to *vis viva* had arisen because kinetic energy was not properly understood.

Thomas Young first used the term *energy* in the modern sense in 1807, incorporating *vis viva*; this after *vis viva* bested the caloric theory as better explaining the potential of heat to generate motion. In 1845 English physicist James Prescott Joule discovered the link between mechanical work and the generation of heat.

Mathematical ponderings about heat led to laws of thermodynamics, based upon the core assumption that energy in a *closed system* is, as an aggregate, a fixed quantity. Our universe has been shown to *not* be a closed system energetically. Hence, these 'laws', while mathematically neat, are fictional.

❧ Laws of Thermodynamics ☙

The development of the steam engine created an urgent need to understand the nature of heat. Early theories had heat emanating from the friction of unseen moving particles.

Heat itself, its essence and quiddity is motion and nothing else. ~ English scientist Francis Bacon in the 17th century

Continuing inquiry into thermodynamics led to laws about energy – specifically, the distribution of energy in the universe. These laws center on *entropy*: the observed tendency of energy to dissipate, and thereby equilibrate.

The laws of thermodynamics are fundamental in Nature, as they do not rely on any specific microscopic theory. ~ Russian American physicist Anatoli Polkovnikov

♂ 1st Law: Conservation of Energy ♀

The 1st law of thermodynamics relates to a reciprocal of entropy in a closed system: that energy endures.

A body of matter cannot disappear completely. It only changes its form, condition, composition, color, and other properties, and turns into a different complex or elementary matter. ~ Persian polymath Nasīr al-Dīn Tūsī in the mid-13th century

Long after Tūsī, Welsh physical scientist William Robert Grove pondered conservation of energy from a holistic viewpoint.

The question of whether there can be absolute motion, or indeed any absolute isolated force, is purely the metaphysical question of idealism or realism. ~ William Robert Grove in 1844

This 1st law of thermodynamics – the conservation of energy – was more firmly put in place by German physicist Hermann von Helmholtz.

The quantity of force which can be brought into action in the whole of Nature is unchangeable and can neither be increased nor diminished. ~ Hermann von Helmholtz in *On the Conservation of Force* (1847)

Kinetic or potential energy may be locally gained or lost during energy transformation. But, according to the 1st law, energy may neither be created nor destroyed.

Energy is just a concept, and the 1st law of thermodynamics is merely an assumption that works well in equations. There is *no evidence* to support conservation of energy.

☡ Conservation Contravened ☡

The universe does not violate the conservation of energy; rather it lies outside that law's jurisdiction. ~ Australian astrophysicist Tamara Davis

Classical thermodynamics adhered to an assumption still widely held: that the universe is a closed system. That is, the cosmos is presumed isolated and self-contained, with all the energy in evidence (theoretically).

All thermodynamics laws rely upon a closed system, but none so much as the 1st: that the quantity of energy is unchangeable.

The phenomena of light shows that no vibrations go outside of three-dimensional space, even in the luminous aether. If there is another universe, or a greater number of universes, outside of our own, we can only say that we have no evidence of their exerting any action upon our own. ~ Canadian astronomer and mathematician Simon Newcomb in 1894

What modern physics has learned is that vast amounts of energy are continuously interchanged between the observable 4 dimensions (4D) and extra spatial dimensions (ED). This has been shown at both the quantum and cosmological scales.

There is a constant flux of so-called *virtual particles* about every 4D subatomic particle. Virtual particles are extremely short-lived energetic quanta that pop in and out of 4D; subatomic popcorn out of empty space that is quickly consumed by a vacuum void. These ED quanta shape the basic properties of 4D particles, including mass.

The 4D energy drained by the singularity sink of a black hole leaks into ED, rendering a net energy loss 4D.

To extend the conservation of energy law to a higher dimensionality (HD) – to include ED – assumes that energy ED behaves the same as it does 4D; an assumption with no evidentiary basis (nor can there be).

Virtual particles and black holes show that 4D and ED are intertwined energy gyres. We can never find out about the confines of holistic dimensionality (HD). While we may experience ED effects in 4D, the contours of existence are beyond our ken.

This universe may be a gyre with others. Our cosmos may be one in a community. However far-fetched that seems, it is entirely consistent with the interconnections that ubiquitously exist within this universe, and so is only an extension of a known paradigm. It is also coincident with some modern cosmology and physics models.

⌀ Vacuum Genesis ⌀

Maybe the universe is a vacuum fluctuation. ~ Edward Tyron

In a 1969 seminar, English physicist Dennis Sciama jokingly suggested that the universe was a supersized virtual particle – having popped into existence for an extended visit before popping back out. American physicist Edward Tyron took the idea seriously. But it was not until after Guth's 1980 cosmic inflation conjecture – how a universe could inflate from a tiny particle – that anyone else took vacuum genesis seriously.

The obvious problem with the vacuum genesis hypothesis is presupposing a background space from which our universe arose. In 2014, Chinese physicists Dongshan He, Dongfeng Gao, and Qing-yu Cai mathematically showed how to get something from nothing; well, not just *something* – everything!

> The universe can be created spontaneously from nothing, where "nothing" means there is neither matter nor space or time, and the problem of singularity can be avoided naturally. ~ Dongshan He, Dongfeng Gao & Qing-yu Ca

In this quantum cosmogony model, "the universe is described by a wave function rather than the classical spacetime."

> The birth of the universe completely depends upon the quantum nature of the theory. ~ Dongshan He, Dongfeng Gao & Qing-yu Cai

Cosmogony theories that do not posit an eternity of universes fail to address where the cosmic energy comes from – a grievously lame omission. But that is small potatoes to the universal failure: cosmogony theories ignoring the critical question of how the coherent diversity of Nature is obtained from a singularity.

> The universe is one of those things that happens from time to time. ~ Edward Tyron

ଶ 2nd Law: Thermalization & Entropy ଛ

With the tendency to thermalize – that is, the inclination of energy to equilibrate – the 2nd law of thermodynamics embraces entropy and crafts a thermodynamic arrow of time. In a nutshell, everything runs down.

French military engineer Nicolas Léonard Sadi Carnot was fascinated with steam engines. Carnot abstracted an idealized heat engine in 1824.

> The production of motive power is therefore due in steam engines not to actual consumption of caloric but to its transportation from a warm body to a cold body. ~ Nicolas Carnot

While Carnot developed an otherwise compelling analysis of how to efficiently convert heat into work, the *Carnot cycle* was grounded in the clumsy *caloric theory*: an obsolete conjecture that heat is a self-repellent fluid that flows from hotter to colder bodies.

French engineer and physicist Benoît Paul Émile Clapeyron conceptually cleaned up the Carnot cycle, presenting it in 1834 in a more acceptable form: as an analytic graph.

Rudolf Clausius formulated the 2nd law of thermodynamics in 1850. Further thinking about thermalization led Clausius to the notion of entropy in 1865.

The 2nd law of thermodynamics is what forbids perpetual motion machines.

✄ Thermalization Thwarted ✄

The 2nd law's edict that systems thermalize is regularly violated. Some quantum systems thermalize; others don't.

One thermalization violation occurs when cooling gas, substantiating *Maxwell's demon*: an 1871 thought experiment by Scottish physicist James Clerk Maxwell, who hypothesized a way to decrease entropy by a method now proven.

In nanoclusters of jostling atoms, some clusters ricochet off each other faster than their collision speed. This violates the 2nd law.

On average, the rebound is less energetic than the collision. Fast bouncers appear 5% of the time; enough to violate the law while leaving it an adequate approximation.

Experiments with entangled atoms have shown that heat may flow from cold to hot, contrary to the 2nd law. This may not be a strict violation of that law, which presumes no correlations between particles – an unrealistic assumption, however commonly it may appear to apply.

☟ 3rd Law: Maximum Entropy ☞

The 3rd law of thermodynamics takes entropy to the max: entropy approaches a constant value as the temperature approaches zero.

German physicist and chemist Walther Nernst developed his theory of the 3rd law 1906–1912, whereupon it became known as *Nernst's postulate*. Lack of contradiction led to its acceptance.

☟ 0th Law: Temperature ☞

Temperature is a single-parameter curve fit to a probability distribution. ~ American physicist Lincoln Carr

The 0th law of thermodynamics defines *temperature* as an absolute measure of heat. We have no conception of how hot energy may get. Cold is altogether another matter.

Scottish mathematical physicist William Thomson, better known as Lord Kelvin, imagined in 1848 a temperature so low as to be absolute zero. That chilling concept became the Kelvin temperature scale.

Absolute zero corresponds to the theoretical state in which particles of a gas have no energy at all: utter entropy. This situation is traditionally characterized as a measure of the disordered motion in a classical idealized gas.

☟ So Bitter Cold That It's Hot ☞

At 0 K, most particles would be at rest, but a few might have higher-than-average energy. This state can be manipulated magnetically to turn positive zero Kelvin to negative.

Quantum particles can be chilled to their lowest possible energy state. Their spins go down. Add energy and some particles' spins go up. When half are down and half are up, maximum disorder (entropy) is reached.

Add more energy after maximum entropy and the quantum system shifts into negative temperature, where a high-energy state is the only way to accommodate the extra energy. In systems with negative temperature, particles prefer to populate high-energy states instead of low-energy ones.

Entropy does a backflip, throwing the laws of thermodynamics into a tizzy. Atoms instantaneously shift from their most stable, lowest energy state to the highest possible energy state.

Absolute zero is not absolute. Negative temperature exists; a state which is almost infinitely hot.

Converse to normal matter, where atoms naturally repel each other, and thus maintain their personal space, atoms in a –K gas are attractive: energetically driven to collapse inwards (and so disappear into a black hole). They do not only because the negative absolute temperature stabilizes them.

These conservation "laws" are global, applying throughout our universe. Any other form of these laws would be so astounding as to force us to look for some more complex explanation. ~ American particle physicist Victor Stenger in 2000, who never bothered to look for "some more complex explanation"

The laws of thermodynamics are a legacy of classical physics left intact. Unlike Newtonian gravitation, which was supplanted by Einstein's general relativity, the tenets of thermodynamics remain undisturbed, with no theoretical replacement.

A place for everything, and everything in its place. ~ Mister Dog

The thermodynamics laws are tidy laws, however messy it might be under the rug of reality. Mister Dog would approve, as do old-school physicists, who continue to grant credence to these 'laws'.

Modern physics has found that all the laws of thermodynamics are violated beyond the confines of ambient existence; that thermodynamics ultimately has no laws which we can ascribe. Alas, for lack of insight, theoretical physicists have not walked through this open door.

It is not at all natural that "laws of Nature" exist, much less that man is able to discover them. ~ Hungarian American physicist and mathematician Eugene Wigner

ஐ From Classical to Modern Physics ଔ

Classical physics accepted what the 5 senses facilely perceived: that observable space and time was all that there is. This philosophical stance is *naïve realism*: the belief that actuality is reality. Universal laws of Nature were built upon that precondition. As physics' perspective of existence expanded, the scope of universal laws shrank.

> The important thing in science is not so much to obtain new facts as to discover new ways of thinking about them. ~ William Henry Bragg

Inquiry into the nature of radiation ushered in modern physics, which is only partly a post-Newtonian conception.

At the turn of the 20th century, Max Planck discovered that energy, while having wavelike properties, only manifests in quantized (particulate) form. Thus arose the *Planck constant*: the smallest possible increment of energy. Space and time also quantize into a minimal *Planck length* and *Planck time* respectively.

Einstein extended Planck's discovery and found that space and time, which Newton had considered absolute, were instead relative.

> The world looks classical because the complex interactions that an object has with its surroundings conspire to conceal quantum effects from our view. ~ Serbian-born British physicist Vlatko Vedral

The onset of modern physics came from poking holes in classical descriptions, finding them lacking when considering the cosmic or infinitesimal. The irony of modern physics has been to create new holes that bring the exploration of physics to its limits; a demonstration of how little can be empirically sussed about the nature of Nature, and how easily theory misleads.

ஐ Toolkit ଔ

> A person with a new idea is a crank until the idea succeeds. ~ American author and humorist Mark Twain

Math is the tool of the trade for theoretical physics, equations the universal language.

> Models are a means of extrapolating from what is known to create proposals for more comprehensive theories with greater explanatory power. ~ American theoretical physicist Lisa Randall

A *physical model* is a mathematical model, typically geometric or algebraic, providing a symbolic description of the embodied phenomena. The quality of a model is how well it agrees with empirical observations and its predictive power. Newton's motion laws came from a physical model.

> All great discoveries in experimental physics have been due to the intuition of men who made free use of models, which were for them not products of the imagination but representatives of real things. ~ Max Born

A *physical theory* describes relationships between various measurable phenomena, often considered as cause and effect. A physical theory may include a model of physical events.

In the late 5th century BCE, Greek philosopher Pythagoras explained the relation between the length of a vibrating string and the musical note it produced. In the early 3rd century BCE, Greek polymath Archimedes understood that a boat floats by displacing the water that would otherwise be there.

> What is especially striking and remarkable is that in fundamental physics a beautiful or elegant theory is more likely to be right than a theory that is inelegant. ~ American particle physicist Murray Gell-Mann

Physical models bias physics. Physicists are understandably fond of mathematical simplicity, termed *elegance*, which comes via reducing independent variables. Symmetry is also essential to simplicity.

Otherwise, models become unwieldy if not insolvable, however better they may reflect Nature, and thereby offer predictability. Complexity is considered a nemesis, as it is a hindrance in workability, and an encumbrance to comprehending what are taken to be fundamental operating principles.

The result has been a strong inclination toward simplifying reduction that is often amended with exceptions when a model is found wanting, as most are. Putting a patch on a model lessens its elegance. Applying multiple patches can bring a model to its knees, as predictive power wobbles on an increasing number of variables and/or contingent conditions.

Just because the results happen to be in agreement with observation does not prove that one's theory is correct. ~ Paul Dirac

Mathematically, any system with 3 or more independent variables is unpredictable. Patches to improve predictability leave room to ponder if something else essential is being left out. Many theories, and the models on which they rely, flounder on these shoals. Such as been the case with the standard cosmological model (ΛCDM) and quantum physics' standard model.

It is impossible to trap modern physics into predicting anything with perfect determinism because it deals with probabilities from the outset. ~ English physicist Arthur Eddington

Beyond description and prediction lays explanation. The most powerful theories go beyond mere mechanics, yielding insight into the nature of phenomenal relations. Ultimately, this is what physics, and every branch of inquiry, strives for: knowledge.

Although we live in a world of constant motion, physicists have focused largely on systems in or near equilibrium. ~ American physicist Michael Kolodrubetz

A *theory* is a statement of how a relationship is presumed to behave, based upon some evidence. In contrast, a *law* is a conclusion of a universal natural tendency.

Whereas a theory is confined to specific relations, a law applies to everything. Laws invariably underpin theories.

While a theory may not necessarily ratify an implied aspect not central to the theory, it at least suggests that any implications of the theory are as valid as the theory's central tenet. After all, if a theory appears to well-describe its intended target, its ancillary implications intrigue.

Sometimes the implications of a theory turn out to be more important than the target phenomena described, in the doors they open, as questions raised of issues previously unconsidered. Maxwell's implicit discovery of wave/particle duality, leading Einstein to his relativity theories, is exemplary.

The open-ended nature of mathematics has engendered the belief that the fundamental constructs of physical existence can be formed into formulas.

> We exist in a universe described by mathematics. But which math? ~ American theoretical physicist Antony Garrett Lisi

While math has paved a remarkable path, ultimate understanding of existence via science remains as easily reached as the end of a rainbow. Behind every model and theory is insight which only opens more doors. The atomistic unraveling of reality is endless.

> The hypotheses we accept ought to explain phenomena which we have observed. But they ought to do more than this: our hypotheses ought to foretell phenomena which have not yet been observed. ~ English polymath William Whewell

ൠ The Inscrutability of Infinity ൠ

> To infinity and beyond! ~ action figure Buzz Lightyear in the movie *Toy Story* (1995)

In the early 20th century, modeling subatomic particles in classical electrodynamics hit a stumbling block: *infinity*. The mass-energy of a charged particle field veered to the infinite as its mass approached zero.

2 divergent preeminent models emerged in modern physics: relativity and quantum mechanics. Attempts to bridge the relativistic cosmological realm with the infinitesimals of quantum theory foundered on the ethereal rocks of infinity. At the quantum level, gravity became an infinite cavity. So too relativity, where gravity culminated in black holes.

The quantum Standard Model, which is the supposed success story of quantum theory, was long beset with pathological infinities. For instance, quantum electrodynamics – the quantum theory of the electromagnetic force – initially posited the mass and charge of an electron as infinite.

Infinity won't do for theoretical glue, so a workaround was developed, along with a euphemism for ejecting infinity from equations: *renormalization* – pitch the infinities with any workaround that works.

Once introduced, renormalization became the norm. If a relative measure could be established in relation to an experimentally known quantity, such as the electron's mass and charge, the infinities popping out the equations could be ignored. The residual finite results from renormalization serve as a workable approximation.

> The horrible "infinity" could be subsumed, hidden from view as if didn't exist, leaving an apparently pristine theory on display. ~ English particle physicist Frank Close

Ubiquitous in their cameo appearances, virtual particles are a phenomenon that smacks of infinity. The model for virtual particles is anomalous in multiple ways. Photons are massless, but virtual photons have mass. Virtual particles also have an energy-momentum that is not allowed by any relativistic equation for a particle of that mass.

Virtual particles are accounted for using integrals. These integrals are often *divergent*: their solution yields infinity. So virtual particles are mathematically ignored, while their physical effects are factored in.

A crucial particle in the quantum Standard Model – the *Higgs boson* – supposedly gives mass to all matter particles (fermions) via the Higgs mechanism, where a universal field imparts just the right amount of juice. The Standard Model states that the Higgs boson has infinite mass. This unacceptable result is washed out by mathematical trickery.

> Infinities, when considered absolutely without any restriction or limitation, are neither equal nor unequal, nor have any certain proportion one to another, and therefore, the principle that all infinities are equal is a precarious one. ~ Isaac Newton

While all infinities are not necessarily created equal, the models that turn up infinity are all making the same statement: that their accounting is incomplete, that something essential is missing.

✁ Fields ✂

Many acts of physics transpire as fields. A *field* is a physical quantity represented by a scalar, a vector, or a tensor.

A *physical quantity* is a mathematical representation of a property in a *physical system*, which is an arbitrary geometric region under examination. The adjective *physical* refers to a study of something which may be observed, as contrasted to utterly imaginary concepts.

A *scalar* is a quantity which is representable as a point on a scale. Scalars are expressed as real numbers.

A *vector* is a geometric quantity with both magnitude and direction. Vectors are typically represented by arrows. Although a vector has a significance and an orientation, it lacks a position. In physics, a vector is a movement of energy at a certain strength.

A *tensor* is a geometric object describing linear relations between other geometric entities (scalars, vectors, tensors). Tensors are typically entangled with other tensors, forming a *tensor network*.

In physics, a *field* is considered a region of space with integral energy. More pointedly, a field is an energy associated with a spacetime point.

Like energy, fields do not exist. Physics fields are just a mathematical modeling technique: a geometrical way of characterizing phenomenal transformations of matter.

Newton's law of universal gravitation was an expression of force fields, though Newton lacked the idea of fields. English physicist Michael Faraday coined the term *field* in 1849, referring to electromagnetism.

Gravity, which is an entropic distortion of spacetime, acts as if it was an attractive force by a body because of its mass. Mathematically, gravity is a monopolar field.

An electrical field has 2 opposite point charges, so is dipolar. The 2 point charges of an electrical field are negative and positive. Electrons carry a negative charge. Protons carry a positive charge. By definition, an electric current flows away from a positive charge and toward the negative charge.

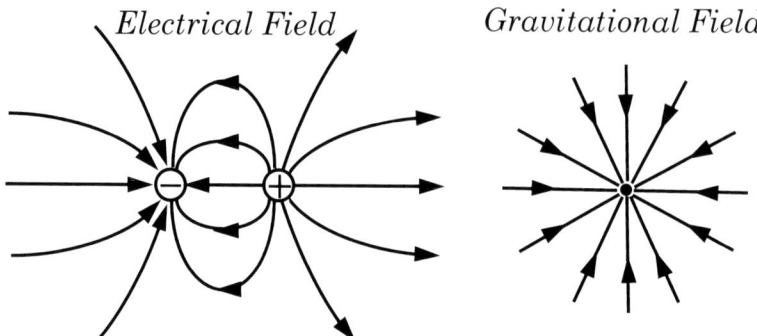

Electrical Field   *Gravitational Field*

All matter is loaded with electric charges. We are usually unaware of this because the opposing charges within matter – between protons and electrons – neutralize one another.

An electric charge creates a field which exerts an outward-radiating force, called the *Coulomb force*. The lines of force of an electric field flow between the oppositely charged dipoles.

Charles-Augustin de Coulomb published his speculations on electricity and magnetism in 1785. The *Coulomb force* came from characterizing static electricity.

The strength of an electric field decreases as the square of the distance from its source. Moving twice the distance from a point charge saps the felt field strength by 1/4th. This inverse-square dynamic is termed *Coulomb's law*, though German physicist Franz Aepinus suspected as much in 1759, before Coulomb published his law.

A moving electric charge – an *electric current* – creates a magnetic field. As everything is always moving, electric and magnetic fields are coincident.

The charges of a magnetic field are like those of an electric field, with the south pole of a magnet analogous to a negative electric charge, and the north pole like a positive electric charge.

The 2 charges in a hydrogen atom, with its single proton (+) and sole electron (–), may be pulled apart to distinguish electric monopoles. That cannot be done with magnets. Magnets are always dipolar. Theorists hypothesize that there may have been magnetic monopoles in the early universe, but none have ever been observed.

ೂ The Imaginary Complexity of Reality ೪

Standard quantum theory is based on a complex Hilbert space. ~ William Wootters *et al*

For over 2,000 years, there was only 1 geometry, devised by Greek mathematician Euclid of Alexandria in the 4th century BCE. What belatedly became termed *Euclidian geometry* was described in the most influential mathematics book of all time: *Elements*, the primary textbook for math, especially geometry, into the early 20th century. The 3-dimensional spatial world has long been described as *Euclidean space*.

With general relativity, Einstein inadvertently introduced an extra spatial dimension. Quantum theory and its poster child, the Standard Model, upped the ante with even more dimensions.

These models required geometric description that exceeded Euclid's conception; a *non-Euclidean geometry*.

Work by German mathematician David Hilbert and others in the 1st decade of the 20th century provided the mathematical means; so arose *Hilbert space*.

In supporting any number of dimensions, Hilbert space generalizes Euclidean space. To do so, complex numbers are employed.

In construing geometric points, *complex numbers* are inherently 2-dimensional, with real and imaginary parts. While the real number is real enough, quantitatively speaking, the imaginary part (*i*) is unworldly in satisfying the equation:

$$i^2 = -1.$$

Despite their surreality, complex numbers have been a mathematical convenience since the 16th century. Waves of all sorts are most easily expressed using complex numbers. The rotations and oscillations of quantum mechanics are neatly described via complex numbers. In contrast, real numbers alone cannot describe the wave/particle duality that is the fuzzy foundation of quantum theory.

There is some saving grace in that the imaginary *i* washes out when measuring a quantum phenomenon. The uncertain complex plane collapses to a real measured point.

If one takes the quantum model to be a map of reality, which physicists most certainly do, the issue is what the imaginary part means. It must somehow represent information that is required for the system as a whole, but not in the perceived instant. In other words, conceptually taking complex numbers as representing something real, the imaginary portion must provide a necessary context that is not apparent when viewing moments of spacetime.

Perhaps the best way to see what the imaginary brings to reality is to try to set it aside. American theoretical physicist William Wootters did so.

Wootters and his colleagues constructed a real-number quantum theory. An extra bit was needed to fill the void of the imaginary; a supposed physical entity that Wootters dubbed the *ubit* (for "universal quantum bit").

The ubit turned out to be a master of entanglement: an information conduit interacting with all the other ubits in the system describing the universe. In a word, the ubit signified *entanglement*.

With the ubit, the modeled world is all real. The same could be said for the imaginary i in complex Hilbert space, which takes the same backseat driver role that the ubit has; a difference with scant distinction. Except, in building the real-only model, the ubit came in as a necessary accouterment, rather than being built-in as part of the complex mathematical fabric.

With his real-number model, Wootters was able to shine a spotlight on an essential element that drops out of view in ever-emergent actuality: the meaning of i in Hilbert space. Wootters' work showed that our world has a complexity which contains a bit of the imaginary, all entangled.

> People always thought of complex numbers just as a tool, but increasingly we are seeing that there is something more to them.
> ~ English mathematician Dorje Brody

ઠ Transmission Optimality ର

Around 60 CE, Egyptian mathematician Hero of Alexandria noted that reflected light takes the shortest path.

During the 160s, Ptolemy characterized perceived properties of light, including reflection, refraction, and color. Though he construed refraction tables, Ptolemy failed to discover the exact math relating angles of incidence and refraction (*Snell's law*).

Persian mathematician Ibn Sahl discovered the law of refraction in 984. Sahl's derivation was unknown to later Europeans who rediscovered the law multiple times. The mathematical rule of refraction is now attributed in name to Dutch astronomer Willebrord Snell, who derived the law in 1621 but never published it in his lifetime.

In 1658 French mathematician Pierre de Fermat proposed an optics *principle of least time*: that light always travels most efficiently: from one point to another in the least time, regardless of being reflected or refracted. *Fermat's principle* (as it is commonly called) was broadened to encompass all wavefront behavior by Dutch physicist Christiaan Huygens in 1678.[*] Augustin Fresnel supplemented Huygens' principle in 1818 with mathematical treatment of interference (diffraction).

In the figure, a ray of light going from *a* to *b* would travel the least distance via the hypothetical straight line. Instead, light actually traverses a longer distance that takes less time, as light moves slower through water than air – the straight-line path would incur longer, sluggish passage in water.

Fermat's Principle

The law of wave transmission optimality (the *Huygens–Fresnel principle*) was generalized for all dynamics in any physical system by Irish physicist and mathematician William Hamilton in 1827. This *principle of least action* is based on a single function: the Lagrangian. *Hamilton's principle* was a rehash of the same discovery independently-made by

---

[*] Huygens' principle applies only with 1 time dimension and an odd number of space dimensions. The principle fails with an even number of spatial dimensions.

Gottfried Leibniz, Leonhard Euler, and Pierre Maupertuis in the first half of the 18th century.

Lagrangian mechanics was a 1788 reformulation of classical mechanics by Italian French mathematician and astronomer Joseph-Louis Lagrange. The Lagrangian is widely used in physics. Lagrangian equations provide that any motion may be calculated by incorporating all the information about the dynamics of the system.

Although originally formulated for classical mechanics, Hamilton's principle also applies to all physics theories, notably playing a key role in quantum mechanics.

As has long been known, light travels optimally, as do all propagating energy waves. The mathematics behind this shows that such matchless motion necessitates omniscience: knowing all the instant information in the universe.

This profundity is no casual conclusion. It is a statement of fact. For light, or any energy wave, to behave as it does, all information about actuality must be instantaneously incorporated. Every physics theory accepts this axiomatically.

Optimal propagation clearly indicates a unified, coherent intelligence from the quantum level on up, and strongly suggests teleology: that the game afoot which we call Nature has intention.

The history of physics has been an unwinding, from describing observed Nature to formulating Nature as an illusion, from equations that made sense of what the senses sensed to formulas that make foolery of what is perceived.

Even if there is only one possible unified theory, it is just a set of rules and equations. What is it that breathes fire into the equations and makes a universe for them to describe? The usual approach of science of constructing a mathematical model cannot answer the questions of why there should be a universe for the model to describe. Why does the universe go to all the bother of existing? ~ Stephen Hawking

♒ Albert Einstein ♒

I am no Einstein. ~ Albert Einstein

In 1895 Albert Einstein failed the entrance exam at the Polytechnic University in Zurich, Switzerland, so his parents sent him to a secondary cantonal school in northern Switzerland. After a year there he made his way into Polytechnic.

After graduating from university, Einstein got a temporary teaching position at a school in Schaffhausen. His 2-year search for a permanent teaching post proved fruitless.

A former classmate's father used his influence to get Einstein a job at the Swiss patent office in 1902, where Einstein became a 3rd-class examiner for patent applications related to electromechanical devices. Although his job became permanent, he was passed over for promotion to 2nd-class until he "fully mastered machine technology."

In the meantime, the lackluster but ambitious Einstein was writing papers. His 1901 paper on capillary action in straws was published in a prestigious physics journal.

1905 was Einstein's jackpot year. In what has been called his *annus mirabilis*, or "miracle year," Einstein completed his thesis – on molecular dimensions – and got his PhD. He also published 4 papers over various topics: the photoelectric effect; Brownian motion; a terribly simple formula equating matter and energy ($E = mc^2$); and special relativity, a mathematical statement striking at the heart of Newtonian absolutist physics.

♋ Mass-Energy Equivalence ♌

The mass of a body is a measure of its energy content.
~ Albert Einstein in 1905

Classical physicists conveniently multiplied an object's mass by the square of its velocity (mv^2) to come up with a useful indicator of its kinetic energy: $E = mv^2$. In his relativistic equation $E = mc^2$, Einstein simply substituted the speed of light (c) for the classical notion of velocity (v).

Mass and energy are both but different manifestations of the same thing. ~ Albert Einstein

Max Planck quickly followed Einstein in expressing mass as a form of energy. Other physicists contemporaneously converged on the same equation.

The relationship between matter and energy is asymmetric. Whereas disbanding matter releases energy, as atomic bombs illustrate, matter cannot construct energy.

The formula is astonishing in its implications. The speed of light squared is a huge number. $m = E/c^2$ means that even the smallest amount of matter locks up an unimaginable amount of energy. In contrast, chemistry, which works profound transformations by tweaking chemical bonds, involves the slightest flutterings.

For his efforts in mastering machine technology, Einstein was promoted to technical expert "second class" at the Swiss patent office in 1906. Einstein left the patent office in 1909, headed to the type of university position he had sought a decade before.

ಶಿ Light Speed ಲ

If you are in a spaceship that is traveling at the speed of light, and you turn the headlights on, does anything happen? ~ American comedian Steven Wright

In the observable dimensions, light seems to travel as fast as anything can. Nonlocality has shown light speed as an apex to be an assumption with a limited domain.

That light even has a speed was something long unconsidered: light just was, pervasive from its source. Studying the movement of Jupiter's moon Io in 1676, Danish astronomer Ole Rømer first demonstrated that light traveled at a finite speed, not instantaneously.

In 1865, with a paper on electromagnetism, James Clerk Maxwell started using V – for *velocity* – as the symbol for light speed. That was the notation adopted by Einstein in his 1905 papers on relativity.

By the end of the 19th century, *c* commonly denoted the speed of electromagnetic waves. This convention began in 1846, with a paper by German physicist Wilhelm Weber that aimed to unify electrostatics with electrodynamics. As light chauffeurs electromagnetism, you can tell where this story is going.

Writing about electrons in an influential paper published in 1904, German physicist Max Abraham used *c* rather than *V* as the speed of light. In 1907, Albert Einstein started using *c* to signify the speed of light, editing his seminal relativity papers to use *c* instead of *V*.

Increasingly precise measurements of light in transit culminated in 1975 with the speed now used: 299,792,458 meters per second (in a vacuum).

℘ The Horizon Problem ℞

Cosmic microwave background (CMB) radiation left a fossilized imprint of the universe. By that time the universe was already quite spread out, to put it mildly.

The CMB indicates that the temperature and other fundamental physical properties of the universe were largely uniform then. Such consistency should not be possible. There is no mechanism to explain how the universe had a consistent temperature long before heat-carrying photons had time to scour the cosmos and deliver uniformity.

American physicist Charles Misner considered this conundrum in the late 1960s and termed it the *horizon problem*. Alan Guth dreamed up cosmic inflation to address this inscrutable homogeneity. But there is another possible explanation: an inconstant light speed; or, as it is commonly called, a varying speed of light (VSL).

Einstein first mentioned VSL in 1907 and seriously considered it until propounding general relativity in 1915. His conclusion was that light speed was subject to gravity (by warping spacetime).

French astrophysicist Jean-Pierre Petit first proposed VSL in 1968 to solve the horizon problem. Others have since variously modeled how VSL might work.

The CMB reflects the speed at which light and gravity propagate as the temperature of the universe changes. Following the ΛCDM model, some astrophysicists propose VSL in the feverous early universe, with light outracing gravity by exceeding the blazing speed it now travels. Though merely speculative, VSL is not contradicted by evidence like cosmic inflation is.

Varying light speed would invalidate Einstein's relativity theories, which anyway were inapplicable in the infant universe before matter took form and mass had much meaning, which meant that gravity would have been nebulous. The physics of the early universe are not understood.

✺ Relativity ✺

When you are courting a nice girl, an hour seems like a second. When you sit on a red-hot cinder, a second seems like an hour. That's relativity. ~ Albert Einstein

In 1632, Galileo Galilei stated his principle of relativity: that the laws of physics are the same for all inertial frames. Galilean invariance would underpin Newtonian classical mechanics, which was the mathematic clockwork of classical physics, with an absolute space sharing a universal vector of time.

Galileo and Newton took for granted that reality was circumscribed to the observable dimensions, and that physics could posit properties that were absolute cornerstones of reliance; whence 'laws of Nature'.

Classical physics modeled the everyday world; a set of rules only for the dimensions we can see. When one's view veers from the ordinary, onto a path partly paved by Einstein, those rules are rent.

✺ Special Relativity ✺

Henceforth space by itself, and time by itself, are doomed to fade away into mere shadows, and only a kind of union of the two will preserve an independent reality. ~ Lithuanian mathematician Hermann Minkowski

Einstein's special relativity confirmed the Galilean relativity that physics laws are inviolable while trashing the Newtonian relativity that space and time are unqualified. Under Einsteinian relativity, only the speed of light is absolute in that it appears the same to any observer, regardless of how fast that observer is traveling. All else is subject to change.

Simply, special relativity put a cap on how fast anything may travel: the speed of light. Time stops there (as it does at the opposite end of the spectrum: in a black hole).

Practically, the speed of light depends upon the medium it travels through. Light can be slowed or bent.[*]

Light propagates through transparent materials, such as air or glass, at less than c. The ratio between c and the speed at which light travels in a material (v (phase velocity)) is called the *refractive index* (n) of the material: $n = c / v$.

Under special relativity, all uniform motion is relative. Passengers in planes, trains, and automobiles get a visceral sense. But special relativity leads to some counterintuitive predictions for objects traveling at obscene speeds.

1st, *relativity of simultaneity*: simultaneity is not absolute; instead, it depends upon an observer's frame of reference. Different observers in relative motion to one another may legitimately disagree as to whether 2 events occurred simultaneously, or one before the other.

2nd, *time dilation*: that time itself is relative to the relative motion of an observer. In other words, clocks tick at different rates depending on their relative motion. Time dilation has been experimentally demonstrated.

3rd, *length contraction*: a moving ruler that appears at rest to an observer will measure shorter than otherwise. Length contraction is noticeable when the frame of reference approaches the speed of light.

This principle of relativity is "special" in that it applies only to inertial reference frames. Under special relativity, the maximal speed of light is the only absolute.

---

[*] Even in a vacuum, structuring light can slow it down. While light is usually approximated as plane waves, its structure is considerably more complex.

As everything is in relative motion, all phenomena are qualified, including space and time. Special relativity posits an adamant bond between space and time: the two are entwined as *spacetime*. General relativity exposed the mutability of this medium in which existence swims.

> The geometry of spacetime is not given. It is determined by matter and its motion. ~ Austrian theoretical physicist Wolfgang Pauli

♎ Gold ♎

> Gold is a difficult system. ~ Indian chemist Sourav Pal

Chemistry is mostly concerned with the reactivity of elements, which owes to the number of electrons in the outer shell; the fewer electrons there, the more reactive.

Cesium and gold both have a single electron in their outer shell (the 6th such shell for them). Cesium is the most alkaline of natural elements, and highly reactive: it explodes if dropped in water, and even reacts to ice.

In contrast, gold is stalwart to a fault. Gold does not react to oxygen at any temperature, nor with ozone. Gold is unaffected by most acids and most bases. Hence gold does not tarnish. Special relativity accounts for gold's stability and its color.

Negatively charged electrons whirl about their atomic nucleus with a speed and tightness corresponding to the intensity of the positively charged protons within, along with the nuclear core's mass. With 79 protons in gold's nucleus versus the 55 protons in cesium, and half again as many neutrons, gold's tightly bound nuclear core has much more pull on its orbiting electrons.

The electrostatic attraction of gold's nucleus relativistically speeds up, and tucks in, gold's electrons, making it less reactive, and increasing its light absorption.* Thus, gold

---

* The subatomic attraction of gold's positively charged protons to the negatively charged electrons in orbit both increases the electrons' speed and increases their mass, causing a relativistic contraction in their orbits because, as an electron's mass increases, the radius of its orbit with constant angular momentum shrinks proportionately.

soaks up blue light; reflecting the reds and greens which combine into the golden hue we see.

Gold is not the only element under the influence of special relativity. Mercury is another heavy atom, with electrons held close to the nucleus. But the bonds between mercury atoms are weak; hence mercury has a low melting point and is liquid at ambient temperature.

Just as Galilean relativity is a slow-motion approximation of special relativity, special relativity approximates general relativity for weak gravitational fields.

Special relativity is violated by quantum entanglement: simultaneity faster than light. Special relativity applies only within certain dimensional constraints.

෨ General Relativity ൠ

There are really four dimensions, three of which we call the three planes of Space, and a fourth, Time. ~ English author H.G. Wells in the novel *The Time Machine* (1895)

Issued by Einstein in 1916, *general relativity* posits gravity as a geometric property of 4-dimensional (4D) spacetime, based upon the mass of objects. General relativity is a simple theory of gravitation; a generalization of special relativity coupled to Newton's law of universal gravitation. Under general relativity, objects are deflected when they pass near a massive body, not because of a force per se, but because spacetime itself around the body is warped.

It took a half-century for the implications of general relativity to be appreciated. After an initial burst of excitement following confirmation of the theory – a 1919 announcement of light-bending by gravity – general relativity was ignored for decades. Only during the 1960s did appreciation of this gravitational theory sink in.

⚡ Mass ⚡

Mass is the fundamental physical property of matter. Mass is measured in terms of its *inertia*: a body's resistance to change in motion (acceleration).

In everyday parlance mass and weight are used interchangeably, but *weight* is a measure of inertia within a specific gravitational field. 2 objects with the same mass have different weights on the surface of Earth and the surface of the Moon, as these celestial bodies have their own gravitational pull.

Mass determines the degree to which an object is affected by or generates a gravitational field. Mass creates gravity and is affected by bodies with greater mass.

In providing both the gravitational power that conducts cosmic motion and the sense of solidity at ambient scale, mass is the maestro of materiality; its effects achieved by entropy and inertia.

Though inertial mass and gravitational mass are conceptually distinct, there is no actual difference between them. Repeated experiments since the 17th century demonstrate their identicalness.

Oddly, the concept of mass in general relativity is amorphous and problematic. Mass is a measure of motion, which means that the mass of a physical system is within a reference frame of momentum. Hence, mass is always relative to the system in which it is observed.

Subatomic particle mass is a homonym to the concept understood in classical physics: same word, different meaning. A practical conception of subatomic particle mass is the threshold energy at which a certain particle may appear. Quantum mass has nothing to do with gravity, and is only tangential to inertia, in that subatomic particles hesitate to display themselves until sufficiently prodded energetically.

There are various theories which attempt to explain how mass is generated. None succeed beyond some fancy mathematics that fit within a system of equations. Some *mass generation mechanisms* (as such theories are called) involve gravity, which presents an insolvable chicken-or-egg problem, as mass and gravity are manifestations of the same thing.

Space is not a lot of points close together; it is a lot of distances interlocked. ~ Arthur Eddington

The first evidence that Einstein was on the right track with general relativity came from space. By the end of the 19th century, slight perturbations in Mercury's orbit indicated that Newton's gravitation theory was off. What Newton could not account for Einstein could.

Atomic clocks on GPS satellites tick faster than they would on Earth by about 45,900 nanoseconds per day. This is because they experience less gravity out in space.

This uptick is somewhat compensated by satellites orbiting the planet rather being stationary on it; a slower tick by ~7,200 ns/day, owing to special relativity.

Cosmological objects themselves experience relativistic effects. Earth's core is 2 1/2 years younger than its crust thanks to gravity.* So too the Sun: its center is some 40,000 years younger than its surface.

General relativity is a description of the whole universe as a closed system. ~ Canadian theoretical physicist Lee Smolin

As general relativity predicted, gravity manifests as a curvature of 4D spacetime. To capture this distortion geometrically requires extra dimensionality (ED), an obvious implication Einstein studiously ignored.

The warpage of spacetime caused by an orbiting body is termed the *geodetic effect*. A feebler distortion, in which a spinning body yanks and twists surrounding spacetime, is the *frame-dragging effect*. These effects have been verified via satellite observations of Earth.

---

* Relative youth has nothing to do with why Earth's crust is wrinkly and its core is not.

Though classical in its assumption of the universe as a 4D closed system, general relativity propelled predictions that differed from classical physics, notably the geometry of space, the passage of time, the motion of bodies in gravity-influenced free-fall, and the propagation of light.

> Spacetime tells matter how to move; matter tells spacetime how to curve. ~ American theoretical physicist John Wheeler

General relativity is not the only relativistic theory of gravity, but it is the most mathematically simple theory to explain the experimental data which validates it.

> General relativity is the basis of our understanding of gravity. But 21st-century work in cosmology and particle theory strengthens the belief that it is an incomplete description. ~ American astronomer Gary Wegner

In his quest for a unified field theory, Einstein spent the back half of his life trying to incorporate electromagnetism into relativistic spacetime. He never succeeded. An unrelenting bias against higher dimensionality and mediocre mathematics acumen made Einstein ill-suited for the task he had set himself.

∼ William Kingdon Clifford ∼

English mathematician and philosopher William Kingdon Clifford literally worked himself to death, succumbing at 34. In his short life, Clifford blazed an illuminating trail.

> He with great ingenuity foresaw in a qualitative fashion that physical matter might be conceived as a curved ripple on a generally flat plane. Many of his ingenious hunches were later realized in Einstein's gravitational theory. ~ Hungarian physicist Cornelius Lanczos

Clifford published in 1870 *On the Space Theory of Matter*, where he advanced the concept of reality as particles in space, though appearing from a higher dimensionality; matter as non-Euclidean disturbances viewed from a perspective of "flat" (noncurved) 3D space.

Clifford envisioned fields (electric, magnetic, gravitational, et cetera) expressed geometrically, and that particles interacted by means of these fields.

Relatively little was known about the composition of matter at the time, and so Clifford's explanations lacked sophistication. Working from received wisdom, Clifford presaged the most advanced theories of modern physics.

Among other musings, Clifford developed the notion of consciousness as being formed from a composite of information ("mind-stuff"); and the basis of moral law as being founded upon social interdependence ("tribal self").

It is wrong always, everywhere, and for anyone, to believe anything upon insufficient evidence. ~ William Clifford

‮ஐ‬ Time ‮ை‬

Time by itself does not exist. It must not be claimed that anyone can sense time apart from the movement of things. ~ Roman philosopher Lucretius

What is time? If nobody asks me, I know; but if I were desirous to explain it to one that should ask me, plainly I know not. ~ Augustine of Hippo

Classical physics regarded time in the way we are familiar: a vector of past, present, and future; subject to subjective interpretation, but ultimately an objective metric.

Absolute, true and mathematical time, of itself, and from its own nature, flows equably without regard to anything external, and by another name is called *duration*. ~ Isaac Newton

Near the end of the 19th century, Austrian physicist Ludwig Boltzmann assaulted the temporal vector, arguing that time had no built-in arrow; but that its application in space gave time meaning, as entropy was irreversible.

The only reason for time is so that everything doesn't happen at once. ~ Albert Einstein

Temporality lost its absoluteness with special relativity, which rendered time as relative to a frame of reference. 2 events may occur simultaneously or sequentially, depending upon point of view.

> The past, present and future are only illusions, even if stubborn ones. ~ Albert Einstein

Under special relativity, causality is a point of view. The implication is that time, as a matter of perspective, is only an apperception.

Equivalently, time has no meaning in the quantum realm, nor does causality. The world simply incessantly *is*. Existence itself as an emergent phenomenon: spacetime itself coming into being. Continuity is a perspective, not a phenomenon.

General relativity posits that gravity warps spacetime: not just space, and not just time. Under general relativity, time stops upon entry into a black hole, and space ceases to exist. A singularity of infinite mass collapses space upon itself and squeezes time to a standstill.

Whereas space comprises equivalent dimensions that must be unified for manifestation, time stands alone. Relativity's twining of spacetime goes only to gravity's effect: inferring elasticity in time as a geometric fluid medium.

Heisenberg's uncertainty principle posits an inherently nondeterministic reality through time. Under uncertainty, space and time are in a flux: *where* and *when* are probabilistic.

> Everything in the future is a wave. Everything in the past is a particle. ~ English physicist William Lawrence Bragg

The only particle of time is the present moment, when wavelike interconnections in space have crystallized. Nonlocal simultaneity simply is; its recognition merely a perception by a mind insisting upon continuity.

> Nothing in known physics corresponds to the passage of time. ~ English physicist Paul Davies

Between past and future, the present is an instantaneous confluence; the only moment of existence. Despite having no accounting for its passage, physics depends entirely upon time.

> You could make different choices of what you mean by time and get any laws of physics you like. ~ American theoretical physicist Andreas Albrecht

As the potentate of entropy, gravity plays with time. There is more to it than that. Spacetime and gravity define each other.

Many physicists believe that on the very smallest scale of size and duration, space and time might lose their separate identities. ~ Paul Davies

At the quantum level, gravity is assumed to exist in granular form. Similarly, time is simply taken for granted as having tiny steps. If the measures become too tiny, time simply stops. This is the Zeno effect.

⚹ Zeno Effect ⚹

The quantum Zeno effect is the inhibition of transitions between quantum states by frequent measurements of the state. The inhibition arises because the measurement causes a collapse of the wave function. If the time between measurements is short enough, the wave function collapses back to the initial state. ~ American physicist Wayne Itano *et al*

Zeno of Elea was a 5th century BCE Greek philosopher and mathematician who had an inordinate fondness for paradoxes. Aristotle credited Zeno with inventing *dialectic*: logic based upon the interaction of juxtaposed ideas.

The Zeno effect is maintaining stasis in a quantum system via continuous observation. Time evolution in a quantum system is completely suppressed simply by continuously observing it.

The term *Zeno effect* comes from Zeno's arrow paradox, which states that because an arrow in flight is not seen to move at any instant, the arrow cannot possibly be moving at all. It is a proposition based on the infinitesimals upon which calculus is built.

The Zeno effect is also called the *Turing paradox*, from Alan Turing's 1954 observation that continuously measuring a system increases to infinity the probability that the system will be in the same state from one observation to the next: "that is, that continual observations will prevent motion."

The Zeno effect has been repeatedly verified in quantum and atomic systems. It remains inexplicable.

Until a trustworthy algorithm is developed to explain the Zeno effect, the completeness of quantum theory must remain in doubt. ~ Indian physicists Baidyanath Misra & George Sudarshan

Awareness makes time stand still. Time's progression is a matter of inattention.

The quantum Zeno effect is real; a watched quantum pot never boils. ~ English astrophysicist John Gribbin

¤ ✧ ¤

The Zeno effect applies to a localized quantum system. The classical effect of time's arrow takes hold because quantum systems are entangled in a larger-scale multisystem. This entanglement induces delocalization and the classical irreversibility of time: an *anti-Zeno effect*. Classical chaotic dynamics exist because awareness cannot cover the world at large.

✿ ✿ ✿

Quantum events that shape moment-by-moment actuality occur in the shortest time (Planck time). At that scale, time has no meaning. This is in stark contrast to the tale time will tell over the eons of cosmic expanse, from what supposedly started as a singular quantum, then spread through time as a gyral web.

The singular force of time in a timescape is equivalent to all the forces influencing landscape at every scale. But whereas empty space is uniform, time has different characters at the extreme scales of existence.

Our perception of time profoundly influences our perception of change. A drastic change at one scale may appear trivial at another. ~ American evolutionary biologist Stephen Jay Gould

Ultimately, for all its abstract foibles, physics treats time as the pesky vector that it is for all concerned, regardless of scale or scope. The bottom line is that time is as slippery as it seems: an illusion that can't be grasped however strongly felt. That, at least, is a unified theory of time.

Time is free, but it's priceless. You can't own it, but you can use it. You can't keep it, but you can spend it. Once you've lost it you can never get it back. ~ American businessman and writer Harvey MacKay

ೞ Matryoshka Reality ೞ

A careful analysis of the process of observation in atomic physics has shown that the subatomic particles have no meaning as isolated entities but can only be understood as interconnections between the preparation of an experiment and the subsequent measurement. ~ Erwin Schrödinger

♎ Matryoshka Dolls ♎

Matryoshka are diminishing self-similar hollow dolls that can be nested one inside another. Matryoshka derives from the Russian peasant name for females: Matriosha or Matryona, derived from the Latin root *mater*, meaning mother.

In 1890, Russian painter Sergey Malyutin designed the first Russian Matryoshka doll, which was carved by Vasily Zvyozdochkin, a Russian craftsman. Malyutin painted the dolls.

Matryoshka dolls are sometimes called *babushka dolls*, the Russian word for grandmother.

Like the dolls themselves, the history of the Matryoshka doll is nested. The concept predates Malyutin, who was inspired by a wooden doll brought to Russia from the Japanese island of Honshu. The Japanese claim that their dolls derived from the earlier work of a Russian monk, who had created a doll to represent a Buddhist sage.

♋ History ♋

Except for enlightened sages, existence was taken at face value for all but a wink of time in human history. As human mathematics evolved, everyday energetics became codified by classical mechanics.

In the late 19th and early 20th century, physicists' theorizing greatly disturbed physics' status quo. With special relativity concluding that perceptions were relative to the observer, Einstein inadvertently forwarded a startling metaphysical triumph of subjectivity over objectivity.

General relativity trashed gravity as a fundamental force, showing it instead as a distortion in the spacetime fabric. Contemporaneously came the unraveling of particulate matter.

Human conception of what comprises the constituents of Nature has evolved: from 4 basic elements (air, fire, water, earth), to atoms, to a smorgasbord of elementary particles so small that they can only be inferred, bifurcated between those that are matter (fermions) and those that keep matter working (bosons).

Getting past Empedocles' purely theoretical construct of there being 4 basic elements, ancient Greek chemists considered that the fundamental particles of matter were chemical elements. The term *atom* comes from the ancient Greek adjective *atomos*, meaning *indivisible*. The philosophical concept of chemical elementalism as the fundamental unit of matter was commonly held in Greece and India by the 4th century BCE.

5th century BCE Greek rationalist philosopher Democritus and the shadowy Leucippus considered atoms as the bedrock of being: physically, but not geometrically indivisible; of infinite number and kinds, with different shapes and sizes; and forever in motion, whirling in empty space. This conceptualization held sway into the 19th century.

In proposing unalterable chemical elements with specific "atomic" weights, English physicist and chemist John Dalton whittled elemental particles to molecules in 1803, but had no conception that molecules consisted of atoms.

Italian physicist Amedeo Avogadro, seeking more accurate estimates of atomic weight, corrected Dalton's flaw in 1811 by distinguishing between molecules and atoms.

⌘ Brownian Motion ⌘

Particle theory jelled by pondering the haphazard movements of floating debris. In 60 BCE, Lucretius described the random motion of dust particles, which he then used as a proof for the existence of atoms.

In 1827, Scottish botanist Robert Brown wondered about dust particles from pollen grains floating in water, constantly jiggling for no apparent reason. Dutch physiologist Jan Ingenhousz had earlier observed the same effect in floating charcoal, but by virtue of broader publication, Brown won the name game. *Brownian motion* is the seemingly random movement of particles suspended in a fluid (gas or liquid).

Particle theory firmed with Einstein's 1905 mathematical model to explain Brownian motion, which allowed him to determine the size of atoms and calculate how many atoms are packed into a mole (the standard molecular weight). French physicist Jean Perrin experimentally validated this facet of atomic theory in 1908.

Atoms were considered the smallest possible division of matter until 1897, when English physicist J.J. Thomson found that there was something smaller: what he termed *corpuscles*, the subatomic particle now called the *electron*. Several scientists before Thomson had suggested that atoms were built up from a more fundamental unit but conjectured that this unit was about the size of the smallest atom, hydrogen.

Thomson found that cathode rays could be deflected electrically, and figured that subatomic corpuscles emerged from gas atoms. He concluded that atoms were divisible into constituent corpuscles, whereby he concocted a plum-pudding model of atomic structure.

To explain the overall neutral charge of an atom, as contrasted to the corpuscle (electron) negative charge, Thomson proposed in 1903 that corpuscles floated in a sea of positive charges, with electrons embedded like plums in a pudding, though Thomson's model posited rapidly moving corpuscles instead of plopped plums.

Japanese physicist Hantaro Nagaoka rejected Thomson's model on the grounds that opposite charges are impenetrable. Nagaoka proposed a planetary model in 1904, in which a positively charged nucleus was surrounded by revolving negatively charged electrons. Nagaoka had in mind Saturn, with its satellite rings.

One of Thomson's pupils, English physicist and chemist Ernest Rutherford, disproved the atomic plum pudding in 1909. At the behest of Rutherford, German physicist Hans Geiger and New Zealand physicist Ernest Marsden performed their famous "gold foil experiment": shooting a beam of radium alpha particles at gold foil, whereupon they measured a widespread deflection of radioactive decay. If Thomson's plum-pudding atomic model had been correct, the deflection would have been at most a few degrees.

Following Nagaoka, Rutherford proposed his planetary atomic model in 1911: a cloud of negatively charged electrons swirling in orbits over a compact positively charged nucleus. Only a concentrated charge could have accounted for the heavy deflection found in the gold foil experiment. This subatomic particle was the proton, which Rutherford identified in 1918.

Rutherford was working with Niels Bohr, who conjectured in 1913 that electrons moved in specific orbits which were regulated via Planck's quantum of action.[*]

In 1919, Rutherford became the first to transmute one element into another, converting nitrogen into oxygen through the nuclear reaction $^{14}N + \alpha \rightarrow {}^{17}O + proton$.

Rutherford speculated in 1921 about how atomic nuclei stayed together rather than flying apart. Rutherford concocted neutrons, which could, via some attractive nuclear force, somehow compensate for the repelling effect of the positively charged protons. Neutrons account for much of the extra mass of atoms heavier than hydrogen.

Rutherford's neutron hypothesis was experimentally shown in 1932 by his associate, English physicist James Chadwick. Chadwick would later join the American Manhattan Project which developed the atomic bombs dropped on Hiroshima and Nagasaki, to close World War II with a most spectacular war crime. For his valiant effort in developing the first weapon of mass destruction, Chadwick was knighted in 1945.

---

[*] Planck's quantum of action, also known as the *Planck constant*, is the essential granularity of existence.

Non-cooperation in military matters should be a vital part of the moral code of basic scientists. ~ Albert Einstein

ঃ০ Quantum Mechanics ৎঃ

I consider the methods of quantum mechanics fundamentally unsatisfactory. ~ Albert Einstein

Nobody understands quantum mechanics. ~ Richard Feynman

Quantum theory cannot be extrapolated to complex systems. ~ Swiss theoretical physicists Daniela Frauchiger & Renato Renner

There is no quantum world. There is only an abstract description. ~ Niels Bohr

The subatomic particle Matryoshka became a babushka when French physicist Louis de Broglie speculated in 1924 that all particles in motion might exhibit wave-like behavior. A fascinated Austrian physicist Erwin Schrödinger took the idea and ran, unknowingly injecting uncertainty into quantum mechanics with his 1926 publication that described an electron as a wave function rather than a particle at a particular point in time. This became known as *Schrödinger's equation*.

Schrödinger formed his fundamental equation during an erotic tryst with a lover, on holiday in Arosa Switzerland. In an equal and opposite reaction, Schrödinger's wife proved that what goes around comes around, by having an amorous relationship with German mathematician and theoretical physicist Hermann Weyl. Or, as Weyl might have said, "it all adds up."

Schrödinger's equation had mathematical elegance and accounted for many spectral phenomena that Bohr's particulate atomic model failed to explain. But Schrödinger's wave function was difficult to visualize and so faced opposition.

The wave did not waver. Instead, the idea of wavy particles matured into *quantum field theory*, also known as *quantum theory* and *quantum mechanics*.

> I do not like it, and I am sorry I ever had anything to do with it. ~ Erwin Schrödinger

In 1925, German physicist and mathematician Max Born formulated a matrix representation of quantum mechanics, based upon interpreting Schrödinger's equation as a probability function for an electron's location. Born's theory formally introduced *wave/particle duality*: an electron had properties of both a particle and a wave, thus reconciling opposite views.

Einstein had essentially come to the same conclusion in 1905, when he argued that radiant energy consisted of quanta. But Einstein did not appreciate the implications of his discovery. It would not be the last time that Einstein failed to fathom the import his own conclusions.

At the heart of quantum field theory (QFT) is wave/particle duality. QFT stuffs the basic bits of Nature into quanta while acknowledging their wavy properties as paramount.

QFT proposes an umbrella for understanding the fundamental nature of existence with a quantum that defies precise characterization as to position, path, and speed. That makes a quantum amenable to being mathematically sketched simultaneously as both a particle *and* a wave function.

> Subatomic particles are far from solid. They are nothing like matter as we know it. Much of the time they seem more like waves than particles. Whatever matter is, it has little, if any substance. ~ English physicist Peter Russell

A *quantum* is not a particle, like some itty-bitty billiard ball, but instead a little localized chunk of ripple in a field that deceptively looks like a particle. Understanding a particle is more about wave interactions than about particulate properties. It is a story of field behavior, not characterizing fantastically fleeting fragments.

> Particles are epiphenomena arising from fields. Unbounded fields, not bounded particles, are fundamental. ~ American theoretical physicist Art Hobson

A *photon*, or particle of light, is more essentially a vibratory wave matching the intensity of the fields that surround an electrically-charged object: *electromagnetic radiation*.

This radiation comprises electric and magnetic field components oscillating in phase perpendicular to each other and perpendicular to the direction of energy propagation.

The oscillation is a wave which yields a quantum. Photons, and everything else, are interconnected energetic vibrations that appear particulate.

Quantum mechanics at its heart is a statistical theory. It predicts probabilities of outcomes. This probabilistic nature of quantum theory is at odds with the determinism inherent in Newtonian physics and relativity, where outcomes can be exactly predicted given sufficient knowledge of a system. Perhaps quantum systems are controlled by hidden variables that determine the outcomes of measurements. ~ American physicist Lynden Shalm *et al*

♂ Uncertainty ♀

The highway is for gamblers, better use your sense. Take what you have gathered from coincidence. ~ American musician Bob Dylan in the song "It's All Over Now, Baby Blue" (1965)

A most significant consequence of describing electrons as waveforms, as Schrödinger had done, was to make it mathematically impossible to state the position and momentum of an electron at any point in time. This observation was first made by German theoretical physicist Werner Heisenberg in 1926. It became known as the *uncertainty principle*: a measurement may be made to get a sense of either a quantum particle's position or momentum, but not both at the same time.

A quantum system – for instance, a photon – may behave either as a particle or a wave. However, the way in which it behaves depends on the kind of experimental apparatus with which it is measured. Hence, both aspects, particle and wave, which appear to be incompatible, are never observed simultaneously. ~ Italian physicist Alberto Peruzzo *et al*

Heisenberg's principle came from a thought experiment: using light to measure the position of an electron (though it applies to any subatomic particle). The necessarily short wavelength that might be used to measure an electron would necessarily give the particle a kick of energy in the process of measurement, thus creating an error by the disturbance.

Trying to get an accurate account is not the core issue, nor even unique to quantum physics. Such measurement disturbances also occur in classical physics.

Quantum particles cannot be described as a point-like object with a well-defined velocity because they inherently behave as a wave. For a wave, momentum and position cannot both be defined accurately at any instant.

Mathematically, uncertainty between position and momentum arises because the expressions of the wave function for these supposedly independent variables are actually Fourier transforms of one another. Position and momentum are conjugate variables, which means they are *symplectic*: interdependent, not independent.

Neil Bohr interpreted the uncertainty principle holistically: the universe is basically an unanalyzable whole, in which the notion of separation of particle and environment is a vacuous abstraction except as an approximation.

To extract information about a certain wave, *interferometry* superimposes one wave upon another. Interferometry is also used to measure subatomic particles.

During interferometry, the measurement wave and subatomic target wave become coherently entangled. The measurement itself decoheres the particle to a definite observable outcome which is the *local state* result.

Meanwhile, entanglement continues. The global *measurement state* (MS) is the context surrounding the local state. Observation of the measurement state creates uncertainty.

> As a consequence of nonlocality, the states we actually observe are the local states. These actually observed local states collapse, whereas the global MS, which can be "observed" only after the fact by collecting coincidence data from both subsystems, continues its unitary evolution. This conclusion implies a refined understanding of the eigenstate principle: following a measurement, the actually observed local state instantly jumps into the observed eigenstate. ~ Art Hobson

An *eigenstate* is a measured state of an object with quantifiable characteristics, such as position and momentum.

Even if a measurement is attempted, but no result read, entanglement ensues in such a way that the target quantum system is disturbed, and information of the event – as a lingering resonance in the affected quantum field – is retained.

Measurement is immaterial to uncertainty. The quantum level exhibits uncertainty regardless of observation.

Uncertainty is a form of potentiality, which is inherent in the nature of the particle field itself owing to unobservable entanglements. Uncertainty is phenomenal, not merely mathematical. Quite simply, as existence is emergent and always in flux, quantum uncertainty is a property of Nature. (The common assumption that Nature is ultimately observer-independent (objective) is a false axiom. See *Spokes 8: The Web of Being*.)

The uncertainty principle unsettled Bohr's neat model of clearly defined circular tracks for electron orbits in atoms. Electrons became clouds without definite path or pinpoint.

Existence at the subatomic level became a juggle of Matryoshka dolls, a dice throw that appears as chance in when and where; but by no means random. Uncertainty does not mean disorder; quite the contrary. The greatest beauty of Nature is its spontaneity in what is obviously a deeply nested order.

It's déjà vu all over again. ~ American baseball player Yogi Berra

¤ ✧ ¤

The math leading to uncertainty was elegant. Einstein admired the formula, but hated the idea, revolted by what empirical uncertainty implied. In a 1926 letter to fellow physicist Max Born, Einstein wrote:

Quantum mechanics is very impressive. But an inner voice tells me that it is not yet the real thing. The theory produces a good deal but hardly brings us closer to the secret of the Old One. I am at all events convinced that He does not play dice.

To this sentiment Niels Bohr retorted:

Do not presume to tell God what to do.

Einstein never did reconcile himself with uncertainty.

I still believe in the possibility of a model of reality, that is to say a theory, which shall represent the events in themselves and not merely the probability of their occurrence.

Einstein was ever uncomfortable with a universe that wasn't predictable or stable, and remained adamant in denying what he had proven: that the cosmos had more dimensions than could be observed.

For all that his physics intimated, Einstein believed in an ethereal presence as a cosmic mystical force. Einstein never reconciled his mysticism with his felt need for determinism and certainty.

¤ ✧ ¤

Interpreting wave/particle duality is particularly vexing for a discipline determined to pin things down. Inherent uncertainty is anathema.

Schrödinger first conceived wave/particle duality as a reality. That was cast aside as fantastic in what emerged at the end of the 1920s as the *Copenhagen interpretation* – a term Heisenberg applied in a 1955 series of lectures. The Copenhagen interpretation considered the wave function a computational tool, giving good results, but not to be taken literally. The Copenhagen interpretation later lost its popularity owing to its intrinsic inconsistency.

> The Copenhagen Interpretation is hopelessly incomplete because of its *a priori* reliance on classical physics as well as a philosophic monstrosity with a "reality" concept for the macroscopic world and denial of the same for the microcosm.
> ~ American physicist Hugh Everett III in 1957

But then the idea that the wave function reflects what we can know about the world, rather than actuality, came back into vogue with the rise of *quantum information theory*, which generalizes *classical information theory* to the quantum level. This arbitrary mix-and-match by physicists of what part of a model is to be considered real and what is not presents a serious problem.

> Our present quantum mechanics formalism is not purely epistemological; it is a peculiar mixture describing in part realities of Nature, in part incomplete human information about Nature – all scrambled up by Heisenberg and Bohr into an omelet that nobody has seen how to unscramble. Yet we think that the unscrambling is a prerequisite for any further advance in basic physical theory. For, if we cannot separate the subjective and

objective aspects of the formalism, we cannot know what we are talking about; it is just that simple. ~ American physicist E.T. Jaynes

Schrödinger's first impression was supported by English particle physicists Matthew Pusey, Jonathan Barrett, and Terry Rudolph (PBR) in 2011.

> A pure quantum state corresponds directly to reality. Any model in which a quantum state represents mere information about an underlying physical state of the system, and in which systems that are prepared independently have independent physical states, must make predictions that contradict those of quantum theory. ~ PBR

As PBR put it, "the statistical view is not compatible with the predictions of quantum theory." PRB's theory derives from a negative proof: if a quantum wave function were merely a computational tool, then even quantum states unconnected across space and time would be able to communicate with each other.

If that is unrealistic, which necessarily is an epistemological assumption, then the wave function must be physically real. Otherwise, as E.T. Jaynes stated, physicists do not know what they are talking about.

Quantum theory is founded upon the premise that so-called particles are actually chunks in fields. Fields are by definition represented as waves. Denying the reality of the wave function, and its inherent uncertainty, eviscerates quantum mechanics by denying the existence of the foundation upon which the theory is built.

> The linearity of quantum mechanics is intimately connected to the strong coupling between the amplitude and phase of a quantum wave. ~ German theoretical physicist Wolfgang Schleich

Uncertainty at the originating level of existence suggests a deeper reality.

> The particles and fields are very, very crude statistical descriptions. Those particles and those fields are not true representatives of what's really going on. ~ Dutch theoretical physicist Gerard 't Hooft

⌖ Pilot Wave Theory ⌖

All particles must be transported by a wave into which it is incorporated. ~ Louis de Broglie

The central problem with the conventional interpretation of the uncertainty principle is that it merely provides a statistical convenience, rather than a representation of Nature. As a tool rather than a characterization, the uncertainty principle explains nothing, while leaving the universe inherently nondeterministic.

This is a case of consensus writing history, and in doing so wiping away good sense, including Schrödinger's first impression of what the uncertainty principle meant.

For Louis de Broglie, wave/particle duality was no abstraction. He assumed a real wave existed that satisfied Schrödinger's equation, with an attendant particle following a definite trajectory.

de Broglie theorized that each particle is guided by a background wave, which he later called a *pilot wave*. Consistent with thermodynamics, the particle is in a thermal bath provided by a background of vacuum fluctuations.

Phase harmony between a wave and its particle, as well as synchrony between particles, is provided by a periodic process, equivalent to a clock. A pilot wave steers its particle by this nonlinear interaction.

The *pilot wave theory* provides a deterministic system that characterizes existence with a cynosure and casts off uncertainty. In its developed form, the theory is also consistent with classical physics, quantum mechanics, and relativity.

Despite its ostensible appeal, the pilot wave theory was repudiated by physicists at the 1927 Solvay Conference. Einstein's failure to speak up for the theory led to its rejection.

Einstein liked the theory's determinism. His objection was the implication that the entire universe was entangled, affording nonlocal interactions between particles.

The pilot wave theory requires the potential for interaction between any and all particles in a system. Distance does not drive interactivity to zero. The instantaneous state of a particle depends upon its overall environment.

In the long run, only the entire universe can be regarded as self-determinate, while any part may be independent in general only for some limited period of time. The very mode of interaction between constituent parts depends on the whole, in a way that cannot be specified without first specifying the state of the whole. ~ American theoretical physicist David Bohm & British quantum physicist Basil Hiley

The next doll down emanated from particle accelerators and their detectors which were first built in the 1950s. This led to further splitting the protons and neutrons in atomic nuclei into smaller, more "elemental" particles: *hadrons.*

Particle accelerators proliferated hadrons into such a prodigious variety that it prompted Wolfgang Pauli to remark:

Had I foreseen this I would have gone into botany.

By the late 1960s, a "depressingly large number" of hadrons had been found. Hadrons were but a nested doll, not the smallest doll in Matryoshka reality. Hadrons are comprised of quarks.

For decades, hadrons were known to be of only 2 families: baryons, composed of 3 quarks, and mesons, comprising 1 quark and 1 antiquark. In 2014, a new meson family was discovered: a tetraquark, with 2 quarks and 2 antiquarks.

Like hadrons, quarks have their own varieties, called flavors, which are determined by their spin and symmetry.

¤ ✧ ¤

Going bottoms-up: quarks combine to form families of hadrons, which join in threesomes to form protons and neutrons (2 types of baryons), which are enslaved by the strong nuclear force to create atomic nuclei, which combine with electrons, bound together by the electromagnetic force, to create atoms. By ionic attraction, atoms congregate into molecules, which make up everyday matter. Bear in mind that all these particles are nothing more than intense interactions of localized coherent energy fields, posing as something solid.

❧ The Standard Model ☙

The Standard Model leaves gaping holes we don't know how to fill. ~ American astrophysicist Tyce DeYoung

Physicists understand matter and energy in terms of kinematics and the interactions of elementary particles. *Kinematics* is a classical mechanics construct which characterizes the motions of bodies without consideration of the forces that cause movement. This conceptual bifurcation – between matter and the forces that move matter – would live on in the standard model of quantum physics.

The *Standard Model* (SM) is a quantum field theory offshoot which addresses the presumed basic building blocks of matter and their subatomic interactions. SM proposes a set of elementary particles which supposedly fabricate Nature at the quantum level.

Embarrassingly, the Standard Model suggests that nothing should exist. ~ German physicist Werner Rodejohann

The Standard Model was formulated in the 1970s. From the early 1980s, experiments verified various facets of SM. But, as numerous observations have often differed from theory, the Standard Model has undergone repeated patchwork: so much so that SM is a set of theorems cobbled together to render an approximate fit to what has been observed.

Whether you can observe a thing or not depends on the theory which you use. It is the theory which decides what can be observed. ~ Pakistani theoretical physicist Abdus Salam

There are many Standard Model deficiencies. One is a failure to explain how the heaviest chemical elements function. Instead, relativity theory accounts for the behavior of the last 21 elements of the periodic table.

The electrons whirling about berkelium and heavier elements do not organize themselves the way they do with lighter elements. As the nuclei of these heavy atoms are highly charged, their electrons are moving at significant fractions of light speed. Under general relativity theory, the faster anything with mass moves, the heavier it gets. Hence, heavy-element electrons are heavier than ordinary electrons.

The SM rules that typically apply to electron behavior break down, replaced by relativistic rules.

¤ ✧ ¤

An *elementary particle* is a particle that supposedly has no constituents; in other words, is not known to be comprised of smaller particles. These fundamental particles are the bottom-up building blocks of existence.

The qualifier *elementary* is a euphemism. Every SM particle is acknowledged as a conglomeration comprising higher-dimensional (HD) constituents: *virtual particles*. An electron, for example, is considered fundamental, but is encased in an entourage of virtual particles: HD influences which practically define an electron's mass and charge.

> In quantum field theory, the electron is surrounded by a thin soup of evanescent particles which wink in and out of existence in fractions of a second. ~ English physicist Francis Farley

Slam electrons into positrons (anti-electrons), both supposedly elementary particles, and the resulting annihilation produces streams of quarks, gluons, muons, tau leptons, photons, and neutrinos. Knowing that rather eviscerates the definition of *elementary* (or *fundamental*) particle as having integral integrity. Energy decomposes what it has composed by energetic interaction. Which is to say that elementary particles are just a snapshot view of a portion of an energy field that only appears to be a chunky particle.

Nature comprises coherent waves of energy which *appear* to take particulate form. Hence, painting elementary particles as points in 4D fields results in an unfinished canvas, as every particle is a composite of itself and virtual particles; and even a fundamental particle may be broken up (or down, depending upon one's perspective).

Virtual particles are acknowledged in SM, but otherwise ignored as ancillary. Therefore, under the Standard Model, an elementary particle is a perspective limited to 4D. That alone renders the SM an incomplete explanation.

The virtuality of existence is largely extra-dimensional (ED). Only a tiny fraction of baryon mass comes from the quarks within. Over 99% comes from being bound into a hadron. The glue that holds hadrons together are a swarming stew of virtual particles: subatomic dark matter. Further, their energetic interactions create the mass that makes up matter.

SM partly explains interactions in the electromagnetic, weak, and strong nuclear forces. These forces mediate the 4D dynamics of known subatomic particles.

SM is but a start; admitted by physicists as incomplete, as it lacks an accounting for gravity and matter's hegemony over antimatter, among other deficiencies.

℘ Fermions & Bosons ℘

Under the Standard Model there are 2 elementary particle types: fermions and bosons. Whereas *fermions* are the particles that comprise matter, *bosons* are force carriers.

Bosons are termed after Indian physicist Satyendra Nath Bose, best known for his work with Einstein in theorizing Bose-Einstein condensate, which is cold, oddly coherent bosonic gas.

Fermions were named after Italian-born physicist Enrico Fermi, who created the first nuclear reactor in 1942.

The SM particle zoo has 17 main characters: 12 fermions and 5 bosons, complemented by an equivalent set of antiparticles, and at least one hanger-on: the *graviton*, which is the elusive (nonexistent) particle representing gravity.

There is also an ancillary mob to cover various observed oddities. To conclude the parade are virtual particles, which are assumed to flit in and out of existence to lend their support to the proceedings.

Photons – the quanta of light – are the best-known boson. Photons are never alone; they sense their neighbors and coordinate their travels.

> Photons are not just light particles, they are also waves, and waves interact with each other. This creates a link between photons, and photons' path through material is not independent from the other photons. ~ Spanish physicist Pedro David García

Bosons are more slippery than fermions in their spacetime characteristics. Whereas 2 fermions cannot occupy the same space at the same time, bosons with the same energy can – it's all in the spin (which is explained shortly).

Quarks, leptons, and their antimatter equivalents are fundamental fermions. As bosons are immaterial, they have no anti-equivalents.

There are 6 flavors (types) of *quark*: up, down, strange, charm, bottom, and top. Quarks do not naturally exist in solitude. They are always found within baryons and mesons, both composite particles (*hadrons*) which are bound together

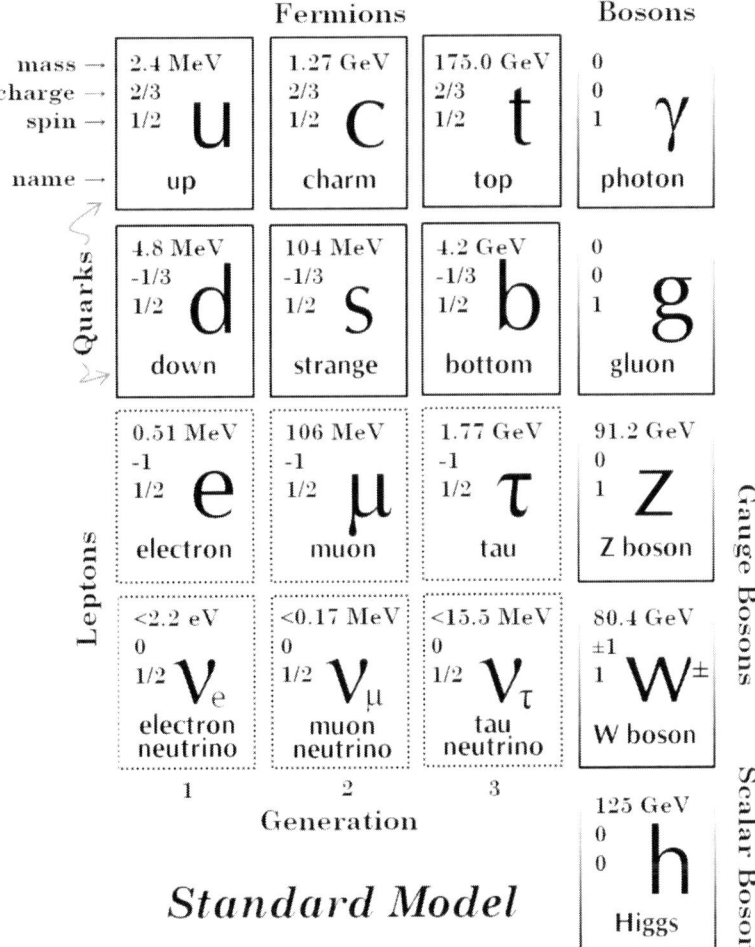

by the nuclear strong force (the physics interaction which binds particles together).

A *baryon* is a hadron with 3 quarks. The most stable hadrons are the baryons that comprise the nuclei of atoms.

Up and down quarks interact via the strong force to form protons and neutrons, as well as other baryons. Nominally, a *proton* has 2 up quarks and 1 down quark, while a *neutron* contains 1 up and 2 down quarks.

> The quark model that developed provided the simple picture that atomic nuclei comprises only up and down quarks. This simple picture has been superseded by one consisting of a range of quarks, antiquarks (the antiparticles of quarks) and gluons (the particles that bind quarks together). In principle, all 6 flavours of quark can be present. ~ Australian physicist Ross Young

Baryons are complex dynamic systems. For example, the size of a proton varies upon its energy state, which depends upon environmental circumstances.

A *meson* is made of 1 quark and 1 antiquark. Mesons appear in nature only as short-lived products of high-energy interactions between quark-based particles.

> A family of 4-quark objects has begun to appear. While the theoretical picture remains to be finalized, more and more clues are suggesting that we are witnessing new forms of matter. ~ American physicist Frederick Harris

Like atoms, quarks may link up to form subatomic molecules. One has been discovered that is a meson-baryon combination. These quark-composite oddities lie outside the Standard Model.

Leptons – electrons and neutrinos – are more ethereal than quarks. The defining quality of leptons is that they are not subject to the strong force, and so never tempted into atomic nuclei.

Electrons are affected by gravity, electromagnetic fields, and the weak force. Neutrinos, lacking charge and largely bereft of heft, are affected only by the weak force, and so subject to decay.

Leptons may be residual particles during radioactive decay. A neutron breaks down into a proton via flavor changing (flipping quark type): a process which emits a virtual W boson, which then converts into an electron and an electron antineutrino. Conversions of fermions into bosons and vice versa lies outside the Standard Model.

Protons are the lightest, and thereby least energetic, baryon. While proton decay is hypothetically possible, there is no experimental evidence that it happens.

The fundamental fermions are grouped into 3 *generations*, each with 2 leptons and 2 quarks. Only 1st-generation fermions exist at everyday energy levels. Hence, only a few elementary particles are 4D stable: the electron, the proton, and the neutrino.

♎ Particle Demolition Derby ♎

Subatomic particles are detected using colliders. A focused beam of specific particles, such as electrons or protons, is accelerated via electromagnetic jostle, then aimed to collide into an opposing beam. The heavier the particle, the more energy it takes to accelerate a beam. Then there is the need for speed. The faster the accelerator, the more spectacular the results in generating bits for boffins.

The latest generation of collider is The Large Hadron Collider (LHC), located near Geneva Switzerland. The LHC was built to hunt the Higgs boson by smacking protons, at a cost of €7.5 billion (euros) ($9 billion US). Like all human endeavors of significant scale, getting the Collider up and running collided with serious glitches and cost overruns.

Particle collisions create detectable tracks of decay. A hammered hadron decays into its constituent quarks and other lower-energy elementals.

Higher-generation particles, created by high-energy collisions, have greater mass and less stability. They decay into lower-generation particles by means of weak interactions.

Fermion decay can be rather directly observed, as these are chunks of matter. Bashed bosons are more mysterious. Their presence must be inferred from what they decay into.

The W boson may decay in many ways, which it does in less than 1 billion-trillionth of a second. If it dissolves into a puddle of neutrinos, it cannot be detected. If instead a W collapses to an electron or muon, it may be measured.

Computer simulations are run based on the equations that characterize a model; typically, the Standard Model. Transformations and decays at different energy levels are predicted.

Repeatedly matching telltale tracks from collisions with predictions builds confidence that a model is on track. Uncertainty propagates in each analysis. Many repetitive runs with similar results are necessary for validation.

♎ Neutrinos ♎

With no electric charge and interacting primarily by the feeble "weak" force, neutrinos are will-o'-the-wisps that can pass through Earth as easily as a bullet through a bank of fog. ~ Frank Close

A *neutrino* is electrically neutral, with scant mass. The Standard Model predicted that neutrinos had no mass. Having been proven wrong, the Model was revised.

Neutrinos interacts gravitationally with other particles and can travel close to the speed of light through ordinary matter with *almost* no effect. While neutrinos are nearly massless and faster than lightning, they pack tremendous mystery.

There is a surfeit of neutrinos. On Earth every second, through every square centimeter, there are 65 billion solar neutrinos flashing by. Most of the neutrinos passing through the Earth come from the Sun.

Wolfgang Pauli theorized neutrinos in 1930 to explain an observed gap in radioactive decay, specifically *beta decay*, which is mediated by weak interactions. To hold to the law of energy conservation, the matter-energy equation must balance between what existed before the decay and after.

Pauli proposed the missing bit being carried off by a hypothetical neutrino. The neutrino was so minuscule and fleeting that Pauli wagered a case of champagne that the little bugger would never be spotted. He lost the bet in 1956.

According to the Standard Model, as neutrinos have mass, there must be 2 varieties, defined by spin. Only the left-handed variety experience beta decay. The right-handed ones may be the Majorana fermion.

Neutrinos come in at least 3 flavors: electron, muon, and tau; all of which can oscillate between flavors spontaneously. That neutrinos can morph into different flavors indicates that they experience change, and thus are subject to time.

Neutrino oscillations in a vacuum are different from those that interact with matter. This *matter effect* occurs because electron neutrinos interact with electrons, which changes the effective mass of the neutrinos.

More than electrons are affected. Neutrinos colliding into an atomic nucleus ricochet away, leaving the nucleus recoiling in response.

Neutrino interactions are born in the shade of high-energy nuclear decay, such as reactions that occur in stars, in nuclear reactors, or when cosmic rays slam into atoms. As they are born of decay, neutrinos are mediated (affected) by the weak force.

According to the Standard Model, all fermions have an antithetical twin. By this reckoning, there are equivalent anti-neutrinos. None have been spotted.

Data suggests the existence of additional flavors of neutrinos. Neutrino flavors were determined from measurements of the width of the Z^0 boson, which mediates the weak force. Z^0 is filled out, so any additional neutrinos that exist would have to be *sterile neutrinos*: not interacting with the weak force, thereby not affecting the width of Z^0.

A non-participant in any 4D particle interactions, sterile neutrinos arise only from non-sterile flavors oscillating into sterile form. 2 sterile neutrino flavors are predicted. There is skepticism about the existence of sterile neutrinos.

> You're trying to prove the existence of something with no interactions. It's like trying to prove the existence of God.
> ~ American particle physicist Patrick Huber

Indirect evidence of sterile neutrinos has been found by observing supernovae, and from a nearby dwarf galaxy, as well as in lab experiments.

> The sterile neutrino is not something bizarre or exotic.
> ~ American particle physicist Paul Langacker

Theoretically, sterile neutrino flavors would help explain why the cosmos is dominated by matter rather than a balance of matter and antimatter. Such insignificant particles, hardly interacting with any matter whatsoever, could go a long way in explaining the most fundamental nature of the universe, at least mathematically from a quantum view.

♎ Higgs Boson ♎

The Higgs gives everything in the universe its mass. ~ physicist David Francis

The *Higgs boson* is a grainy chunk of the *Higgs field*, which permeates all space. Via the *Higgs mechanism* – bathing in the ubiquitous Higgs field – gauge bosons (W & Z) nab mass by absorption of Nambu-Goldstone bosons, which arise from spontaneous symmetry breaking (a mathematical artifice).

The Higgs was hard to find. Along the way, the hunt for it created an extra-dimensional mystery. Some of the particles created by proton collisions synchronize their flight paths, "like flocks of birds," indicating interconnections that can only be explained by HD coherence.

Finding the Higgs filled in the Standard Model, which remains an incomplete explanation of known quantum particles. The discovery of the Higgs also lent support to *supersymmetry*, a rival theory to SM.

According to the Standard Model, all fermions get their mass from interacting with the Higgs field by way of interactions with gauge bosons, and by traveling with a cloud of virtual particles. There is no evidence of the Higgs mechanism conveying mass.

99% of the mass of the visible universe isn't explained by the Higgs. ~ American physicist Peter Steinberg

The mathematical symmetries inherent in SM largely cancel out the influence that virtual particles have on fermion mass, leaving a modest contribution. That same symmetry predicts that the mass of the Higgs itself is infinite. This points out that SM is nothing more than a convenient fiction.

Supersymmetry predicted the Higgs at ~125 GeV, where it duly appeared. That puts the Higgs at 133 times the heft of a proton.

¤ ◇ ¤

The Higgs holds the key to explaining why fermions have mass but bosons don't. The photon and gluon, both bosons, are presumed to be nothing but massless energy.

The *photon*, lover of light, carries electromagnetism. Yet photons do not interact with one another; at least, not unless terribly excited. Light hitting atoms so potently that they ionize causes participating photons to couple into polarization-entangled pairs.

The *gluon* puts the strong force to quarks, hardening them into hadrons. Besides their quark interactions, gluons do interact among themselves.

Gluons are mysterious in many ways. How their supposed interactions create the properties of quarks is not understood.

The way the strong force works implies that gluons are massive, and quarks supposedly get most of their mass from interacting with gluons. But gluons are massless; so, how gluons glue remains inscrutable.

The *W & Z bosons*, which mediate weak interactions, are theoretically massless too, but are actually massive. That discrepancy between theory and observation is explained away in the Standard Model by the mathematical trickery of *spontaneous symmetry breaking*.

The decay of a Higgs boson births fermions, particularly bottom quarks and tau leptons. This illustrates the intertwining of quantum particles, regardless of pedigree, and that matter is merely energy transposed.

♎ Graviton ♎

SM's accounting for gravity has been a tentative positing of a gravity particle: the *graviton*. The graviton is a boson; massless, to give it unlimited range, with a spin of 2: the highest stress-energy tensor of the particle set, twice as fast as a photon (which travels at the speed of light).

The graviton is generally disregarded as existing, as quantum field theories fall apart at high energies when gravitons are factored into the equations. Gravitons result in ineradicable infinities due to quantum effects. In other words, gravitons don't mathematically play well with other particles in the Standard Model, and so the graviton is typically ignored.

> Since the graviton theory made no testable new predictions, no practical reason could be given to prefer the graviton theory over general relativity. ~ Victor Stenger

℘ Particle Properties ჸ

3 properties are typically used to characterize elementary particles: mass, charge, and spin. These properties are not what someone with a knowledge of classical physics would intuitively expect; hence, all 3 terms are homonyms to their classical usage.*

♎ Mass ♎

In the everyday world, *mass* is considered a measure of an object at rest. Special relativity shows that rest mass and rest energy are essentially equivalent ($E = mc^2$). But *rest mass* (invariant mass) does not apply to subatomic particles.

For a subatomic particle, which is really an infinitesimal localized field, *mass* is something of a euphemism. Subatomic particle mass is formally a mathematical outcome, not an actual measurement.

Particle mass is a representation of the isotropy in the Poincaré group that a particle transforms; a nutshell reminder that the constructs of quantum mechanics are entirely mathematical models, and quite complex ones at that.

A practical conception of *subatomic particle mass* is the threshold energy at which a certain particle may appear; put another way, the energy required for a certain quantum to make an appearance.

---

* Beyond the jiggy math, comprehending the concepts of quantum mechanics is hindered by confusing terminology.

Negative mass is possible, in the same sense that an electric charge can be negative or positive. Push something with negative mass and it doesn't accelerate in the direction it was pushed. Instead, a negative-mass object accelerates backward.

A negative effective mass can be realized in quantum systems by engineering the dispersion relation. ~ American physicist Peter Engels *et al*

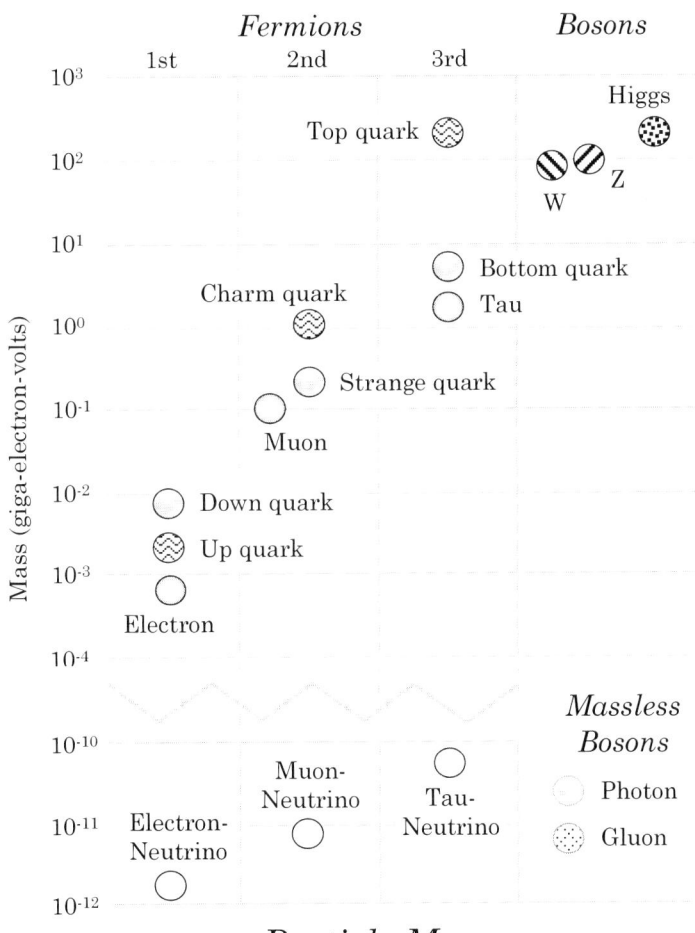

Particle Mass

⳽ The Hierarchy Problem ⳽

There have been numerous approaches to calculating the observed spectrum of particle masses from theory, but they have
not been successful. ~ English theoretical physicist Paul Wesson

The Standard Model having nothing to say about what
the mass of the Higgs bosom should be is just one facet of
what is called the *hierarchy problem*. Higgs' mass is not the
only problem. The masses of all fundamental particles are
100 quadrillion times less than they should be.

A hierarchy problem arises when the fundamental value
of a physical parameter, such as mass or a coupling constant
(such as the cosmological constant) is vastly different from its
effective (measured) value. When it has arisen, this problem
has repeatedly been whitewashed using a mathematical ad-
justment technique called *renormalization*.

All hierarchy problems grapple with relations to matter
(such as mass), which is understandable when you consider
that all of physics is concerned with explaining how matter
behaves. Since matter is all that is observable, it remains the
starting and end point of all physics models. All the transformations in between involve energy.

The most poignant hierarchy problem in theoretical physics is the enormous discrepancy between the weak force and
gravity. There is no consensus as to why the atomic weak
force is 10^{24} times stronger than gravity.

♎ Color Charge ♎

The *color charge* of a particle is an abstracted indication
of a particle's strong interaction according to quantum chromodynamics theory. Color charge is a property of a subatomic
particle's field interaction with the strong nuclear force.

Though distinct, color charge is analogous to electrical
charge. Color charge is not a simple strength measure (such
as with electric charge); far from it.

The term *color* in this context is itself a metaphorical abstraction, related to the nature of additive color, but color as
normally thought of (i.e., red, green, blue) has nothing to do

with color charge, which is a statement of the state of strong interaction at the subatomic particle scale.

A quark can have 1 of 3 charges (colors): red, green, or blue. An antiquark may be either anti-red (cyan), anti-green (magenta) or anti-blue (yellow).

A gluon has a mixed color charge, such as red and anti-green.

♎ Spin ♎

The quantum *spin-statistics theorem* categorizes particle types by their *spin*: the direction of internal angular momentum relative to the direction of linear momentum. The theorem characterizes the wave function of particles according to their symmetry.

The 12 flavors of fermions have a half-integer spin, whereas bosons twirl with an integer spin. Spin is the property that distinguishes fermions from bosons.

The boson integer spin versus fermion half-integer spin means that exchanging the positions of 2 bosons does not change their wave functions: they are symmetrical.

Contrastingly, fermions have asymmetric wave functions. Swapping fermions causes a reversal in their spin (wave function) sign, flipping between positive and negative.

The implication is that the amplitude of 2 identical fermions occupying the same space must be zero. Because of this, 2 fermions cannot simultaneously occupy the same quantum state, whereas symmetrical bosons can.

This fermion phenomenon is termed the *Pauli exclusion principle*, formulated in 1925 by Wolfgang Pauli. Pauli was trying to envision all the possible properties that electrons might have. He realized that the data all pointed to each electron occupying only 1 of a fixed number of energy states; what is now called *spin*. Pauli's exclusion principle started with electrons and was then applied across the board to all fermions. As all matter is made of fermions, the Pauli exclusion principle requires that atoms take up space; whence existence as a fulsome experience.

Electrons cannot congregate in a cloud at the lowest energy state. Instead, they must space out, into orbital shells, with higher-energy electrons at a distance from a shell of lower energy electrons. The Pauli exclusion principle underpins the fundamental properties of chemistry, including atomic stability and the segregation of atoms according to the *periodic table of the elements.*

Quantum spin is conceptually different than classical physics' spin, yet the term stuck.

In classical physics, the spin of a charged particle is associated with a *magnetic dipole moment:* the potential exertion force of magnetism upon the particle. Because of this, when the property was discovered classically, particles were thought to literally rotate to create the magnetic moment; an unproven assumption. Whether subatomic particles actually spin is unknown.

To be clear, quantum spin is not spin as in a spinning ball. If it were, the surface of electrons would spin at several times the speed of light. That supposedly is not so.

The direction of particle spin can change, but an elementary particle supposedly spins at a speed which cannot be changed, either faster or slower.

The nuclei of atoms also exhibit spin. Nuclear spin affects the strength of atomic interactions. If 2 atoms have identical nuclear spin, they interact weakly. In contrast, atoms with different nuclear spin states interact much more strongly.

A theoretically symmetrical system is one where outcomes are equally likely. The Standard Model presumes a symmetrical system, but the hard facts of existence shatter that pristine symmetry. This incongruity between physical model and actuality is patched with math; a clear indicator that the map is decidedly not the territory.

☡ Spontaneous Symmetry Breaking ☡

Spontaneous symmetry breaking (SSB) is ubiquitous in Nature. The examples include magnets, superfluids, phonons, Bose-Einstein condensates, and neutron stars. ~ Japanese physicists Haruki Watanabe & Hitoshi Murayama

Spontaneous symmetry breaking (SSB) is a simple concept: nothing more than stating that actualization breaks an idealized (mathematical) symmetry. SSB is a way of explaining how a perfectly symmetrical physical model can appear broken in view of physical manifestation, yet, paradoxically, the model still be presumed valid.

SSB smashes mathematical symmetry on the stones of sampling. If a symmetrical system is acted upon, a specific outcome arises out of the wave of possibilities. The symmetry breaks. That does not necessarily discredit the underlying symmetry, which by manifestation appears broken, but is simply a hidden symmetry.

A ball sitting on top of a conic hill is in a symmetric state: it could roll down any which way. When the ball actually moves by some force, the symmetry is broken – SSB in action.

A mathematical ideal is balanced until it actuates. By manifesting, perfection becomes imperfect. Phenomena arise from defects.

Particle physics pilfered the concept of SSB from solid-state physics: a discipline that particle physicist Murray Gell-Mann called "squalid-state physics" (an ironic deprecation: for a Standard-Model man to deride the crutch upon which SM depends). Solid-state physics is the study of the intense atomic interactions in solids.

Modeling solids evinced equations that characterized their lowest energy state. The model results were rotationally symmetric, but the solids were not; hence spontaneous symmetry breaking.

SSB is emblematic of the handedness that occurs throughout Nature. Chirality is essential in the basic molecular interactions of life. That asymmetry is also fundamental to physics is unsurprising.

The strong nuclear force, electromagnetism, and gravity all respect symmetry. The weak force, responsible for nuclear decays and neutrino interactions, does not.

The charged $W^{\pm}$ boson, which mediates weak interactions, is responsible for the broken parity symmetry. By their cross-influences, the troika of $W^{\pm}$, Z^0, and Higgs^0 bosons provide the theoretical patchwork by which the Standard Model is reputedly redeemed despite SSB.

Via the *Higgs mechanism* hypothesis, W & Z bosons acquire non-vanishing mass through SSB. SSB is invoked to explain the massive mass discrepancy between theory and observed actuality of these bosons.

Supposedly weightless until caught in the act, W & Z manifest with an immense presence. SSB is also the basis upon which the Higgs particle is predicated in the Standard Model.

A sidekick in the Standard Model – Nambu-Goldstone bosons (NGBs) – facilitate coherent collective behavior in a material. NGBs mystically appear whenever symmetry is spontaneously broken.

The nominal case in SSB is that the number of Nambu-Goldstone bosons equals the number of broken symmetries. But in exotic materials, such as neutron stars, Bose-Einstein condensates, and superfluids, the number of NGBs is less than the number of broken symmetries. A deficit of Nambu-Goldstone bosons makes matter go crazy, as it does in these outlandish coherent constructs.

SSB is *not* ubiquitous in Nature but is instead commonplace in *symbolic representations* of Nature. Spontaneous symmetry breaking is a necessary artifice for physical models that provide an inadequate approximation of intricacy in Nature that is beyond human mathematical skill to capture.

℘ Ghost Fields ଅ

In the Standard Model (SM), the masses of bosons are modified via interrelations with other bosons and fermions. This creates what are called *ghost fields*, which are necessary to maintain mathematical consistency in SM.

Boson-fermion interactions are called ghost fields because they are presumed to not exist. They are instead treated as a computational tool.

Then again, ghost fields are hypothesized to create the *virtual particles* that appear 4D out of the HD ground state that supposedly comprises only vacuum energy. Virtual particles are now taken for granted as existing.

There is a paradox in granting virtual particles existence but considering the originator of virtual particles – ghost fields – to be a fictional construct (i.e., purely mathematical mumbo-jumbo).

Ghost fields play a role in producing a loopy hierarchy of particles in SM, thus creating considerable complexity in the Standard Model construct. This *hierarchy problem* prompted theoretical physicists to derive a more elegant mathematical solution: supersymmetry.

℘ Supersymmetry ꙮ

The necessity for spontaneous breaking of supersymmetry is a disaster. The whole setup is highly baroque and not very plausible. ~ American mathematician Peter Woit

Supersymmetry (SUSY) posits a symmetry between particles by dint of their spin. SUSY brings together all quantum particles as components of a single master superfield.

Mathematically, SUSY is much easier to work with than the Standard Model. Several SM loose ends become exactly solvable in SUSY.

In SUSY, each fermion flavor has a boson shadow and vice versa. Thus, supersymmetry necessitates a *"shadow" partner* for every elementary particle. This shadow partner is also termed a *superpartner* or *sparticle.*

Sparticles are entirely hypothetical. There is no evidence to support their existence, and at least some sparticles should have been spotted by now.

Further, what is known about electrons, which are perfectly round, runs contrary to SUSY, which predicts that electrons have a slightly oval deformation, owing to their having an electric dipole moment which has yet to be found.

Like the Standard Model, unbroken SUSY requires the ruse of spontaneous symmetry breaking (SSB). That makes SUSY as unlikely as SM to decently represent Nature.

Supersymmetry also features in most versions of string theory.

That's not science. That's pathetic. ~ German astrophysicist Sabine Hossenfelder on the frequent tinkering of SUSY precepts

❧ Antimatter ☙

> The discovery of antimatter was perhaps the biggest jump of all the big jumps in physics in our century. ~ Werner Heisenberg

Matter theories have long proposed a negative twin. English physicist William Hicks, via the then-popular vortex theory of gravity, proposed negative gravity in the 1880s.

Along the same lines, in 1886 English mathematician Karl Pearson posited that the gyral flow of cosmic aether had sinks and sources ("squirts"). Squirts were normal matter, whereas sinks represented negative matter.

German-born British physicist Arthur Schuster whimsically proposed anti-atoms and antimatter solar systems. Schuster coined the term *antimatter* in 1898 and hypothesized matter and antimatter as mutual annihilators. Schuster conjectured that antimatter possessed negative gravity.

Paul Dirac formulated *quantum electrodynamics* (QED) in 1920. A relativistic quantum field theory of electrodynamics, QED modeled how radiation and matter interact. Essentially, QED theorizes perturbation of the *ground state*, which, instead of being nothing, as the term implies, formulates to have massive latent energy.

QED theory was the first to harmonize between the otherwise incongruous schools of relativity and quantum mechanics. This bridged a huge theoretical gap. But the infinity issue cropped up. Dirac's early equations led to predictions of infinity, which were considered unacceptable by other physicists. Dirac denied adjustments that washed infinity out of his equations.

> This is just not sensible mathematics. Sensible mathematics involves neglecting a quantity when it is small – not neglecting it just because it is infinitely great and you do not want it! ~ Paul Dirac

In 1928 Dirac produced a relativistic quantum mechanical wave equation, now termed *the Dirac equation*, which characterizes the spin of normal fermions (those with mass and charge). Dirac cast this equation to explain the behavior of a moving electron, thereby allowing an atom's quantum behaviors to be treated in a manner consistent with special relativity.

Yet the Dirac equation created conditions expanding the natures of both material existence and time. Dirac's formulations of quantum mechanics led to a perspective that allowed each subatomic particle its own proper time, escaping relativistic coordinate time.

At the time, this created an apparent dichotomy between quantum and relativity equations. But any theory of relativistic kinematics allows a particle to have an energy such that $E = -mc^2$ as a complement to $E = +mc^2$, Einstein's original equation. Dirac necessarily found that his model gave negative as well as positive energy solutions.

A *many-body* reinterpretation by Dirac of his basic 1928 equation founded quantum field theory (QFT). Many-body problems attempt to characterize a physical system comprising a stupendous number of interacting particles.

Dirac's many-body interpretation involved quadratic equations, which often have 2 solutions: 1 positive and 1 negative. These quadratics predicted *antimatter*: every particle of matter having a mirror antimatter particle.

There was no evidence for the existence of antimatter at the time Dirac constructed his QFT model, but the concept of antimatter was less objectionable to theoretical physicists than infinity.

American physicist Carl Anderson liked to play with cosmic rays. In doing so, in 1932, 4 years after Dirac equation formulation, Anderson discovered antimatter. His discovery was the *positron*, the antimatter equivalent of the electron. The term *positron* is a contraction of "positive electron" (electrons carry a negative electrical charge).

Interpretation of the solutions presented by the Dirac equation has always been controversial. Dirac's own idea was that all the negative energy levels are physically filled in the ground state, while the positive energy states are empty. This is the physics equivalent of double-entry accounting: matter only appears 4D if balanced by energetic opposites extra-dimensionally; existence as a conjuring trick.

This *Dirac sea* idea contradicts the common-sense view of vacuum as a state in which matter is absent. But the ground state has been found to be unimaginably energetic, with substantive material contributions.

Condensed matter physics sails on the Dirac sea as reality. And QFT – which has substantial evidentiary backing – is scuppered if the Dirac sea does not exist.

Another phenomenon depends on the Dirac sea: chirality, which has been observed experimentally in the decay of subatomic particles. *Chirality* is the handedness (left or right) of a particle's spin.

In 1949, American theoretical physicist Richard Feynman mused that positrons were not like holes in a negative energy sea of electrons, like the holes in semiconductors. Instead, Feynman proposed positrons as electrons moving backward in time. An electron moving backward in time would be indistinguishable from a positron puttering forward in time; a negative charge bouncing backward equivalent to a positive charge flying forward.

Though the Dirac equation, and indeed any relativistic theory, requires negative energies and time in reverse, the convention of time vectored in a forward direction was too formidable for Feynman's flaunting of time to be acceptable. Antimatter became treated as a raft of temporally progressive particles; mirrors of matter.

℘ Matter-Antimatter Asymmetry ℘

On the big Bang theory: For every 1 billion particles of antimatter there were 1 billion and 1 particles of matter. And when the mutual annihilation was complete, one billionth remained – and that's our present universe. ~ Albert Einstein

It was long held that matter and antimatter comprised mirror-perfect particles that acted as energetic opposites, in an antagonistic relationship of matter-antimatter mutual annihilation whenever opposing particles encountered one another. Einstein's view, that matter crowded out antimatter, was mainstream.

That presumption has been shown wrong. There is a belt of antiprotons around Earth.

Antineutrons are generated when energetic cosmic rays strike the upper atmosphere. The antineutrons escape the atmosphere, decaying into antiprotons at higher altitudes.

Antiprotons congregate from several hundred to 2,000 kilometers above Earth's surface. Ordinary matter is so scarce there that they seldom meet their particle counterparts – protons – and annihilate each other on contact.

The antimatter is trapped by Earth's magnetic field, forming a thin belt of particles gyrating around magnetic field lines: bouncing back and forth between the planet's north and south magnetic poles.

Lightning storms generate positrons. Solar flares sprout positrons.

Matter and antimatter coexist 4D because they are not mirror-perfect opposites. An asymmetry exists in their behaviors.

For one, mesons go from their antimatter state to their matter state more quickly than going the other way; an asymmetry indicative of why there is more matter than antimatter.

There is one known particle that is also its own antiparticle: the photon. But then, a photon is a boson, which is no matter at all; only a trick of light.

The Standard Model offers no accounting for antimatter; one of the more obvious indications that SM is merely a stopgap story.

℘ CP Violation ℘

QFT physical models create mathematically symmetrical relationships. As with particle symmetry and its breaking (SSB), Dirac's model for matter-antimatter is broken.

CP is an acronym for 2 supposed symmetries: charge conjugation (C) and parity (P). *Charge conjugation* transforms a quantum particle into its antiparticle. *Parity* creates a mirror image of a physical system. Both C & P posit mathematically symmetrical relationships.

C symmetry relates to physical forces: that replacing an interaction with its negative would result in an equivalent dynamic. While the strong interaction, electromagnetism, and gravity obey C symmetry, the weak force violates C symmetry.

Parity transformation is an inversion of a spatial coordinate system, hence also termed *parity inversion*. Parity transformation hypothetically flips between space (x, y, z coordinates) and anti-space (–x, –y, –z coordinates).

Under parity inversion, a particle in the matter realm turns into an antiparticle in the antimatter realm. Charge (C) transpires within a system defined by parity (P), as the interactions of a charged particle are within the context of the physical system (parity). Hence CP is a natural combinatorial construct.

The mirror implicit in parity means that the equations of particle physics are *invariant*: a mirror reaction occurs at the same rate as the original reaction. Under parity symmetry, reaction type doesn't matter, whether chemical or of radioactive decay.

The one consistency in all QFT models has been that Nature is messier than the math readily allows. Hence the necessity of symmetry breaking as a bandage.

From the dawn of quantum theory, *parity symmetry* was held as one of the fundamental geometric conservation laws, along with conservation of energy and conservation of momentum (which are merely mathematical edicts). Parity symmetry seems to hold for electromagnetic and strong interactions. But weak interactions, those responsible for radiation, violate P symmetry (as with C symmetry).

A 1928 experiment showed P symmetry violation, but its results were ignored, as the concepts necessary to understand the experiment's significance were undeveloped. One simply cannot understand what is beyond one's worldview.

In 1956, parity as asymmetric was cracked by observing the beta decay of Cobalt-60. Some reactions did not occur as frequently as their mirror image.

With theory in hand, but upon staring at undeniable proof, CP symmetry became something of a polite fiction. Physicists simply accept their models as imperfect approximations of inscrutably intricate Nature. But the theoretical cloud has a silver lining. Behavioral differences from *CP violation* explain certain incongruities in the Standard Model, and excuse matter and antimatter from mutual annihilation in every instance.

ॐ Forces ॐ

We have to remember that what we observe is not Nature in itself, but Nature exposed to our method of questioning. ~ Werner Heisenberg

A *force* is an effect on an object resulting from an interaction. Forces only manifest as a result of an interaction.

Force is a quantity measured in the standard metric unit known as the *newton*. A 1-newton push accelerates a 1-kg mass 1 meter per second per second.

$$1 \text{ newton} = 1 \text{ kg x m/s}^2$$

There are 4 observable fundamental physics forces in the universe: strong & weak nuclear, electromagnetism, and gravity. As these forces are now interpreted to be impositions of bosons upon fermions, the term *interactions* has become more common. Interactions vary by strength and scale: the distance in which a force may apply.

The nuclear forces are quantum interactions which manifest. Electromagnetism defines the ambient world. While gravity keeps planetary inhabitants grounded, it exerts its real power cosmologically.

Forces are commonly characterized as either by contact or at a distance. Normal forces, such as friction, are considered contact forces. In contrast, electricity, magnetism, and gravity exert their influence at a distance.

Space always exists between interactions of matter, even among atomic nucleons. While proximity is readily comprehended by our everyday experience, quantum nonlocality shows that distance involves a perceptual reference frame.

In physics, every observation is made with respect to a frame of reference. The state of a physical system constitutes a reference frame. ~ Italian quantum physicist Flaminia Giacomini

Per the Standard Model, 3 of the 4 forces represent fundamental interactions accompanying emission or absorption of gauge bosons. But gravity isn't really a force at all; only a spacetime distortion caused by material mass.

✄ Nuclear Forces ✄

There are agents in Nature able to make the particles of bodies stick together by very strong attractions. ~ Isaac Newton

The *strong force* binds protons and neutrons in the nucleus of an atom. It is by far the strongest force: 100 times the tug of the electromagnetic force, 10^6 times that of the weak force, and 10^{39} times that of gravity. But then, the strong force applies only to irascible atomic nuclei, whereas gravity affects entire galaxies. The strength numeric is therefore an apples-and-oranges comparison, as the scales involved, while mathematically figurable, are practically incomparable.

Within the context of the Standard Model, the strong force is a gluon shotgun: forcing quarks to marry each other, and so form *nucleons*: the protons and neutrons which comprise atoms.

The strong force is overwhelming at distances the size of a nucleon (10^{-15} m): squeezing quarks together to form hadrons. It rapidly weakens beyond that range.

Protons are the only hadrons that are stable. All other hadrons, of which there are many, are ready prey to particle decay under sway of the weak force. Neutrons are stable only when inside atomic nuclei.

Protons nominally comprise 2 up quarks and 1 down quark, all different colors. Neutrons are 1 up quark and 2 down quarks, also different colors.

The *weak force* causes particle (*beta*) decay, a form of radioactivity, and initiates hydrogen fusion in stars. Under the Standard Model, the weak force is invoked by interaction between $W^{\pm}$ and Z^0 bosons.

Weak interaction forces quarks to change flavor. Changing flavor means changing into a different type of quark.

As up and down quarks have the lowest mass, they are the most stable. Heavier quarks (strange, charm, bottom, and top) decay by weak interaction into a less energetic (massive) flavor.

The weak force also breaks the symmetry between matter and antimatter. While the strong force is about marriage, the weak force is about divorce.

A typical atomic nucleus has a spherical or watermelon profile, depending upon the nucleons within. But at high energies, nuclei become pear-shaped, as protons are pushed away from the center by an unknown force.

We've found these pear-shaped nuclei literally point toward a direction in space. This relates to a direction in time, proving there's a well-defined direction in time and we will always travel from past to present.

Further, the protons enrich in the bump of the pear and create a specific charge distribution in the nucleus. This violates the theory of mirror symmetry and relates to the violation shown in the distribution of matter and antimatter in our universe. ~ Scottish nuclear physicist Marcus Scheck

Such asymmetry shows that there is yet another nuclear force besides strong and weak; one about which the Standard Model has nothing to say.

✍ Electromagnetism ✍

Electricity and magnetism were long thought distinct. Then, in 1831, Michael Faraday discovered a magnetic field about a wire conducting DC current. Faraday also established that light is affected by magnetism.

In 1865, James Clerk Maxwell published equations that equated electricity, magnetism, and light as manifestations of the same phenomenon; that electric and magnetic fields were waves that traveled at the speed of light, and that light was also wavy.

Maxwell's unified model of electromagnetism was a milestone in physics. The theoretical implications of electromagnetism inspired Einstein to formulate special relativity.

Electromagnetism acts between electrically charged particles. Via attraction of negatively charged electrons to positively charged protons, electromagnetism creates atoms.

The vacuum fluctuations of the electromagnetic field have clearly visible consequences, and, among other things, are responsible for the fact that an atom can spontaneously emit light. ~ Swiss physicist Ileana-Cristina Benea-Chelmus

Electricity and magnetism are dual manifestations of a single interactive force. A changing electric field generates a magnetic field and vice versa. This electromagnetic induction is the basis for electric generators, transformers, and induction motors.

> Anyone who uses electricity is experiencing the effects of relativity. ~ American physicist Thomas Moore

Electromagnetism involves relativistic effects. Moving a loop of wire through a magnetic field generates an electric current. The charged particles in the wire – electrons and protons – are affected by the changing magnetic field: forcing some of them to harmonically sway, thereby creating an electrical current.

Imagine the wire at rest with the magnet moving. In this instance, the wire's charged particles aren't moving, so the magnetic field should not affect them. But it does. Current still flows. This shows that there is no privileged frame of reference; exactly the point that Einstein was making with special relativity: that all frames of reference are relative, and that their relations exhibit phenomenal effect.

> Without relativity, neither magnetism nor light would exist, because relativity requires that changes in an electromagnetic field move at a finite speed instead of instantaneously. If relativity did not enforce this requirement, changes in electric fields would be communicated instantaneously instead of through electromagnetic waves, and both magnetism and light would be unnecessary. ~ Thomas Moore

¤ ✧ ¤

Besides being the source of light by dint of energetic glow, photons are the force carrier of electromagnetism. This is unobvious, as photons do not nominally interact with matter.

Magnets attract each other because they exchange virtual photons. Each virtual photon has its own frame of reference. In their supposed interaction, photons exchange momentum, thereby producing attraction or repulsion, depending upon relative energetic orientation of the object which they encounter. This is a relativistic effect.

¤ ✧ ¤

At an energy level of 246 GeV, the electromagnetic and weak forces unite into electroweak interaction. 246 GeV is estimated upon the calculated value of the Higgs field in a vacuum. Room temperature has a thermal energy of 0.025 eV.

♎ Magnetism ♎

Magnetism is a powerful force that causes certain items to be attracted to refrigerators. ~ American writer Dave Barry

Magnetism is a field of attraction between particles. Most materials are influenced to some extent by magnetic fields.

The magnetic behavior of crystalline materials is highly sensitive to the lattice constant.* ~ Japanese physicist Hideaki Sakai

There are 3 known magnetic states. All appear in crystals. *Ferromagnetism* – the magnetism of magnets and compass needles – has been known for over a millennium.

The 2nd state of magnetism is *antiferromagnetism*: where the ionic magnetic fields of metals cancel each other out, owing to complementary electron spins. Antiferromagnetism was discovered in the 1950s. It is the basis for read-heads in computer hard-disk drives.

Having no fondness for heat, both ferromagnetism and antiferromagnetism exhibit their talents only when cooled below a critical temperature.

The 3rd state of magnetism is in *quantum spin liquids* (QSL), discovered in 2012. QSL is the liquid-like magnetism of quantum entanglement. The state is called *liquid* because it is disordered compared to the spin state of crystalline ferromagnetism. QSL and ferromagnetism are analogous to the states of water and ice.

Quantum spin liquids cannot be described by the broken symmetries associated with conventional ground states. In fact, the interacting magnetic moments in these systems do not order but are highly entangled with one another over long ranges.

A key feature of spin liquids is that they support exotic spin excitations carrying fractional quantum numbers. In a spin liquid, the atomic magnetic moments are strongly correlated, but

---

* The *lattice constant* characterizes the physical dimensions of unit cells in crystal.

do not order or freeze even as the temperature goes to zero.
~ Chinese American physicist Tian-Heng Han *et al*

♎ Lorentz Symmetry ♎

Now we know that time and space are not the vessel for the universe but could not exist at all if there were no contents, namely, no Sun, Earth, and other celestial bodies. ~ Hendrik Lorentz

The idea of Nature having laws is inherent in science; especially physics, which more intently scrutinizes the nature of existence than other disciplines, which generally take materiality for granted (that is, assume naïve realism).

The essence of such laws is consistency. After all, a law is not a law if Nature violates it. In physics, this concept is embodied in symmetry: that physical laws are invariable.

In 1895 Dutch physicist Hendrik Lorentz derived the transformation equations which formed the basis of Einstein's special relativity theory. Behind these equations, and special relativity, is the idea of an *inertial reference frame*, which is inviolable. That the laws of physics are the same for all observers is termed *Lorentz symmetry*.

Paradoxically, symmetry breaking is as important in physics as symmetry. To generate mass in subatomic particles, the Standard Model relies upon breaking electroweak symmetry.

3 distinct classes of fermions have been identified: Dirac (with mass and charge), Weyl (massless, charged), and Majorana (massless, chargeless). Dirac fermions are the stuff of ordinary matter. Majorana fermions were mathematically conjectured in 1937 but were experimentally elusive until the mid-2010s.

Also long shy were Weyl fermions. In 2015, they made an appearance in certain crystalline semimetals made of tantalum and arsenic (TaAs). Sort of. Weyl fermions were discerned via physical effects which can be inferred through the collective excitation of their quasiparticles.

A Weyl fermion can emerge as a quasiparticle in certain crystals: Weyl fermion semimetals. ~ Chinese physicist Su-Yang Xu *et al*

Another sort of Weyl fermion (type 2) showed up in a crystalline solid in 2017. This Weyl fermion broke Lorentz symmetry with an astonishing display of asymmetrical magnetism.

Put a normal material in a magnetic field and its resistance to electrical conduction grows. But in a solid larded with type-1 Weyl fermions, a magnetic field enhances electrical current flow.

In violating Lorentz symmetry, type-2 Weyl fermions are even stranger. In a material with these particles, a magnetic field in one direction increases conductivity; but when magnetized in another direction, electrical flow drops.

Symmetry breaking shows that our supposed 'laws' of Nature are nothing more than conditional codicils to something more fundamental. Nature keeps secrets.

⚡ Gravity ⚡

Before Newton, gravity was considered related to the motions of celestial objects; a theoretical construct along the lines of an Aristotelian view of things, with some modification. Aristotle believed that heavier objects fell faster.

Galileo reputedly dropped balls from the Tower of Pisa to conclude that all objects obey gravity's tug at the same rate, but that's a tall tale. What Galileo actually did was roll differently weighted balls down an incline.

Newton's 1687 publication of *Principia* formulated the gravitational attraction of planets as following a universal *inverse-square law*: the gravitational attraction between 2 bodies is directly proportional to the product of their masses, and inversely proportional to the square of the distance between them. Under Newton's law, gravity is a force. More massive objects create more gravity. Gravity weakens with distance.

Einstein's 1917 theory of general relativity gravitated gravity away from being a fundamental force and into a curvature of spacetime. Einstein's conception taught away from

Newton's construct and headed back to Galileo: that gravitation gives the appearance of acceleration independent of the mass perceived to be accelerating in free-fall.

Einstein began his theory of gravity with the *equivalence principle*: there is no way to distinguish the effects of acceleration (inertial mass) from the effects of gravity (gravitational mass). The key to the equivalence principle is the idea of a *reference frame*. In a proverbial nutshell, within a spacetime reference frame, you couldn't tell the difference between going to hell in a bucket under constant acceleration and merely sitting in the bucket under the influence of a hellish gravitational field.

The equivalence principle can be also got to via the precepts posited by Galileo and Newton. No force but gravity depends on mass. Galileo showed that acceleration is independent of the mass of the object being accelerated.

Newton's 2nd law of motion (acceleration) and his gravitational force law depend on mass in the same way. Newton's laws were consistent. Which is to say that mass cancels out when calculating acceleration. Hence, acceleration doesn't depend on mass.

While the equivalence principle can be deduced from classical physics, it fails to explain how or why the principle works.

Einstein deduced that free-fall is actually inertial motion. Free-falling objects do not really accelerate, but rather, the closer they get to the gravitationally attractive object, the more time stretches due to spacetime distortion around the massive object. This spacetime distortion is gravity.

While gravity may be perceived as a force, that is a statement from a limited reference frame. Gravity is a relation between objects according to their relative masses.

Gravity is not a force, but instead an entropic distortion, an emanating consequence of mass. Gravity warps the spacetime reference frame we call existence. Unsurprisingly, gravity manifests in waveform.

Gravity remains an ontological enigma. The beauty and success of general relativity seems to imply a reality of a curved spacetime framework to the universe. On the other hand, this

curved framework is not at all required, and is indeed a hindrance, for the rest of physics. Furthermore, general relativity is not a quantum theory, and so it must break down at some level.
~ Victor Stenger

℘ Quantum Gravity ᘒ

The fundamental laws necessary for the mathematical treatment of a large part of physics and the whole of chemistry are thus completely known, and the difficulty lies only in the fact that application of these laws leads to equations that are too complex to be solved. ~ Paul Dirac

Quantum gravity is the label for the search to create a conceptual and mathematical reality continuum, ranging from the vast expanses of space to the ineffably small.

Quantum gravity represents a strained effort to marry general relativity and quantum mechanics; specifically, explaining how the gravity of general relativity works in the realm of the infinitesimal. What's missing are the wedding rings. Meshing the two theories flounders over fundamental assumptions of how the universe works.

These assumptions are encompassed in equations that don't fit together, as they result in useless infinities. This mathematical quandary led to applying renormalization.

Renormalization boils down to addressing the resultant infinities that leap out of a physical model addressing how existence is structured. Renormalization scrubs the infinities out, providing usable approximations.

It is difficult to describe Nature when the most elegant equations object with obscenities of unusable infinities. That is an inescapable problem in trying to model the universe in 4D when there are more than 3 spatial dimensions.

Renormalization was first developed for quantum electrodynamics (QED), to get a grip on the infinite integrals that arose in perturbation theory. *Perturbation theory* involves mathematical techniques to squeeze an approximate answer from equations that refuse to resolve to an exact solution. In other words, perturbation theory is a mathematical fudge.

> The lack of understanding of time seems to be one of the chief impediments to developing a quantum theory of gravity. ~ Canadian physicist William Unruh

In 1986, Indian theoretical physicist Abhay Ashtekar reformulated Einstein's general relativity to have it more closely correspond with the rest of physics. Follow-on work led to *loop quantum gravity* (LQG): a theory solely aimed at explaining quantum gravity as a way to marry quantum mechanics to relativity without resorting to renormalization. Unlike string theory, LQG requires no additional dimensions.

The dimensional modesty of loop quantum gravity belies its radical assumption that space itself is quantized at Planck scale as loops. Under LQG, loops of space are linked, forming a network of relations (a tensor network).

> Once again, the world seems to be less about objects than about interactive relationships. ~ Italian theoretical physicist Carlo Rovelli

One consequence of loop quantum gravity is the dispensation of time. LQG equations have no need of temporality. Instead, time's passage is internal to the world, born in the relationships between the quantum events that define existence and are themselves the source of time.

Loop quantum gravity theory has something to say about the origin of the universe. LQG finds that a severely compressed cosmos generates a repulsive force. Thus, LQG supports cyclic cosmology.

Whether gravity at the quantum level is a modeling problem or a misapprehension of spacetime appears to leave the nature of quantum gravity up in the air. But the problem with quantum gravity may be much simpler.

> Gravity may not even exist at the quantum level. ~ Vlatko Vedral

Recall that subatomic particle mass is not classical, which would be a measure of an object at rest. Instead, quantum particle mass is a euphemism for an energy measurement.

Gravity is an entropic distortion caused by mass. At the quantum level, mass is so insignificant as to render gravity negligible. General relativity doesn't break down at the quantum level so much as evaporate for lack of heft.

⚹ Quantum Superposition ⚹

> The ability to live in coherent superpositions is a signature trait of quantum systems. ~ Brazilian physicist Isabela Silva

Quantum superposition assumes that the uncertainty principle is real. At every moment, any physical quantum, such as an electron, exists in all possible states simultaneously until it manifests, whereupon its result is only 1 configuration.

While single particle superpositions can be fairly stable, macroscopic objects never are. The formation of macroscopic superpositions, in which numerous quantum components must maintain a precise relationship with each other, are disrupted by continual environmental influences.

> Gravity as an environment induces the rapid decoherence of stationary matter superposition states when the energy differences in the superposition exceed the Planck energy scale. ~ English physicist Miles Blencowe

A disruption of superposition decoheres a system into a specific state. Gravity is a spacetime disturbance that pushes the quantum components of a system out of sync as they travel across a superposition.

Quantum particles are largely beyond the reach of gravity. An atom is touched by it. A molecule feels gravity, however slightly.

Decoherence rate rises by the square of the energy difference between 2 states in a superposition. The more there is to a physical system, such as a proton versus an electron, the greater the energy differences in superposition states.

Gravitational waves are pervasive and inescapable. They are part of the cosmic background, an echo of inception, and a cousin to the electromagnetic radiation which also pervades. Superposition loses its grip on the macroscopic world via gravity, giving rise to the predictable realm described by classical physics.

Various experiments have lent credence to quantum superposition.

❧ Nonlocality ☙

The statistical predictions of quantum mechanics are incompatible with separable predetermination. ~ John Stewart Bell

Quantum mechanics has an obvious deficiency: its mechanics. Quantifying quantum phenomena is the elephant in the room of interpreting quantum theory.

Measuring fundamental particles is an existential oxymoron. Watching a wave function collapse is a probabilistic event. The math itself is nontrivial, and the appropriateness of the bandied equations contentious.

Nonetheless, some quantum field phenomena have been seen. The most inexplicable is *nonlocality*; what Einstein called "spooky action at a distance."

The principle of *locality* states that an object can be influenced directly only by its immediate surroundings. *Nonlocality* is the notion that distance is ultimately an illusion. Nonlocality and entanglement are synonyms.

A 1935 paper by Einstein, Boris Podolsky, and Nathan Rosen (EPR) posited a paradox over quantum uncertainty, called the *EPR paradox*: either locality or uncertainty must be true. EPR opted for locality, thereby concluding that the wave function must be an incomplete description of actuality.

In response, Irish physicist John Stewart Bell tackled the quantum measurement problem in 1964; whence *Bell's theorem*.

Science in general, and physics in particular, long assumed that locality and objectivity were both true. Locality means that distance affects the probability of interactions. Locality is colloquially codified in everyday *cause and effect*. Locality is an embrace of naïve realism: the idea that actuality is reality, independent of observation, but observable; what is commonly called *objectivity*.

Bell's theorem stated that either locality or objectivity was not true. In opting for the uniformity of objective reality, Bell pitched locality.

Einstein struggled to the end of his days for a theory to uphold causality and objectivity. While Einstein had a woolly spirituality, he was a naïve realist.

Bell's theorem went the other way, stating that some quantum effects travel faster than light ever can, thus violating locality. Bell's theorem painted special relativity into a corner; applicable only at the macro scale; irrelevant at the quantum level.

Regarding causality, *counterfactual definiteness* (CFD) goes to measurement repeatability: whether what has happened in the past is a statistical indicator of the future. Adhering to causality, locality considers events predictable, and thereby certain.

At the quantum level, CFD butts heads with locality by stating that past probability as indicative of the future is a chimera. Instead, uncertainty always reigns.

In accepting objectivity, certainty and uncertainty are mutually exclusive.

Bell's theorem insisted that quantum uncertainty was a certain reality. The principle of locality breaks down at the quantum level.

Nothing can travel faster than light speed. ~ American theoretical physicist Brian Greene

A lot of things travel faster than the speed of light. Nonlocality has been repeatedly confirmed at both quantum and macroscopic scales. Oddly, the larger the system, the greater the odds of nonlocality. At a threshold of about 200 subatomic particles, entanglement becomes the norm rather than the exception. Smaller particle groups are less likely to demonstrate nonlocality.

Entanglement is hard to create from a small system, but much easier in a large system. ~ Indian physicist Harsh Mathur

With spooky-action-at-a-distance a reality, superluminal (faster than light) effects exist. Bell's theorem of nonlocality/entanglement is considered a fundamental principle of quantum mechanics, having been supported by a substantial body of evidence.

Nonlocality is so fundamental and so important for our worldview of quantum mechanics. ~ Swiss quantum physicist Nicolas Gisin

The supposed tradeoff between locality and objectivity is a false one. While the restriction of quantum locality has been lifted, there is no proof that existence is objective. It just appears that way by social consensus, and so is taken as an axiomatic assumption, just as locality was for so long.

What is taken for objectivity is instead *showtivity*: shared subjectivity via consciousnesses within the same enveloping reference frame (consciousness relativity). Each individual consciousnesses exists within a universal field of Consciousness.

⌀ Bose-Einstein Condensate ⌀

In 1924, Indian polymath Satyendra Bose sent Albert Einstein one of his papers. Einstein was impressed. He translated the article from English for publication in a German physics journal, adding material which expanded upon Bose's ideas.

The upshot: supercooled bosons become a new form of coherent matter, a gas that coalesces into a single super-particle with overlapping wave functions. Such behavioral singularity owes to the uncertainty principle, which counterintuitively holds, among other things, that particle positions become increasingly uncertain as their velocity slows, which happens when they are chilled to the core.

Bose-Einstein condensate (BEC) was first demonstrated in 1995. BEC exhibits extraordinary quantum mechanical properties at a macroscopic scale. Further, entanglement between 2 nonlocal BEC clouds was observed in 2011.

The irony is that BEC, named after Einstein, demonstrates nonlocality, an extra-dimensional property which Einstein expressly disbelieved.

⌀ Solitons ⌀

A *soliton* is a self-reinforcing solitary wave that maintains its shape as it travels through a medium at a constant speed. Solitons arise via cancellation of nonlinear and dispersive effects in a gas or fluid.

In 1834, Scottish engineer John Scott Russell saw a solitary wave in a canal travel for over 8 miles without changing shape or amplitude. He then managed to reproduce solitons in a wave tank.

Solitons exhibit startling robustness in their coherence. Solitons can encounter each other and still maintain their integrity. Soliton dynamics vary depending upon the medium in which they appear.

A *dark soliton* is a standing dip in the density distribution of a medium; the opposite of a *light soliton*, which creates a wave of greater density than the surrounding medium.

Superfluidity – frictionless flow – arises in a Bose-Einstein condensate. Dark solitons can arise in a BEC.

Unlike bosons, fermions follow the Pauli exclusion principle, and so cannot occupy the same quantum state simultaneously.

To condense and form a superfluid, fermions must turn into bosons. They can do so by forming entangled pairs that have the requisite integer spin (each fermion has a half spin).

The size of an entangled fermion pair critically depends upon the interaction between pair members. An entangled pair may be tightly bound (a Cooper pair) or be at some distance. This determines the underlying physics of the condensate. Entangled pairs at a distance are relatively weakly bound, and yet more readily given to superfluidity.

A condensate may transition from close-knit to a greater inter-particle spacing or vice versa. In a condensate progressing to greater pair spacing, a dark soliton becomes more filled with non-condensed-gas atoms, making it heavier, and slowing it down.

This wave change occurs because quantum fluctuations have a more pronounced effect on the dark soliton. Solitons, which arise from coherence, are heavily influenced by the degree of fluctuations in the ground state.

The Ground State

The idea that nothing can exist has been controversial throughout history. Ancient Greek philosophers debated the possibility of a void in the context of atomism.

Plato found the idea inconceivable. Following Plato, a featureless void faced skepticism – how could something exist that could not be perceived? Aristotle considered vacuum impossible: nothing could not be something. In the 1st century BCE, Lucretius thought that a vacuum was possible, but his argument went nowhere.

Medieval Christians held the idea of a void to be heretical. The absence of anything implied the absence of God; harking back to the void prior to creation, as described in the book Genesis in the *The Bible*. This led to the commonly held view that Nature abhorred a vacuum (*horror vacui*); an axiom carried forward during the Scientific Revolution. The concept of a cosmic aether reflected this belief in substance even when nothing was manifest.

¤ ✧ ¤

The *ground state* is the lowest energy state of a quantum mechanical system, with supposedly zero-point energy. In quantum field theory, the ground state is called the *vacuum state*, or simply *vacuum*.

> Vacuum is not empty. Particles appear out of nothing.
> ~ Russian quantum physicist Andrey Moskalenko & Australian quantum physicist Timothy Ralph

In the first iteration of general relativity, the ground state appeared as the *cosmological constant*: a construct Einstein coined to create a stationary universe. He then abandoned the constant as a bad idea, as the universe appeared to be not as stationary as he first supposed.

The 1998 faux discovery of an accelerating expansion of the universe renewed interest in a cosmological constant that characterizes the ground state.[*] That interest hit a cosmological conundrum when calculating the vacuum energy that signifies the ground state, which came out as 10^{120} times too much.

---

[*] Accelerating cosmic expansion has been discounted by observations since its 1998 false discovery. The paradox of enormous vacuum energy remains.

Quantum electrodynamics (QED) calculates that vacuum energy is 10^{113} joules per cubic meter; unimaginably enormous power. Yet this is a comedown from the early universe, when the ground state was even more energetic. Some of the earliest proto-hadrons formed with strange quarks, with a greater mass than the baryons that now comprise ordinary matter.

It seems the boundary between 4D and ED shifted and settled as the cosmos diffused and average temperature lowered. But then, as the energy of the ground state demonstrates, dimensionality is something of an artifice.

♎ The Dance of Spacetime ♎

In 2017, Chinese physicist Qingdi Wang and colleagues investigated "the gravitational property of the quantum vacuum by treating its large energy density predicted by quantum field theory seriously, and assuming that it does gravitate to obey the equivalence principle of general relativity." What they found was that "spacetime itself is constantly moving."

> It's similar to the waves we see on the ocean. They are not affected by the intense dance of the individual atoms that make up the water on which those waves ride. ~ William Unruh

From a fluctuating m of spacetime emerges the illusion of a stable cosmos.

According to the 3rd law of thermodynamics, the ground state is supposed to be at absolute zero temperature (0 K). That is a theoretical fiction, as the ground state is not a void, or empty space.

According to a QED hypothesis, the ground state is continually perturbed by ghost fields of tremendous vacuum energy, making matter radiate over it, in what has been termed *quantum foam*.

The conventional comprehension is that fleeting virtual particles and electromagnetic waves pop in and out of "existence" from the ground state.

The notion of matter/energy popping in and out of existence is ridiculous. The popping is between the perceivable 4D and ED; a dimensional phase shift. Our experience of actuality may be largely 4D, but the dimensionality of existence is more expansive (HD).

Virtual particles are an HD phenomenon with 4D cameo appearances. The ground state is simply a limit boundary to phenomenal space.

You can expect what you inspect. ~ American statistician Edwards Deming

Deming's comment to "expect what you inspect" was a statement referring to quantitative quality control in manufacturing, but it also applies to biased conception in a realm requiring open inquiry. In the case of dimensions, physicists expect only 4 because they can only inspect 4, even though their physical models tell them there are more.

℘ Quasiparticles ℞

These particles are just smoke and mirrors, handy mathematical tricks and nothing more. Or are they? ~ English physicist Andrea Taroni

Quasiparticles are emergent phenomena that behave as quantum particles but are not considered legitimate in the sense of being a fermion or boson. Quasiparticles are to quantum mechanics what epigenetics is to genetics: potent, but not quite kosher. Both illustrate the deep, entangled intricacy that characterizes Nature.

Emergent quanta of momentum and charge, called *quasiparticles*, govern many properties of materials. ~ Dutch physicist Dirk van der Marel

Formally, whereas a *quasiparticle* is related to a fermion, a *collective excitation* is related to a boson. Both types of emergent energies are casually referred to as *quasiparticles*.

In solids, many-body correlations lead to characteristic resonances – quasiparticles. ~ German physicist Fabian Langer

♒ Ettore Majorana ♒

Italian physicist Ettore Majorana first proposed the existence of neutrons. Enrico Fermi urged him to write an article on it. Majorana demurred, considering his own work banal. The credit was instead given to James Chadwick, who won a Nobel prize for it.

In 1937, Majorana discovered a hitherto unknown solution to the equations from which quantum particles are deduced. Out of it came a prediction for an exotic fermion, initially thought as perhaps a type of neutrino: the Majorana.

On 27 March 1938, Majorana took a boat trip from Palermo to Naples. He disappeared. His body was never found.

Majorana had emptied his bank account prior to the trip. 2 days before he left, Majorana wrote a note to the Director of the Naples Physics Institute, apologizing for the inconvenience that his disappearance would cause.

♎ Majorana Fermions ♎

The *Majorana fermion* is a charge-neutral, zero-energy quasiparticle. The Majorana is unique in being its own antiparticle, hence existing on the shadowy border between matter and antimatter; hence its designation as *quasiparticle.* That may not make the Majorana novel.

The nature of neutrinos is unsettled. As neutrinos have no charge, they too may be *Majorana fermions*, rather than *Dirac fermions*, where a particle has a mirror image antimatter partner.

Mathematically, neutral spin 1/2 particles, such as neutrinos and the Majorana fermion, can be characterized by a *real wave equation* (the *Majorana equation*), instead of the more typical wave function that predicts a particle and antiparticle via *complex conjugation.*

A *complex conjugate* is a complex-number pair, where the real components are identical, but the imaginary parts, though of equal magnitude, have opposite signs. *1 + 2i* and *1–2i* are exemplary complex conjugates.

The Majorana is not included in the Standard Model particle clique even though its existence is certain. Majoranas do not comfortably fit within SM owing to their uniqueness, especially the mathematics that characterize them.

Whereas Majoranas have no charge, and do not interact very strongly with light or other electromagnetic radiation, they can be detected by electrical measurements, and are affected by the electrical environment. Majorana fermions facilitate superconductivity.

When 2 Majoranas are repositioned relative to each other in a superconducting region, they essentially remember their previous position. This property has been suggested as valuable in constructing a quantum computer, with Majoranas as a memory mechanism.

The oddity of Majoranas as chargeless, and yet subtlety interacting electromagnetically, may be explained by Majoranas having a magnetic anapole moment. An *anapole* is a toroidal dipole: a solenoid field bent into a torus.

Particles with electrical and magnetic dipole moments interact with electromagnetic fields regardless of their momentum. In contrast, electromagnetic interaction with an anapole particle strengthens with speed. The Majorana having an anapole moment fits well with its mass, speed, and electromagnetic properties.

Neutrinos travel near light speed because they are theoretically nearly massless. An accurate measurement of neutrino mass has eluded experimental physicists. Majoranas are assumed to be like neutrinos, albeit slower moving.

♎ Plasmons ♎

A *plasmon* is a quasiparticle quantum of plasma oscillation. Plasmons appear as an oscillation in conducting electrons, and so are an exhibition of electromagnetism. The visual effect of a plasmonic object can be different colors, depending on how light strikes the object.

The 4th century Lycurgus Cup was made of dichroic glass. It displays a different color depending upon the direction of light passing through it. The cup appears red when lit from behind, and green when lit from the front. The Lycurgus light show is an example of quasiparticle plasmons at work.

Plasmons can be produced in a metal, whereupon they dance along its surface. If the piece of metal is less than 10 nanometers (nm) thick, quantum effects emerge, changing a plasmon's oscillation frequency and 4D lifetime.

Plasmons at one scale behave differently than at another scale. The optical absorption of plasmons is proportional to their volume.

In particles larger than 10 nm, plasmons respond to electromagnetic fields as a classical electron gas. This is done by adhering to the uncertainty principle.

Smaller than 10 nm, and a plasmonic field defies its supposed basic nature by interacting only weakly with light. There are so few conduction electrons participating in the plasmons (~250 electrons in a 2-nm particle) that the electrons appear at a discrete set of energy levels, which are increasingly separated from one another as particle size shrinks.

This produces individual electron jumps between occupied and unoccupied electron energy levels. The separation and jumpiness increase uncertainty and reduce plasmon 4D life.

This is exemplary of quantum effects, which are essentially a breaking down of the reality construct by surging uncertainty into events. Nonlocality can also occur with plasmons at this small scale. Distant individual electrons within the plasma become entangled in an HD dance.

♎ Phonons ♎

A *phonon* is a collective excitation that shuttles heat around solids, and herds electrons into the coherence that affords superconductivity. Just as plasmons are quantized plasma oscillations, phonons are quanta of mechanical vibrations.

> Phonons are not actually real. They are really just a way of simplifying a very complicated problem. ~ English physicist Jon Goff

As sound is a mechanical vibration, it quantizes as phonons. Sound waves have negative effective gravitational mass.

A sound wave not only is affected by gravity but also generates a tiny gravitational field. ~ Italian physicist Alberto Nicolis

♎ Magnons ♎

A *magnon* is a collective excitation that quantizes the spin wave which characterizes the spin property of all quanta. Magnons emerge from waves of flipping spin. They explain electron behavior in a crystal lattice. Swiss physicist Felix Bloch introduced magnons in 1930 to elucidate abrupt, spontaneous changes in magnetism at low temperatures; whence their name.

♎ Polarons ♎

Electrical conductivity is more than merely the flow of negatively charged electrons. Positively charged atomic protons play a critical part in cooperating with or countering electrons on their merry way.

Atoms are more than the gatekeepers of electricity. The relation between electrons and nucleons is entangled. The movement of electrons has a direct effect on atomic arrangements.

Water conducts electricity. In doing so, flowing electrons tug on water molecules' hydrogen atoms, moving protons. This process, known as the *Grotthuss mechanism*, also occurs in vision, when light hits the eye's retina.

A *polaron* is a quasiparticle that characterizes electron mobility. Polarons were proposed by Lev Landau in 1933.

> Polaron transport, in which electron motion is strongly coupled to the underlying lattice deformation or phonons, is crucial for understanding electrical and optical conductivities in many solids. ~ Chinese physicist Junjie Li *et al*

When an electron cloud (polaron) enters an atomic lattice, the two try to accommodate one another by modifying their shapes.

> Lattice vibrations are interacting with the electrons; proof that polarons exist. ~ Chinese quantum physicist Weiguo Yin

♎ Excitons ♎

Quasiparticles exist even when nothing is there. American physicist William Shockley Jr. was working with semiconductors when he had an epiphany that permitted the perfection of the transistor in 1947.

It had been known for a decade that electrons moving through semiconductors left gaps of nothingness. But no one thought of these "holes" as anything more than an electron's absence.

Shockley proposed to treat holes as particles in their own right: like an electron but with a positive charge. This crucial paradigm shift led to better understanding the flow of energy in semiconductors, and so fashion the junctures and switches that characterize transistors.

Since then physicists have conceived that electrons and holes can combine, yielding a whole new quasiparticle: the *exciton*. Plants were way ahead of us on this. The light-harvesting proteins responsible for photosynthesis use electrons to absorb photons of sunlight. The resultant energy kick knocks an electron out of position, creating a hole.

The electron and hole link up to form an exciton, which is shuttled about the photosynthetic machinery. When the exciton gets to where it's needed to do its bit, the electron and hole recombine, releasing energy employed to split water into constituent hydrogen and oxygen; a key step in making sugar from sunlight, air, and water.

⌀ Quasiparticles Forever ⌀

Quantum states of matter typically exhibit collective excitations. These involve the motion of many particles in the system, yet, remarkably, act like a single emergent entity – a quasiparticle. Known to be long lived at the lowest energies, quasiparticles are expected to become unstable when encountering the inevitable continuum of many-particle excited states at high energies, where decay is kinematically allowed. Although this is correct for weak interactions, strong interactions generically stabilize quasiparticles by pushing them out of the continuum.
~ German quantum physicist Ruben Verresen *et al*

Decay reigns in the ambient domain. The quantum world is something else.

> The assumption was that quasiparticles in interacting quantum systems decay after a certain time. The opposite can be the case: strong interactions can even stop decay entirely. ~ German physicist Frank Polimann

Quasiparticles may decay and then reorganize themselves, becoming virtually immortal.

> Quasiparticles do decay, but new, identical particle entities emerge from the debris. This process can recur endlessly. A sustained oscillation between decay and rebirth emerges. ~ Ruben Verresen

ಏ String Theory ಞ

> I would not even be prepared to call string theory a "theory." Just a hunch. ~ Gerard 't Hooft

String theory postulates subatomic particles as infinitesimally thin strings vibrating through a holistic dimensionality (HD) that has more than 4 dimensions.

> We still don't know what string theory is. ~ American particle physicist and string theorist David Gross

The "string" in string theory seems somewhat misleading, as the significance is that particle fields have resonances at different frequencies, harmonically interacting with their brethren. *Vibe theory* sounds more appropriate.

ଶ History ଧ

> String theory is 21st-century physics that accidentally found its way into the 20th century. ~ Ed Witten

String theory was presaged by Einstein's 1907 generalization of photons: a suggestion that solids came as vibrating particles, now termed *phonons*. Einstein was guessing. The structure of atoms was not discovered until 1911.

Yet Einstein's phonon serendipity hit the right note. Phonons are relevant to characterizing Bose-Einstein condensate and other exotic thermodynamic phenomena. A phonon is a quasiparticle that represents the excited state which brings electrons together into an entangled Cooper pair.

Phonons were more formally conceptualized by Russian physicist Igor Tamm in 1932 as the particle form of wave/particle fields performing a certain vibration.

¤ ✧ ¤

The 1st string theory was proposed in 1926, in the swirl of the quantum revolution, then lost, only to be rediscovered decades later.

Early string theories worked for bosons, leaving fermions out. Bosonic string theory posited 26D. The extra spatial dimensions were necessary to remedy inconsistencies.

In 1968, Italian physicist Gabriele Veneziano was working with the *Euler Beta function*: an equation used to characterize scattering amplitude. He noticed that it could explain particle reactions involving the strong nuclear force. Others then realized the equation made sense to them when they thought of subatomic particles as connected by little strings, vibrating their very little hearts out.

The concept was controversial. Shortly thereafter, the Standard Model swept aside strings as the great explainer of particle interactions.

But interest in strings abided. Variant models were devised, all constrained to bosonic strings. A major flaw of bosonic string theory was it positing the *tachyon*: a superluminal particle with imaginary mass.

In 1971, French physicist Pierre Ramond worked a 2D model that included fermions by generalizing the Dirac particle equation into string form. As reality is known to have more than 2 dimensions, Ramond had built a starter kit.

With André Neveu and John Schwarz in tow, Ramond's concept was extended into a 10D supersymmetry string theory, aka *superstring theory*.

> I think all this superstring stuff is crazy and is in the wrong direction. ~ Richard Feynman

In 1995, American particle theorist Edward Witten, who had been fiddling strings for over a decade, had a vision of unifying the variant quantum field theories. The result was

Witten's *M-theory*, which postulates 11 dimensions of spacetime: 10 of space and 1 of time. *'M'* stood for *membrane.*

M-theory is naturally extensible in the number of dimensions. In M-theory, a single brane string may be a membrane of greater dimensions.

American string theorist Joseph Polchinski and Czech string theorist Petr Horava independently extended M-strings into higher-dimensional objects: *D-branes* (a Horava term). Among other things, D-brane theory attempts to characterize string endpoints.

D-branes add rich mathematical texture to M-theory, paving the way for constructing more intricate cosmological models with greater explanatory power. Numerous *brane-world* (brane cosmology) models have emerged.

♎ Stringy Liquids ♎

String theory has been derided by partisans for its lack of track record.

> String theory cannot give any definite explanations of existing knowledge of the real world and cannot make any definite predictions. ~ American theoretical physicist Daniel Friedan

One experiment tried to simulate the trillion-Kelvin conditions just after the supposed birth of the universe, by smashing gold ions together at 99.99% of the speed of light. Instead of the expected gaseous plasma, a hot quark soup with liquid-like behavior was produced.

Another experiment confined lithium atoms and cooled them to as cold as practically possible: within 1×10^{-8} K, barely above absolute zero. The behavior was also liquid-like.

Both these liquids at opposite extremes exhibited collective behavior: flowing with the lowest possible viscosity. String theory successfully modeled these phenomena as strongly coupled particles linked by ripples traveling extra-dimensionally. In contrast, the Standard Model cannot account for these stringy liquids.

> A new truth always has to contend with many difficulties. If it were not so, it would have been discovered much sooner.
> ~ Max Planck

☙ Planck Units ❧

> All matter originates and exists only by virtue of a force which brings the particle of an atom to vibration and holds this most minute solar system of the atom together. We must assume behind this force the existence of a conscious and intelligent mind. This mind is the matrix of all matter. ~ Max Planck

At the heart of string theory and quantum mechanics lies an inherent ambiguity: the impossibility of probing space when the distance considered heads to the infinitesimally small. The idea of space becomes meaningless when it becomes smaller than what physicists term *Planck length*.

Under contract to find a way to get the most luminance out of a light bulb using the least energy, German physicist Max Planck originated quantum field theory in 1900. Studying thermal radiation, Planck discovered that energy was always emitted or absorbed in discrete units: *quanta*. Planck expressed this relation in the simple equation: $E = hv$, where E is the energy of a wave, v is the frequency of the radiation, and h is a very small number that came to be called the *Planck constant* (aka *Planck's action quantum*), which is 6.626×10^{-34} joule/second in meter/kilogram/second units, with just a bit of uncertainty.

Though energy is utterly wavelike, the work of energy only manifests in the discrete amounts specified by Planck's quantum of action. Phenomenality is confined to granular form – particulate puppets on the strings of waves.

Planck length is a measure derived from Newton's gravitational constant, the speed of light in a vacuum (c), and Planck's constant. Planck length is 1.616199×10^{-35} meters.

Comparing Planck length to the size of a bacterium is like comparing a bacterium to the size of the known universe. Essentially, Planck length is the theoretical minimal spatial distance.

Planck time is the time required for light in a vacuum to travel a single Planck length; the shortest sprint imaginable, at 5.391×10^{-44} seconds. Present physics theories have nothing to say about the universe younger than a Planck time instant. Physicists hold out hope that a theory of quantum gravity might illuminate that moment.

Planck length and Planck time are one system of *natural units* used in physics, known as *Planck units*. Planck units serve to mathematically normalize the fundamental quantities of matter, thereby elegantly simplifying algebraic expressions which express properties of elementary particles.

If matter's basic nature is of tiny resonances, beyond Planck's length, the nature of space cannot be probed. Under string theory, space itself appears an emergent property, appearing from spacelessness.

If quantum physics hasn't profoundly shocked you, you haven't understood it yet. ~ Niels Bohr

✄ Entanglement ✄

That one body may act upon another at a distance through a vacuum, without the mediation of anything else, by and through which their action and force may be conveyed from one to another, is to me so great an absurdity that I believe no man who has in philosophical matters a competent faculty of thinking can ever fall into it. ~ Isaac Newton

Basic notions in physics depend upon a time continuum: cause preceding effect. A principle of locality must exist for cause and effect to work.

If causality is kicked aside, such as with simultaneous ("spooky") action at a distance, locality is violated.

Nonlocality is a well-established fact. Quantum entanglement at a distance has repeatedly been demonstrated. In one experiment, a single photon entangled nearly 3,000 atoms.

The fundamental properties of chemistry rely upon entanglement. Solids form, and retain their solidity, via quantum entanglement of the electrons in the material.

Quantum entanglement is a strange and non-intuitive aspect of the quantum theory of matter, which has puzzled and intrigued physicists since the earliest days of the quantum theory.
~ American physicist Leon Balents

Superluminal communication presents a challenge to theoretical physics that has not been resolved. It is a challenge that can never be met by insisting upon the universe as a 4D closed system; the basic axiom which Newton and Einstein were so confident of.

Entanglement demonstrates that time, as well as space, is *emergent*: constantly coming into being, as contrasted to preexisting and incrementally evolving, as it appears to us.

Space and time will end up being emergent concepts; i.e. they will not be present in the fundamental formulation of the theory and will appear as approximate semiclassical notions in the macroscopic world. This point of view is widely held in the string community. ~ Israeli theoretical physicist Nathan Seiberg

String theory implies countless possible vacuum states; that is, spatial constructs with different properties. From vacuum a spacetime emerges into the universe which we experience. This seeming science fiction is a science fact.

℀ Entanglement Out of Time ℀

A practical pointer to time as an emergent property occurs by entangling particles that don't exist at the same time. Nonlocality can also be nontemporal.

A scheme termed *entanglement swapping* – chaining entanglement through time between subatomic particle pairs – has been experimentally demonstrated using 4 photons.

First, entangled photons 1 and 2 are created by zapping a special crystal with laser light. The polarization of photon 1 is measured while 1 and 2 are entangled. Photon 1 is destroyed by the measurement.

Then the entangled pair of 3 and 4 are created. Next, an entangling measurement of photons 2 and 3 is made even as it absorbs and destroys them. Finally, the polarization of photon 4 is measured.

Thanks to unavoidable uncertainty, unobserved photons on the fly are simultaneously polarized vertically and horizontally. Measuring a photon collapses its uncertainty wave function such that it will always be found to be either horizontally or vertically polarized.

Even though there is no moment in time when photons 1 and 4 coexist, they show entanglement by their measured polarization matching.

> There is no moment in time in which the 2 photons coexist, so you cannot say that the system is entangled at this or that moment. ~ Israeli physicist Hagai Eisenberg

ೞ The Information Paradigm ෬

> A century of physics has taught us that information is a crucial player in physical systems and processes. Regard the physical world as made of information, with energy and matter as incidentals. ~ Israeli theoretical physicist Jacob Bekenstein

As matter is made of energy, and energy is nothing more than an immaterial concept, theoretical physicists progressed to considering whether there is a meaningful essence of existence. Their answer: information.

♉ Information Theory ♌

> The fundamental problem of communication is that of reproducing at one point, either exactly or approximately, a message selected at another point. ~ Claude E. Shannon

American mathematician Claude E. Shannon founded information theory in 1948 with a seminal paper on "a mathematical theory of communication." Shannon sought to comprehend the fundamental limits of signal processing and communications operations, such as data compression. Information theory has since been applied to several sciences, including physics, genetics, evolutionary biology, intelligence physiology, and ecology.

Shannon treated *information* as meaningful content received from a message transmission. The potential problem is noise. Shannon used the term *entropy* to ascribe the inherent uncertainty of received (destination) information equating to transmitted (source) information.

Bizarrely, information theory ignores the most important aspect of information: that there must be both a source and a perceiver of it. With their ersatz information paradigm, physicists cannot pick the lock of significance because they overlook the key.

¤ ✧ ¤

Everything in our reality is made up of information. ~ Vlatko Vedral

Slamming into a dead end in trying to suss the nature of energy, Shannon's information theory washed up on the shores of physics, with the notion that the universals of the universe were themselves *information*. Gerard 't Hooft proposed the idea in 1993. Instead of a law where energy was only transmuted in a closed cosmos, never created nor destroyed, the dicta morphed into a universe where no information is lost – a conceptual absurdity taken seriously.

Bytes of reality may have their bits scrambled beyond practical redemption, but alchemic physicists believe it is at least theoretically conceivable that some accounting trick might right a digital Humpty Dumpty. Otherwise, all is lost.

The whole structure of everything we know would disintegrate if you opened the door even a tiny bit for the notion of information to be lost. ~ American theoretical physicist Leonard Susskind

⚮ The Holographic Principle ⚭

Holography is a huge leap forward in the way we think about the structure and creation of the universe. ~ Dutch theoretical physicist Kostas Skenderis

With energy cast aside for sheer data as the source of cosmic construction, the *holographic principle* emerged, with existence as an information structure painted on a cosmological canvas; like a hologram, where information is both distributed and entangled. Gerard 't Hooft concocted the concept in 1993. Leonard Susskind gave it wings via strings in 1995 with a string theory model. Inscrutably, all that is needed to encode the HD richness of the 3D world is a mere 2 dimensions, as if Nature could be written on a piece of paper.

> Most physicists believe that the degrees of freedom of the world consist of fields filling space. Instead of a 3-dimensional lattice, a full description of Nature requires only a 2-dimensional lattice at the spatial boundaries of the world. The world is 2-dimensional and not 3-dimensional as previously supposed. ~ Leonard Susskind

Remarkably, what goes unremarked is *energy–data equivalence*. The ordered patterns that energy take are inherently the informational content of existence. As energy creates forms greater than 3D spatially (e.g., virtual particles), the holographic principle is hooey. Treatment of black holes under this hypothetical regime illustrates the folly.

> Singularities imply information loss. ~ Gerard 't Hooft *et al*

Radiation from black holes, predicted by Stephen Hawking in 1974, has been shown to exist. The randomness of Hawking radiation obliterates information. For years, Hawking had no problem with that idea. But once quantum information theory became popular among physicists, Hawking contradicted himself, with the hedge that information is preserved if one waits for the black hole to completely evaporate. In Hawking's imagination, the 'information' rent by a black hole reintegrates in a Humpty Dumpty manner once it dries up.

> The physics of black holes – immensely dense concentrations of mass – provides a hint that the principle might be true. Studies of black holes show that, although it defies common sense, the maximum entropy or information content of any region of space is defined not by its volume but by its surface area. ~ Jacob Bekenstein

In bounding the universe to 2 spatial dimensions under the holographic principle, by definition, no energy can actually go into a black hole; doing so would defy the 2D limit, as well as information being lost. So, the matter drawn to a black hole simply piles up on its horizon (surface) as a sheet of increasingly dense information entropy.

When matter falls into a black hole, the increase in black hole entropy always compensates or overcompensates for the "lost" entropy of the matter. ~ Jacob Bekenstein

⌑ ✧ ⌑

The holographic principle requires a medium for the canvas. The concept of a cosmic canvas is not new.

⌖ Aether ⌖

Aether has a long history. 2,300 YA, Aristotle proposed *aether* as a divine substance that makes up the heavenly spheres and bodies. Aether was the 5th element; the basic 4 being earth, water, fire, and air.

As physics solidified through the centuries aether held its own, being the medium by which electromagnetism propagated and through which light traveled.

The concept of luminiferous aether had its heyday in the 19th century. Particularly popular with British physicists and mathematicians was the *Victorian Theory of Everything*, whereby every atom was soaked in aether. Lord Kelvin developed the *Vortex Theory*, based upon mathematical knots, whereby atoms were vortices in the aether.

There can be no doubt that the interplanetary and interstellar spaces are occupied by a substance. ~ James Clerk Maxwell on cosmic aether in 1870

In 1887, American scientists Albert Michelson and Edward Morley set out to show the aether flow by measuring light patterns. In one of the most famous failed experiments of all time, the aether wind didn't blow. Nevertheless, the *Michelson-Morley experiment* shed some light for Albert Einstein, who came up with special relativity in its wake.

With the holographic principle, the aether is back, though shrunk to a fantastically flimsy sheet in a 3D universe, with information digitally tucked away in a fluctuating foam, with the bits miraculously shorter than Planck length, which is the theoretical limit of spatial measurement.[*] Foam at this

---

[*] An electron is 10^{15} larger than Planck length.

resolution is eminently convenient, as it puts any prospect of proof out of reach.

At Planck length, the structure of spacetime becomes dominated by quantum effects. It is theoretically impossible to determine the difference between 2 locations less than 1 Planck length apart.

The human imagination incessantly demonstrates that actuality is no bar to abstraction. Freewheeling minds roam all the way to the foam, where physics gives way to Dada philosophy.

> At a very, very small scale, there are these little foamlike fluctuations. ~ American astronomer Nicholas Suntzeff

ᔆ Entropy ᔆ

> In classical physics, concepts of entropy quantify the extent to which we are uncertain about the exact state of a physical system at hand or, in other words, the amount of information that is lacking to identify the microstate of a system from all possibilities compatible with the macrostate of the system. If we are not quite sure what microstate of a system to expect, notions of entropy will reflect this lack of knowledge. Randomness, after all, is always and necessarily related to ignorance about state.
>
> In quantum mechanics, positive entropies may arise even without an objective lack of information. This entropy arises because of a very fundamental property of quantum mechanics: *entanglement*. ~ German physicist Jens Eisert *et al*

The *volume* of a thermodynamic state is an intensive property that equals an examined system's volume per unit of mass. Volume is a function of state, and is interdependent with other intensive properties, such as temperature and pressure. But the entropy of a region within a thermodynamic system – its *volume scaling* – is an extensive property.

Whereas an *intensive property* is independent of system's size or materiality, an *extensive property* is dependent upon such system characteristics. The basic difference is how a property is stated. Whereas extensive properties are (absolute) measures of a system, intensive properties tend to be ratios of certain system attributes.

Intriguingly, for typical ground states, volume scaling follows a simple, often logarithmic, *area law*. The scaling of ground state in a designated region is merely linear to its boundary area. This owes to inherent entanglement in a ground state.

¤ ✧ ¤

By terminating existence at its singularity (past the event horizon), a black hole can be said to hide information. Following the holographic principle, *entropy* is a measure of missing information, and so involves a level of uncertainty. From this perspective a black hole is chock full of entropy.

Black hole entropy is considered the amount of entropy that a black hole must have for it to comply with laws of thermodynamics, as interpreted by an observer outside the black hole. But the entropy of a black hole follows the volume scaling of a typical ground state, not that of a thermodynamic system. Black hole entropy increases only as fast as its surface area increases, not its volume.[*]

From an information theory standpoint, the area law for black hole horizon entropy is analogous to measuring how many files are in a filing cabinet drawer based upon the surface area of a drawer, rather than how deep the cabinet goes.

Cool helium to 2.17 K and it becomes a superfluid. Owing to the entanglement incurred in this state, measuring the entropy of a puddle of supercooled helium follows the same area law as for a black hole or a ground state: sussing surface area is sufficient.

So-called area laws for quantum entanglement are widespread. ~ Italian theoretical physicists Paolo Zanardi & Lorenzo Campos Venuti

---

[*] Note that the volume of a black hole, in being a singularity of mass/energy, is necessarily infinite, which means that black holes are beyond thermodynamics, and everything else in existence, for that matter.

⚄ Unity ⚄

By shifting viewpoint to information instead of energy, physicists vest hope in the holographic principle to point the way to a unified theory of everything, especially reconciling relativity with quantum mechanics.

> The holographic principle is a signpost to quantum gravity.
> ~ American string theorist Raphael Bousso

That signpost is illegible.

> The general consensus is that the amount of information that Nature can store in a very tiny volume of space and time is gigantic, it is so tremendously big that there is no hope whatsoever to follow this thing with any rigorous mathematics at all.
> ~ Gerard 't Hooft

In 1997, Argentinian theoretical physicist Juan Maldacena developed, via dazzling math, a version of the holographic principle which reconciles the paradox between black holes and string theory, via M-theory D-branes. A mathematical minimalist, he was able to do so by upping the number of necessary spatial dimensions to 5, a cardinal violation of the holographic principle's minimal spatial dimensionality (a 3D world from a 2D construct). Maldacena admits that the holographic principle is just flashy hokum.

> It is not clear how to define a holographic theory for our universe; there is no convenient place to put the hologram. So far, no example of the holographic correspondence has been rigorously proved – the mathematics is too difficult. ~ Juan Maldacena

⚄ Gravity & Thermodynamics ⚄

Dutch theoretical physicist Erik Verlinde suggests that gravity arises once spacetime has emerged.

> Gravity is explained as an entropic force caused by a change in the amount of information associated with the positions of bodies of matter. ~ Erik Verlinde

Working out gravity using the holographic principle, Verlinde mathematically demonstrated that Newtonian thermodynamic and gravitational laws, and Einstein's general relativity, naturally arise at appropriate scales of observation.

Indian theoretical physicist Thanu Padmanabhan agrees. Padmanabhan showed how Einstein's equations describing gravity can be rewritten in a form that makes them identical to the laws of thermodynamics. Gravity turns out to be an emergent metric of spacetime. Considering that gravity dimensionally defines the characteristics of spacetime, Padmanabhan's common-sense conclusions proverbially sewed silk from the sow's ear of holography.

The underlying description of gravity may lie in a microstructure made up of "atoms of spacetime." ~ Thanu Padmanabhan

The explanatory power of the holographic principle in deriving observed forces does give pause to wonder about what is behind the veil of Nature. If existence is information in action, the cosmos must be a coherent illusion coming from an intelligent source.

The whole 3-dimensional physical world is an illusion born from information encoded elsewhere. ~ Canadian theoretical physicist Mark Van Raamsdonk

Special relativity creates a universal speed limit: nothing can travel faster than the speed of light. In other words, causality is universal. This is *locality*.

Electromagnetism and gravity work at a distance but in concert; hence *fields*. All the known forces of physics work at a distance, including the nuclear forces, though the distances there are subatomically short. Still, locality is maintained.

But entanglement of particles is also known. Entangled quanta respond to each other instantaneously. Alter an entangled photon, and its twin instantly changes with it. This is n*onlocality*. What can explain it? Only hidden dimensions.

೫ Holistic Dimensionality ೪

If it is possible that there are extensions with other dimensions. Spaces of this kind, however, cannot stand in connection with those of a quite different constitution. Accordingly, such spaces would not belong to our world, but must form separate worlds. ~ German philosopher Immanuel Kant

Our experience of Nature is ostensibly limited to 4 dimensions: 3D space and a time vector. 4D is a natural province for chemistry, as elemental and molecular interactions may be sufficiently understood in the ambient; *sufficient* in the sense of workable predictability.

Historically, 4D had been the home stomping grounds for physics. But physics is a discontented discipline in its striving to be a know-it-all. Modern physics peeled back the 4D veneer and delved deeper, where it encountered uncomfortable unknowns.

There are known knowns; there are things we know we know. We also know there are known unknowns; that is to say, we know there are some things we do not know. But there are also unknown unknowns: there are things we do not know we don't know. ~ American bureaucrat Donald Rumsfeld

4D exists in a higher-dimensional envelope. Extra dimensions (ED) exist. *Holistic dimensionality* (HD) is the realm of actuality as it is, not as we perceive it.

$$HD = 4D + ED$$

ED has long been regarded as the "spirit plane." There are many spiritual attributions throughout history for an HD existence, focused on ED as a piece not to be left out of the picture in perceiving Nature.

Some live an HD actuality. The spiritually sensitive feel the energy of their chakras flow. ED-resident presences may be sensed 4D. Those less aware easily dismiss such.

¤ ✧ ¤

There is nothing sacrosanct about dimensionality. ~ Paul Wesson

In declaring gravity as a curvature of spacetime, Einstein unwittingly made the case for extra dimensionality. If gravity distorts spacetime, there must be more than 4 dimensions. Warping 3D space geometrically requires at least 1 extra spatial dimension; an obvious implication which Einstein refused to acknowledge out of religious conviction for naïve realism.

Further, general relativity predicted black holes; a most massive curvature that appears planar only because visibility cannot extend beyond a black hole's event horizon. Surely the massive volumes of matter/energy sucked into a black hole go somewhere. Such singular sinks must be beyond 4D.

Even on the rim of black holes are HD dynamics. *Hawking radiation* is the predicted black-body radiation emitted from a black hole's event horizon. The gravitational force is so strong as to spontaneously produce 4D photons (light energy), and even mass-laden particles. Small-scale experimentation has demonstrated this dynamic. This *black hole evaporation* is necessarily an example of HD interaction.

> Dimensions must not be multiplied beyond necessity, but a multiplicity of dimensions is absolutely unavoidable. ~ French - American mathematician Benoît B. Mandelbrot

In 1921, German mathematician and physicist Theodor Kaluza extended general relativity to a 5D spacetime, thereby unifying electromagnetism and gravitation, the forces relevant to general relativity. Einstein was impressed by the math, but was discomforted with the idea of unseen dimensions, as were many of his contemporaries. To them the notion of ED smacked of mysticism. Welding time to 3D space was as much tinkering as Einstein could abide.

> The non-mathematician is seized by a mysterious shuddering when he hears of "four-dimensional" things, by a feeling not unlike that awakened by thoughts of the occult. And yet there is no more commonplace statement than that the world in which we live is a 4-dimensional spacetime continuum. ~ Albert Einstein

¤ ✧ ¤

A cosmological issue is how the universe developed its HD existence with 4D as a predominant scale. Somehow our 4D diverged from ED while leaving an integrated HD.

One proposal has been that a quantum energy fluctuation near the start of the emerging universe broke the symmetry of the dimensions, with 3 of space to blossom in scale, resigning other spatial dimensions to less than Planck length; meanwhile, leaving time as the eternal illusion of passage.

In 1926, Swedish physicist Oskar Klein extended Kaluza's 5D concept with *compactification*. Klein proposed that the 4th space dimension was *curled up*: existing at a wavelength beyond measure (i.e., shorter than Planck length).

The *Kaluza–Klein theory* is exemplary of models extending the number of dimensions and defining ED characteristics, such as whether the extra dimensions are curled or flat.

In their search for a unified field theory, Lisa Randall and fellow theoretical physicist Raman Sundrum developed 2 braneworld models in 1999 (the *Randall–Sundrum models*). Both braneworld models construe a universe of 5 dimensions using warped geometry, with the force of gravity (via gravitons) emanating from the 5th dimension. RS1 has a small $(10^{-31}$ cm) 5th dimension, but not compact; instead, merely phase-shifted from 3D space. RS2 was bolder, with a warped 5th dimension that is infinitely large.

As Randall–Sundrum suggests, the path of peeling ED from 4D does not necessitate compactification. The problem remains in properly characterizing the geometry of HD.

¤ ✧ ¤

However HD may be shaped – with ED either compact, small, or large – all modern physics' models require more than 4 dimensions. Though relativity and the Standard Model (SM) of quantum physics require extra dimensions, theorists of both those models have often stayed silent about ED, spooked by the implications of supporting spooky action dimensionally.

SM is replete with examples of higher dimensionality. The separation of matter (fermions) from fundamental interaction forces (bosons) requires extra dimensions.

3 dimensions of space provide only sufficient mathematical room for fermions. No dimensional room is left over for bosons to instantaneously act. If limited to 4D, bosonic fields have no means to instantaneously impose their coherent forces on fermions.

The uncertainty principle itself is an extra-dimensional cry for help. Explaining nondeterministic quanta in a 4D harness can't be done. Actuality as a probability matrix must be a constrained view, leaving the basic gyre unexplained. In other words, exhibition of inherent uncertainty suggests an underlying determinism, as light refraction illustrates.

Wave-based interpretation of particle momentum is hard to adjudge as a mere 4D mathematical construct when quantum entanglement, as well as much less exotic effects, demand an extra-dimensional vehicle. Nonlocality is sheer magic if existence is constrained to 4D.

Virtual particles present a most pressing case for HD. These little bits make constant cameo appearances 4D. There is no other way to explain their comings and goings except via extra-dimensional shifts. Virtual particles must have active HD lives.

Virtuality as fleeting is a peephole 4D perspective. The notion that particles pop in and out of existence is ridiculous, beginning with the issue of where they could possibly come from. Yet that is the stance that SM modelers typically take for virtual particles.

Further, SM posits ghost fields that give rise to virtual particles – a mathematical necessity treated with contempt by not granting the possibility that those fields are actual.

Ghost fields are as unsavory to SM physicists now as uncertainty was to Einstein. The incongruity of begrudging acceptance of virtual particles, while maintaining denial for the ghost fields from which the particles supposedly emanate, is virtual irrationality: self-contradiction between cause and effect.

The energy of virtual particles – vacuum energy – is figured to be enormous. Yet vacuum energy appears not to impact 4D in a rip-roaring way. That can only make sense if vacuum energy is mostly tucked away ED, with fleeting 4D appearances via virtual particles. But virtual particles do carry considerable 4D heft.

An electron is an elemental fermion, but it is overstatement to say that an electron has no constituents. An electron travels with an entourage, dragging a cloud of virtual particles around. The virtual cloud literally defines the electron, contributing to its mass, volume, and charge.

That the electron has an HD existence is known for several electromagnetic quantum phenomena, including near-field effects, such as induction, the Casimir effect and van der Waals force, and the quantum energy fluctuation known as the Lamb shift.

The photon is another self-standing elemental, a light-giving boson well known for its HD qualities. Virtual photons are behind static electricity (*Coulomb force*).

Bodies in motion create quantum vacuum fluctuations, causing photons to be emitted while dissipating the body's motional energy. This is an example of the *Casimir effect*, which characterizes physical forces arising from a quantized field.

> The Casimir force can, as a function of distance, oscillate between attractive and repulsive, and it can be tuned by application of an external magnetic field. ~ American theoretical physicist Frank Wilczek

Magnetic fields are caused by the exchange of virtual photons via vacuum polarization. Magnetricity flows like electricity.

When vacuum polarization happens, a background electromagnetic field produces virtual electron–positron pairs which change the charge and current distributions that generated the original electromagnetic field. Vacuum polarization is the self-energy of the photon. Generally, a particle's self-energy is its contribution of energy or effective mass in HD interactions.

The strong nuclear force between quarks is an interaction outcome of virtual gluons. Likewise, exchange by virtual W & Z bosons results in the weak force.

All these expressions of virtuality poignantly prove that Nature is a front propped up by extra dimensions.

¤ ✧ ¤

Spontaneous emission is the process where an atom or molecule transitions from an excited state to one with a lower energy, emitting a photon as a tell. This everyday experience is another HD event.

> Extra dimensions really exist. They're part of Nature. ~ Edward Witten

In the macroscopic world, photosynthesis and animal sensory perception work with astonishing efficiencies, as do other natural processes at the subcellular level, including genetics. Such energy economies in organic processes are HD dynamics.

> There is something beyond that which we call the visible universe; and that individual consciousness is in some mysterious manner related to, or dependent upon, the interaction of the seen and unseen. ~ Scottish physicists Balfour Stewart & Peter Tait in 1874

♋ 3D ♌

> Space is nothing else than the form of all phenomena of the external sense, that is, the subjective condition of the sensibility, under which alone external intuition is possible. ~ Immanuel Kant

No physics theory demands 3 dimensions as the nominal spatial construct. Why then does existence appear in 3D?

Nature's love of diversity withstanding, economy is a cardinal facet of existence. All of physics pronounces it: from classical thermodynamics to quantum mechanics. Chemistry cogently coheres about energy parsimony guiding reactivity, despite the plethora of ways in which reactions may transpire.

Time is a single vector. No one questions why there are not 2 dimensions of time, because it is inconceivable how that would work.

Following Occam's razor, the snappiest answer for 3D space as spot-on is adequacy: 3 dimensions are the fewest that can create a rich life experience.

Extra dimensions may be necessary to explain quantum and cosmological mechanics, but ambient 3D is opulently textured in a way that flatland (2D) could never approach, at the expense of only a single extra dimension.

Conversely, experiencing more than 3 spatial dimensions would create an entirely new level of complexity in experiential existence, and in so doing contravene Nature's proclivity for economical sufficiency.

ಬಿ Theory of Everything ಜ

From time immemorial, man has desired to comprehend the complexity of Nature in terms of as few elementary concepts as possible. ~ Abdus Salam

Physicists have long sought an umbrella theory that explains and connects all known physical phenomena, along with the predictive power to anticipate the outcome of any experiment within the physical realm – in other words, the formula to make the world deterministic.

Archimedes was perhaps first in describing Nature by axiomatic principles and using them to deduce new results. The path of theoretical unification has been dreamt by many of his successors: Democritus, Newton, Laplace, Einstein, and others.

Some of the most recent attempts are string theory and M-theory, though those are more particularly oriented toward an approach for producing a unified theory, rather than being a completed theory unto themselves.

A pivotal struggle has been unifying the 3 forces understood at the quantum level (strong, weak, electromagnetism) with gravity, which is incidental at the quantum scale, but long proved tricky to bottle in equation form at that level, owing to the infinities that arise. Relativity reveals a bendable 4D spacetime, whereas quantum mechanics has treated spacetime as unrealistically rigid.

All unification theories suffer from the same philosophical problem: the barrier between the continuity of the field and the discreteness of the quantum. Any quantum-based theory of physical reality must explain everything in terms of the discrete nature of particles, while any relativity-based theory must proceed from the continuity of the spacetime structure. ~ American theoretical physicist James Beichler

৶ Wormhole Entanglement ৬

The intrinsically quantum phenomenon of entanglement appears to be crucial for the emergence of classical spacetime geometry. Spacetime is just a geometrical manifestation of entanglement. ~ Mark Van Raamsdonk

Canadian theoretical physicist Mark Van Raamsdonk first proposed a way to connect quantum theory with relativity in 2009. The key insight involved entanglement of quantum fields, which has been observed.

Neighboring fields are typically more entangled than those farther away. This relative locality intimates that entanglement plays a role in the geometry of spacetime.

The continuity of spacetime, which seems to be something very solid, could come from the ghostly properties of entanglement. If you change the pattern of entanglement, you also change the geometry of spacetime. ~ Juan Maldacena

In the large, entanglement of quantum fields via flexible spacetime reconciles relativity with quantum mechanics. But explaining entanglement specifically entails synchronicity, which requires an instantaneous channel through spacetime to allow "spooky action at a distance." Such a shortcut is now known as a *wormhole*; a term coined by American theoretical physicist John Archibald Wheeler in 1957. The idea of wormholes is older than the word.

Hermann Weyl posited a wormhole theory in 1921, as a way to reconcile relativity with electromagnetic field theory. Einstein and Rosen proposed wormholes in 1935 as a means by which disparate black holes were connected, either in this universe or to other universes. This theoretical observation was an extension of general relativity, which renders the fabric of spacetime riddled with wormholes.

Wormholes act as the conceptual conduit of entanglement. That they exist in actuality has been repeatedly shown in experiments demonstrating quantum nonlocality.

Wormholes demonstrated by quantum nonlocality and hypothetical black hole wormholes are the same thing, just on a vastly different scale. This is simply scale-invariant uniformity; the physics of the universe behaving consistently.

The flexible geometry of spacetime at the macroscopic level emerges from quantum entanglement, which is mathematically characterized as a tensor network. A *tensor* is a geometric object that embodies a localized locus of relations. Tensor networks are used to describe various systemic entanglements, from quantum mechanics to holographic information storage to brain function. The upshot: entanglement defines the geometry of spacetime.

> Our best description of the past is not a fixed chronology but multiple chronologies that are intertwined with each other.
> ~ American theoretical physicist Jordan Cotler

ஐ Matryoshka Transition ରଃ

That Nature imposes an intricate order is obvious. How it does so is the quest of science. Much mystery remains.

Existence incessantly emerges from the quantum realm to the everyday, where classical physics applies, through 2 mechanisms: decoherence and perception.

The type of measurement determines whether a phenomenon will appear as quantum or classical. Quantum effects may be inferred using equipment. To the naked eye, such small-scale specificity is beyond observation.* Hence, the world always appears classical.

Existence at the quantum scale is always coherently entangled: waves of possibilities (superpositions) which emerge at every instant via *decoherence*: environmental interactions that are thermodynamically irreversible. The emergence of observable phenomena requires localization and quantization, which appears as an absence of entanglement. Decoherence generates diversity.

௸ Superconductivity Seam ௹

The emergence of complexity is often tied to novel forms of collective behavior driven by strong interactions. Unconventional superconductors are archetypical examples for materials in which the competition and entwining of collective forms of

---

* Humans cannot consciously perceive quantum phenomena. Whether other life can is perhaps unknowable.

behavior give rise not only to a complex phase diagram, but also to unexpected properties that have resisted all attempts of a theoretical explanation. ~ German American physicist Dirk Morr

Whereas *superconductivity* is of coherently-paired electrons with opposite spins and momenta, *charge order* involves orderliness in the arrangement of electrons and holes (electron absences) that have the same spin and momenta. Antiferromagnetism relies upon charge ordering.

Charge order and superconductivity are 2 competing pairing tendencies which cannot be simultaneously satisfied. This mutual exclusivity is on display when an antiferromagnetic material gives way to superconductivity.

In between these highly-ordered states is a transition phase which defies characterization, as it does not correspond with the *Landau–Fermi liquid theory*, which is the accepted model of how metallic fermions interact at low temperatures. Quantum states cannot be accounted for in this transitional gap.

Discrete discontinuities exist between spatial scales, and at certain phase transitions, which defy theoretical explanation. The gap between the quantum and classical worlds, and the energetic path to superconductivity, are exemplary illustrations of inscrutable seams in the fabric of existence.

ଖ Paradigms ଓ

Our minds are filled with concepts. Templates for conceptual organization are *paradigms*. It is not that paradigms are in any sense "true" – no more than concepts are anything but convenient fictions. Instead, our minds strike a pose with paradigms and we come to believe that they are true because they are useful. So, we perceive that Nature follows patterns which couple time and space to produce systematic confluences of dynamic and domain which can be concisely conceptualized. Hence, the world before us appears comprehensible.

ℬ Systems ℭ

One of the basic assumptions of general systems theory embraces the concept of order – an expression of man's general need for imagining his world as an ordered cosmos within an unordered chaos. ~ Swedish systems theorist Lars Skyttner

Existence appears ordered into systems. A *system* is a set of interdependent or interacting components that are considered an integral whole. It is the connections or interactions of components within that determine a system's salient characteristics. Properties emerge from the connections themselves. A clock provides a measure of time; something none of its individual parts can.

A system is a theoretical incarnation of a tensor network. The relations between components in a system are as significant, if not more so, than the components themselves.

Biology is replete with synergies. Proteins interact with each other to form complex modules and express functionality that they individually are incapable of.

Hierarchy is inherent in all systems. While modules themselves may express emergent properties, their connective arrangements produce further novel synergies.

In appreciating that coherent patterns of energy underlie every appearance of physicality, each body constitutes a system. Perception is a system, in bringing order to disparate, disorganized inputs.

Systems are not confined to phenomena. Skills and beliefs are synergistic thought systems, built through conceptualization, which itself is a system.

The hierarchy of systems is endless. The proximate ecologies of bodies produce ecosystems with their own characteristics and dynamics. Thus, all systems are ultimately entangled, with emergent properties at every level.

Constructal law construes the relation between form and function – the evolution of tangible systems. *Gyres* revolve around dynamic interactions in systems. *Self-organized criticality* comes into play when cause and effect come a cropper

to threaten and disrupt systems. All involve the appearance of *causality*.

ॐ Causality ॐ

Causality lies at the interface between quantum mechanics and general relativity. ~ Austrian theoretical physicist Caslav Brukner

One event occurs, and another after it. The two are related or not; coincidence or cause and effect. Sussing circumstance is the mind's constant entertainment.

Causality is considered a philosophical concept but it is also an ever-present issue in physics, chemistry, and biology, where its interpretation forms the foundation of science.

Causation can be of many kinds. They form our ways of ordering our scientific understanding of the world, all the way from the reductive concept of cause as elementary objects exerting forces on each other, through to the more holistic concept of attractors toward which whole systems move, and to adaptive selection taking place in the context of an ecosystem. ~ South African logician George Ellis *et al*

Reductionism is the philosophical stance that even complex systems are simply the sum of their parts. An accounting of a reductionist system may be made solely by reference to its individual constituents.

It is difficult enough understanding something in isolation; comprehending ecology or a dynamic system is practically impossible. Hence, empirical science is inherently reductionist in its approach. Science embraces a methodology of simplification.

There are no reductionist systems in Nature. Hence, the simplification inherent in science is prone to false conclusions, biased by the limits of our perception and mentation.

Emergence embraces ways in which systems and patterns stem from a multiplicity of relatively simple interactions. Emergence is a top-down causality: a perspective of a larger order arising from atomic actions.

Emergence is reductionist only if a dynamic system can be accounted for by a tally of its constituents. That is never the instance.

If something is more than the sum of its parts, its coming into being is emergent but not reductionist. In this case, a cascade of cause and effect create more than the individual reactions involved would produce. Such is life, the ambient world, and the cosmos.

Interactions may occur at the same scale. Otherwise, in sussing the ultimate source of an action, order of incitement determines whether causality is bottom-up or top-down.

To perceive a flow from micro to macro scale is bottom-up. To instead sense a larger force working smaller bits is top-down: a teleology.

> You, your joys and your sorrows, your memories and your ambitions, your sense of personal identity and free will, are in fact no more than the behaviour of a vast assembly of nerve cells and their associated molecules. ~ English biologist Francis Crick

If, as Crick claimed, behavior is bottom-up, one cannot assign causality to intermediates. The causal originator would not be nerve cells or their minion molecules, but the fundamental building blocks of matter.

> Neuroscientists believe that neurons do indeed do real work. This is only possible if they act to channel and control the flow of electrons in neural axons – that is, if top-down causation takes place from the neuron to the electron level. ~ George Ellis

To admit that matter complexes – intermediates – determine events is to confess that the door of causality swings both ways. If all is bottom-up, quantum interactions determine everything that goes on.

There a serious problem arises when it comes to causality. Uncertainty pervades the quantum world, including the sequence of events. The best that can be done is to consider correlation a surrogate for cause.

> The weirdness of quantum mechanics means that events can happen without a set order. ~ Australian quantum physicist Jacqui Romero

> If quantum mechanics governs all phenomena, the order of events could be indefinite. ~ quantum physicist Fabio Costa

Evolution is a process where environmental interaction shapes genetic modification. Its causality is top-down.

As an example of causality, vision is blurry. At first glance, it may seem that photons strike light receptors that lead to mental integration of perceived patterns: a bottom-up approach. Instead, what is in view is mostly mental anticipation. We mostly see what we expect to see: top-down.

Top-down causality belabors the imagination to conceive of the emanating force; which is to say that coherence is at play in having events go a certain way.

Implicit in top-down causality is *determinism*: the doctrine that all events exemplify natural laws. The idea of scientific determinism led to *Laplace's demon*.

> We may regard the present state of the universe as the effect of its past and the cause of its future. An intellect which at a certain moment would know all forces that set nature in motion, and all positions of all items of which nature is composed, if this intellect were also vast enough to submit these data to analysis, it would embrace in a single formula the movements of the greatest bodies of the universe and those of the tiniest atom; for such an intellect nothing would be uncertain and the future just like the past would be present before its eyes. ~ Pierre-Simon Laplace in 1814

Laplace's demon bears more than passing resemblance to the acts of God espoused by sects of Protestant Christians who consider predeterminism de rigueur and free will illusory.

Scientifically, Laplace's demon was presumed banished by developments in thermodynamics, particularly process irreversibility as exemplified by entropy. But the demon dodged this bullet because of time's one-way arrow: cause and effect and reversibility are coincidental. Going back in time is irrelevant.

Determinism is damned by *chaos theory*, which proposes divergent outcomes from an initial condition in dynamic systems, where *distributed causality* reigns. Entanglement of events makes pinpointing initial cause practically impossible.

The *butterfly effect*, an example of chaos theory, illustrates distributed causality. The butterfly effect occurs when a small variation at a relatively small scale produces large perturbations in the behavior of a nonlinear system through a cascade of changes. The butterfly effect stems from a sensitive dependence on initial conditions. American meteorologist Edward Lorenz coined the term *butterfly effect* in 1963 from his study of weather patterns.

We attribute experienced event sequences as having cause and effect. This ambient sense of causality helps us understand how the world operates, and so anticipate events by their harbingers. In contrast, the energy of the universe is interwoven; an entanglement that renders causality a casualty of chaotic complexity. When it comes to causality, what works locally falters globally.

ℬ Constructal Law ℛ

The Constructal Law is a universal tendency toward design in Nature, in the physics of everything. This tendency occurs because all of Nature is composed of flow systems that change and evolve their configurations over time so that they flow more easily, to create greater access to the currents they move. ~ Romanian American mechanical engineer Adrian Bejan

Constructal law states that flow favors the path of least resistance. Any system, whether inanimate or alive, evolves to optimize efficient transport.

Constructal law is a tenet of all design and evolution in Nature. It holds that form arises to facilitate flow. Essentially, constructal law is a statement of energy economy shaping structure.

Tree structures – from lightning bolts to circulatory systems – are ubiquitous because they optimize the flow of energy or material from source to destination.* Alternately, as with plants, a tree structure affords ideal access flow – of sunlight and air.

The essential properties of water create the affinity for drops to coalesce and flow together in following gravity. So, streams flow into rivers, which invariably lead to the sea. The flow is more than mere topography.

---

* The concept of destination involves a teleology when actions have an intended end. Lightning, for instance, seeks an outlet of discharge.

Constructal law can be construed as a patterning mechanism for gyres. Constructal law is as readily applied to optimizing the arrangement of cells and organs as it is to cosmological components, from planets and star systems to galactic structures.

Constructal law also applies to economic and social systems, providing the basis for viewing exploitative corruption as a natural tendency: that wealth accumulation, and sustained poverty, are Nature's way of just desserts.

Nested hierarchical networks of systems are pervasive: from cells to ecosystems to social interactions. Hierarchies afford optimization of communication and controlled energy flow.

Both biological evolution and cultural evolution operate under a number of deep constraints. The majority of webs display a balance between integration of multiple signals and control over multiple targets under a bow-tie structural pattern. ~ Adrian Corominas-Murtra *et al*

In a *bow-tie* structure, a diversity of inputs (fan-in) are processed via a limited set of protocols, with various resultant outputs (fan out). Bow-tie architectures often appear in complex and self-organizing systems ranging from biology to technology. Bow ties provide heterogeneous stimulus-response via orderly processing. In operation, bow ties often mediate tradeoffs between efficiency and robustness.

Constructal design applies at every scale. Hence, constructal law stipulates a hierarchy of coherence at every level of existence.

For a finite-size system to persist in time (to live), it must evolve in such a way that it provides easier access to the imposed currents that flow through it. ~ Adrian Bejan

ᔆᓭ Gyres ᓂᖃ

Most systems in Nature are not in equilibrium; they exchange fluxes of matter or energy with their surroundings or undergo chemical reactions. ~ French physicist Jacques Prost & French biophysicist Jean-Francois Rupprecht

In the 5th century BCE, Greek philosopher Leucippus developed a theory of atomism. Less than a century later, Greek philosopher Democritus posited vortex motion as a law of Nature.

Copernicus's cosmic heliocentric model had a gyre flavor. In the mid-17th century, French philosopher René Descartes proposed planetary motion as occurring in whorls. At the smallest scale, Descartes declared that no empty space could exist, and so it must be filled with matter, in vortices of aether.

A conceptual convolution that kept coming around, the notion of gyres as explaining fundamental physical dynamics kept reappearing. Lord Kelvin's *Vortex Theory* had atoms swirling about in an aether soup. Maxwell used the gyre in formulating different theories of electromagnetism. Feynman correctly predicted quantum vortices in 1955.

In 2011, American microbiologist Erik Andrulis reintroduced the gyre as a paradigmatic framework for understanding the basic mechanics of Nature at every scale, from quantum to organic to cosmic.

> The gyre appears, a posteriori, to be a prime candidate for a core model of natural systems. ~ Erik Andrulis

A *gyre* is a vortex that forms a system interacting with its environment. A gyre is characterized by its structure, qualities, thermodynamics, and interactions.

A gyre's core is the position about which its matter and energy revolve. The shape of the swirl about a core may be various: spherical, conically cylindrical, or disk-like.

Gyres possess chirality: a handedness or certain direction. Chiral gyres may be observed at the quantum level, in various organic molecular structures, and at the galactic level. DNA is a chiral helix. Chirality is significant in several facets of biology.

Andrulis attributes organic qualities to gyres, including spontaneous self-organization, and a life cycle which originates when conditions are favorable. As an interacting thermodynamic entity, a gyre consumes matter and energy and exudes it in turn. A gyre selectively attracts and repels. If it can, a gyre self-regulates, adjusting itself to an internal balance (homeostasis).

Any cycle that exists in Nature – in physical, chemical, or biological systems – may be viewed as a gyre. ~ Erik Andrulis

Unlike the physical models preferred by physicists, gyres are inherently nondeterministic in their reliance upon environmental interaction; hence, not given to modeling, though facets of a gyre can be mathematically characterized. There are too many variables flowing in vortex behavior for a gyre to be entirely predictable. Mathematically, any set of equations with more than 3 independent variables yields unpredictable results.

∞ Self-Organized Criticality ∞

Consider a collection of electrons, or a pile of sand grains, a bucket of fluid, an elastic network of springs, an ecosystem, or the community of stock-market dealers. Each of these systems consists of many components that interact through some kind of exchange of forces or information... Is there some simplifying mechanism that produces a typical behavior shared by large classes of systems? ~ Dutch economist Henrik Jensen

Self-organized criticality (SOC) is a property of dynamic systems where a threshold exists that when passed sets off a substantial reaction. In having a *critical point* that unleashes a cascade consequence, SOC demonstrates the butterfly effect.

Self-organized criticality showed itself to researchers who looked at growing piles of rice grains in 1995 experiments at the University of Oslo (Norway). A rice pile was made by dropping grains of rice one on top of another.

In long-grained rice, SOC was shown by a single dropped grain triggering a sudden avalanche. Just before, the rice pile had reached a *stationary critical state*: stable, but on edge. That grain that launched an avalanche was the proverbial straw that broke the camel's back.

In a pile of short-grain rice, with a smaller aspect ratio than longer grain, SOC was not apparent. Thus, SOC is sensitive to system features. Systems with SOC tend to be slowly driven, without fixed equilibrium, but extended degrees of freedom, and are highly nonlinear.

Avalanches of snow show SOC. Layers inside snow form a fragile network of ice grains with lots of space in between. Some arrangements are an avalanche waiting to happen.

A key element of SOC involves *power-law distributions.* When the frequency of an event varies as a power of some attribute of the system, the frequency follows a *power law,* which is a consistent mathematical relationship between system attributes or behaviors. A growing rice pile follows a power law by having far fewer large avalanches than small ones.

SOC is scale-invariant: the laws that apply to a system's behaviors are not dependent on system size. SOC systems have a fractal aspect, of self-similarity in dynamic regardless of scale.

Another intrinsic facet of systems subject to self-organized criticality is *interdependence*: one feature dynamic may affect another. Whereas SOC is an observation about a system, interdependence is the inherent nature from which SOC arises. A cascade event can only occur because a system possesses interlocked features. Whereas the degrees of freedom a system has characterizes its malleability, interdependence creates conducive conditions by which a critical point may be exceeded via feedback loops.

Physics, geology, biology, and sociology are all entrenched with SOC systems. In a vast variety of contexts, complexity in structures and behaviors emerge from roots bound by relatively simple rules, however difficult discerning those rules may be. Self-organized criticality pervades Nature in myriad forms, as well as in complex man-made systems, such as economics and politics.

While prediction is an elusive oracle to find, mathematical models can give insights into the workings of SOC systems by correlating crucial factors, which can give a rough idea as to how close a system may be to a tipping point. Critical points are often presaged by behavioral changes that indicate a phase shift coming.

Stable systems tend to quickly recover after minor turbulences. There is a robustness that tends to equilibrium: *homeostasis* in organisms or biological systems/populations.

Recovery is engendered by intrinsic compensatory mechanisms awakened by feedback. But exceeding criticality in multiple facets can result in systemic stress, where recovery becomes problematic.

> Animal social groups are complex systems that are likely to exhibit tipping points – which are defined as drastic shifts in the dynamics of systems that arise from small changes in environmental conditions. ~ American ecologist Jonathan Pruitt *et al*

Species populations can experience catastrophic collapse in response to small changes in environmental conditions. Recovery can prove impossible.

A notable example of self-organized criticality in Nature is overfishing: the collapse of sardine stocks in California and Japan in the late 1940s, and Canadian cod in the 1990s. At low population densities, group dynamics falter. Difficulty in finding mates, and the breakdown of cooperative behaviors, such as forming schools for hunting or predator avoidance, can preclude recovery.

SOC in ecosystems is often broad-based, as populations of different species interrelate in a network of interactions, most obviously the local food web.

> Climate change often leads to local extinctions and declines by influencing interactions between species, such as reducing prey populations for predators. These shifting interactions may make even small climatic changes dangerous for the survival of plant and animal species. ~ American ecologist John Wiens

℘ Symmetry ℘

> A thing is symmetrical if there is something you can do to it so that after you have finished doing it, it looks the same as before. ~ Hermann Weyl

Aristotle argued that the stars were pasted on celestial spheres which moved in circular orbits. He was wrong, but the assumption behind the idea pervades physics: symmetry.

Through meticulous observation in the early 17th century, German astronomer Johannes Kepler discovered the elliptical orbits of planets about the Sun, albeit moving at varying speeds: faster closer in, slower further out. Yet all balances out. An imaginary line connecting planets to the

Sun traces out equal areas in equal times: what is now called the *conservation of angular momentum.*

Newton explained why this happens with his *universal law of gravitation*; a behavior grounded in symmetry: the force of gravity acting equally in all directions. Einstein's refinement of gravity was also founded upon a symmetry: the *equivalence principle.*

The cosmological principle and all of physics' conservation laws are built upon an assumption of symmetry within a closed system; something for which, in both instances (symmetry and system), there is no evidence.

In the 1960s, theoretical physicists Sheldon Lee Glashow, Abdus Salam, and Steven Weinberg independently discovered that they could unify electromagnetism with the weak force through a symmetry which became a keystone of quantum physics' Standard Model; but SM is reliant on violation of symmetry, not on its adherence. Breaking electroweak symmetry produced a prediction of the Higgs boson, which was discovered nearly a half-century later.

In producing quantum color charges, the symmetry of the strong force is what makes particles of matter possible. Yet all particles in Nature exist in colorless states: effectively white. Such equanimity breaks symmetry.

Symmetry underlies causality: that an action produces a reaction. Newton's 3rd law of motion – of equal and opposite reaction – applies as much to psychology and politics as it does to physics. Yet such symmetry is belied by the butterfly effect, from which great movement may be had from the most minuscule initial fluttering.

Chemistry is rife with butterfly effects and other asymmetries. Handedness plays a vital role in molecular interactions and associations. All organic molecules are chiral. Life depends upon asymmetry.

The basic laws of physics supposedly do not vary over time or space. Forces emanate with energetic equanimity. Yet such equality is not the story of existence.

Symmetry may give order to the universe, but breaking it is essential to anything happening. With matter ruling the roost, spacetime is defined by distortion – such is the gravity of the situation in Nature.

≈ Synopsis ≈

The most incomprehensible thing about the world is that it is comprehensible. ~ Albert Einstein

Einstein was simpleminded in considering the cosmos comprehensible. The human mind makes sense of what it can and leaves the rest behind; rendering an illusion of apprehension by omission of what is not understood or not believed.

Physicists have come to see that all their theories of natural phenomena, including the 'laws' they describe, are creations of the human mind; properties of our conceptual map of reality, rather than reality itself. ~ Austrian physicist Fritjof Capra

All theories amount to nothing more than mental maps. Whereas physics has progressed by questioning assumptions, the failure to understand the nature of existence lies in mental maps which take for granted what should not be assumed.

Physics Paradigms

➢ A *paradigm* is a conceptual template for comprehending related phenomena.

➢ A *system* is an integral whole which has internal, related components with various degrees of interaction or interdependency.

➢ *Causality* may be bottom-up or top-down. Perceiving micro-to-macro cause and effect is bottom-up causality. Conversely, a conclusion that an overarching force effects changes at lower levels is top-down causality.

Top-down and bottom-up causality are not mutually exclusive. As existence is constantly emergent, Nature proceeds using bottom-up causality. By contrast, that Nature exhibits order by operating via algorithms is top-down causality.* Top-down causality necessitates *coherence* as a force of Nature. The teleology of evolutionary adaptation is an example of top-down causality.

---

* Such natural algorithms are often called *laws of Nature.*

Constructal law claims that the design and evolution of everything in Nature – alive or inorganic – aims to optimize flow. Constructal law embodies top-down causality.

➤ *Reductionism* is the idea that a system is nothing more than the sum of its parts. While a piecemeal approach is convenient for basic understanding, no dynamic system is reductionist. Instead, dynamic systems are synergistic gyres. *Synergy* occurs in a system when its movements or effects cannot be explained via reductionism.

➤ A *gyre* is a vortex that forms a system interacting with its environment. The gyre is an apt paradigm for the dynamics of existence at every scale, from quantum to organic to cosmic. A gyre is beyond mathematical modeling, indicating that the vicissitudes of Nature cannot be bottled by equations.

➤ Some systems undergoing change possess *self-organized criticality* (SOC), where crossing a threshold sets off a nonlinear reaction. Gyral systems which practice *homeostasis* – feedback-based self-correction toward a balance – aim to avoid a point of breakdown criticality. Healthy organisms have homeostatic tendencies.

➤ Assumptions of *symmetry* underpin theoretical physics, yet phenomena manifest by breaking symmetry via decoherence.

➤ Theoretical physicists have begun to look at the universe as an encoding of *information*, while failing to appreciate that the coherent localized patterns of energy which encode existence are themselves informational. Energy is intrinsically a medium of information.

The Nature of Physics

Physics is experience, arranged in economical order.
~ Austrian physicist Ernst Mach

➤ Physics is an inquiry into the fundamentals of matter and *energy*, which is the impetus necessary to put matter into motion. Until the 20th century, that inquiry – *classical mechanics* – was confined to the ambient environment, in the 4 observable dimensions (4D): 3D space and time.

➤ A *physical model* is a mathematical characterization of a system. A *physical theory* expresses relationships among phenomena.

➤ Modern physics discovered that existence involves extra dimensions (ED), yielding a *holistic dimensionality* (HD) : 4D + ED = HD. Oddly, this key discovery is ignored in many models.

➤ Physicists insist that *time* is an illusion but treat it the same way we all do: a vector from past to future.

➤ In propagating most expeditiously, light waves typify all moving energy. *Energy wave optimization*, such as occurs during refraction, is only possible by there being an omniscient force of coherence behind Nature.

Astrophysics

➤ In 1905, Albert Einstein posited *special relativity*, which twined space and time, and found equivalence between mass and energy. Einstein's 1916 *general relativity theory* characterized gravity as an HD phenomenon, in warping space and slowing time. Gravity is not a force, like electromagnetism, but an entropic distortion caused by mass. In producing gravity matter shapes its spacetime environment.

Particle Physics

> According to quantum mechanics, what we can observe about the world is only a tiny subset of what actually exists.
> ~ Sean Carroll

➤ Breaking matter down into its constituent parts has turned into ratchets of scaling down, from atoms into an abyss of the infinitesimal; a Matryoshka doll of incrementally smaller constituents. *Atoms* were the smallest thing until 1897, when J.J. Thomson sussed *electrons*.

This led to discovering that an atom has a *nucleus*, comprising *nucleons* (protons and neutrons), with electrons whirling about in an orbital cloud. Ernest Rutherford played a leading role in this discovery.

In the next doll down, nucleons were found to be made of even smaller bits: *quarks*, which are supposedly elementary quanta. The quark model was independently proposed by Murray Gell-Mann and George Zweig in 1964.

➢ At the beginning of the 20th century, Max Planck discovered the limits of phenomena at the smallest scale. The *Planck quantum of action*, commonly called the *Planck constant*, characterizes the level at which Nature quantizes fields into phenomenal existence.

➢ Fields naturally localize and granulate, creating the seeming particles which compose Nature. A *quantum* is not a particle, like some infinitesimally tiny billiard ball, but instead a segmented field phenomenon that deceptively appears particulate.

➢ Quantum mechanics is bedeviled by inherent *uncertainty*. Localized fields emergently quantize. Before they do, their actualization is uncertain. Quantum uncertainty is a property of Nature, not just a measurement difficulty.

➢ The *Standard Model* (SM) is the conventional model of quantum mechanics, focused on interactions of *elementary particles*, which are particles supposedly lacking constituents, not comprised of smaller particles. According to SM, *fermions* are the elementary particles of matter, while immaterial *bosons* are force carriers. Bosons impart on fermions the attributes that make matter function as it does.

> The Standard Model cannot possibly be right because it cannot predict why the universe exists. That's a pretty big loophole. ~ American physicist Gerald Gabrielse

SM is a 4D fictional account. Every subatomic particle entrains an entourage of *virtual particles* which define particle properties. Virtual particles are HD particles that constantly make transitory 4D appearances in near-Planck time.

> Many independent lines of evidence now suggest that most of the matter in the universe is in a form outside the Standard Model of particle physics. ~ English astrophysicist David Harvey *et al*

➢ Particle physics models all posit symmetries which do not exist. To account for actuality, quantum theories apply a patch: *spontaneous symmetry breaking* (SSB). Resorting to SSB as a bandage indicates that quantum theory is at best a coarse approximation.

> When quantum mechanics was discovered, we realized that classical mechanics was just an approximation. We'll see that continuous quantum mechanics is itself just an approximation to some deeper theory. ~ Swedish American cosmologist Max Tegmar

	Relativity	*Quantum theory*
Explains:	Gravity	Fundamental interactions
Concept:	Spacetime: spacetime guides matter, matter distorts spacetime	Fields: universal fields localize Quanta: local fields naturally granulate
Properties of matter:	Definite	Probabilistic
View of spacetime:	Dynamic	Static

➢ Under the principle of *locality*, interactions between quanta are bound by the speed of light. Relativity forbids superluminal effects. But *nonlocality* occurs. What Einstein called "spooky action at a distance" has been demonstrated by *entanglement*: quanta far apart behaving synchronously. Entanglement has even been shown in subatomic particles that do not exist at the same time.

This demonstrates that time, as well as space, is *emergent*: spontaneously coming into existence coherently.

➤ The non-mainstream quantum *pilot wave theory* postulates pilot waves which guide the emergence and activities of quanta. This apt theory invokes a coherence in Nature, provides for deterministic order, and is consistent with known physics.

Theory of Everything

> One ring to bring them all and in the darkness bind them.
> ~ English novelist J.R.R. Tolkien in the novel *The Lord of the Rings* (1955)

➤ Natural philosophers and physicists have long sought a *theory of everything*, which is realized by recognizing that matter is energy transposed, and that all energy is entangled. That energy is nothing but an abstraction means that any theory of everything explains nothing about the nature of existence. Like energy, ironies never cease.

Existence incessantly emerges from the quantum realm via decoherence of natural entanglement. The world appears only through perception – a product of the mind, made possible by each individual consciousness entangled within an enveloping field of Ĉonsciousness. In simplistically explaining some basic mechanics, physics provides a rough sketch of Nature's sophistication.

> I am now convinced that theoretical physics is actually philosophy. ~ Max Born in 1954

☙ Chemistry ☙

No inanimate object is ever fully determined by the laws of physics and chemistry. ~ Hungarian–English polymath Michael Polanyi

Chemistry is the science of matter and its interaction. Physics bounds what exists. Chemistry gives it form.

ঙ History ୯

By 2,000 years ago, humans had developed diverse technologies that would eventually lead to an understanding of chemistry. The technologies included the making of pottery and glazes, extracting metal from ore and chemicals from plants, fermentation, dying cloth, tanning leather, rendering soap from fat, making glass, and attaining alloys such as bronze. These were all practical skills that drove those involved to greater mastery.

The progenitor of chemistry was *alchemy*: seekers of knowledge over the power of the elements, some of whom sought the *philosopher's stone*, which was the legendary substance capable of transmuting base metals of scant worth into gold, the most precious metal. Written mention of the philosopher's stone dates to at least 300 CE. Its alchemical symbol is shown.

Alchemy failed in its craven direction, but its methodology – experiment and recordkeeping – set the stage for modern chemistry.

I have always looked upon alchemy in natural philosophy to be like enthusiasm in divinity, and to have troubled the world much to the same purpose. ~ English Anglican clergyman William Temple

♒ Antoine Laurent Lavoisier ♒

We must trust to nothing but facts: these are presented to us by Nature and cannot deceive. We ought, in every instance, to submit our reasoning to the test of experiment, and never to search for truth but by the natural road of experiment and observation. ~ Antoine Lavoisier

18th-century French nobleman Antoine Laurent Lavoisier is considered the father of modern chemistry. He was an exacting experimenter and an observant man with a keen sense of reason.

In addition to his other considerable accomplishments in chemistry and biology, Lavoisier, by his meticulous measurements and quantitative observations of chemical phenomena, was critically important in creating a paradigm shift from acquisitive alchemy to the natural science of chemistry.

Lavoisier collated the first extensive list of elements, revising chemical nomenclature in the process. He discovered that the mass of matter is constant, even though it may change shape or form. Lavoisier was also instrumental in devising the metric system.

༄ Atoms ༅

Nothing exists except atoms and empty space. Everything else is opinion. ~ Democritus

All matter is made of atoms. An atom is a conglomeration of a nuclear core (*nucleus*) covered by a swirling cloud of electrons at a distant remove.

With its single proton and solitary electron, monatomic hydrogen (hydrogen-1) is the progenitor of all other matter. Unsurprisingly, hydrogen is by far the most abundant chemical element, constituting ~75% of material mass in the universe.

That atoms are spherical is a crude approximation. Only the simplest atoms come close to being round. Heavier atoms have more complex shapes.

There are 92 distinct configurations of atoms (*species*). Men with ungodly machines have created new heavy elements which rapidly decay for lack of natural stability – a gamey alchemy at the atomic level.

The maximum size of an atomic nucleus is determined by its tendency to decay. Obese nuclei decay by shucking off excess neutrons – particulate radioactivity.

> There is no evil in the atom; only in men's souls. ~ American politician Adlai Stevenson II

℘ Protons ℭ

> The proton is such a complicated system. ~ German physicists Jan Bernauer and Randolf Pohl

Every atom has a *nucleus* that carries a positive charge by virtue of possessing at least 1 proton. The *atomic number* of an atom, its *nuclide*, refers to the number of protons in the nucleus.

¤ ✧ ¤

Like the base of a pyramid, the physics of protons provide the foundation of what is understood about matter. It is a tenuous base. Researchers cannot agree about the radius of the proton; somewhere around 0.88 femtometers, ±4%. 4% sounds small, but the discrepancy cannot ostensibly be explained by either experimental method or error. It may be that the uncertainty principle is exercising itself in a grandiose way, in that how big protons are depends upon how they are observed. But that is less uncertain than it is unlikely.

The situation is a bit tight inside a proton. Pressure in the center of a proton is 10 times greater than in the heart of a neutron star, which is as packed as atomic matter may be.

The quark-trio picture by which protons are painted is simplistic. In addition to these ever-present constituents, a swarm of transient particles churn within a proton. Meantime, gluons – the bosonic glue that holds protons together – ceaselessly careen between quarks.

The upshot of this bustle is that the properties of protons, and their neutral cousins, neutrons, are hard to get a handle on. Spin exemplifies the problem. Physicists have studied subatomic spin for decades, but it's still not sorted out.

Like the Earth rotating on its axis, quantum particles act as if they are whirling at blistering speed. Because a rotating charge creates a magnetic field, this spin makes protons behave like tiny magnets. This property is key to the medical imaging procedure called *magnetic resonance imaging*, popularly known by its acronym: *MRI*.

But there's no actual spin going on. Because fundamental particles like quarks don't have a finite physical size, they can't twirl. They just give the appearance that they do, yielding a proton spin of 1/2.

In 1987, physicists discovered that only a small fraction of the spin owed to the quarks inside. They then suspected gluons as being spin-meisters. No such luck.

The current tally is that gluons are responsible for only ~35% a proton's spin. Quarks make up ~25%, leaving 40% unaccounted for.

> We have absolutely no idea how the entire spin is made up. We maybe have understood a small fraction of it. ~ American nuclear physicist Elke-Caroline Aschenauer

Experimental physicists get little help from their theoretical counterparts when trying to unravel the proton's perplexities. Quantum chromodynamics – the theory of the strong force transmitted by gluons – is a mathematical marvel of such complexity that its equations cannot be solved. Instead, theorists rely upon an approximate technique that roughly quantizes the quantum actors and their actions. Results only coarsely correspond with experimental measurements, which indicates that rough is not good enough.

> The proton is not something you can calculate from first principles. ~ Elke-Caroline Aschenauer

¤ ✧ ¤

> We have a lot of circumstantial evidence that something like unification must be happening. ~ Indian nuclear physicist Kaladi Babu

Protons seemingly live forever: a fact that physicists are reluctant to accept. The rub is that all of the various physical models which unify electromagnetism with the nuclear forces demand that protons give up the ghost some time.

Experiments show that a proton has a life expectancy of at least 1.6×10^{34} years. That may understate the situation, as proton decay has never been seen.

The proton is the most fundamental building block of everything, and until we understand that, we can't say we understand anything else. ~ Scottish nuclear physicist Evangeline Downie

¤ ✧ ¤

Hydrogen-1 is the only instance of an atom comprising only a proton and an electron, as protons will not bind with each other: their electromagnetic repulsion is stronger than their nuclear strong force attraction.

Enter *neutrons*, the strong-force glue in atoms, holding protons together in a nucleus. The number of neutrons in an atom determines the *isotope* of an element.

Besides having selfsame spin, protons and neutrons have equivalent girth: 10^{-14} meters; and about the same mass: 1.7×10^{-24} grams.[*]

℘ Neutrons ℞

Experimenters have shown that, to very high precision, the neutron has no electric dipole moment. ~ American physicist Adrian Cho

Neutrons bound into an atomic nucleus are homey hadrons, but when set free they quickly become unstable, radiating away, undergoing beta decay with an average lifetime of less than 15 minutes. A decaying neutron turns into a proton, letting fly an electron (the beta particle) and a ghostlike antineutrino.

---

[*] A 2010 measurement of the proton, using a muon racing around it as a metric, put the proton 4% smaller than previously found. The measurement was supposedly more accurate than those used before. The different result put quantum electrodynamics (QED), which describes how light and matter interact, in deep trouble, as it predicts a fatter proton. At of 2018, whether QED is off-base, or using a muon somehow mucked the result, or the experiments themselves were faulty, remains unknown.

Decay rate can be affected by magnetic fields, a phenomenon without explanation in known physics. Extra-dimensional dynamics explain this oddity: symmetry via ED mirror particles.

A neutron is made up of 2 charged quarks: 2 down quarks, each with a negative charge 1/3rd that of an electron, and an up quark that carries a 2/3rds positive charge. The arrangement leaves a neutron electrically neutral. Hence the particle's name.

❧ Nuclear Clusters ☙

The internal structure of a nucleon – a proton or a neutron – depends on its environment. The structure of a nucleon in empty space is different from its structure when it is embedded inside an atomic nucleus. There is a variation in the momentum distribution of quarks inside the nucleons embedded in nuclei. Specific nucelons are altered by interacting in pairs over brief time periods. Similar nucleons are less likely to pair up than are dissimilar nucleons. ~ American nuclear physicist Gerald Feldman

The textbook view of an atomic nucleus is that of a largely homogenous collection of protons and neutrons as a spherical dollop of nuclear matter. But nucleons are predisposed to cluster into certain configurations according to relative stabilities.

Neutrons bring the bosonic strong force to atoms, binding with protons that are otherwise ill-disposed to being in close quarters with other protons. The bosonic character of a nucleus, based upon quantum spin, determines cluster stability. A helium-4 nucleus, with 2 protons and 2 neutrons, is exceedingly stable. This owes to its clustering quality.

Such nucleon quartets – with even and equal number of protons and neutrons – make the most stable atomic nuclei. This includes beryllium-8, carbon-12, oxygen-16 and neon-20.

Nuclear clustering has cosmological significance. Supposedly, within a few minutes after the Big Bang, a soup of free protons and neutrons had formed in a 6 to 1 ratio.

When the cosmos had cooled enough to permit nuclear binding, almost all the earliest nuclei were helium-4. Hence, the early universe had a hydrogen/helium ratio of 3 to 1, like today, with nearly all the neutrons in the universe trapped in helium-4.

Carbon-12 is formed by triple fusion of helium-4; a process termed *triple-alpha*. A star feeds itself during its red-giant phase through the triple-α process, fusing helium into carbon.

Without nuclear clustering, there would be little carbon in the universe, and no organic life.

⚡ Magic Nuclei ⚡

Nucleons are always antsy: racing about each other in an intricate orbital dance orchestrated by the strong force. They fill orbital slots in a sequential manner according to energy levels.

Spin-orbit interaction is the interplay between a proton's orbital momentum and its angular momentum (spin). It is critical in keeping nuclei steadfast. Radioactive beta decay is a product of instability in spin-orbit interaction.

Most atomic nuclei have a fairly constant density, regardless of the number of nucleons they contain. But some exhibit orbital oddities.

Silicon-34, a radioactive isotope with a life expectancy of less than 3 seconds, is one such heretic. It has a largely empty orbital, creating a bubble in its nucleus.

The orbital shells of protons and neutrons are independent of each other. The number of nucleons (either protons or neutrons) forming a complete nuclear shell, and so promoting maximal stability, is its *magic number*. Silicon-34 is doubly magic, in that both its protons and neutrons are at a magic level.

A magic number means that the energy needed to boost a nucleon into the next orbital is particularly high. This explains a nuclear bubble's origin.

'Doubly magic' nuclei have fully occupied shells of protons and neutrons. Such nuclei are therefore more strongly bound together and more difficult to excite than their neighbors.

~ American nuclear physicists Gaute Hagen & Thomas Papenbrock

For a proton in silicon-34 to jump into the unfilled central orbital, it needs an energetic boost. Hence, its center proton shell remains sparsely populated.

♎ Eternal Element ♎

Named after a villain in Greek mythology, *tantalum* is an uncommon heavy metal found in Earth's crust. By weight, 1.5 parts per million of Earth's crust is tantalum. Most tantalum is in the form tantalum-181. But 0.01% is in the isomer form of tantalum-180m.

An *isomer* is a molecule with the same molecular formula as another, but with a different chemical structure. Tantalum-180m is an isomer with a naturally agitated nucleus.

Normally, excited nuclei quickly calm down, dropping to a lower energy state and emitting a photon in the process. Somehow tantalum-180m is stuck in its frenzied state.

After extensive observation, physicists decided that tantalum-180m could not have a half-life shorter than 45 million billion years.* No other known element has anywhere near such a determined buzz.

The oddest thing is that tantalum-180m even exists. The element-forging processes that transpire in stars and supernovas seem to bypass this nuclide of 73 protons.

We don't understand how it is created. ~ American nuclear physicist Eric Norman

---

* *Half-life* is the duration required for a material to decay to half of its initial mass. The term is commonly used in nuclear physics to state the radioactive decay rate of atoms. Because of the uncertainty principle, it is impossible to predict when radioactive decay may occur. So, decay rate is expressed in terms of probability, represented as a half-life.

The fusion of lighter elements into heavier ones is termed *nucleosynthesis*. Heavy elements are created via nucleosynthesis during supernova explosions. But there have been too few exploding stars to account for all the corpulent chemicals that exist.

Some forms of nucleosynthesis, such as the fusion of hydrogen into helium, provide the energy that keeps a star from collapsing and generate its luminosity. Others do not, such as the transformation of gold into mercury by adding a neutron. Some elements, once fused, remain locked in the dead cores of stars and are not released into the surrounding galaxy. ~ Jennifer Johnson

Supernovas enrich the universe with matter in 3 ways. 1st, they eject the products of nucleosynthesis built up over the star's lifetime. Most oxygen, carbon, and magnesium are made during star time. The explosion simply hurls these elements into space.

2nd, the extreme densities and temperatures caused by the shock wave of a supernova drives additional nucleosynthesis. The iron ejected by core-collapse supernovae does not come from the core but from explosive fusion of material in the silicon shell during the supernova.

3rd, the explosive shock of ejected material plowing into ambient gas generates cosmic rays by accelerating some particles to near light speed. Cosmic rays are energetic enough to break apart heavier nuclei, producing lighter elements through fission. Cosmic ray disassembly is responsible for large fractions of the lithium, beryllium, and boron made. Cosmic ray fission also renders elements such as carbon and oxygen, though the abundance of those elements is dominated by other modes of production.

Black holes occasionally collide with neutron stars. The black hole chews on the star and spits out neutron-rich material which forms heavy elements once swept up in another star.

❀ Periodic Table of Elements ❧

In the late 5th century BCE, Empedocles originated the cosmogenic theory of there being 4 basic substances: earth, water, wind, and fire. He also proposed that love and strife could mix and separate these substances, which Plato termed *elements*.* Aristotle furthered Empedocles' elementalism with the idea that all matter was made from a mixture of 1 or more elements (which he called *roots*).

The periodic table is intertwined with the quest to discover new elements. The 1st element discovered since ancient times was furtively made by German merchant and alchemist Hennig Brand, who in 1669 distilled human urine to reveal a glowing white substance which he named *phosphorous*. Brand's reeking distillation was part of his continuing quest to find the philosopher's stone.

Brand kept his discovery of phosphorus secret until its published rediscovery by Anglo Irish chemist Robert Boyle in 1680. That same year, Boyle put a phosphorous paste with a sulfur tip on wooden sticks and invented the forerunner of modern matches.†

Boyle defined a *chemical element* as "a substance that cannot be broken down into a simpler substance by a chemical reaction." This definition served for 3 centuries, until subatomic particles were discovered, whereupon the definition shifted to characterizing an element by proton count.

Lavoisier's 1789 *Elementary Treatise of Chemistry*, the first modern chemistry textbook, listed the "simple substances" that Plato and Boyle had termed *elements*, bifurcating them into metals and nonmetals. Lavoisier also listed *light* and *caloric* as elements, which were believed at the time to be material substances.

German chemist Johann Wolfgang Döbereiner began classifying elements into groups by atomic weight in 1817. In arranging the chemical elements by atomic weight, French geologist and mineralogist Alexandre-Emile Béguyer de Chancourtois first noticed their periodicity. de Chancourtois failed to publish the irregularly arranged table that he had

---

* Ancient traditions throughout the world had also found these 4 basic elements.
† Phosphorus' fiery reactivity earned it the distinction of being "the Devil's element."

constructed. It did not matter. His 1862 article was ignored by chemists, having been written in terms of geology.

Based upon accumulated understanding of chemical properties, Russian chemist Dmitry Mendeleyev published the modern form of the periodic table in 1869, having noticed the regularities that form the basis of the table, and having the foresight to complete the table by predicting the characteristics of elements not yet discovered.

The periodic table is a tabular arrangement of the 118 chemical elements (94 of which occur naturally on Earth), organized on the basis of their atomic numbers, electron configurations, and recurring chemical properties. Elements are ordered by atomic number, which is a product of protonic girth.

Group ›1	2		3	4	5	6	7	8	9	10	11	12	13	14	15	16	17	18
↓Period																		
1 1 H						*Periodic Table*												2 He
2 3 Li	4 Be					*of Elements*							5 B	6 C	7 N	8 O	9 F	10 Ne
3 11 Na	12 Mg												13 Al	14 Si	15 P	16 S	17 Cl	18 Ar
4 19 K	20 Ca		21 Sc	22 Ti	23 V	24 Cr	25 Mn	26 Fe	27 Co	28 Ni	29 Cu	30 Zn	31 Ga	32 Ge	33 As	34 Se	35 Br	36 Kr
5 37 Rb	38 Sr		39 Y	40 Zr	41 Nb	42 Mo	43 Tc	44 Ru	45 Rh	46 Pd	47 Ag	48 Cd	49 In	50 Sn	51 Sb	52 Te	53 I	54 Xe
6 55 Cs	56 Ba	*	71 Lu	72 Hf	73 Ta	74 W	75 Re	76 Os	77 Ir	78 Pt	79 Au	80 Hg	81 Tl	82 Pb	83 Bi	84 Po	85 At	86 Rn
7 87 Fr	88 Ra	*	103 Lr	104 Rf	105 Db	106 Sg	107 Bh	108 Hs	109 Mt	110 Ds	111 Rg	112 Cn	113 Uut	114 Fl	115 Uup	116 Lv	117 Uus	118 Uuo

	57 La	58 Ce	59 Pr	60 Nd	61 Pm	62 Sm	63 Eu	64 Gd	65 Tb	66 Dy	67 Ho	68 Er	69 Tm	70 Yb
*	89 Ac	90 Th	91 Pa	92 U	93 Np	94 Pu	95 Am	96 Cm	97 Bk	98 Cf	99 Es	100 Fm	101 Md	102 No

The table is typically a grid of elements, with rows called *periods* and columns called *groups*. Element groups have the same number of electrons in their outer shell. A new row (period) begins when an element sprouts a new electron shell.

The distinction between metals and nonmetals is one of the most fundamental in chemistry. Good conductors of heat and electricity, most elements are metal.[*] They may be pulled into wires because they are *ductile*; hammered into sheets because they are *malleable*; and most are *lustrous* (shiny).

---

[*] Hydrogen (H) is the only nonmetal on the left side of the table (at least at ambient temperature and pressure). The chemical properties of elements 109–111 and 113–118 are unknown.

Toward the right of the table in the figure above, metalloids (semimetals) are shown darkly shaded. To the right of the semimetals are nonmetals. The lightly shaded, far-right group (18) comprise the noble (inert) gases.*

Electrons

An electron is a particle and a wave; it is ideally simple and unimaginably complex; it is precisely understood and utterly mysterious; it is rigid and subject to creative disassembly. No single answer does justice to reality. ~ Frank Wilczek

Swirling in an orbital cloud around a nucleus, at a distance 10,000 times that of the nucleus' diameter, are perfectly round, negatively charged *electrons*. An electron has a mass estimated at 1/1,836th of a proton.

Individual electron orbitals are limited to pairs. An *electron pair* comprises 2 electrons in the same orbital, albeit with opposite spins.

The Pauli exclusion principle forbids fermions from simultaneously occupying the same space. Electrons are fermions. Despite the Pauli exclusion principle, there is a distinct probability of an electron being inside the nucleus of an atom.

To fly together, members of an electron pair have different angular momentums – a different spin makes for an orbital twin.

Electrons commonly escape their atomic bond and become free electrons, at least temporarily. Electron transitions between bound and free characterize chemical reactions.

An electron in motion generates a magnetic field in its wake; whence *electromagnetism*. Electrons possess a *magnetic dipole moment*: a polarity like a bar magnet. Supposedly, that is because an electron is a spinning smear of charge. Elementary electromagnetism stipulates that this creates a magnetic dipole field. The more complex actuality is not understood.

---

* The term *noble* refers to low reactivity. It was first used in 1898 by German chemist Hugo Erdmann, as an analogy to "noble metals," which also have hesitant reactivity.

That electrons have an *electric dipole moment* is strongly suspected. Confirming that by isolating an electron has proven tricky.

A free electron will accelerate under the influence of an electric field and crash into whatever is in its path. This is handy for recruiting electrons into employment, but self-defeating for measuring elusive electron properties. So, experiments to date aimed at determining whether electrons have an electric dipole moment have been thwarted.

An electron having an electric dipole moment might doom the Standard Model. Consider time-reversal symmetry.

Supposedly, the laws of physics stay the same if time ran backward. But for a spinning electron, the north and south poles would swap. An electric dipole would accumulate charge at one pole; the inverse of which does not happen with time running forward. So, an electron with an electric dipole moment would violate time-reversal symmetry.

That would not be a first. Mesons are known CP violators and do not respect time-reversal symmetry. But to have the star of electromagnetism and the cornerstone of chemistry blithely skirt the fundamental principles of the Standard Model would bring the model's reign to an end.

The strength of the electron's magnetic field provides perhaps the most stringent and brilliantly successful comparison of theory and experiment in all of physical science, whereas the value of the electric field has never been measured. It is a mystery even to theory. ~ Frank Wilczek

There is a great irony at the heart of electronics. At a practical level, electrons provide a steady charge. At a more fundamental level, that charge is immeasurable.

There are also limits to electron reliability. Semiconductor fabrication has reached the point of *nano wires* only a hundred atoms wide. At that point, electrons behave like quantum waves. Electrodes to control electron flow create a mountainous terrain that electrons struggle with.

They bash against the walls, and sometimes reflect from the flanks of the mountain pass. They also sense each other's presence. ~ Dutch quantum physicist Casper van der Wal

At the quantum level, electron flow is an incredibly complex interaction of various physical phenomena: indescribable by any single formula. Depending upon the environment, electrons may scurry about, with eddies that create resistance among themselves, or may flow with no friction whatsoever.

Maxwell's unified field theory of electromagnetism is neat packaging that does not always apply. In some environs, the electrical and magnetic properties of electrons are divisible. Electrons may behave as fractional particles: splitting into a magnet and an electrical charge which can move freely and independently of each other. Electrons' magnetic moment may split into 2 halves and move apart, albeit with extra-dimensional linkage.

Electrons amply illustrate how little we know about how existence is constructed.

The bottom line of a 'normal' atom is no charge: the positive protons and negative electrons balance out. Thankfully, ionic heretics are everywhere.

☿ Ions ☋

> The removal of an electron from the surface of an atom – that is, the ionization of the atom – means a fundamental structural change in its surface layer. ~ German physicist Johannes Stark

An *ion* is an atom or molecule with an electrical imbalance owing to an unequal number of protons and electrons. Ions are promoters of chemical reactions.

Gaseous ions are highly reactive. They relatively rare on Earth: appearing only in flames, lightning, sparks, and other plasmas. The energy required to remove electrons from gaseous atoms or ions is termed *ionization energy* or *ionization potential*.

Liquid or solid-state ions naturally occur when salts interact with a solvent, such as water, and are more stable than gaseous ions, owing to energy and entropy changes as ions move away from each other to interact with the liquid (or solid). Stabilized ions are commonly found at low temperatures, such as in dissolved salts in cold seawater.

ᛞ Shells ᛉ

Electrons are construed to swirl in orbital layers – *shells* – defined by their relative quantum energy state. The energy ranges of shells can overlap. Generally, more energetic electrons move in *orbitals* farther from an atom's nucleus. But electrons can store some energy before getting so excited that they feel obliged to migrate to another shell.

Like an electronic Matryoshka doll, a shell may have *subshells*. While it is commonly stated that same-shell electrons have the same energy, that is an approximation. But electrons in the same subshell are equally energetic.

Shell layering is related to *atomic spectral lines* (electron energy level changes). These energetic relations are a product of quantum interactions and are an HD phenomenon. Electrons in different shells exchange status information to favor a coherent spin alignment of all electrons in their atom.

As atom size increases, more electrons whiz about it. Some electrons in heavier elements attain velocities approaching the speed of light.* This increases relativistic mass, causing certain inner orbitals to contract and stabilize; which, in turn, destabilizes outer orbitals and provokes their expansion.

The relativistic effects of electrons in heavier atoms contribute to their attributes. Gold has its characteristic amber color because of it. Mercury is a liquid at room temperature owing to the relativistic speed of its electrons.

Electron orbitals have a wavelike existence. Their position at any point in time is only a probability.

Today, instead of thinking of electrons as microscopic planets circling a nucleus, we now see them as probability waves sloshing around the orbits like water in some kind of doughnut-

---

* More specifically, electron velocity is correlated with its *effective nuclear charge*, which is the net positive charge experienced by an electron in any atom with multiple electrons.

shaped tidal pool governed by Schrödinger's equation.
~ American physicist James Trefil

℘ The Atomic Void ଅ

By convention sweet, by convention bitter, by convention
hot, by convention cold, by convention color, but in reality atoms and void. ~ Democritus

An atom's electron swirl is a long way away from its nucleus. In 1913, Niels Bohr constructed a model of hydrogen-1
and figured that its radius – the distance from proton to electron – was 100,000 times the size of the nucleus. Heavier atoms are more compact: nuclei gain girth but the orbits of
electrons are constrained.

Measured across the electron cloud, a typical atom, such
as carbon, is 10^{-8} cm: 1/10-millionth of a millimeter. The diameter of a nucleus is about 10^{-12} cm. This is 1/10,000th of
the diameter of the atom in which the nucleus resides. Most
atoms have electrons whirling in a cloud at a distance somewhat over 10,000 times the radius their nuclei. These numbers are proximate, as electrons move in probabilistic orbits,
and nuclei themselves restlessly skitter about.

An atom is 99.999+% empty space, albeit seething with
ED energy. If a hydrogen proton were a tennis ball, 6.35 centimeters in diameter, its electron would be 6.35 kilometers
away. If devoid of void, all the matter in the universe would
fit within the size of a single pea, less than 1 cm in diameter.

ଷ Conductivity ଇ

Electrons coursing through materials normally encounter
electrical resistance: opposition to the flow of an electric current. The resistance is typically only partly successful.

How well a substance allows electron flow falls into 2 cat-
egories: *conductor* and *resistor*. Temperature affects conductivity and resistance.

In being a lattice of atoms, metals are inherently good
conductors. A *positive ionic lattice* engenders outer shell electrons to disassociate from their parent atoms, creating an
electron sea that flows with the force: an electrical *current*.

The flow in conductors is not smooth, as individual electrons suffer scatter due to destructive interference of free electron waves to non-correlating ion potentials – a 4D phenomenon with ED interactions behind it. These collisions transfer energy from an electron to a metal ion, causing the ion to vibrate more vigorously. The result is resistance, manifest as heat.

✠ Bad Metals ✠

Metals conduct electricity because they contain electrons that are free to move through the material. In bad metals the electrons seem to reversibly disappear and reappear. ~ American solid-state physicist Rafael Jaramillo

Theoretically, the hotter the metal, the worse it conducts electricity. Most metals obey this inverse relationship between temperature and conductivity. But some do not. These conductive miscreants are termed *bad metals*.

At high temperatures, the electrons in bad metals ought to violate Heisenberg's uncertainty principle, rendering them nonconductive. But they refuse to bow to uncertainty. Instead, bad metals invoke extra-dimensional effects to keep their electrons flowing.

Electron-electron interactions are so strong in bad metals. ~ American physicist Ray Osborn

✠ Superconductivity ✠

Electrons lose their individual identities in superconductors, in which electrons pair up to form a pervasive sea. Thus, electrons become their own antiparticles. ~ Frank Wilczek

Helium was first liquefied in 1908 by Dutch physicist Heike Kamerlingh Onnes, who studied the resistance of solid mercury when supercooled. Mercury is the only metal that is liquid at room temperature; hence its nom de plume: *quicksilver*. In 1911, Onnes discovered that at 4.2 K, quicksilver got quicker: shedding electric resistance entirely.

Normally, the pull of one passing electron is drowned out by ion vibrations. Cooling a metal down can lessen ionic wiggles enough that an electron's gentle tug can be felt by another, and so the 2 electrons form a *Cooper pair*: entranced to dance together via a pied-piper phonon. The temperature below which a material becomes superconducting is called its *critical temperature*.

A Cooper pair passing through are attractive to the ions in a superconductive metal. The metal ions stray as far toward the electron's wake as their lattice structure allows. This distorts the crystal for a short time, creating a concentrated positive charge that encourages free electrons in the vicinity to pair up and join the flow.*

Once paired, electrons stop behaving as ordinary matter particles and enter a collective HD quantum state, where they become entangled; oblivious to the ions, and so lose no energy bumping into them. Current passes through the crystal without resistance.

Superconductivity is zero electrical resistance, resulting from electrons overcoming their mutual repulsion and pairing up, creating a coherent flow. Electrons act oddly on their way to flowing freely. In crystals that become superconductive when cooled, electrons become 1,000 times more massive than normal as they become entangled. These hot-to-trot heavy electrons are actually composite objects, mixing opposite behaviors together: localized bonding to individual atoms and freely flitting between atoms in the lattice.

Entangled electrons can become so obese that they refuse to budge: freezing into a magnetized state while stuck at an atom, spinning in unison. Tweaking the crystal composition can charm the fat electrons to dance in superconducting entanglement as the crystal is cooled.

Superconductivity is a full-blown quantum field party, characterized by the *Meissner effect*: the complete expulsion of magnetic field lines from the inside of a superconductor as it transitions to a superconducting state. Owing to the Meissner effect, a magnet put over a superconducting substance levitates. The electromagnetic free energy of the superconductor approaches zero, creating a repulsion of objects with

---

* Most inorganic solids, including metals, are *polycrystals*: a plenitude of microscopic crystals (crystallites) fused together.

intense magnetic field lines, like the repulsion between a magnet and a diamagnetic object.

Absence of resistance does not make a magnet grow fonder; quite the opposite. Magnetism surrenders to super-conductivity.

As antiferromagnetism breaks down, superconductivity appears, encouraging electrons to pair up and flow freely. Long-range magnetism – where atoms align their magnetic moments – ceases in giving way to superconductivity. This abdication belies a deeper connection.

Magnetism is the quantum glue underlying the emergence of superconductivity. ~ Dirk Morr

The electromagnetism in the electrons involved in super-conductive substances bifurcates into an electric charge and a magnet. American physicist Peter Anderson suggested the fractionalization of electron properties ("fractional particles") as facilitating superconductivity in 1987. Altering the inter-activity of magnetism is key to superconductivity.

Whereas atomic magnetism vanishes with superconduc-tivity, local magnetic moments become more powerful, as the electrons in individual atoms synchronize. Superconductivity is a transformation of the turbulent flow of an electron sea into coherence by ordering quantum spin.

Relatively high-temperature superconductivity has been shown in insulators. Intuitive expectation would be that con-ductors would become more conductive when cooled. One would not expect that insulators would reverse themselves, flipping from resisting to superconducting.

The search for high-temperature superconductivity in hydro-gen-rich compounds hinges on a theory that, under certain cir-cumstances, elements that have low atomic masses can contribute to high critical temperatures. Hydrogen, being the lightest element, is optimal for high critical temperatures. ~ American physicist James Hamlin

Temperature is not the only variable affecting potential superconductive materials. Applied pressure matters. Under pressure (specifically, 16,000 times atmospheric pressure (1.6 Gpa)), iron selenides superconduct at 30–32 K. Pressed further, superconductivity disappears at 9 Gpa, only to reappear at 12.4 Gpa. Yet the basic structure of the compound is retained throughout the pressuring.

More recent research has focused on the internal structure of Cooper pairs. Cooper pairs in "conventional" superconductors spin in opposite directions, resulting in the pair having zero spin.

In more exotic "triplet" superconductors, spins line up, such that a Cooper pair carries some spin of its own. Lanthanum-nickel-carbon ($LaNiC_2$) and lanthanum-nickel-gallium ($LaNiGa_2$) are exemplary triplet superconductors.

Unlike conventional superconductors, triplet superconductors are not ferromagnetic. The magnetic moments of the Cooper pairs themselves create the magnetism favorable to superconductivity.

Non-unitary triplet pairing bootstraps itself into superconductivity via quantum spin liquidity generated by emergent superconductivity. This is a superconducting analogue of how magnetism develops in ferromagnetic metals.

Superconductivity appears in divergent materials at different temperatures owing to their magnetic properties and tendency to become a better conductor of electricity in certain directions. Subatomic alignment matters.

Superconductivity shows how the electromagnetic force is highly variable, depending upon material and ambient conditions. Superconductivity remains a mysterious demonstration of 4D effects from HD interactions.

Emergence, the coming into being through evolution, is an important concept in modern condensed-matter physics. Superconductivity is a classic example of emergence in the realm of quantum matter: as the energy scale decreases, the effective electron–electron interactions responsible for Cooper pairing, and thus superconductivity, evolves from the elementary microscopic Hamiltonian through unanticipated modifications.

This evolution is why it is so difficult to derive superconductivity from first principles. Finding the microscopic mechanism of Cooper pairing means discovering the nature of the ultimate effective electron–electron interaction at the lowest energy scales. ~ Scottish Irish American physicist Séamus Davis & Taiwanese American physicist Dung-Hai Lee

♎ Fractal Superconductivity ♎

There are various classes of superconducting substances, each with a different molecular structure at the atomic bonding level. Researchers create novel superconductors by manipulating atomic arrangements.

The oxygen in a copper-oxide semiconductor settles into a fractal pattern when superconducting. The more fractal, the better the superconductor's performance.

Fractals are a mathematical construct, displaying a self-similar pattern that is scale indifferent. In fractals, similar patterns may be discerned regardless of how closely a fractal is examined.

Fractals repeatedly appear in Nature, from lightning bolts to shorelines to cauliflower heads. A snapshot of the solar wind shows a fractal signature, imposed by the magnetic field of the Sun.

♎ Superinsulators ♎

Insulators are materials with inherent resistance. In contrast, *superinsulators*, which were discovered in 2008, absolutely resist electrical conductivity.

Cooper pairs avoid each other in superinsulators. They instead generate enormous electrical forces that oppose penetration of current into the material. A superinsulator is the exact opposite of a superconductor.

ଞ Radiation ଛ

It is probable that all heavy matter possesses – latent and bound up with the structure of the atom – a similar quantity of energy to that possessed by radium. If it could be tapped and controlled, what an agent it would be in shaping the world's destiny! The man who puts his hand on the lever by which a parsimonious Nature regulates so jealously the output of this store of energy would possess a weapon by which he could destroy the Earth if he chose. ~ English radiochemist Frederick Soddy in 1904, anticipating atomic weaponry

French physicist Henri Becquerel accidentally discovered radioactivity in 1896 while investigating the glow of phosphorescence. These radiations were, for a while, called *Becquerel Rays*.

Radiation pervades the universe, differing only in source and energy intensity. Radiation may be an energetic display of electromagnetism. It is otherwise a decay of atomic stability: the weak force overcoming the strong.

> Weak interaction effects can be observed in decays and in collisions only when they are not hidden by the presence of strong or electromagnetic forces. ~ Italian particle physicist Alessandro Bettini

There are 2 forms of radiative emission, corresponding with radiation types: waves and particles. *Electromagnetic radiation* is massless, traveling as photonic waves of energy at some frequency. Light is an electromagnetic radiation.

Particulate radiation consists of high-speed particles. Such radioactivity emanates from *atomic decay*: neutrons shedding something of themselves while heading toward entropy.[*] Atomic decay is an interplay of the strong and weak nuclear forces, along with the electrostatic force, with HD quantum phenomena conducting.

Atomic radiation rearranges a nucleus, something normally hindered. HD forces are thought to kick off the reaction, as a quantum fluctuation forces a nucleus to a lower energy state. This relaxation is a *decay event*, caused by quantum tunneling. A particle under duress overcomes its 4D classical confines, and cuts loose a particle or energetic wave.

✣ Electromagnetic Radiation ✣

An atom receives a burst of energy which may come in a flush of heat. An electron absorbs the energy bit, boosting it to a higher orbital, farther from the nucleus. But the electron cannot hold that position for more than a fraction of a second, and so it falls back to its original shell.

---

[*] The lightest of baryons, protons do not decay.

An electron losing energy, as happens when dropping to a lower orbital shell, results in releasing that energy as a photon. How much energy a released photon has depends upon how far the electron dropped between orbitals. That determines the photonic wavelength.

In this case, *wavelength* is spatial period of an energy wave: the point-to-point distance between wave repetitions. Mathematically, wavelength (λ) is the spatial period of a sine wave.

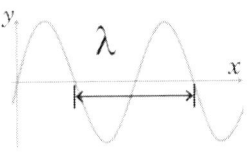

The energy of a single wavelength is a product of its mass, angular frequency, and amplitude (height). The *mass* of a wave defines its momentum.

$$E\lambda = \tfrac{1}{2}\mu\omega^2 A^2\lambda$$

angular frequency
mass amplitude

Wavelength Energy

The wavelength gives the *angular frequency*, which is the rate of change in the wave. The shorter the wavelength, the higher its frequency and greater its energy.

Electromagnetic radiation is a wave of radiant energy. The photon is its quantum poster child.

☪ Electromagnetic Spectrum ☪

Newton coined the term *spectrum* to describe the rainbow of colors when sunlight is split by a prism. The *electromagnetic spectrum* is a continuum of increasing energy intensity. High-energy wavelengths are shorter than low-energy wavelengths.

Radio waves have the longest wavelength in the EM spectrum, and so are the least energetic. A single radio wave may be up to 100 kilometers long. The shortest radio wave is 1 millimeter.

Befitting their name, *microwaves* are more energetic than radio waves, with a shorter wavelength: 30 centimeters to 3 millimeters. The entire universe has a faint background radiation of microwaves (cosmic background radiation (CMB)), generally considered a lingering effect of the universe's origin.*

Microwaves can penetrate smoke, haze, or clouds, making them useful for transmitting information. Radar, a common tool for weather forecasting, uses microwaves.

At a shorter wavelength lies *infrared* (IR), which touches upon the EM spectrum detectable by living organisms. IR extends from 300 micrometers (μm) down to 0.74 μm.

Humans feel infrared radiation from the Sun as heat, both in the eyes and skin. Snakes can sense infrared radiation, allowing them to locate endothermic prey in total darkness.

The small slice of the EM spectrum that can be seen by humans is anthropomorphically referred to as *visible light*. EM radiation between 400–700 nanometers is visible. The Sun is the natural source of most visible light.

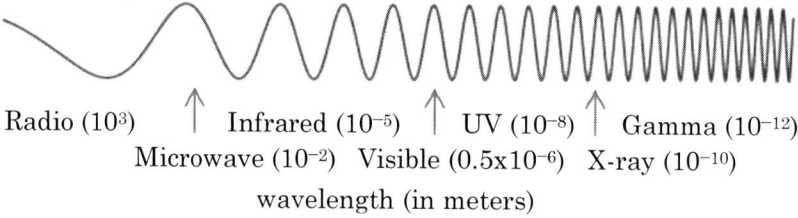

Radio (10^3)   Infrared (10^{-5})   UV (10^{-8})   Gamma (10^{-12})
Microwave (10^{-2})   Visible (0.5×10^{-6})   X-ray (10^{-10})
wavelength (in meters)

♎ Ultraviolet ♎

Just beyond our visibility is *ultraviolet* (UV) light, which ranges between 10–400 nanometers. Some insects, including spiders and honeybees, see UV. Consequently, flowers look much different to a bee than a human, with intricate patterns and subtle colorations undetectable by us; even more beautiful to their intended audience than they are to our eyes.

Ultraviolet is biologically significant in several ways. UV irradiation of interstellar ice, such as in comets, assisted creation of organic compounds in prebiotic times. Ultraviolet radiation may have produced a disequilibrium in biomolecular chirality, creating an evolutionary impetus to the L-amino acids and D-sugars common in terrestrial life forms.

---

* This assertion of CMB being an artifact of cosmic origination is a conjecture connected with the Big Bang hypothesis.

At 290 to 400 nanometers, the ultraviolet radiation reaching Earth's surface now is non-ionizing. Wavelengths shorter than 200 nm are invariably ionizing.

Solar radiation fluxes reaching Earth's surface have changed over geologic time. Part of the flux has been from photochemical reactions, most notably in recent decades from the depletion of stratospheric ozone.

Ultraviolet radiation (UVR) played an important, almost ironic role in prebiotic chemistry and habitat creation before life arose. In energizing O_2 to O_3, UVR initiated the photochemical processes that led to the formation of ozone (O_3) in the paleoatmosphere (the atmosphere before life arose).

The formation of a stratospheric ozone layer was critical to life evolving, as ozone strongly absorbs 220–330 nm solar radiation. These wavelengths can be hazardous to life.

Ultraviolet radiation is a photochemical catalyst for the formation of hydrogen peroxide (H_2O_2), which produces reactive oxygen species (ROS) that can damage DNA. DNA naturally absorbs UV.

Ultraviolet photochemistry might have been essential in developing the biosynthetic pathway of chlorophyll. Photosynthesis may have initially evolved from a mechanism to protect cells from UVR, at a time when atmospheric oxygen was scant, and the ozone shield not in place.

Anaerobic prokaryotes (single-celled organisms not respiring oxygen) are intrinsically resistant to UV radiation, and are able to repair DNA damage from ultraviolet radiation. They arose when such sunscreen protection was essential.

Biochemical UV protection mechanisms may have been instrumental in the evolution of eukaryotic life via *ploidy*: the number of sets of chromosomes in a biological cell. Diploid cells, as in animals, are more resistant to radiation damage than haploid bacteria and yeast cells.[*]

Early life not only learned to deal with reactive oxygen species (ROS); it was even harnessed as a biochemical tool.

---

[*] A *haploid* organism has only 1 set of chromosomes (genetic packaging). A *diploid* has 2.

Macrophages, a key actor in vertebrate immune systems, employ ROS bursts to blast microbial invaders to oblivion.

In a subtler application, ROS is used in inter-cell signaling. It is an axiom of evolution that adaptive mechanisms are often redeployed for applications different from those to which an initial response developed (what is termed *pre-adaptation*).

ROS accumulates in animal cells with age. Cells in older animals have a harder time warding off oxidative stress; hence, *senescence*: growing old.

UV may have provided the pressure for evolution of human skin color. Skin color closely correlates to latitude, as does solar ultraviolet. Human skin is protected from ultraviolet radiation by 2 means: the pigment melanin, and the thickness of the uppermost skin layer (the stratum corneum).

Plants too can suffer senescence via ROS. But some plants ward off oxidative stress and live for hundreds or thousands of years.

By depleting the ozone layer via pollution, especially chlorofluorocarbons (CFCs), surface-level UV radiation at damaging wavelengths has increased. Skin cancer is stimulated by overexposure to ultraviolet light.

♎ X-Rays ♎

X-rays range from 10 to 0.01 nanometers. X-rays naturally occur in space, but generally don't travel to the Earth's surface in the present eon, as they are absorbed in the upper atmosphere. Lightning on Earth generates X-rays. Peeling back clear office-supply sticky tape emits X-rays.

♒ Wilhelm Röntgen ♒

In 1895, German physicist Wilhelm Röntgen accidentally created X-rays in the lab while experimenting with electron beams in a gas discharge (vacuum) tube. In the process, he accidentally founded *radiology*, by seeing an image of the bones of his hand illuminated by X-rays.

Not long on imagination or ego, Röntgen coined the term *X-ray* and stuck to it, even as the penetrating new rays would come to bear his name in several languages: Röntgen rays.

Röntgen died of intestinal carcinoma, but his X-ray work is not generally considered causal, because he was one of the few pioneers in the field who routinely used protective lead shields during experimentation.

♎ Pedoscopes ♎

Goofing with X-rays continued well past Röntgen. From the 1920s to the 1960s in the United States, and into the mid-1970s in Britain, X-ray fluoroscopes were employed to fit shoes. Called pedoscopes, their use was a sales gimmick: claiming a better fit, and more fun for the kids at the shoe store. The traditional method was equally effective, more convenient, and much less hazardous to health.

The danger of the pedoscope was belatedly revealed in the US in 1949. The machines were quietly phased out as the next decade wore on. Pedoscopes may not have resulted in gross overexposure to customers but did take a toll on shoe clerks who regularly used them.

♎ Gamma Rays ♎

The EM spectrum energetically ends with gamma rays, with wavelengths less than 10 picometers (10^{-12} meters): less than the diameter of an atom.

In deep space, gamma ray bursts occur regularly. The Sun powers low-energy gamma rays (up to 10^{10} eV). Mid-range rays are energized by the shock waves of supernovae. The origin of high-energy rays, above 10^{15} eV, is a mystery.

Planet-side, radioactive atoms and nuclear blasts create gamma rays. Gamma rays readily kill living cells.

♋ Particulate Radiation ♋

Particulate radiation comes from subatomic particles carousing about. An atom with too many or too few neutrons makes for an unstable nucleus. The nucleus becomes radioactive. A disintegrating atom produces particulate radiation, including alpha and beta particles.

When radium, uranium, and polonium decay, they effervesce *alpha particles*: 2 protons and 2 neutrons. Alpha particles are relatively chunky and slow. Protons and neutrons that can travel only short distances.

Alpha particles can be stopped by a piece of paper, or skin. Inhaling or ingesting alpha particles is dangerous, in exposing sensitive internal tissues to radiation.

Beta particles are hightailing electrons or positrons. Beta decay may be β^-: where a neutron converts into a proton, an electron, and an electron-type antineutrino, or β^+: where a proton transforms into a neutron, a positron, and an electron-type neutrino.

Lower-energy beta particles can be slowed, or even stopped, by clothing or aluminum foil. Higher-energy beta decay can reach speeds that are *ultrarelativistic*: approaching the speed of light.

When passing through matter, particulate radiation ionizes whatever it encounters, incrementally losing energy in the process.

The distance to where a charged particle is spent is called the *range* of the particle. Range depends on particle type, initial energy, and what matter is in the way.

Similarly, energy loss per path length, *stopping power*, depends on the particle's type and energy, and the material bombarded. Stopping power, which equates to density of ionization, increases until it reaches an apex, the *Bragg peak*, just before the energy drops to zero. English physicist William Henry Bragg discovered the elemental dynamics of ionizing radiation in 1903.

♎ Earth's Age Via Isotope ♎

The heaviest elements are naturally radioactive. They readily shed to get down to *lead*, the heaviest steady-state metal.

For his 1948 dissertation project, American geochemist Clair Cameron Patterson determined the duration of Earth by measuring the age of meteorites found on the planet, using a technique developed by his academic mentor, American chemist Harrison Brown: counting lead isotopes in igneous rocks.

Patterson calculated Earth to be 4.55 billion years old, give or take 70 million years. The number stands still, though the margin of error has been whittled down to 20 million years.

⚗ Radiation Exposure ⚗

Ionization is the energetic process of converting an atom or molecule into an ion with randy electrons.

Radiation may be ionizing or non-ionizing. Both can be harmful to living organisms and change the natural environment. Non-ionizing forms of radiation are at the low end of the electromagnetic spectrum. They include radio waves and visible light. While non-ionizing radiation is considered less dangerous than ionizing radiation, overexposure or long-term exposure to non-ionizing radiation can degrade health.

Long-term exposure to transmission power lines affects health, especially infants and youngsters during early development, but it took decades of research for scientists to come to even a tentative acceptance of that. Cell phones alter brain function on the side of the head used.

Overexposure to infrared radiation can result in burns. Ultraviolet radiation overexposure can be insidious, because there are no immediate symptoms of overexposure. Sunburn is exemplary.

Besides sunlight, sources of UV include black lights and welding tools. UV overdoses can lead to cataracts and skin cancer, as well as compromising the immune system.

But nothing is simple. Ultraviolet radiation in the eyes, and on the skin, is essential to stimulate the body to producing vitamin D. As little as 5 to 15 minutes, 3 times a day, is more than enough.

Higher-energy radiation can ionize atoms: strip electrons off by knocking them out of their shells. This leaves a *cation*: an atom with a net positive charge. At higher energy, the nucleus of an atom can be destroyed.

Ionizing radiation may be particulate or by high-energy electromagnetic rays.

The noble gas *radon*, found underground, is a natural ionizing radioactive material. Radon is one of the densest elements that remains gaseous under atmospherically ambient conditions.

X-rays and gamma rays are a common man-made ionizing radiation. Bones are examined using X-ray photography.

As a shotgun therapy, gamma rays are blasted for radiation treatment to kill cancer cells. The stopping power Bragg peak effect allows gamma radiation therapy to work on tumors without demolishing a large swath of nearby healthy tissue.

X-rays and gamma rays are much the same, but from a different origin. Whereas gamma rays emanate from inside an atom's nucleus, X-rays are stirred from discombobulating the electrons of an atom. X-rays are less penetrating, and so can be stopped by a few millimeters of lead, which absorb the energy.

Overexposure to ionizing radiation can cause genetic mutations, raising the risk of cancer. At higher exposures, burns or radiation sickness results.

✄ Cosmic Radiation ✄

Besides the gentle echo of EM background radiation from the universe's distant past, cosmic radiation continues. Highly charged particles are ejected from supernova explosions: ersatz atoms, stripped of their electrons by the extreme temperatures within these giant furnaces.

Supernova-borne particle types vary: primarily hydrogen nuclei (protons) (85%) and helium nuclei (alpha particles) (12.5%), but also heavier nuclei, such as iron and nickel (1%). There is also a smattering of electrons (1.5%) in the cosmic radiation broth. Their scurry is ultrarelativistic (near the speed of light). The radiation is isotropic: coming from all around, and so constantly bombarding Earth's vicinity.

Solar wind is the constant, fluxing flow of particulate released from the Sun's atmosphere. A portion of the galactic radiation is deviated by the magnetic field carried by the solar wind. So, the more intense the solar wind, which runs in an 11-year cycle, the less the galactic cosmic radiation in the neighborhood of the Sun.

Between 10 and 100 keV, in a stew of mostly electrons and protons, the emissions from the solar wind are less energetic than cosmic radiation. Few solar wind emissions reach Earth's surface and they are not evenly distributed. The rotating ball with an iron core called Earth creates a magnetosphere that deviates most of the solar wind and cosmic radiation. But determined particles breach the magnetosphere and reach the upper layers of the atmosphere, where the ions socialize with local atoms. These collisions create less energetic secondary radiation that sometimes manages to reach the ground.

There are also solar flares which dust up the solar wind and can cause disruption in communications satellites in orbit. The largest solar flares originate in complex formations of sunspots, the dynamics of which are little understood.

On average, only ~11% of the ionizing radiation that strikes Earth's surface is cosmic in origin.

✂ Earthbound Radiation ✂

More radiation emanates from natural land-based sources than from the sky. Radon, the gaseous descendant of earthbound natural uranium, concentrates in enclosed areas, such as houses.

Soil-based radiation comes from surface rocks, granite in particular, which totes radioactive elements dating from the formation of the planet. Water and foods consumed have bits of radioactive elements. Finally, potassium-40, naturally present in human tissue, completes the story of radiation coming from everywhere. It's a miracle anything lives, what with all the atomic noise going on.

♎ Watch That Dial ♎

Wrist watches became popular during World War I. A few years later, quite a few companies came up with the bright idea of rendering watches easier to read, and even glow in the dark, using radium-laced paint. The Radium Dial Company of Ottawa, Illinois was one; U. S. Radium Corporation of Orange, New Jersey another.

All told, 4,000 young women were hired for the painstaking painting by hand. They were encouraged to make a fine point on their brushes by rolling the tips on their tongues before dipping them in the paint.

Their bosses told them "not to worry. If you swallow any radium, it'll make your cheeks rosy." All the while, staff doctors routinely checked dial painters for radioactivity exposure, though the employees weren't informed.

Many died, and the companies lied. Workers were exposed to more than 1,000 times the amount of radiation scientists considered safe at the time.

In a Chicago court in 1938, after a worker sued, a supervisor explained why the company didn't post the results of the physical exams: "My dear girls, if we were to give a medical report to you girls, there would be a riot in the place."

The radium scattered from the Ottawa factories when the buildings were bulldozed in 1969 and 1984. No precautions were taken.

ಬ Molecules ಛ

A *molecule* is a bonded conglomeration of atoms in tentative relationships. 37 distinct types of chemical reactions are known.

When atoms and molecules undergo chemical reactions, the electrons are the ones that do the heavy lifting. They regroup and move to allow new bonds between molecules to be created or destroyed. ~ Swedish atomic physicist Marcus Isinger

Empedocles considered chemical changes as emotional relations: substances combine in love to birth something new, or discordantly divorce. No wonder it's called chemistry.

Nature does what must be done to achieve equilibrium, which is a minimum-energy configurational space. ~ American chemist Scott Chambers

❧ Bonding ❧

Atoms join into molecules by sharing electron bonds. In becoming a molecule, atoms give or take electrons, depending upon their species. Atoms with few electrons in their outermost shell are (electron) *donors*, while those in want of a few are *acceptors*.

In forming molecules, atoms tend to find the most stable relationships among themselves. This takes the form of bonding that requires the least energy to sustain the association.

Without its outer shell of electrons filled to the brim, an atom feels lonely, and readily mingles with another. The mingling may be inscrutable.

Glass seems solid, but at the molecular level, glass resembles a liquid. Its molecules do not make a tidy lattice, like crystals do. Instead, glass is a molecular jumble. How glass settles into its dual state – solid with a liquid structure – is a mystery.

♋ Valence ♋

Each electron shell of an atom has a limited capacity. The 1st shell layer (shell 1) is stable at 2 electrons. The 2-electron atom helium (He) is relatively stable, and so reluctant to join with other atoms to make a molecule.

♎ Hydrogen ♎

Helium's lightweight cousin, hydrogen (H), is the most abundant chemical element, and the simplest. With only 1 electron, hydrogen is a randy joiner: willing to donate its electron to make a molecule. Hydrogen's enthusiasm plays an indelible importance in cosmic construction, and in the dynamics of life.

The simplest molecule is diatomic hydrogen gas (H_2), which creates a closed shell configuration with 2 electrons. *Diatomic* refers to 2 nuclides of the same type of atom (*atomic species*) forming a molecule. Most diatomic molecules are gases. Hence H_2, O_2, and N_2 are molecularly happy campers, mated with each other.

Normally, hydrogen is a gas of diatomic molecules. Pressure and temperature can change the situation. When cooled enough, hydrogen solidifies. Compressed hydrogen has 3 phases as a quantum crystal, consisting of rotating molecules in tightly packed hexagonal lattices.

Another phase appears at 2.2–3.4 million times normal atmospheric pressure, with 2 divergent types of hydrogen molecules in its structure. One type interacts very weakly with its neighbors, which is unusual for molecules under high compression. The other type of molecule tightly bonds with its neighbors, forming planar sheets. Solid hydrogen under these conditions is on the borderline between a semiconductor, like silicon, and a semimetal, like graphite.

At tens of millions times pressure, hydrogen turns into a metal. The interior of Jupiter is largely cold, metallic hydrogen; some of it as a liquid metal.

> With hydrogen, nothing is simple. ~ condensed matter physicist Eugene Gregoryanz

The 1st shell of an atom, as with helium, has only 2 electrons. The 2nd shell is full at 8 electrons. Since negatively charged electrons are attracted to their positively charged nucleus, electrons in an atom will generally occupy an outermost shell only when the shells within have been filled.

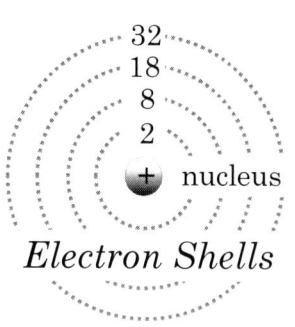

Electron Shells

The general formula for filling shells is *2 x 2ⁿ*, where n is the shell number. Hence: $2 = 2 \times 1^2$; $8 = 2 \times 2^2$; $18 = 2 \times 3^2$; $32 = 2 \times 4^2$.

Atoms with an incomplete outer shell naturally bond with other atoms to form molecules. These bonds involve sharing electrons.

The *valence* of an atom tells how many electrons an atom needs to fill its outermost shell. The chemical properties of an atom are defined by its outermost shell: the *valence shell*.

The simple story is that the valence shell determines how reactive an atom is. More accurately, the electrons traveling farthest from the nucleus – the most energetic ones – determine how an atom reacts chemically. These *valence electrons* are necessarily in the valence shell.

Oxygen, at 2|6, has a valence of 2, and so readily combines with 2 atoms of hydrogen gas to make H_2O: water (6 (from oxygen) + 2 (from hydrogen) = 8 (stable valence shell)).

The 5 most abundant elements in the solar system are: hydrogen (1), helium (2), oxygen (2|6), carbon (2|4), and nitrogen (2|5). Only helium is not prone to forming molecules.

Helium is rare on Earth and has no role in organic chemistry. The other abundant atomic species – carbon, nitrogen, oxygen, and hydrogen – all play active roles in organic chemistry precisely because they are reactive.

✂ Covalent Bonds ✁

In forming molecules, electron shells fill themselves out by sharing pairs of electrons. This is basic chemical affinity: forging a *covalent bond*, with electrons portioning their orbits among atoms.

There are several types of covalent bonds, which vary by the nature of electron sharing. Sigma, pi, and 3-center 2-electron bonds are illustrative.

A *sigma bond* (σ) is the strongest type of covalent bond, corresponding to valence shell sharing as previously described.

A *pi bond* (π) is formed by overlapping atomic orbital lobes. Pi bonds are more diffuse than sigma bonds, and so somewhat weaker.

In a *3-center 2-electron bond* (3c-2e), 3 atoms share 2 electrons. This odd bonding comes up an electron short. Typically, the bonding orbital is not equally allocated, but skewed toward 2 of the 3 atoms in the molecule. The simplest example of 3c-2e bonding is H_3^+: the trihydrogen cation.

¤ ✧ ¤

Covalent bonding takes energy. The strength of a chemical bond – *bond energy* – is measured by the energy needed to separate the bonded atoms of a molecule.

Electronegativity is electron sex appeal: the ability of an atom or molecule to attract electrons. Electronegativity is an inelegant term, and confusing, because it applies to the positive appeal that an element has for an electron, albeit an electron is negatively charged.

On the other end of the electron exchange program, some elements are willing to donate electrons. The measure of electron donation willingness is termed *electropositivity*.

ß Bond Orders ৯

A covalent bond may be of a single, double, or triple electron pair. Quintuple bonds exist, but are rare, confined to certain metals. A quintuple bond involves 10 electrons bonding between 2 metal centers.

Diatomic fluorine (F_2) is an exemplary single bond. Molecular oxygen (O_2) is cozy as a double bond, while molecular nitrogen (N_2) forms a triple bond.

Bond	F_2	O_2	N_2
Order	Single	Double	Triple
Length	1.42	1.21	1.10
Energy (10^{-19} Joules)	2.6	8.3	15.6

The higher the bond order, the tighter the bond – literally: bond length shortens. Bond energy increases with bond order. Tighter bonds are energetically stronger. Hence, the triple bond of dinitrogen is a particularly strong one.

৯ Ionic Bonds ৯

Electrostatic attraction results in 2 oppositely charged ions coupling. This is *ionic bonding.*

Ionic bonds are formed between a cation and an anion. A *cation* is an ion with a positive charge, as it has fewer electrons than protons. Conversely, an *anion* has a surfeit of electrons, and so is negatively charged. A cation is commonly a metal, while the average anion is a nonmetal.

All ionic compounds also have some degree of covalent bonding. The larger the electronegativity between the atoms involved in bonding, the more ionic (polar) the bonding is.

Ionic molecules are electrically conductive when molten or in solution but not when they are a solid. Ionic compounds tend to be water-soluble and have a high melting point.

While compounds with ionic bonds intact are electrically neutral, they are subject to *ionization* upon encountering a solvent: the partner atoms disassociate back into ions. *Electrolytes* are substances, such as salts, acids, and bases, that release ions when dissolved in water.

⚡ Hydrogen Bonds ⚡

A *hydrogen bond* is the attraction between a hydrogen atom and an electronegative atom, such as oxygen, nitrogen, or fluorine. Hydrogen bonds may be within a single molecule or between molecules.

A hydrogen bond is weaker than a covalent or ionic bond, but stronger than the van der Waals interaction, which couples 2 dipoles via HD connection.

Commonly, hydrogen bonds form between the ends of polar molecules. Water molecules attain their fluidity through hydrogen bonds. The high boiling point of water (100 °C) owes to intermolecular hydrogen bonding, as does water's surface tension.

Hydrogen bonds hold the 2 halves of DNA together. Hydrogen bonds facilitate the complex 3D structures that proteins take. More generally, hydrogen bonding plays an important role in the structure of polymers.

⚡ Bonds Beyond ⚡

There is a standard law that says as you lower the temperature, the rates of reactions should slow down. ~ English chemist Dwayne Heard

The vast expanse of interstellar space is supposed to be too cold for most chemical reactions to occur. The more frigid it gets, the harder it is to spark a chemical reaction for lack of energy, which is the very definition of *cold*.

Yet a vast variety of complex organic molecules are formed in space. Some reactions transpire on the surface of cosmic dust grains, or with a little help from gamma rays or stray high-energy electrons. But most happen beyond the laws of chemistry.

Colossal clouds of alcohol float in Sagittarius B2: a giant molecular basin of gas and dust, 120 parsecs from the center of the Milky Way. These clouds are at a balmy 40 K or less: far too cold to explain such ample booze in space. Interstellar spirits brew despite chemistry's cardinal rules giving them the cold shoulder. They do so by *quantum tunneling*, which is an HD trick that allows a particle to surmount a barrier that it cannot breach classically.

> The tunneling of a particle through a potential barrier is a key feature of quantum mechanics that goes to the core of wave–particle duality. The phenomenon has no counterpart in classical physics. Tunneling events are only as 'instantaneous' as the electron wavefunction collapse that orthodox interpretations of quantum mechanics associate with the appearance of continuum electrons.* ~ Russian physicist Igor Litvinyuk *et al*

The label *quantum tunneling* names, but does not explain, one of Nature's most mysterious mechanisms. Despite the intense energy already invested, nuclear fusion nevertheless depends upon quantum tunneling.

The method for making methoxy comes in combining methanol gas with a hydroxyl radical. Both are found in the cold expanse of space, but the energy to put the two together is not. Nonetheless, reactions happen, and prodigiously so. Quantum tunneling bestirs interstellar spirits 50 times faster than would occur sitting at room temperature.

> There is organic chemistry in space of the type of reactions where it was assumed these just wouldn't happen. Scientists have been severely underestimating the rates of formation and destruction of complex molecules, such as alcohols, in space. ~ Dwayne Heard

---

* *Continuum mechanics* is the study of matter as a process.

✍ HD Harmonically Bound ✍

Bonding can also occur by the non-conservative forces responsible for interaction-induced coherent population trapping. The bound state arises in a dissipative process and manifests itself as a stationary state at a preordained interatomic distance. Remarkably, such a dissipative bonding is present even when the interactions among the atoms are purely repulsive. ~ Russian physicist Mikhail Lemeshko & German quantum physicist Hendrik Weimer

A bond between atoms typically forms when it is energetically more favorable for atoms to stick together than stay apart. This requires an attractive force.

Even energetic repulsion between atoms can be overcome by local HD harmonic fluctuations of the ground state, which generate a quantum interference that nullifies repulsion, trapping atoms into union at a distance which sets the bond length.

The nature of an HD harmonic bond is strikingly different than ordinary chemical bonds, most notably in being remarkably robust. Even applying a constant amount of energy may not break a harmonic bond.

✍ Molecular Geometry ✍

Shape is destiny in the world of molecules. ~ American philosopher Daniel Dennett

The position of each atom in a molecule is determined by the nature of bonding with its neighbors. Molecular geometry is determined by the quantum behaviors of atomically bound electrons. Molecular geometry characterizes the shape of molecules by the positions of constituent atoms in space – evoking the bond length between 2 atoms, the bond angles among 3 atoms across 2 bonds, and the torsional angle of atoms chained together, such as carbon chains.

Torsional angle is the angle between 2 planes. In molecular geometry, bonded atoms in a chain form 1 or more planes.

Molecules form a tremendous variety of 3D shapes. The geometry of a molecule affects and reflects its reactivity, polarity, phase, color, magnetism, and biological activity.

Isomers are compounds with the same molecular formula but different shapes. As molecular structure is functionally significant, isomers may have different properties than their cousins with the same atomic composition.

Each molecular bond is a region of negatively charged electrons. These regions repel each other. Individual bonds want to stay as far away from others as possible.

Molecules assume a geometry that minimizes the repulsive energies of electron orbital regions. These regions always include shared electron pairs between atoms in a molecule, but valence shell electrons that stay at home also get into the act.

Most of the modern understanding of chemistry, including the very notion of a well-defined molecular structure, rests on the concept of a potential energy surface – a $3N$-dimensional 'landscape' that plots the total energy of a collection of N atoms as a function of the atomic positions. ~ American chemist Todd Martinez

☿ Lone Pairs ☿

The electrons in the valence shell of an atom in a molecule may not be shared. A nonbonding valence electron pair is termed a *lone pair*.

Lone pairs fashion a high electron density in their region of orbital space. So, even though they are not bonded, lone pairs influence the shape of a molecule.

In bonding, shared electrons are concentrated between 2 atoms. Lone pairs, without the attention of a 2nd atom, are cozy homebodies: typically located a bit closer to the atomic nucleus than bonding pairs.

Lone pairs often exhibit a negative polar character. As electron regions repel each other, a lone pair nudges its brethren bonding pair away, reducing molecular bond angle.

☿ Polarity ☿

When atoms with different electronegativity join by covalent bonding, electrons may not be equally shared: electrons are pulled more toward one atom than another. When this happens, the force of alignment causes one end of a molecule to have an overall negative charge, leaving the other end to assume a positive charge. A molecule with such an unequal charge distribution is *polar*.

Polarity means that a molecule has positive and negative poles; in other words, the molecule possesses an electric dipole moment. Polarity underlies several physical properties, including surface tension, solubility, and critical thermodynamic points (melting and boiling). Water (H_2O) is the poster child of polar molecules.

Polar molecules interact through dipole–dipole intermolecular forces and hydrogen bonds. A polar molecule with multiple polar bonds must have an asymmetric geometry so that the bond dipoles do not cancel each other.

Nonpolar molecules share electrons equally, rendering them electrically neutral. Oxygen (O_2) and methane (CH_4) are nonpolar, as are lipids: fat not being so easily excited makes it a relatively contented energy storage medium.

෪ Phases ෨

> While all matter is characterized by mass, a gas has only mass. A liquid has, in addition, a specifiable volume. A solid has, besides mass and volume, some specific shape. ~ Steven Vogel

The states of matter may be positively classified in various ways. Gases and liquids are *fluids*, whereas crystals and glasses are *solids*. Alternate views emphasize density or atomic arrangement.

Gases naturally expand to occupy available volume. A *plasma* is an ionized gas. Liquids maintain their volume but adapt their shape when confined. Solids maintain their volume and shape.

99% of the matter in the universe is plasma, including stars and the medium that permeates in between. Most plasmas are in a turbulent state, threaded with magnetic fields that generate magnetic reconnection.

A *crystal* is characterized by an orderly, repeating 3D pattern. A lattice is a typical crystal.

In contrast, *glass* has an amorphous structure that undergoes a *glass transition* while being heated toward a liquid state. The glass transition temperature is always lower than the melting temperature.

The glass transition of a liquid toward a solid may be inspired by cooling or compression. This transition is a smooth increase in viscosity, by as much as 17 orders of magnitude before any pronounced change in structure.

✄ Triple Point ✄

At any temperature and pressure, the thermodynamically stable phase of a compound is the one with the lowest free energy – that is, compounds are most stable when then have acquired the maximal amount of local energy. The arrangement, motion, and interactions among chemical constituents determines this.

When 2 phases coexist stably, their free energies must be equal. This condition forms a coexistence line in a phase diagram.

Ice and water coexist in equilibrium at 0 °C and atmospheric pressure. Increasing pressure decreases melting point.

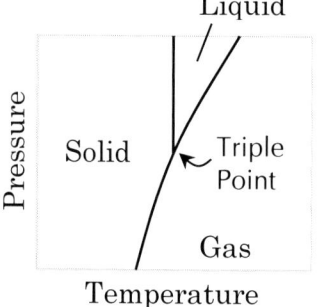

Similarly, water vapor and liquid coexist stably at 100 °C and atmospheric pressure. Decreasing pressure drops that temperature. Hence water boils at a lower temperature on top of a mountain than at sea level.

These 2 coexistence curves can only intersect at a single value of temperature and pressure. The *triple point* is the temperature and pressure at which a specific molecular structure coexists in the 3 phases: gas, liquid, and solid.

The singular stability of *water* – where water is simultaneously vapor, liquid, and ice – is at 273.16 K (0.01 °C) and 611.73 pascals. This particular triple point defines the Kelvin temperature scale. From its triple point, slight changes in pressure and/or temperature transforms water alternately to ice, liquid, or vapor.

The triple point of H_2O is also the minimum pressure at which water is liquid. At pressures below the triple point, such as in outer space, a rise in temperature (at constant pressure) turns ice into vapor; a process termed *sublimation*. A temperature rise at constant pressure above triple point melts ice to liquid.

Water is peculiar in many ways, including its triple point behavior. Whereas the melting point of ice to liquid water decreases as a function of pressure, for most substances, the triple point is the minimum temperature at which a liquid can exist. For water, at a constant temperature just below the triple point, compression turns water vapor to ice, then to liquid form.

In addition to phase transition, for solid *polymorphs* – materials that may exist in different forms – there can be multiple solid triple points. Helium-4 is a unique polymorph in having a triple point for 2 different fluid phases (known as its *lambda point*).

✀ Wettability ✄

The internal coherence of a liquid makes a surface something unattractive which the system will arrange itself to minimize. A droplet will spontaneously round up into a sphere. ~ Steven Vogel

Although the surfaces of liquids and crystals differ, they both have a tendency toward minimizing surface area. This owes to the energy associated with such a boundary.

A surface atom lacks neighbors on one side and so has a deficit of interatomic bonds. Bond energies are negative, so the missing surface bonds correspond to a positive energy contribution.

For a crystal, *surface tension* arises from stretching interatomic bonds, whereas liquid surface tension is more about the extra atoms introduced when spreading out in increased surface area.

Interesting dynamics transpire at the interface of different substances. Cocktails come with complete *miscibility*: a thorough water-alcohol mixture. This passage of molecules between species is known as *diffusion*.

If instead the interphase bonding is competitive, the liquid *wets* the solid. Solders and brazes function as they do by readily wetting metallic surfaces.

Water is especially sensitive to its environment: reacting distinctly depending on what it interfaces with. This reactive versatility is called *wettability*: spreading when a material is wettable or beading when not. Wetting occurs when the adhesion of liquid to solid gains the advantage over internal cohesion.

Mercury (Hg) does not wet glass. Hence its *meniscus* (surface) in a thermometer is concave: bulging upward, lower along the edges.

In contrast, water wets glass, owing to the attraction of H_2O to the silica in glass. Hence water in a glass is convex: bulging downward, higher at the edges.

Capillary action – the ability of a liquid to readily flow when narrowly confined in a solid tube, essentially ignoring gravity – owes to wetting. Vascular plants are major employers of capillary action.

ϼ Superfluidity ϟ

Helium is the 2nd lightest and 2nd most abundant element, right behind hydrogen. The two could not be more different. Helium's stability is the opposite of hydrogen's volatility.

Helium is a gas except under extreme conditions. Helium has the lowest boiling and melting points. Helium becomes a liquid at 4.2 K. Below 4 K, helium fiercely boils.

Below 2.172 K, the boiling stops; helium becomes a superfluid. A superfluid exhibits zero viscosity and zero entropy. Superfluid helium flows without friction; through tiny holes as small as a molecule, up and out of a tube, over the edge of a cup.

More particularly, helium-4, the ubiquitous isotope comprising 2 neutrons with 2 protons, becomes a superfluid more readily than helium-3, which is a rare isotope, with only 1 neutron per helium atom. Helium-3 only becomes a superfluid when chilled to 2 millikelvins.

The reason is that helium-4 atoms are bosons, whereas helium-3 atoms are fermions. Hence helium-4 corresponds to the characteristics of Bose-Einstein condensation, whereas helium-3 becomes a fermionic condensate.

A fermionic condensate interacts by Cooper pairing between atoms, as contrasted to the electron Cooper pairs that facilitate superconductivity.

Helium-4 is bosonic via the subatomic components (protons, neutrons, and electrons) canceling each other's complementary spins, resulting in zero spin for the helium-4 atom as a whole. This facilitates the stability of helium-4.

Only neon is less reactive than helium, and it needs a cluster of buddies to be so. Neon is most commonly molecular as ^{20}Ne, whereas helium is monoatomic (^{1}He).

Helium is not the only element capable of superfluidity. Rubidium, a highly reactive silvery-white metal, becomes a superfluid at 500 nanokelvins (500 10^{-9} Kelvin).

৯ Supersolids ৶

Whereas a liquid conforms to its surroundings, a solid rigidly retains its shape. A *supersolid* is a state of matter that marries solidity with superfluidity.

Like a solid, a supersolid is rigid. Its atoms snap back into place if displaced. But, owing to its highly ordered state, particles can flow through a supersolid without viscosity, aided by quantum mechanics which smear out the positions of quanta being shoved aside.

From the 1950s, theoretical physicists argued whether a supersolid could exist. Failed experiments into the 2110s confirmed doubters. But then in 2016 came experimental proof that supersolids can exist in ultracold quantum matter.

৯ Energy Transitions ৶

The energy of an atom or molecule changes not only when it loses or gains electrons, but also when electron orbits change. It takes energy to move electrons from an inner shell outward. Conversely, some energy is released if electrons move to an inner shell.

The energy to effect shell transitions may be meager. For example, some solids naturally fluctuate between 2 different structural forms, each with a different oscillation frequency. Slight differences at the subatomic level influence the dynamics of transformation.

Intermolecular forces determine temperature magnitude of melting and boiling as well as the solubility of molecules. The greater the attraction between the elements of a specific compound, the higher the melting and boiling points are likely to be.

Nonpolar molecules are most soluble in nonpolar solvents. Likewise, polar molecules are most soluble in polar solvents.

At the molecular level, life transpires as a channeling of energy flows; the chemistry of *metabolism*. The elements that comprise the mainstay of organic life readily combine in a variety of ways.

As a network of about a thousand enzymatic reactions, metabolism fuels growth by converting nutrients into building blocks and energy. ~ Swiss microbiologist Robert Schuetz

ༀ   Catalysts & Enzymes   ༃

Existence would be much different if chemical reactions transpired at their nominal rate. Instead, Nature quickens the pace with catalysts, which arise in a wide variety of chemical contexts. If not for catalysis, life would not be possible.

ༀ Catalysts ༃

Catalysts promote chemical reactions. The term *promote* is used because it is possible for a reaction to happen without a catalyst, but the likelihood of non-catalytic reaction is analogous to a coin flip resulting in the coin standing on edge: not likely.

Catalysts lower the activation energy required for a reaction to occur. In organic chemistry, that role is played by proteins known as *enzymes*.

☙ Enzymes ❧

> Alcoholic fermentation is an act correlated with life. ~ Louis Pasteur

French chemist and microbiologist Louis Pasteur, thirsty for knowledge, concluded in 1857 that some vital force catalyzed yeast in its fermentation of sugar into alcohol. He termed that vitality *ferments*. What he discovered were enzymes in action. German physiologist Wilhelm Kühne coined the term *enzyme* in 1877.

> An enzyme is able to speed up a chemical reaction by as much as 10 million times. It had to do this by lowering the energy of activation – the energy of forming the activated complex. It could do this by forming strong bonds with the activated complex, but only weak bonds with the reactants or products. ~ Linus Pauling

An *enzyme* is commonly defined as a protein that acts as a catalyst in reactions involving proteins or other substrates, typically polymers. A *polymer* is a large molecule (macromolecule) comprising repeating molecular units (monomers).

Enzymatic action is the karmic wheel of organic life in polymeric construction (anabolism) and deconstruction (catabolism). Enzymes enable the manufacture of macromolecules which may be consumed. Conversely, consumption requires enzymes to break big molecules down to release the energy within.

Because proteins are picky about being prodded to perform, every biochemical reaction is promoted by a specific enzyme.

> Every organism contains a vast number of different enzymes, involved in a complex web of metabolic interactions. ~ Andrew Clarke

Enzymes are often quite specific in their binding, but some are also produced to be non-specific, with specificity generated by controlling access to specific substrates – a learning mechanism. Proteins can thus be made to act like switches, by having 2 or more conformations, thus able to interact differently with signaling proteins. This is a facet of intracellular communication.

> An enzyme can use a rare and transient conformational state in its substrate to direct an outcome. ~ Canadian geneticist David Pulleyblank

Enzymes dance to do their job. Subtle changes in shape play a crucial role in enzyme function. Enzyme conformational changes are highly dynamic.

2 enzymes with virtually identical molecular shapes may catalyze reactions at very different rates. One may regulate insulin production in humans, while its evil twin gives a bacterium the power of bubonic plague.

The rate of molecular motions is critical to enzyme function. Context matters.

The enzyme responsible for insulin regulation moves slowly, ensuring certainty in cellular processing. Conversely, the plague enzyme is carefree; moving 30 times faster. While proper construction requires meticulousity, the power for destruction can be swiftly rendered.

The exuberance of enzymes often needs to be adjusted to suit cellular needs. The rate of enzymatic activity is affected by a variety of conditions: temperature; ambient chemical environ, such as pH; and chemical concentrations. Whereas *activators* increase activity rate, *inhibitors* slow enzymes down. Many drugs and poisons are enzyme inhibitors.

♎ Humble Agent ♎

The Onion reported a 2013 interview with an α-amylase enzyme involved in speeding up the breakdown of starch into maltose. Calling itself "just a catalyst, nothing more," the enzyme remarked: "all I did was lower the activation energy required for the reaction to take place, but if I don't have an amazing substrate to act upon, there is no reaction, period."

α-amylase commented on all the factors necessary for success. "Say the pH isn't slightly acidic, or the ions are not properly aligned. Are we left with a simple sugar that can be used as an immediate energy source? Absolutely not. You need teamwork for that, and thankfully, that's what we had today."*

℘ Pseudoenzymes ଖ

Every cell produces a significant number of enzymes which appear to be inert. Of the 518 enzymes that are supposed to catalyze proteins involved in phosphorylation – protein *kinases* – about 10% seem to be duds, as they lack key amino acids to act as catalysts.

The researchers who discovered these pseudoenzymes in 2002 were shocked. "We thought we must have got it wrong," fumed American geneticist Gerard Manning.

The DNA which delivers dead enzymes are not degraded. Those protein-producing sequences have hardly changed over millions of years.

Instead of catalysts, pseudoenzymes are employed in a variety of other roles. Some help 'true' enzymes do their job by shoving them into the correct shape. Some provide a venue where proteins can mingle. Some latch onto receptors to help cells communicate. Some act as bodyguards, escorting another protein to a work site.

The lesson of pseudoenzymes is that there are many other jobs which enzymes do besides catalyzing reactions. A better definition of an *enzyme* is a protein that facilitates the activities of other proteins or substrates.

ഏ Water ര

If there is magic on the planet, it is contained in water.
~ American anthropologist Loren Eiseley

Hydrogen and oxygen are among the most abundant elements in the universe. Water exists in outer space. Much of the water in space is a byproduct of star formation.

---

* The interview was, of course, fictitious. But the point that biochemical reactions are often daedal behaviors is true.

Water formed early in the history of the universe; as quickly as supernovas spread oxygen into space, allowing it to combine with ubiquitous hydrogen gas.

In most planets, including Earth, vast reservoirs underground belie a more modest surface presence. The Moon appears bone dry but stores considerable H_2O within.

Even Mercury, with daytime surface temperatures reaching 400 °C, hot enough to melt lead, has ice tucked within.

☿ History ☿

Philosophers from antiquity regarded water as a basic element, typifying all liquids. Scientists did not discard that view until the back half of the 18th century.

English chemist Henry Cavendish discovered hydrogen in 1766, calling it "factitious air." Cavendish synthesized water in 1781 by detonating hydrogen, which combined with the oxygen in the air to form water as a byproduct.

Lavoisier termed both *oxygen* (1779) and *hydrogen* (1783) and determined water as a combination of oxygen and hydrogen, thus dispelling water as an element unto itself.

☿ Properties ☿

Water shows exceptional properties, different from all other liquids. ~ Austrian chemist Katrin Amann-Winkel *et al*

Pure water is an odorless, tasteless liquid, with a bluish tint observable only at a deep volume. Water is transparent in the visible spectrum, but strongly absorbs both infrared and ultraviolet light.

Water is so commonplace that it is regarded as a typical liquid. Water is anything but.

Water is one of lightest gases. Liquid water is much denser than expectable. Water's solid phase – ice – is surprisingly featherweight in light of its liquid form.

Water can be extremely slippery and exceedingly sticky simultaneously. This slippery/sticky behavior is how the feel of water is sussed. There is no direct sensation of moisture.

Every body of water, from droplets on to more voluminous realms, is an intricate viscoelastic system, with an underlying extended structure governed by quantum forces; whence water's enduring mystery.

Although water is one of the simplest molecules, it is in reality a very complex liquid displaying more than 64 counterintuitive anomalies, most of which have not been adequately explained. ~ Italian physicist Francesco Mallamace *et al*

℘ Water & Life ℈

Life as we know it depends absolutely on the presence of liquid water, and the existence of the liquid state sets the boundary conditions for life itself. ~ Andrew Clarke

The quirks of water are what makes life possible. The high cohesion of water molecules gives it high freezing and melting points, making its liquid phase predominate: the phase upon which life directly relies. Water's thermodynamics – its large heat capacity and high thermal conductivity – vitally help organisms control body temperature.

The high latent heat of water defines an evaporation which resists dehydration while yielding considerable evaporative cooling. Water's unique hydration properties critically affect important biological macromolecules – proteins and nucleic acids – that determine their 3D structures, and hence their biological functions.

Water being a small molecule, an excellent solvent, and having high relative permittivity, all contribute biological utility. The easy ionization and proton exchange of H_2O contributes to a rich repertoire of biological interactions.

Water's thermal connectivity and density characteristics means that all of a body of freshwater, not just its surface, must be close to 4 °C before freezing can start. The freezing of lakes, rivers, and seas is top-down. This insulates water from further freezing, permitting bottom dwellers to survive.

Water reflectivity aids rapid thawing. Water's density-driven thermal convection affords seasonal mixing in deep temperate waters, carrying life-essential oxygen into the depths.

The compressibility of water lowers sea level ~40 meters, yielding 5% more land on Earth. Water's high surface tension, and its expansion on freezing, engenders the erosion of rocks, providing soil upon which life on land may thrive.

The tremendous heat capacity of oceans allows them to act as global thermal reservoirs, such that sea temperatures vary only 1/3rd as much as land temperatures, and so moderate the world's climate.

The quantum properties of water facilitate life in innumerable ways, from the molecular to the planetary scale.

☙ Polarity ❧

A water molecule is in sum electrically neutral, but the atomic arrangement of a water molecule is not linear, so there is a skew of electrical charges: a slight negative charge at the oxygen atom interface, while the hydrogen atoms are slightly positive. Hence, water is a polar molecule, with an electric dipole moment. This creates a weak and fleeting affinity for hydrogen bonding. That affinity has a cumulative effect, in making water molecules slightly cohesive. Water's flowing effect owes to hydrogen bonding, which is somewhat like static cling.

By having one hydrogen atom in the drink and the other in the breeze, about a quarter of the water at the interface between a water body and atmosphere straddles the liquid-gas phases. This layer of molecular ambiguity at the interface is extremely thin – 0.3 nanometers – and has no effect on the behavior of water molecules below because of their submerged location.

At extremely low temperatures, a water molecule has quantum tunneling ability forbidden in the classical world. The oxygen and hydrogen atoms are *delocalized*: present in all 6 symmetrically equivalent positions simultaneously. This only occurs at the quantum level, with no parallel in everyday experience.

> The average kinetic energy of water protons at almost absolute zero temperature is about 30% less than it is in bulk liquid or solid water. This is in complete disagreement with accepted

models based on the energies of its vibrational modes. ~ Russian nuclear physicist Alexander Kolesnikov

℘ Bonding ℘

What makes water molecules unique is that they can form hydrogen bridge bonds. ~ English physicist Gareth Parkinson

The molecular bonding of water has a strangely strong looseness: constantly rearranging restless hydrogen into irregular molecular arrangements owing to the polar alignment of H_2O. Water molecules have the unique ability to form short-lived connections that create fleeting networks. These transitory networks are crucial to water's peculiarity in many ways.

The oxygen in H_2O seeks connections with the hydrogen in neighboring water molecules, consistently maintaining tetrahedral coordination between close neighbors. At any given time, an oxygen atom in water is bonded to roughly 3.6 hydrogen atoms, with ever-changing interaction patterns on a picosecond time scale. The hydrogen bonds that link water molecules together break and form several thousands of billions of times per second.

Yet, measured at the level of attoseconds (10^{-18} seconds), water is not H_2O. It is instead $H_{1.5}O$. This was demonstrated in a 1995 British neutron-scattering experiment that found 25% fewer protons in water than expected.

The protons in hydrogen were sometimes not detectable by neutron probes. The reason is quantum entanglement. This phenomenon also occurs in other hydrogen compounds.

Water's bonding network is extensive. A single ion can tweak the bonds of several million water molecules over a distance exceeding 20 nanometers, causing the liquid to stiffen. This networked entanglement is central to many of water's unique properties, beginning with surface tension.

℘ Surface Tension ℘

You may have inner tranquility, but you can't escape surface tension. ~ American evolutionary biologist Louise Roth

Water's high surface tension owes generally to H_2O polarity, which creates molecular bonding affinity, but specifically results from hydrogen bonding at the water-air interface, where adsorption of hydroxide ions occurs. Rounded raindrops reflect it. Ice is an unusually strong molecular solid because of it. Earth's rapid water cycle is abetted by surface tension.

Nonpolar volatile molecules like carbon dioxide (CO_2) and methane (CH_4) can't form droplets. Instead, they float airborne as a fine mist, unable to rain.

Surface tension aids *capillary action*: the ability to flow against gravity when confined, such as in a thin tube. Owing to hydrogen bonding networks that form when water is squeezed, water molecules move faster as their density increases.

Vascular plants (tracheophytes) rely upon water's facile capillary action for liquid transport between roots, wood, stems, and leaves. Without capillary action, tracheophytes could not have evolved.

For thousands of years it was known that oil may float on water, but that the converse was not true. But it is possible for water to float on oil, depending upon the size of the water droplet and the surface tension of the oil. Mineral oils have a low surface tension, and so will not support water; but commercial vegetable oil will.

ꟿ Solvency ꟿ

Water is an ionizing agent. Because of its polarity, the positive and negative poles exert forces that can pull apart other molecules. As such, water is a solvent.

Rocks are slightly soluble. Hence, water erodes solid landscapes, fashioning valleys and canyons, ferrying inorganic nutrients in its flow.

Sugar dissolves in water because sugar's slight polarity is matched to covalent hydrogen bonds. This polarity prevents sugar molecules from regaining integrity. Because water is a polar molecule, other polar molecules are generally water-soluble.

Salt (sodium chloride (NaCl)) crystals enveloped by water dissolve because its 2 elements succumb to water's polarity: positive sodium ionization and negative chlorine ionization. This is seduction by electrical deception. It takes several water molecules to simulate the ionic appeal that divorces sodium from chloride in salt's natural crystal ionized form.

Insoluble ions are not so easily seduced: their ionic bonds remain faithful, unpersuaded by H_2O's partial charges. By such chemical stoicism, insoluble substances resist water's siren song of solvency.

Salt and sugar are the leading actor and actress in life's metabolism theater. A lot of organic molecules besides salt and sugar are water-soluble. Others, such as oils, are not.

The difference in organic solubility comes from attachments. If atoms other than carbon (C) and hydrogen (H) are present, a molecule has partial charges that can be emulated by water.

Hydrocarbon oil, naturally occurring as petroleum (crude oil), is steadily insoluble. But the sugar glucose ($C_6H_{12}O_6$), thanks to ionic oxygen in the mix, readily succumbs to dissolution.

Positively charged ions (cations) are loath to pair up, strongly preferring anions for their partner. But when liquored up by water, cations do couple.

❧ Forms ☙

Water's seeming simplicity is deceptive. The pairing of an oxygen atom with 2 of hydrogen has variants which abstrusely affect water's behaviors. Further, the thermodynamics of water are entangled with the molecular constructions which this sublime fluid takes. At both the atomic and molecular levels, water is a system like no other.

❧ Atomic Forms ☙

Water has different forms, both as isomers and isotopes.

♎ Isomers ♎

Water has 2 distinct nuclear-spin isomers of its hydrogen atoms: para-water and ortho-water. The isomers are stable, though interconversion does mysteriously occur sometimes. Ambient-temperature water is a mixture of these 2 isomers.

Whereas the nuclear spin of the hydrogen atoms in para-water sum to 0 via asymmetric wavefunctions, ortho-water nuclear spins symmetrically sum to 1. The atomic twists and turns of para-water let it react 23% faster than the stolider ortho-water. Para-water is also able to more strongly attract its reaction partner than ortho-water, which leads to enhanced chemical reactivity.

♎ Isotopes ♎

Water has several isotopes, especially hydrogen, though oxygen too has isotopes. Known oxygen isotopes range in mass number from 12 to 28. Oxygen has 3 stable isotopes: ^{16}O, ^{17}O, and ^{18}O.

^{16}O is the most abundant: 99.762%. ^{16}O is the principal form by stellar evolution, as ^{16}O can be made by stars that were initially fueled entirely by hydrogen.

^{17}O and ^{18}O are produced later in the stellar life cycle. ^{17}O comes by burning hydrogen into helium during the CNO cycle.

^{18}O is common in the helium-rich zones of stars. ^{18}O is typically produced when ^{14}N (nitrogen), abundantly derived from CNO combustion, captures a ^{4}He (helium) nucleus.

Hydrogen has 2 stable isotopes: *protium* and *deuterium*. Protium has a nucleus of a single proton. Deuterium (aka *heavy hydrogen*) has a nucleus comprising a proton and a neutron. Both protium and deuterium sport a single electron.

Water is typically a mixture of protium and deuterium, naturally varying by source. More than 99.98+% of the hydrogen in ocean water is protium.

Recently evaporated ocean water, including rainwater, river water, and snow, tends to have the lighter isotopes of hydrogen and oxygen. Such waters evaporate faster than heavier varieties.

Heavy water, properly termed *deuterium oxide* (D_2O), has a richer deuterium content. 0.0156% of the hydrogen in ocean water is deuterium.

Rats avoid heavy water by its smell. Humans are generally unaware of the difference (from ordinary water), other than heavy water sometimes has a sweet flavor, or causes a burning sensation.

Conversely, *light water* is deuterium depleted. Light water has been found beneficial for mammals with cancer.

There are differences in the bonds between light and heavy water, at the atomic and molecular levels.

The distance between deuterium and oxygen nuclei in D_2O is 3% shorter than the distance between hydrogen and oxygen in an H_2O molecule. Conversely, the hydrogen bonds between one molecule of water and another are 4% longer in heavy water than light water.

Despite their differences, light and heavy water behave much alike. Their melting temperatures differ by less than 4 °C, and they boil at even closer temperatures. This owes to competing nuclear quantum effects (NQEs) being offset along different molecular axes.

One NQE is associated with motion perpendicular to the plane of a water molecule. Another NQE relates to motion parallel to the hydrogen bond.

The difference in nuclear quantum effects between H_2 and D_2 is that one NQE increases while the other lessens. Hence, a small quantum net effect between the different hydrogen forms, and there is similarity in phase transitions.

Another oddity of water is its lack of chemical purity. Water is never just H_2O. A small fraction of any body of H_2O splits into positively charged *hydrons*: hydrogen ions (H^+), which are protons without bonded electrons, and negatively charged *hydroxyl* groups (OH^-).

Hydrons latch onto water molecules, forming *hydronium* ions (H_3O^+). So-called "pure water" at room temperature balances equal numbers of positive hydronium ions and negative hydroxyl groups, creating a neutral solution (pH = 7).

Molecular Forms

> Water has the remarkable property of being more compressible in winter than in summer. ~ English chemist John Canton in 1764

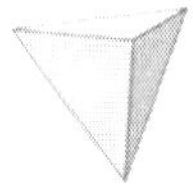

Water molecules take 2 structural forms, based upon the instant inclinations of the hydrogen bonds. One structure is tetrahedral, the other disordered, in a myriad of ways.

> Despite its simplicity, water tends to form tetrahedral order locally by directional hydrogen bonding. This structuring is known to be responsible for a vast array of unusual properties. ~ Japanese physicist Hajime Tanaka *et al*

Water is densest at 4 °C, when its tetrahedral ordering is at its apex. Heating reduces the number of tetrahedral structures, resulting in more disorder.

In water cooler than 4 °C, disordered areas are more densely packed than the tetrahedrons. Agitated by heat, warming above 4 °C spreads out molecules in disordered regions, making the water less dense.

In 1762, Scottish chemist Joseph Black discovered *latent heat*: that materials may absorb heat without changing temperature. Black also noted that different substances have dissimilar heat capacities.

Directly related to latent heat, *heat capacity* is the ratio of heat absorbed by a substance to its temperature change. In other words, heat capacity is the measure of how much the temperature of a certain substance rises in response to heating: a ratio of heat energy/temperature. The term *specific heat capacity* is a measure of heat capacity per unit mass of a material.

Heat capacity is an extensive property: proportional to the size of the system. In the instance of H_2O, heat capacity proportionally varies by how big the body of water is.

Water has an exceptionally high specific heat capacity: it takes a lot of energy to heat water. Water is laden with latent heat. This is because much of the extra heat applied to H_2O converts molecular regions from tetrahedral form to disordered structures, rather than increasing the kinetic energy of the water molecules, which would raise the temperature.

Water's specific heat capacity is at a minimum at 35 °C, increasing as its temperature falls or rises. The heat capacity of most other liquids rises continuously with temperature.

Between 0–35 °C, increasing water temperature steadily removes regions of ordered, tetrahedral structure, reducing the ability to absorb heat. Above 35 °C, so few of the tetrahedral regions are left that water behaves like a regular liquid.

The strong attraction between water molecules keeps them more tightly packed than most other liquids. As such, water is especially difficult to compress. Compressibility is at its lowest when the higher-density disordered structures dominate.

With most liquids, compressibility rises continuously with temperature. Water is anomalous. As water hots up, the dense, disordered regions become more prevalent; areas which are more difficult to compress. Still, heat forces molecules within disordered regions further apart, making them more compressible. Disordered region expansion takes precedence over structural shifts beyond 46 °C.

The speed of sound in water increases up to 74 °C, after which it starts to fall again. This owes to the interplay between water's density and compressibility profiles, which stem from the changing balance between the 2 structural forms.

Exceptionally, water molecules diffuse more easily at higher pressures. That is because high pressure converts more molecules to a compressed disordered structure, making them more mobile. Water becomes less viscous at higher pressure.

Increasing pressure increases the amount by which water expands when heated. Rising temperature expands disordered regions more rapidly than ordered, tetrahedral ones, and high pressure forces fluctuations in disordered regions.

Properties such as viscosity, melting point, and boiling point are significantly different in heavy water (D_2O) from H_2O. The heavier isotopes change the quantum mechanical properties of water molecules, altering the balancing act between tetrahedral and disordered regions.

༄ Phases ༅

States of matter are also called *phases*. Whereas *state* stresses stasis, *phase* suggests fluidity via energy level.

A molecular system may be arranged in a certain variety of configurations depending upon total energy in the system. Constituent molecules have the liberty of greater possible configurations at higher temperatures/energies and can freely move about. This is the *gaseous* phase.

At lower temperatures, configurations are more constrained. At this more ordered phase, the material system is *liquid*.

If the temperature/energy drops to a certain threshold, the molecules are confined to a specific configuration, forming a *solid*.

This picture is common for relatively simple molecules, which have 3 distinct states: gaseous, liquid, and solid. such CO_2 is exemplary. For more complex or slippery molecules, including H_2O, there are greater possibilities for combinations, giving rise to more phases.

Liquid crystals are a beautiful illustration of this: complex organic molecules which can liquidly flow within a solid-like crystalline structure.

Water is one of the few chemical compounds that exists alternately as a gas, liquid, or solid within the temperature range of ordinary life, with an intricate quantum dance in transitioning from one state to another. Whereas water exhibits its phase range within 100 ºC, most other compounds have a much broader range for their temperature transitions.

According to its structural properties, water should be a gas at the temperature and pressure where water is liquid. Consequently, liquid water behaves much differently than other liquids.

Water has more than the 3 well-known phases (gas, liquid, solid), and may selectively exist in more than 1 phase simultaneously.

Liquid water uniquely exhibits 2 distinct states, with different rates of thermal expansion, surface tension, and refractive index (a measure of how light travels through it), as well as other characteristics. The thermal crossover for these states occurs at ~50 ±10 °C.

> Water at room temperature can't decide in which of the two forms it should be, high or low density, which results in local fluctuations between the two. Water is not a complicated liquid, but two simple liquids with a complicated relationship. ~ Swedish physiochemist Lars Pettersson

Water boils at 100 °C. At a critical temperature and pressure (374 °C and 22.064 MPa), found naturally only in the hottest regions of hydrothermal vents, water becomes *supercritical*: the gas and liquid phases merge into a homogeneous fluid phase, with properties from both phases.

Water's thermal conductivity is not constant. Water cooled to 225 K starts to more efficiently conduct heat. Below this temperature, liquid water undergoes sharp but continuous structural changes, becoming highly ordered: very much like ice.

Water cooled to less than –135 °C turns syrupy but clear: glassy water. This phase is well past freezing.

ϐ Freezing ϙ

> The fact that water has previously been warmed contributes to its freezing quickly. ~ Aristotle

Water nominally freezes at 0 °C, but, like most other liquids, may exist as a supercooled liquid: staying fluid below its freezing point. Liquid water at –40 °C has been found in clouds. Water may remain a liquid to –48 °C. Supercooled water freezes if disturbed.

The freezing of water is controlled not only by its temperature, but also by its size. The nucleation of ice in small droplets is strongly size-dependent.

Freezing is a process of water molecules forming a bonded network. In transition to freezing, water physically changes into tetrahedrons, with each water molecule loosely bonded to 4 others.

Freezing water droplets garb themselves with pointy tips, sporting fractal shapes as they crystallize. Depending upon temperature and pressure, ice may take various crystalline forms while water molecules cling to each other via hydrogen bonds.

Aristotle, and others after him, noted that hot water freezes faster than cold water. Yet namesake credit is given for its 1963 rediscovery by Tanzanian scientist Erasto Mpemba, who first noticed the *Mpemba effect* when he was 13, while making ice cream in haste: mixing boiling milk with sugar, rather than waiting for the milk to cool first, as instructed.

Mpemba asked his physics teacher for an explanation. The teacher told Mpemba that he must have been confused, as what he supposedly saw was impossible. Mpemba persisted in his inquiry.

Though the mechanics of the Mpemba effect remain something of a mystery, the likely explanation lies in the strength and arrangement of hydrogen bonds, which are affected by temperature, and partly depend upon nearby water molecules.

৶ Ice ৹

There are 17 known forms of ice. Many of them form under extreme pressure, such as in the interiors of frozen planets; 2 take on a quartz-like crystalline structure.

Whereas most solids sink into their liquid forms because the solids are denser than the liquids, ice floats on top of water. Ice, like water, has unusual structural properties.

Water attains its maximum density at 4 °C. Liquids typically condense as they cool but water expands upon freezing.

Whereas molecules of water vapor behave rather independently, ice forms a strong structural lattice. Liquid water molecules have weaker interactions with each other than they do in ice. Water also exists in a liquid-crystal state when adjacent to hydrophilic surfaces, which tend to interact with water or be dissolved by it.

Ice possesses quantum properties opposite those of other crystals. Typically, crystalline molecules shrink as they cool, but the shrinking stops before reaching absolute zero (0 K), owing to zero-point energy.

In theoretical physics, *zero-point energy* is the lowest possible energy that a particulate matter may have: the energy of its ground state. Zero-point energy is inherently tied to the uncertainty principle.

Typically, less massive atoms have more zero-point energy. Lighter nuclei need more room to move than heavier elements, which translates into larger crystals for lighter elements as they cool.

At higher temperatures, this quantum effect becomes less pronounced. Hence volume differences decrease as temperature rises.

The opposite occurs with ice. D_2O (heavy water) occupies more volume than H_2O ice. This volume difference increases as temperature rises.

> In order to access and measure quantum mechanical effects in matter, we usually need to go to very low temperatures, but in water ice some zero-point effects actually become more relevant as the temperature increases. ~ physicist Marivi Fernández-Serra

The structural integrity of ice is illustrated in its melting. Ice melt is not continuous. Instead, ice discontinuously liquefies layer by layer.

Ice's structure varies at different scales. Macroscopically, ice crystals form a hexagonal lattice, which is why all snowflakes are 6-sided. In contrast, in crystallites of up to 100,000 molecules, ice crystals form from water in layers of hexagonal and cubic arrays.

Ice formation in clouds – nucleation – is a critical aspect of precipitation development. The semi-disordered cubic-hex stacks of ice that form from cloud droplets facilitate faster development of rain than if ice nucleation occurred solely like snowflakes (hexagonally). Earth would be much different if ice formation physics was otherwise, as the water cycle would be hindered.

℘ Evaporation ୪

Evaporation turns out to be a process driven by very small temperature differences. Often, only 10/10,000th parts of Kelvin are enough to make it happen. ~ Polish physicist Daniel Jakubczyk

Evaporation is critical to life, and fundamental to the world's climate by way of the water cycle.

For evaporating droplets, the temperature before evaporation starts and after it is accomplished is the same. Tiny temperature fluctuations between these moments tells evaporation's tale.

The heat flux between a droplet and its surroundings plays a key role in evaporation. If the thermal flux between a hot pan and water drops were efficient, the water would instantly evaporate in boiling off. Instead, this flux is hindered, as droplets slide on an insulating layer of water vapor. This thermal layer is thick enough to restrain the heat flux. The thickness of the thermal layer depends upon surrounding conditions, not droplet size.

୪ Organic Role ୨

Water acts as a catalyst in many biochemical reactions by stabilizing charge states on a surface. Via condensation, large organic molecules, which have a skeleton of carbon and hydrogen, are built up by forming covalent bonds and eliminating water. These organic molecules can be broken down by adding water, as covalent bonds are split by hydrolysis. In the hydrolysis ionization reaction, both the organic molecule and the water molecule are split.

Water's stickiness is instrumental in engendering life, as water's properties assist both the formation and concentration of organic molecules (*anabolism*), and their transition to simpler forms (*catabolism*). Thus, water abets life at the molecular level by facilitating metabolism.

Cells are 60% water and 40% macromolecules: proteins, carbohydrates, and other components. Water confined within a cell slows by a factor of 10 compared to pure water alone.

This impacts the intricate cellular machinery that is self-organized and precisely choreographed to transform energy, manufacture molecular products, and be responsive to its environment.

> While it is tempting to focus on the macromolecules, it is becoming increasingly clear that the water is not simply an innocent bystander, but instead plays an active and sometimes decisive role in mediating the processes carried out by the cell's machinery. ~ Canadian chemist Kevin Kubarych

Water has a high heat capacity, so it buffers aqueous systems from dramatic temperature changes. As water evaporates, it cools the liquid that remains. Frozen water forms an ice lattice, taking up more space. Hence, ice floats, providing a blanket of insulation.

At the planetary scale, water offers some stability by mediating temperature extremes; another benefit to life.

⚡ Fear of Water ⚡

> The hydrophobic interaction is one of the most important driving forces in Nature and is key to processes such as protein folding and the self-assembly of lipid membranes. ~ Dutch chemical physicist Huib Bakker

Water-repellent molecules are termed *hydrophobic*; literally: "fear of water." This molecular fear puts the surrounding water on guard.

Hydrophobic groups in molecules enhance the ordering of the surrounding hydrogen-bond network of water. These fearful groups form ideal templates around which a water network can fold, leading to greater local tetrahedral order, like water on its way to freezing.

A hydration shell forms around a hydrophobic group. Water network folding creates ridges 0.3 nm high, which is the intermolecular distance of water molecules. Ridge angles are 104.5°, corresponding to the bond angle of a water molecule. This is typical of hydrogen-bond networks in water.

Temperature matters. The hydration shell surrounding a hydrophobic group melts as water warms. The hydrogen-bond network gradually weakens. Disorder gains the upper hand as water hots up.

Size matters. Alcohols with hydrophobic chains longer than 1 nm are contrary in their effect on hydrogen bonding from hydrophobic groups half that size.

While small hydrophobic structures enhance order, large ones cause surrounding water molecules to have more broken hydrogen bonds, as the curvature of the large hydrophobic surface does not fit the 3D spatial arrangement of water hydrogen bonds. At temperatures above 80 °C, water molecules surrounding such plump hydrophobic groups are less ordered than their nearby brethren in bulk liquid.

ꙮ Water Memory ꙮ

Homeopathy is wholly capable of satisfying the therapeutic demands of this age better than any other system or school of medicine. ~ American physician Charles Menninger

In 1796, German physician Samuel Hahnemann claimed a medicinal treatment based on a "law of similars," where "like cures like." This notion gave rise to *homeopathy*: medicines made from substances causing similar symptoms in healthy people, prepared by serial dilutions until little or none of the supposed active ingredient remains.

In 1988, French immunologist Jacques Benveniste published an article in the magazine *Nature*, to the open skepticism of the publisher, of an experiment reportedly showing that water has a "memory" of the compound last diluted in it, even after dilution is repeated until no molecule of the diluted compound remains. Such a claim would provide scientific support for homeopathy, a treatment based upon such dilution. But Benveniste's results were irreproducible.

Nature magazine called the idea of water memory "scientifically unacceptable, although this doesn't yet seem to have affected the commercial success of homeopathy." Nor perhaps should it.

Homeopathy as a purely medicinal treatment may be harmlessly ineffective, but placebos can be powerful medicine. A skilled, seemingly knowledgeable speaker imparts to someone willing to believe that healing power is within the patient's grasp. That life-affirming belief alone positively affects the spirit, and thus the immune system. (While adults may self-deceive themselves to health, babies cannot. Treating infants homeopathically has an abysmal track record.) Beyond belief, homeopathy is bunk.

Homeopathy not only doesn't work; it couldn't possibly work. It is inconsistent with our basic knowledge of physics, chemistry, and biology. ~ American physician Harriet Hall

๛ Organic Chemistry ൽ

All life is chemistry. ~ Belgian chemist Jan van Helmon

Organic chemistry is the chemistry of carbon compounds. Biochemistry is the study of carbon compounds that crawl. ~ Mike Adam

Life sustains itself by chemical energy. Transformative sustenance transpires in water.

Clean water is vital for all known life. Mars may have never evolved life because the salinity of its water has long exceeded levels by which life could arise, survive, or thrive.

Humans are 65–90% water, scaffolded by carbon-containing organic molecules. 99% of human body mass comprises 6 elements: oxygen (65%), carbon (18%), hydrogen (10%), nitrogen (3%), calcium (1.4%), and phosphorous (1.1%). By contrast, aluminum and silicon are abundant in Earth's crust but rare in life forms.

Carbon's unsurpassed flexibility lends itself to complexity. Sturdy nitrogen provides stability. Oxygen and hydrogen readily react, and so make excellent elements for cellular activity. Being readily reactive, calcium ions play a central role in many cellular functions, and in intercellular communication. As a store of energy, phosphorous is vital to metabolism. 2 more elements are noteworthy for their organic roles: potassium and sulfur. Potassium (0.4%) assists in homeostasis and cellular communication. Through its surfeit of electrons which willingly shuffle, sulfur (0.3%) helps catalyze reactions.

All told, cells make a living using just 30 different monomers (molecule types).

The characteristics of a cell rest on the structure of its molecular components. ~ French biologist François Jacob

The molecular components of organisms are remarkably uniform in the nature of the components, as well as in the ways in which they are assembled and used. ~ Spanish American biologist Francisco Ayala

᪣ Functional Groups ᪤

A *functional group* comprises slightly different configurations of the same organic compound, all with similar behaviors. A functional group is typically a molecular subset of a larger molecule. The same functional group produces self-same or similar reactions regardless of the molecule of which it is part.

An organic compound's functional group is indicative of the entire molecule's reactive properties. Reactions of an organic compound can be predicted by knowing the kind(s) of functional group(s) it has.

Carbon molecules are commonly classified based upon their functional group. *Proteins*, the building blocks of life, are complex carbon-based macromolecules comprising many functional groups. In doing their job, proteins selectively employ a certain functional group as a tool.

The term *moiety* is often used as a synonym for functional group, but a moiety is a specific segment of a molecule, regardless of functional group. A moiety may be part of a functional group or encompass parts of different functional groups.

᪣ pH ᪤

The meaning of the "p" in "pH" is unknown. ~ Wikipedia

pH is a measure of how base or acidic an aqueous solution is. In a chemical reaction, an *acid* is a molecule or ion capable of donating a cation of protium ($^1H^+$); a *base* is accepting of $^1H^+$.[*]

---

[*] The term *proton* is sometimes used for $^1H^+$. This usage comes from the 1923 concept of acids and bases independently developed by chemists Johannes Brønsted and Martin Lowry. Of course, using *proton* here is inexact. The term *hydron* is also sometimes used,

Water is neutral (pH = 7). Acids have a pH < 7, while bases have a > 7 pH.

pH was coined in 1924. The "H" in pH stands for hydrogen, but the meaning of "p" is a long-standing mystery. *p = power* is a common guess. pH: the power of hydrogen.

While acidity is related to the concentration of hydrogen ions, it is not ionic concentration per se that confers pH, but instead the *activity factor* of a solution, which is the tendency of hydrogen ions to interact with other elements in a solution. Acid-base reactions are tangled enterprises.

☡ Neutralization ☡

pH is a major aspect of organic chemistry. Acids or bases promote most organic reactions or are at least involved at some stage in a reaction pathway.

After the needed action is accomplished, pH hyperactivity needs calming down; in a word: *neutralized.*

Acids and bases are ionic go-getters. It would make sense that neutralizing them has something to do with taking a charge off: *deionization* (the removal of ions). The ease of neutralization is related to the electronegativity of the elements involved. But something much fatter than electrons easing up is involved. A proton jumps between ions to neutralize. In aqueous neutralization involving an acid and salt, the salt is but a spectator. Proton hops turn acidic ions (H_3O^+) into water (H_2O).

More generally, *proton transfer* either turns an acidic H_3O^+ (hydronium) ion into a water molecule or ionizes a water molecule to the base HO^- (hydroxide). Transfer takes 1–2

_____

though generally disfavored, as hydron is the general name for H^+, which is the cationic form of hydrogen, regardless of isotope. Hence, hydron represents the nucleus of $^1H^+$, $^2H^+$ (D^+) (deuterium isotope), and $^3H^+$ (T^+) (tritium isotope). It may seem strange that chemistry has no exact term for $^1H^+$. There have been numerous characterizations of acids and bases, including a 1923 electronic theory of acid-base reactions by Gilbert Lewis. The Brønsted–Lowry conceptualization is a theory for which exceptions are known. Much of the intricacy of chemistry remains to be discovered.

picoseconds, with the proton hopping along a particular pathway.

The local H-bond water network provides a pathway for proton transfer; or not. Without a permissive aqueous dynamic and structure, proton transfer is stymied. A proton may hop its way to a dead end, where it is locally trapped for an extended period.

In a gas reaction, ions clump by attraction. Without any solvent at all, proton transfers neutralize.

Acids and bases mixed in the right proportions neutralize by proton transfer: acids donate protons that bases accept.

ஐ Carbon ଔ

Life exists in the universe only because the carbon atom possesses certain exceptional properties. ~ English physicist James Jeans

Carbon (C) is the only element that can form chains and rings on its own, using single or double bonds. Thus, carbon acts as a backbone in manifold molecular construction.

4 of carbon's 6 electrons are available to form strong covalent bonds. This uniquely flagrant friendliness makes carbon the most flexible and stable element for complex compound construction. Hence, life is built on carbon. Carbon provides the chemical skeleton of the 4 major types of organic compounds: nucleic acids (aka polynucleotides (RNA, DNA)), carbohydrates (sugars), proteins, and lipids (fats).

An organic compound, which necessarily has a carbon backbone, is an assemblage (polymer) of smaller molecular subunits (monomers), altogether forming a *macromolecule*. *Polymerization* is the process by which repeating subunits (monomers) are bound into chains of various lengths, forming *polymers*. *Amino acids* (monomers) are polymerized into *proteins*.

Coupled to their energetically ready construction, the complexity and flexibility of macromolecules lets them work in a wide variety of roles: functioning as structural components, molecular messengers, enzymes, energy sources, nutrient storage, and storehouses of genetic information.

¤ ✧ ¤

While carbon normally bonds with 4 other atoms, other configurations are possible. *Mellitene* ($C_{12}H_{18}$) is based upon a hexagonal ring of 6 carbon atoms offering 6 arms for 6 carbon hydrogen atoms that each have 3 hydrogen atoms attached. Leftover electrons zip around the middle of the ring, 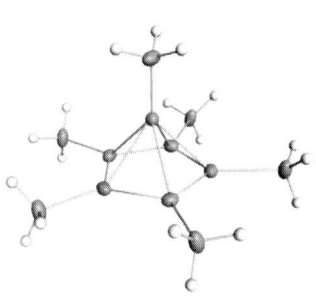 strengthening core carbon bonds and thereby stabilizing the molecule.

Pure carbon in crystalline form appears as *diamonds*: a lattice allotrope with remarkable optical characteristics, and the hardest natural material known.

An *allotrope* is an element existing in multiple forms, with atoms bonded together in different ways. The structural variations characterizing *allotropy* apply only to distinct forms of an element within the same phase, and only to elements, not compounds.

A more pedestrian carbon allotrope than mellitene is *graphite*, the stuff of pencil lead. While diamond won't conduct electricity, graphite will. Yet graphite is the most stable form of carbon.

Because of carbon's inherent promiscuity, there is a greater variety of carbon compounds than all other elements combined. Plastics and petrol, which are derived from erstwhile plant matter, are carbon polymers.

✑ Atmospheric Carbon ✍

Atmospheric carbon most naturally occurs as carbon dioxide (CO_2). While carbon dioxide is currently a relatively minuscule component of the atmosphere, at about 0.039%, Earth's early atmosphere was rich in CO_2.

CO_2 is transparent to visible light, as are N_2 and O_2, the other notable atmospheric gases. But whereas nitrogen and oxygen are also insensitive in the infrared spectrum, carbon dioxide absorbs infrared light.

In trapping infrared, CO_2 prevents some of that radiation from escaping into space. Hence atmospheric carbon dioxide acts as a greenhouse gas, helping keep the planet below warm.

Water vapor has a powerful greenhouse effect, absorbing light in a broader spectrum than carbon dioxide. H_2O absorbs essentially everything at wavelengths longer than 500 cm^{-1}. CO_2 centers its absorption around 667 cm^{-1}.

Absorption dynamics depend upon quantum mechanical properties specific to a molecule's geometry. Radiation at specific frequency ranges jiggles the bond angles of a molecule, affecting molecular vibrational mode, allowing a molecule to hold on to extra energy.

It was the carbon in carbon dioxide, not the oxygen, that was the ready fuel for emergent life. The magic trick was transforming the carbon in CO_2 into energy.

⚘ Photosynthesis ⚘

Photosynthesis powers life on our planet. ~ American physicist Franklin Fuller

The chemical equation for *photosynthesis* is: $6\ CO_2 + 12\ H_2O$ + sunlight energy → $C_6H_{12}O_6 + 6\ H_2O + 6\ O_2$.

A photosynthesizing organism derives energy from sunlight via charge separation; forging atmospheric carbon dioxide and water into energy-storing sugars by freeing electrons to work chemical transformations via photonic energy. *Charge separation* is the process of an atomic electron being excited to a higher energy level by absorbing a photon, and thereby leaving its home to join a nearby electron acceptor molecule.

Chloroplasts – plant organelles that conduct photosynthesis – have chlorophyll that absorb various wavelengths of light, with peak inputs around 430 nanometers (blue light) and 660 nm (red). Chlorophyll is green because it shies from absorbing mid-spectrum green light (526–606 nm), which is thereby reflected.

O_2 is a byproduct of photosynthesis; an oddity considering that oxygen is itself energy-rich by its easy reactivity. But throwing off water's O_2 makes sense in light of the photosynthesis system.

In plant photosynthesis, oxygen stubbornly sticks to the carbon in CO_2, which becomes the backbone molecule upon which the target sugar (glucose: $C_6H_{12}O_6$) is built. In the same reaction as evaporation, water molecules are split to liberate oxygen (O_2) into the atmosphere.

The rate of charge separation for evaporation in photosynthesis is enhanced by coherent dynamics that are vibronic in nature. This allows photosynthesis to be optimal at the quantum level.

Hydrogen is the watery element elicited in photosynthesis; valued for its donation in the *electron transport chain*: the molecular processing system for deriving energy.

Carbon has a dual role in photosynthesis. It is the proton acceptor (HCO_3^-) in the very beginning of the photosynthetic reaction, and the terminal electron acceptor (CO_2) at the end. This varied employment of different species in a reaction series illustrates the wondrous flexibility of carbon.

৶ Hydrocarbons ৶

A hydrocarbon is any variety of organic compound consisting entirely of carbon and hydrogen. Owing to the inherent molecular flexibilities of carbon and hydrogen, hydrocarbons assume a vast variety of structures.

Saturated hydrocarbons (*alkanes*) are the simplest hydrocarbon species, composed entirely of single bonds. Petroleum fuels are derived from alkanes. Alkanes with 1 or more rings of carbon atoms are *cycloalkanes*.

Unsaturated hydrocarbons have 1 or more double or triple bonds between carbon atoms. *Alkenes* are double-bonded hydrocarbons. Hydrocarbons with triple bonds are *alkynes*.

Hydrocarbons come in a variety of phases at room temperature: gaseous (e.g. methane (CH_4)), liquid (e.g. hexane (C_6H_{14}) and octane (C_8H_{18})), wax or low-melting solid (e.g. paraffin (C_nH_{2n+2}, from $C_{20}H_{42}$ to $C_{40}H_{82}$)), and polymers (e.g.

the ubiquitous plastic polyethylene (always following the formula: $(C_2H_4)_nH_2$), which is so durable as to resist biodegradation).

The more complex the hydrocarbon, the more stable.

Hydrocarbons are hydrophobic, as are lipids.

☑ Lipids ☒

Lipids are a broad molecular group, engorged with nomenclature. The lipid group includes fats, waxes, sterols, glycerides, phospholipids, and fat-soluble vitamins.

Although the terms *lipid* and *fat* are sometimes used as synonyms, fats are a lipid subgroup.

☒ Fats ☒

All fats are derivatives of glycerol, combined with fatty acids. *Glycerol* is a simple alcohol (*polyol*) compound, comprising 3 hydroxyl groups. A *hydroxyl* is a functional group comprising an oxygen atom covalently bonded to a single hydrogen atom. *Fatty acids* comprise a carboxylic acid with a long aliphatic tail (chain). An *aliphatic compound is a* group of hydrocarbons that do *not* link together to form a ring.

Triglycerides – the common form of fat in animals – have 3 fatty acids. Humans store unused energy (*calories*) in triglycerides. Overeating high-carbohydrate foods packs on saturated triglycerides.

Chemical composition complexities aside, the biologically relevant facts of fats are that they act as an energy store, and that fats come in 2 forms: saturated and unsaturated.

A *saturated fat* has only single bonds between carbon atoms. The term *saturated* is used because the fat is chock full of hydrogen atoms.

In contrast, *unsaturated fat* has at least 1 double bond within the fatty acid chain. A fat molecule with only 1 double bond is *monounsaturated*. Molecules of fat with more than 1 double bond are *polyunsaturated*.

Hydrogen atoms are eliminated where double bonds are formed. Thus, metabolically, unsaturated fats hold a bit less energy (i.e., fewer calories) than an equivalent portion of saturated fat.

Unsaturated fats are regarded as healthier for human consumption; the more unsaturated, the better. Unsaturated fats metabolize more cleanly.

Nature equipped biological systems to recognize the shapes and other characteristics of molecules and process them accordingly. Metabolically, a saturated fat is treated to a different set of biochemical reactions than an unsaturated fat.

Each form of fat has a different shape. Fats are metabolized by geometry.

The more double bonds there are, the more readily a fat will react with oxygen. The tendency to rancidity is greater the more unsaturated the fat is. So, to preserve shelf life, fatty foods are commercially processed by *hydrogenation*: saturating the fat at high temperature, thus breaking the double bonds and slimming them to single bonds, with hydrogen attached.

✄ Carbon Cycle ✄

You will die but the carbon will not; its career does not end with you. It will return to the soil, and there a plant may take it up again in time, sending it once more on a cycle of plant and animal life. ~ Polish polymath Jacob Bronowski

The *carbon cycle* is the gaseous cycling of carbon exchange among the pedosphere (soil), hydrosphere (water bodies), atmosphere, and biosphere (living ecological systems) of the Earth. Carbon deposits and exchanges among the different spheres differ.

The atmosphere has a small active exchange of CO_2. The pedosphere is both a carbon sink and primary cycler of carbon.

Autotrophs, predominantly plants, move carbon through the cycle, fixing CO_2 into organic compounds. These organic compounds are consumed by hungry heterotrophs (e.g. animals), and then decompose into a carbon sink. Carbon is returned to the atmosphere as CO_2 by heterotrophic cellular respiration, such as by breathing.

Recesses below ground and in the deep ocean are vast carbon storehouses, with exchange becoming active only by violent disturbances: volcanic eruptions and larger-scale geological movements (metamorphism), such as plate tectonics.

⌘ History ⌘

Long before life emerged, carbon was cycled from the atmosphere into the geosphere as carbonate sediments. The evolution of photosynthesis accelerated the carbon cycle, which gained an even greater boost as plants colonized land.

Organic carbon from plant matter became buried as coal and in marine sediments, resulting in a dramatic decline in atmospheric CO_2, while plant exhaust (oxygen) became prominent.

The interplay of carbon and oxygen have been a central driver in the emergence and development of life on Earth, and their cycles remain pivotal in the biological fortune of the planet.

Between 24,000 years ago and today, large ice sheets covering most of Canada, and parts of Europe and Asia, melted away. Global sea level rose by 120 meters. Earth warmed 5 °C. Rainfall and vegetation patterns shifted, sometimes abruptly.

Milankovitch cycling, including orbital shifts, set these changes into motion. But a complex feedback system, in which the carbon cycle was integral, governed the transition from glacial to interglacial state.

17,500 YA, CO_2 levels started to rise from the ice age level of 180 parts per million (ppm), reaching 265 ppm 10,000 YA. Over the next 10,000 years, atmospheric carbon dioxide rose another 20 ppm, until the rapid ramp originating with the onset of the industrial revolution in the 1850s.

In 2010, atmospheric CO_2 was 380 ppm. It rose to 405 ppm in 2017. By 2080, atmospheric CO_2 should be over 700 ppm.

⚘ In the Ocean ⚘

The ocean has the largest store of actively exchanged carbon. The seas near the surface constantly exchange carbon with the atmosphere. The depths are not so yielding.

The deep ocean holds 50 times more CO_2 than the atmosphere. It does so by a *biological pump*.

Phytoplankton fix CO_2 at the sea surface by photosynthesis, forming organic carbon. When they die, the phytoplankton sink. The dissolved organic matter (DOM) – including carbon – is consumed by other organisms, and respired back into CO_2, which forms dissolved carbonate.

The biological pump also depletes nutrients such as nitrogen and silica from the surface waters. Nitrogen is needed by all phytoplankton, while silica is in demand by diatoms. All these nutrients become trapped in the deep sea by the shallow surface layer.

Ocean circulation returns this nutrient-rich water to the surface in the Southern Ocean, and at upwelling regions. When this happens, biological productivity at the surface is enhanced, and CO2 is released into the atmosphere, if degassing outpaces the biological pump.

⚘ Seagrass Meadows ⚘

Seagrass meadows are concentrated stores of carbon dioxide. A typical forest stores 30,000 tonnes of CO_2 per square kilometer. Coastal seagrass beds store up to 83,000 tonnes per km^2.

Although seagrass meadows are less than 0.2% of the world's oceans, they hold more than 10% of all carbon buried annually in the sea. 90% of the carbon storage is in the soil, which accumulates as the meadow thrives.

Seagrass meadows can last for centuries; but not any longer. Seagrass beds are among the most threatened ecosystem worldwide, thanks to dredging and degrading water quality.

29% of all seagrass meadows have already been destroyed, with further yearly loss of at least 1.5% of the historic total. CO_2 emissions for seagrass meadow destruction could be as much as 25% of that which comes from deforestation.

☡ On Land ☡

Dryland systems have high rates of carbon turnover compared to other biomes. ~ French ecologist Philippe Ciais

Semi-arid biomes, from deserts to shrublands, drive variability in the carbon cycle. The greening of these lands from sporadic or seasonal rainfall has an outsized effect on carbon cycle dynamics. As the climate warms, drylands become particularly sensitive to precipitation.

Forests comprise the great storehouse of terrestrial carbon: 86% of the aboveground carbon. Forest soil has 73% of the Earth's soil carbon.

Forests are extremely active in the carbon cycle. Forests store massive quantities of carbon in the soil, through the root systems of trees and the microbes which support vegetation in mutualistic relationships.

The quality of the soil determines how much carbon is retained. Fertile soils sequester 30% of the carbon they take in during photosynthesis. In contrast, nutrient-poor soils may only retain 6% of that carbon.

Logging accelerates carbon release into the atmosphere in multiple ways. Besides killing the trees that act as carbon keepers, logging cuts off the food supply for subterranean life in the soil. Deforestation bankrupts the prospect of the ground acting as a carbon store.

The organic matter in soil is 60% carbon. If the withdrawal of carbon from the soil went up by just 0.3%, the release into the atmosphere would equal a year's worth of emissions worldwide from human use of fossil fuels.

Scientists long thought that the soil locked in carbon; a comforting fiction. The supposedly stable carbon keepers in soil are instead relatively volatile.

For over a century, it was supposed that soil carbon became locked in large compound molecules, highly resistant to microbial assault. Only recently has it been discovered that those presumed molecules – so-called humic soil substances – simply don't exist in any significant quantity. Instead, the persistence of organic matter in soil is controlled by the surrounding microbial ecosystem.

Plants thrive on CO_2. The more there is in the air, the faster and larger plants can grow, if other conditions conducive to growth exists (such as quality soil and proper rainfall).

Plants, especially ones with extensive roots, such as trees, have steadfast friends in the soil: symbiotic microbes. Soil microbial ecosystems rely upon plant life. Plant loss releases CO_2, as soil microbes lose their livelihoods, and their lives.

The carbon cycle is also invigorated by the soil in forests. As trees thrive from more CO_2, so too soil life, so much so that the microbes start consuming and releasing the carbon long locked up in the accumulated plant matter from generations long past.

At 700 ppm CO_2, trees are pumping a bit more oxygen into the air is grossly outweighed by soil carbon release, which is as much as 15% at the surface. That's 50 years of worldwide fossil fuel burning at the current rate.

While the outlines are clear, the carbon cycle is not well understood at the detail level. This is because isotopic variations of CO_2 have distinct significances in the carbon cycle. Different carbon cycle processes affect different carbon dioxide isotopes, and the different isotopes have varying impacts on global warming.

In 2010 there were 1 billion vehicles on the road. 300 million more were added by 2018. The number of vehicles is projected to rise to at least 3 billion by 2050. That will put air pollution, and atmospheric CO_2 release, into overdrive.

It is a safe bet that deforestation worldwide will continue apace if not accelerate. With more cars doing their part to increase atmospheric CO_2, and fewer trees, the forest carbon bank will be paying its interest in a positive feedback cycle to accelerate atmospheric CO_2 well beyond current projections, which do not take into account the actual dynamics of soil in the carbon cycle. Human mass intervention in the carbon cycle is clearly an experiment of global consequence.

ഔ Nitrogen ൶

All we are is a lot of talking nitrogen. ~ American playwright Arthur Miller[*]

Scottish physician Daniel Rutherford is credited with discovering nitrogen in 1772, though he did not identify it. He simply killed a mouse to show that it couldn't breathe once the ambient oxygen had been used up.

Rutherford's discovery was the leftover gas besides oxygen: mostly nitrogen, with residual carbon dioxide; what he called *noxious* air (fixed air). Rutherford's noxious air provided him further proof of his conviction to phlogiston theory.

Imaginative German alchemist Johann Joachim Becher concocted the *phlogiston theory* in 1667. By his account, *phlogiston* was an odorless, tasteless, colorless, massless fire element, contained within combustible substances, and released during combustion.

Once burned, a *dephlogisticated* substance was held to be in its "true" form, the *calx*. Phlogiston theory purported to explain both combustion and the rusting of metals, processes now collectively known as *oxidation*.

If carbon is the king of organic chemistry, nitrogen is queen. Nitrogen (N) is an essential element in building amino acids and nucleic acids, which are respectively the building blocks of proteins and genomes. Nitrogen is the 4th most abundant element in organisms, and the 7th most common in the cosmos.

While essential and abundant, nitrogen only reluctantly plays its organic role. Nitrogen has 5 electrons in its outer shell, and so is *trivalent* in most compounds. The triple bond of molecular nitrogen is a tough bond to break.

---

[*] We are also a lot of farting nitrogen. Nitrogen is the main constituent of flatus (20–90%), followed by carbon dioxide (10–30%).

✍ Fixing Nitrogen ✍

For nitrogen to become organically employed, it must be "fixed." *Nitrogen fixation* is the ability of an organism to transform atmospheric N_2 into usable nitrogen species, such as ammonia (NH_3). Only prokaryotes, supposedly simple soil bacteria, are capable of nitrogen fixation.

Any supposition of simplicity is deceptive. Wholesale nitrogen fixation occurs only after bacteria become intimate with a plant by a welcomed invasion. The process is the most complex coordination between 2 species that is known.

Diazotroph bacteria mastered the knack of fixing nitrogen thanks to *nitrogenase*, an enzyme that assists in the necessary chemical reactions. Some higher plants and termites joined with diazotrophs in symbiotic relationships, providing a home in return for the ability to fix nitrogen.

The primary role of plants in the nitrogen cycle is assimilation of biologically-accessible nitrogen (NO_{-3} & $NO+_4$) into plant biomass, which is locked up until the plant dies.

Protists and fungi incorporate organic nitrogen into their biomass, but do not actively excrete nitrogen compounds; but animals do.

The primary nitrogen excretory product from animals is ammonia (NH_3), derived from the digestion of proteins. Because ammonia is highly toxic, the urinary and excretory systems of animals are complex regulatory mechanisms that afford quick and efficient elimination.

All known life needs fixed nitrogen. Because nitrogen is physiologically essential, organisms evolved various mechanisms for regulating its uptake and excretion. These mechanisms vary widely in different life forms.

✍ Nitrogen Cycle ✍

In many ecosystems worldwide, nitrogen is the element whose supply rate from the environment is most limited. Because competition is fierce for this resource, nitrogen supply controls the behaviour of many organisms and shapes the structure and function of whole ecosystems. ~ American ecologist Edward Schuur

The *nitrogen cycle* is the gaseous cycling of nitrogen between the environment and life, from elemental to biologically accessible. Much of the nitrogen that organisms need to grow is supplied by recycling. New deposits of nitrogen are minuscule compared to quantities recycled. But nitrogen infusion is vital for newly forming ecosystems, and for balancing natural nitrogen loss from an ecosystem into streams or back into the atmosphere.

Nitrogen gas (N_2) comprises 78% of the atmosphere, and so the air forms a large reservoir pool. The soil makes a small exchange pool.

This is quite a contrast to the carbon cycle, where the atmosphere is the exchange pool, and the soil a sink. What is much the same between the 2 cycles is the prominent part played by plants, which, in collaboration with soil bacteria, fix nitrogen into an organically usable form.

On land, nitrogen cycling comes in the organic back end: from the decomposition of organic matter. Denitrifying bacteria convert soil nitrates into gaseous N_2.

The largest pool of fixed nitrogen is dissolved in the oceans. Dissolved organic nitrogen (DON) accounts for 60% of the reactive nitrogen in the ocean.

DON fuels formation of organic molecules from carbon dioxide, especially in oceanic regions low in inorganic nutrients, such as open-ocean gyres: large systems of rotating currents. Most DON is cycled by phytoplankton in less than a year.

Marine sedimentary rock is a rich source of nitrogen: 10 times that found in igneous rock, which explains why plants have such a tough time establishing themselves after a volcanic eruption. 75% of Earth's crust is covered by sedimentary and related rock types.

The nitrogen cycle is naturally a well-orchestrated system, balanced on a planetary scale. For Earth as a whole, the nitrogen fixed equals the nitrogen returned to the atmosphere. If this were not true, atmospheric nitrogen would become depleted.

It is crucial for healthy ecosystems that a balanced nitrogen cycle be maintained. Too much new nitrogen introduced into the cycle – such as excessive application of nitrogen-based fertilizers and other pollution that releases nitrogen into the atmosphere – and world ecology is profoundly affected. This has been happening at least since the onset of the Industrial Revolution in the mid-19th century, beginning with widespread industrial burning, the subsequent commercialization of internal combustion engines, and capped off with nitrogen-rich fertilizers.

Nitrogen fertilizer to promote plant growth has increased crop yields while severely polluting the environment with excess nitrogen. Because of the inefficiencies of nitrogen uptake, only about 10% to 15% of the nitrogen applied via fertilizer is used by crops. The rest is released into the environment as pollution. Burning *fossil* fuels – the fossils being those of plants – also releases nitrogen.

Excess nitrogen from human intervention has altered the Earth's nitrogen cycle in the soil, oceans, and atmosphere. Atmospheric depositing of reactive nitrogen into terrestrial ecosystems globally doubled from 1968 to 2007. The pace is accelerating.

Excess reactive nitrogen in water depletes oxygen supply by redox reactions. Runoff from fertilizer into bodies of water has created a growing number of "dead zones" around the world. Only primitive anaerobic life, which needs no oxygen, can survive in these aquatic dead zones.

As a result of fertilizer runoff, half of the lakes in the United States are *eutrophic*: have excess nutrients that result in poor water quality, including turbidity and oxygen depletion, from the bottom of the water body on up. Eutrophic water is conducive to algal blooms, which further reduce the oxygen in the water.

♒ Fritz Haber ♒

> During peace time, a scientist belongs to the world, but during war time, he belongs to his country. ~ Fritz Haber

German chemist Fritz Haber invented the *Haber process* in 1909, allowing nitrogen fixation by reaction of nitrogen gas and hydrogen gas, catalyzed by enriched iron or ruthenium; a process used industrially to produce ammonia. The Haber process was an important step for the industrial production of both fertilizers and explosives.

For his role in creating chemical weapons in World War I, Haber is considered the father of chemical warfare: developing and deploying chlorine and other poisonous gases on the enemy. Haber personally supervised the first gas attacks. The Kaiser promoted him to captain; a rare gift for a scientist too old to enlist.

Haber's wife of almost 2 decades, also a chemist, committed suicide over his enthusiasm for his work. His son would later commit suicide, in shame over his father's accomplishments.

To further his career prospects, Haber, born into a Hasidic family, renounced Judaism and became a Lutheran.

It did him no good when the Nazis came to power. Haber fled Germany in 1933 but left a legacy. Haber developed the gas Zyklon B, which was later used to exterminate Jews in the Nazi death camps.

ಖ Oxygen ಞ

Oxygen (O) is a colorless, odorless gas, though molecular oxygen appears pale blue. Like nitrogen, oxygen's most common molecular form is diatomic (O_2). But unlike sturdy nitrogen, oxygen is highly reactive: forming *oxide* compounds with almost all other elements. The electron orbitals of ground state dioxygen (O_2) have 2 degenerates: 2 unpaired electrons with antibonding orbitals. The O–O molecular bond is weaker than the N–N molecular bond.

☙ Molecular Forms ❧

Diradicals are molecular species with 2 electrons occupying 2 equal energy (degenerate) orbits. Diradicals vary by the spins of covalently bonded electrons. O_2 and H_2C (methylene) are exemplary diradicals.

Diradicals have 3 possible arrangements: singlet, doublet, and triplet.

The oxygen molecule's ground state is triplet. Both electrons in the degenerate orbitals spin up. The parallel spins of the unpaired electrons make gaseous dioxygen paramagnetic (reactive to magnetic fields), which is quite unusual for a gas. Liquid oxygen is magnetic.

Singlet oxygen has 1 spin-up electron and 1 spin-down electron in 1 orbital, with an equal energy orbital empty. The singlet configuration has several species, all relatively high energy. Hence singlet O_2 is much more reactive than triplet O_2.

If the O_2's ground state was singlet instead of triplet, life would be impossible, and any accumulation of organic matter unlikely. Singlet oxygen is used as an industrial-strength pesticide in buildings, exterminating even the most persistent critters.

The human immune system produces singlet O_2 for weaponry: *reactive oxygen species* (*ROS*). Plants have an ROS response to attacking pathogens: strengthening the cell wall with superoxide or hydrogen peroxide to imprison the infection.

In a reaction powered by sunshine, singlet oxygen is formed from water during photosynthesis. Carotenoids in chloroplasts absorb energy from singlet oxygen using *tetraterpenoids*: a molecular skeletal structure comprising 40 carbon atoms. Carotenoids convert the highly reactive singlet to triplet ground state before it inflicts harm on tissues.

The ability to detoxify ROS evolved in the earliest life, prior to photosynthesis and aerobic respiration; otherwise, aerobic organisms would have poisoned themselves.

Photolysis of ozone by short-wavelength (high-energy) light in the troposphere produces singlet O_2. The troposphere is the lowest layer of the Earth's atmosphere, 17 kilometers thick in the middle latitudes, with about 75% of the atmosphere's mass and 99% of its water vapor and aerosols.

Doublet O_2, with one electron unpaired, is a simple free radical, and highly reactive.

O_2's odd triplet ground state prevents molecular oxygen from reacting directly with many other molecules, which are often in the singlet state. But triplet oxygen will readily react with doublet molecules, such as radicals, to form a new radical. And the univalent pathway of adding electrons 1 at a time in series reactions is quite common.

✂ Oxygen Toxicity ✂

Oxygen's ready reactivity explains its toxicity, by too-enthusiastic production of intermediates, especially singlet oxygen, and, from water, hydrogen peroxide (H_2O_2) and the hydroxyl radical (*OH). Free-radical oxygen also enhances the lethality of ionizing radiation.

There are biochemical defenses against oxygen toxicity. Antioxidants are a rear-guard to minimize tissue damage. A free-radical chain reaction triggered by singlet O_2 can be broken by antioxidants that react with the chain-propagating radicals.

Vitamin E (α-tocopherol) is one antioxidant biocompound. Vitamin E deficiency can cause muscular dystrophy and reproductive failure in humans.

Chronic, low-level, cumulative oxygen toxicity is the biochemical cause of aging.

✂ Discovery of Oxygen ✂

English theologian Joseph Priestley is often credited with discovering oxygen, though claims can be made for Antoine Lavoisier and Carl Wilhelm Scheele.

Priestley gained a scientific reputation for inventing soda water. He also wrote on electricity, and discovered several *airs*, including "*dephlogisticated air,*" now known as oxygen. Priestley was a staunch adherent of Johann Joachim Becher's phlogiston theory: that combustible matter contained a hidden fire element, phlogiston.

Priestley's attribution of oxygen as dephlogisticated air was confused, because, under the theory of phlogiston, oxygen would be the gaseous venue by which phlogiston can be released: a phlogisticating air, not dephlogisticated air.

Priestley's discovery was made by heating mercury oxide by sunlight, and having mice, and later himself, breath the results (vaporous mercury and oxygen). His 1776 published description of isolated dephlogisticated air did not identify oxygen per se.

Antoine Lavoisier determined air as a mixture of gases, primarily nitrogen and oxygen. Lavoisier demonstrated oxygen as the agent of rusting metals and explained oxygen's role in plant and animal respiration.

Lavoisier's explanation of combustion disproved the phlogiston theory that Priestley held dear. Priestley never accepted Lavoisier's outrageous speculations on oxygen, respiration, and rust. Instead, Priestley religiously defended phlogiston theory for the rest of his life.

Priestley, with a Presbyterian cast of mind, also maintained that humans have no free will, that instead conditions create dynamics with inevitable outcomes (predeterminism). According to Priestley, everything in Nature, including men's minds, are subject to the law of causation, but because a benevolent God created all, perfection of man and the world would come in due time. For Priestley, evil arose only from an imperfect understanding of the world.

Swedish pharmaceutical chemist Carl Wilhelm Scheele was called "hard-luck Scheele" by American writer and biochemistry professor Isaac Asimov because Scheele made several chemical discoveries before others who are generally given the credit. Slow to publish, Scheele was scooped on oxygen, hydrogen, chlorine, barium, tungsten, and molybdenum. Scheele isolated oxygen about 2 years before Priestly, but Priestly published first.

Like Priestley, Scheele, who called oxygen "fire air," described the gas using phlogistical terms, as he considered his discovery as confirming, not overturning, phlogiston theory.

Though Scheele did not comprehend the import of his oxygen discovery, others did. Scheele wrote Lavoisier about his findings. Lavoisier grasped the significance. Ironically, Scheele's report was pivotal in invalidating the long-held theory of phlogiston.

♎ Apollo 1 ♎

One giant leap for mankind. ~ American astronaut Neal Armstrong while standing on the Moon in 1969

The Apollo space program was NASA's last step to having an American walk on the moon before the 1960s expired; a national goal set in 1961 by President Kennedy at the height of the Cold War, as a rallying distraction to the diplomatic follies of his and the Soviet Union's administrations.

The distraction proved expensive. $24 billion was spent getting there; at the time, by far the largest commitment of resources ever made by any modern nation in peacetime. At its apex, the Apollo program employed 400,000 people, and required support from over 20,000 corporations and universities.

NASA management was always concerned about cutting payload. To this end, only pure oxygen was circulated in a spacecraft. Nitrogen, 78% of ordinary air, was considered deadweight.

NASA partly acknowledged the risk of using pure O_2 in a 1966 technical report: "in pure oxygen [flames] will burn faster and hotter without the dilution of atmospheric nitrogen to absorb some of the heat or otherwise interfere."

As soon as O_2 absorbs heat, the molecule atomizes. Each atom raises hell by stealing electrons from nearby atoms; sizzling larceny that hots up any fire.

Worse, it takes but the slightest stimulation to spark an O_2 orgy. Some NASA engineers fretted that static electricity from the Velcro on the astronauts' suits might cause spontaneous combustion.

Yet NASA's 1966 report concluded: "inert gas has been considered as a means of suppressing flammability... Inert additives are not only unnecessary but also increasingly complicated."

In space, where there is no atmospheric pressure, just enough gas to breath is all that is needed. But on the ground, owing to atmospheric pressure, technicians had to pump the simulators with prodigious quantities of pure oxygen to keep the simulator walls from crumpling.

Mere months after the O_2 report, during a capsule training simulation, an unexplained spark ignited a fire that cremated the 3 astronauts inside within seconds; whereupon NASA management decided that inert gases had something going for them after all.

⚡ Atmospheric Oxygen ⚡

Oxygen has been the volatile actor in life's evolution. Before the origin of life, in the Hadean eon, atmospheric oxygen was negligible. Similarly, oxygen also exists in space, scattered about in minuscule quantities.

Anaerobic organisms arose that produced oxygen as exhaust, gaining usable carbon in a CO_2 conversion catalyzed by sunshine: photosynthesis. Such phototropic life appeared within a billion years of Earth's formation. Planetary oxidation had begun by 3.4 BYA.

But not without counterforce. Methanogens thrived in the nickel-rich seas billions of years ago, belching methane into the air. The methane reacted with atmospheric oxygen, creating carbon dioxide and water.

Eventually, the methane party wound down. Oceanic nickel levels began to drop 2.7 BYA. Nickel levels halved by 2.5 BYA. The heyday of the methanogens had passed.

The mantle of the early Earth was so hot that dynamic tectonic plate flow with subduction did not begin until 3.0–2.7 BYA. Oxidized material from Earth's surface began being recycled into the mantle. Pressure at depth resulted in oxygen release into the atmosphere.

Relatively rapid oxidation, albeit fractional, resulted in Earth's first extinction event. Much microbial life had evolved to survive in an oxygen-poor environment. The ascent of O_2 spurred adaptation to greater oxygen tolerance.

Over 2.5 BYA, cyanobacteria evolved, and begat the slow oxygenation of the planet, by inhaling carbon dioxide and exhaling oxygen.

It took 500 million years for the atmosphere to begin oxidizing. The oxygen produced by photosynthesis readily reacts with ferrous iron and other elements to form precipitates, such as insoluble ferric oxide (e.g., rust).

1,500 different minerals were found on Earth prior to life arising, generated by dynamic mantle and crust processes during the first 2 billion years. Oxidation of Earth created 2,500 new minerals, many of those being oxidized and weathered products of predecessor minerals.

As the atmospheric oxygen level rose, every mineral that could be oxidized was. Once the weathering of iron-rich rocks abated, photosynthetic cyanobacteria belched so much oxygen so quickly as to overwhelm the planet's ability to soak it all in.

Free oxygen poured into the air and oceans, radically altering geochemical dynamics, as well as the biochemical evolution of life. Rising atmospheric oxygen facilitated the evolution of eukaryotes: a significant step from purely prokaryotic life.

Life added to Earth's mineral stock. Over 4,400 different mineral species have been cataloged. 400 have been added since eukaryotes arose.

Aerobic respiration may have presaged the oxygen surplus from photosynthesizers. 2.9 BYA, the most ancient aerobic process produced pyridoxal ($C_8H_9NO_3$), the active form of vitamin B_6, and an oxygen-based enzyme, manganese catalase. The enzyme detoxifies hydrogen peroxide by breaking it down into oxygen and water. These early aerobic organisms may have got the oxygen needed for pyridoxal production by busting up hydrogen peroxide (H_2O_2), which might have come from glacial ice being bombarded by ultraviolet radiation, which generates generous amounts of H_2O_2.

UV levels were very high at the time, as the atmosphere had yet to form its later ultraviolet shield. The UV shield that eventually formed was an ozone layer in the upper atmosphere, a byproduct of atmospheric oxygen proliferation.

Despite unfiltered radiation from the Sun, early life prodigiously evolved. A few factors were in its favor.

1st, the Sun burned less brightly. 2nd, early life was in the oceans, which provided some protection from UV. 3rd, primordial bacteria developed protective mechanisms to limit DNA damage from UV radiation.

Without atmospheric O_2, there is no O_3 (ozone). Though the ozone layer accounts for only 0.00001% of the volume of atmospheric gas, its accumulation in the upper stratosphere, and its ability to absorb 99% of incoming ultraviolet rays, provides a blanket of protection for life on the surface.

Atmospheric oxygen stops water loss from the planet. Hydrogen released from water bumps into oxygen in the air before it wafts into space and is recaptured in rain droplets.

What started as a primitive organism waste product became organic fuel. The energy that can be derived from fermentation, or the early-evolved methane-sulfate reactions, are puny compared to the potency of aerobic respiration.

Nothing else could have powered multicellular life. All plants and animals depend upon oxygen for at least part of their life cycle. Only the early risers, microscopic life, managed to eke out an existence before the oxygen bloom.

The proof is prehistoric. 300 MYA, atmospheric oxygen levels were 66% higher than today. The higher oxygen levels greatly affected the species that adapted to intake oxygen as an energy source. Paleozoic amoebas were 100 to 1,000 times larger than they are now.

During the atmospheric oxygen bloom, insects supersized. Dragonflies had wingspans of 70 centimeters. There were millipedes over a meter long.

Oxygen transport in vertebrates is through the bloodstream. But insects move air through their bodies by *trachea*: an internal network of channels. So, for insects, a higher oxygen level readily supports rapid evolution of a larger body.

While life adjusted to an oxygenated atmosphere, its toxicity at the cellular level remains a challenge. Both plant and animal tissues create anoxic conditions to generate stem cells.

Oxygen is a diffusible signal involved in the control of stem cell activity. ~ Italian botanist Francesco Licausi *et al*

ஐ Proteins ௸

To a large extent, the structure, behavior and unique qualities of each living being are a consequence of the proteins they contain. ~ American molecular biologists Kathleen Park Talaro & Barry Chess

A *protein* is a linear polymer comprising amino acids; an intricate, living macromolecule, typically folded into a globe or fiber as a product of energetic economy. Hyperactive proteins are called *enzymes*.

The assembly of amino acids into a protein is done one at a time in linear and specific order. ~ American biologist Peter Ward & American geobiologist Joe Kirschvink

Organisms are largely built of proteins. The concept of the set of proteins in a cell comprises a cell's *proteome*.

Proteins are the workhorses of biochemistry, the chief actors for cells: for structure (e.g. membranes) and internal organization, cell identification, transportation (of molecules within, and in and out of cells), communication (signaling, both within and between cells), enzymatic action, gene regulation, defense, and cell movement.

Proteins are commonly capable of a variety of functions. But they often need to be focused to a specific task, else they might create cellular havoc, possibly leading to disease.

This is where cofactors come in. A *cofactor* is a chemical compound that binds to a protein as a requisite for tailored activity. Enzymes are typically activated by cofactors, which essentially act as helper molecules.

Proteins are characterized by 4 levels of structure: chemical (amino acid composition) (aka primary); fold pattern (aka secondary); 3D shape (aka tertiary); and assembly (aka quaternary): an assemblage of several protein molecules, such as with polypeptide chains. These characteristics are genetically defined. Each is significant in protein functioning.

Proteins are machines that have structures and motions. ~ American molecular biologist Peter Wright

Proteins remember their situation, and so are prepared for what needs to be done next. For example, by binding to a certain ion, a protein activates. In this way, the state of a protein at any instant embodies a memory of its past.

¤ ✧ ¤

Even tiny changes in structure can greatly affect the properties of a compound. ~ German chemist Jochen Küpper

Most molecules occur in several shapes, each of which may behave differently from another. Spatial arrangement can also affect reaction rate.

Such diversity of functionality by topology is at the heart of protein functioning. These multifaceted macromolecules evolved to perform different complex functions. One aspect of this comes in the way that proteins fold and unfold to activate and perform tasks.

♎ Cell Size ♎

Cells achieve their steady-state size by adding a constant volume between birth and division, regardless of their size at birth. ~ Korean Canadian molecular biologist Suckjoon Jun *et al*

To realize their biological potential cells must attain a specific size. They do so through certain proteins responsible for growth which are also involved in cell division. The proteins know exactly how large a cell should be by understanding the genetic blueprints. Some eukaryotic cells edit the instructions that these proteins get to alter desired cell size.

It's a very robust mechanism. ~ Suckjoon Jun

♐ Protein Folding ♑

Over time, Nature improved protein folding so that more complex structures were able to develop. ~ German molecular biomechanist Frauke Gräter

Proteins are produced in a linear sequence. The last step in protein production is an elaborate folding to a shape that is energetically in repose.

Chaperonin are proteins that provide scaffolding for the first folding of new proteins. In a high-speed origami, a protein assumes its final resting structure in a series of rapid incremental steps. This process takes only a few seconds, thanks to active guidance by chaperonin.

A protein partly unfolds to work. How a protein folds/unfolds, and at what speed, has much to do with its performance. The intricate 3D conformation of a protein is essential to proper functioning.

A major force behind protein folding and polypeptide interaction is the avoidance of water by hydrophobic amino acids. Internal friction, which reflects the energy landscape of the protein, plays an important role in the dynamics of folding, and the ability of a protein to function properly. The folding process of a protein is regulated by a formidable network of other proteins, and other, smaller molecules.

There is a sea of small molecules – ligands – with which proteins live in a cell. These ligands, which are very small molecules only about 100 daltons in size, are critical in determining the behavior of folding macromolecules on the order of 100 kilodaltons in size, that is 1,000 times larger. It's like the mouse telling the elephant what to do. ~ American biochemists Lila Gierasch & Scott Garman

Many proteins have varying degrees of folding. 30% of human proteins have unfolded portions.

Whether a portion of a protein is folded or not, or even appears disordered, is scant indication of what it may do. A protein operates orchestrally; all portions have some part to play. The odd bits may be the piece that give a protein its versatility.

Many diseases result from misfolded proteins (*prions*), notably neurodegenerative diseases associated with aging, such as Parkinson's, Huntington's, and Alzheimer's. Prions refold from a harmless form into one that is malicious and contagious, initiating a chain reaction that creates a prion aggregation.

RfaH is a protein which activates genes that allow *E. coli* bacteria to launch a successful attack on a host, inciting disease. It does so by folding itself into different shapes, and by doing so is able to perform vastly different tasks. Specifically, RfaH can fiddle with both genetic transcription and translation – coupling the two together – by smartly changing its shape.

Dihydrofolate reductase (DHFR) is an enzyme common to *E. coli* bacteria and humans and everything in between. DHFR structure is almost identical throughout all life forms; conserved through evolution, as its function is critical for the synthesis of DNA. But how DHFR unfolds to expose amino acids differs between bacteria and primates. This distinction in atomic dynamics makes a significant difference.

DHFR motion evolved to fit the cellular environment. Human DHFR is so well-tuned for its own cells that it won't work in bacteria: the product molecules in *E. coli* are packed too tight for human DHFR to function.

✄ Protein Regulation ✄

   Man's health and well-being depends upon, among many things, the proper functioning of the myriad proteins that participate in the intricate synergisms of living systems. ~ American biochemist Stanford Moore

The activities of proteins are regulated in a variety of ways, though the agent almost always involves a *ribozyme*, which is an RNA-based enzyme.

Proteins have an *active site*, where substrates bind and undergo a chemical reaction. The active site is not the only place which affects protein functionality.

Protein activity can be regulated by an *effector molecule* binding at a protein's *allosteric site*: a site that is not the active site. Enzymes are typical effector molecules at allosteric sites, though an enzyme too is subject to allosteric regulation, as enzymes are themselves proteins.

Allostery refers to protein regulation at an allosteric site. *Allosteric regulation* is the action of effectors at the allosteric site, either enhancing a protein's activity (*allosteric activator*), or tamping it down (*allosteric inhibitor*). Allostery is especially important in cellular communication, where getting the right message across is critical.

Because eukaryotic cells are highly compartmentalized, localized control of allosteric regulation is widespread in eukaryotes. Enzymatic activity is itself controlled in a process called *localization*.

Enzymes are continually active. They evolved for flexibility. An enzyme's effect can be controlled by small binding sequences that localize its activity. Enzymes can be very selective about the 3D configuration of the molecules they interact with.

Localization by proximity control is readily engineered because binding sequences to effect such control are typically simple. Because the biomechanics are so easy, much genetic regulation is allosteric.

✄ Generalists & Specialists ✄

Functionally, proteins are either specialists or generalists. Evolution begat specialization. The earliest cells were chock full of jack-of-all-trades proteins.

Generalists dabble in various chemical reactions. They carry out tasks which are less crucial, such as making vitamins (B_{12}), which cells need in limited quantities.

Specialist enzymes stick to specific substrates. Their devotion means that they are more trusted to essential functions, such as translating genetic instructions into new proteins, and the constant chore of producing cellular energy.

A specialist protein senses when a mammal's lungs are full of air, and so plays a critical role in regulating breathing. The same protein is instrumental in the sense of touch and proprioception.

The importance of generalists is not to be understated, particularly in their having a broader vista than their dedicated brethren. In eliminating toxic substances, a cell is better off with enzymes that can recognize and deal with more than one kind of danger.

✄ Pain ✄

Due to specific demands, sensory systems evolve independently in different species. ~ American neurobiologist Marco Gallio *et al*

An animal's tolerances and perils vary greatly, depending upon a creature's adaptation to its environment. Some cells can withstand intense heat, or, conversely, cold, which would damage the average cell; likewise for mechanical stresses and chemical agents.

All animals feel pain, and they do so via the same protein, termed *TRPA*. TRPA acts as a stress detector (nociceptor). Nociception is detection of stimuli – whether chemical, mechanical, or thermal – which are hazardous to a cell. Via TRPA activity, nociception initiates pain to warn an animal.

TRPA is remarkably conserved across animal evolution.
~ Marco Gallio

While TRPA is essentially the same for all animals, its detection acumen varies vastly, depending upon what defines danger for the host in which it works. While the consummate generalist in sensing danger and sounding an alarm, TRPA is specifically adapted to know exactly what tolerances a cell has. How such exquisite expertise can be attained or held by a protein is mysterious.

Even proteins that share the same topology may perform a different range of tasks. The orientation of a protein's amino acids affects electron transfer efficiency.

Channeling energy differentially alters the branching to neighboring amino acids. A simple electron channel change can cascade into a considerable functional difference.

⚭ Amino Acids ⚮

A *protein* is a complex polymer built up from chains of polypeptide molecules. A *polypeptide* is formed from numerous peptides.

A *peptide* is a short polymer of amino acid monomers, linked by peptide bonds. The shortest peptide is 2 amino acids joined by a single peptide bond: a *dipeptide*.

An *amino acid* is a molecule comprising: a *carboxylic acid group*, an *amine group*, and a *side chain* specific to the amino acid. Amino acids have as key elements: carbon, hydrogen, oxygen, and nitrogen.

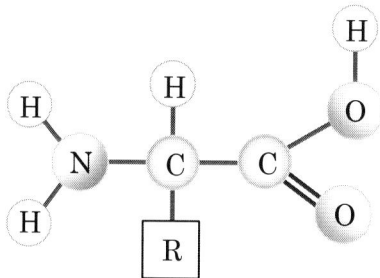

alpha-amino acid (un-ionized)

An alpha-amino acid has both the amine and the carboxylic acid groups attached to the first (alpha-) carbon atom. The 22 proteinogenic (protein-building) amino acids are all α-amino acids.

The *side chain* of an amino acid is a hydrocarbon. A *hydrocarbon* is a molecule comprising only carbon and hydrogen. The side chain alkyl group is historically designated by *R*. Methane (CH_4) is the simplest alkane.

An amino acid side chain may have a string of carbon atoms. The number of carbon atoms defines the size of an alkane.

A *carboxylic acid* is a polar molecule: $-CO_2H$, connected to a hydrocarbon. A carboxylic acid completes itself with a side chain. The minus in $-CO_2H$ indicates a negative charge.

An *amine* is a derivative of ammonia (NH_3). To be an amine, 1 or more of nitrogen's links to hydrogen is replaced by a bond to an *alkane* (an amino acid side chain). The nitrogen atom has a lone pair. An *amine group* is thus: NH_2R, NHR_2, or even NR_3.

Amino acids are simple and easily constructed. The building blocks of life are chemically quite accessible.

20 different amino acids form the basic building blocks of peptides. Some polypeptides have all 20 amino acids, some fewer.

There are 22 known amino acids, but only 20 are commonly used in extant life forms. Amino acid number 22, the most recently discovered, is very ancient; used by methane-producing archaea (methanogens) for energy conversion.

The human body can synthesize 12 of the 20 standard amino acids but cannot make the other 8. These 8 must be supplied by dietary intake, and so are termed *essential amino acids*.

There are also *conditionally essential amino acids*, which some humans can sufficiently synthesize but others cannot, and so must get adequate amounts from their diet. The difference between essential and non-essential amino acids is not fully understood, as some amino acids can be derived from others. For example, the various human amino acids containing sulfur can be converted internally, but not synthesized de novo in humans.

There are specific proteins for many of the hundreds of millions of species that have lived on Earth, and for different plant parts and body organs. The human body synthesizes ~100,000 different proteins.

The vast diversity of proteins comes primarily from variety in amino acid R groups, which yield markedly different chemical and functional properties. Proteins often have over 100 amino acid units.

There are more ways to arrange 20 units in a chain of 100 than there are atoms in the universe; whence the practically unlimited information storage capacity of proteins.

≈ Synopsis ≈

Matter

➤ *Matter* is anything that occupies space and has mass.

➤ The *atom* is a basic unit of matter, comprising a nucleus and electrons. An atomic *nucleus* typically comprises protons and neutrons.

Only the simplest form of hydrogen, with a single proton, can have a nucleus without neutrons. Protons on their own repulse each other. So, any stable atom with multiple protons also has neutrons. Neutrons are necessary to hold atomic nuclei together.

➤ *Matter* involves a hierarchy of combinations. Subatomic particles form atoms. Atoms form molecules. Molecules combine to form a vast variety of compounds.

➤ From atoms up, electrical properties provide guidance for the energetic transactions (*reactions*) of building up more complex macromolecules or breaking them down.

➤ *Electrons* swirl about atomic nuclei in an orbital cloud, equalizing the positive charge of protons via the negative charge possessed by electrons. Electron clouds self-organize into layers (*shells*) which generally differentiate by energy levels. Shell layering is an HD dynamic.

➤ The *Zen of matter* is emptiness. An atom's electron cloud whorls far above its nucleus. An atom is 99.99+% space.

Chemistry

➤ *Chemistry* is the study of matter and the changes which substances undergo. Chemistry is extensively entangled with other fields of study, notably physics. The disciplines concerned with life rely heavily on *biochemistry*: the application of chemistry to biological processes.

➤ *Molecules* are the functional unit of chemistry, the basic building block of matter in an everyday sense.

➤ Seeking stability in the electrical charge created by the outer shell of an electron cloud, atoms combine into molecules by forming *covalent bonds*.

> The internal vibrations of molecules drive the structural transformations that underpin chemistry and cellular function.
> ~ Indian American physical chemist Ara Apkarian *et al*

➤ The *energy to effect chemical reactions* may be meager. Slight differences at the subatomic level influence the dynamics of molecular transformation. Chemical butterfly effects are common.

➤ Chemical reactions are *promoted* by *catalysts* which lower the activation energy required for a reaction to occur. Without catalysts, chemical reactions would be much harder to come by.

Yet reactions readily occur where they should not. According to the classical laws of chemistry, the vast expanses of interstellar space are too frigidly forbidding for organic molecules to form. Via *quantum tunneling* – a term that labels but does not explain – prodigious amounts of alcohol and other organic compounds are formed in space.

➤ *Radiation* arises from energy fluctuations, which are pervasive throughout the universe. *Electromagnetic radiation* emits photonic wave energy. *Particulate radiation* arises from atomic decay. Radiation is an HD process initiated by *quantum tunneling*, where a subatomic particle escapes 4D confines.

➤ *Molecular geometry* characterizes molecules by the positions of their constituent atoms in space. Many molecular properties are affected by shape. The vibrational qualities of atoms also greatly affect their chemistry, but little is known of this.

➤ *Water* is a unique molecule, with unusual chemical and quantum properties. H_2O behaves like $H_{3.6}O$ in its fluid atomic bonding, but a single snapshot of a water molecule appears as $H_{1.5}O$.

Though molecularly stable and supposedly electrically neutral, water is an ionizing agent, and so is a universal *solvent*. Water acts as a catalyst in many biochemical reactions.

Water is one of the few chemical compounds that exists alternately as a gas, liquid, or solid within the temperature range of ordinary life. Water enacts an intricate quantum dance to transition from one state to another.

Water may selectively exist in multiple phases simultaneously. This occurs at the surface of water, in the nanometer interface between air and H_2O.

All of life depends on water. The Earth's hydrological cycle (water cycle) is a primary biospheric cycle. Polluted water ravages life.

➤ *pH* is a measure of aqueous acidity. Pure water is neutral (pH = 7). Acids have a pH < 7, while bases have a > 7 pH.

Neutralization is the process of an aqueous solution approaching neutral pH. This is largely accomplished by proton transfer, not just deionization.

Organic Chemistry

➢ *Carbon* is the life's chemical king, owing to its self-chaining and flexible compounding with other elements.

Nitrogen is queen, as it provides a scaffold for amino acids and thus acts as a key constituent in proteins and nucleic acids (RNA, DNA). Unlike carbon's easy bonding, naturally trivalent nitrogen is not so easily swayed into molecular configurations. So, whereas carbon happily plays its part, nitrogen reluctantly joins in. Yet this perfectly explains why these 2 elements play their respective roles so well.

Carbon's pliability makes it ideal for readily constructing complex compounds and effecting transformations. Nitrogen's inherent stability renders it as the best chemical basis for storing information.

➢ *Oxygen* also plays a key role in life, though as a chemical wild card: biochemistry's joker. Oxygen is highly reactive, hence a fabulous fuel source, if abundantly available. But oxygen, because of its reactive nature, requires special handling, and so is not well-suited as a fuel for the more simply structured life.

Abundant oxygen was not available for early life. The planet transformers that would turn into plants relied instead upon then-abundant CO_2 and H_2O, along with sunlight as a catalyst, to create energy-rich sugar, with extra oxygen and water left over. Their success oxygenated the atmosphere.

➢ Both carbon and nitrogen are critical in the *cycles* by which the planet precariously supports life. Befitting the element's respective characters, Earth's carbon cycle is readily altered, while the nitrogen cycle is more reluctantly swayed.

Balance in the carbon cycle is one of abundance, most notably the level of atmospheric CO_2, while the nitrogen cycle spins on its scarcity in bioavailability. Planetary imbalance in either of the carbon or nitrogen cycles portends serious consequences to all life on Earth.

➢ *Plants* have long held a central regulatory role in the carbon cycle, and thus climate control. The fate of all animals, from air to food to habitat, has always depended upon plants.

➢ *Life* is entangled at the molecular level. The environment that biomolecules live in determines their levels of activity and rates of interactions.

Proteins

➢ *Proteins* are living macromolecules of varying complexity. They are the workhorse molecule of life, participating in virtually every process that takes place within cells, as well as acting as messengers between cells.

Protein folding is essential to proper functionality. Misfolded proteins – *prions* – can cause debilitating diseases.

A protein is a complex polymer built up from chains of polypeptide molecules. A *polypeptide* is formed from numerous peptides. A *peptide* is a short polymer of amino acid monomers linked by peptide bonds.

➢ An *amino acid* is a compound of simple *functional groups*. The key elements of the amino acid functional groups are carbon, hydrogen, oxygen, and nitrogen.

➢ The *active site* of a protein is where substrates bind and undergo chemical reactions. The active site is not the only place affecting protein functioning. Protein activity can be regulated by an *enzyme* or other *effector molecule* binding at a protein's *allosteric site*: a site that is not the active site.

➢ An *enzyme* is a protein that acts as a catalyst in reactions involving proteins or other organic substrates. Every biochemical reaction is promoted by a specific enzyme, though some enzymes learn specificity rather than being produced with an intended specificity.

❧ Life Begins ❦

The intimate relationship between the vital phenomena with chemistry and its laws makes the idea of spontaneous generation conceivable. ~ English naturalist Charles Darwin

Life originating depends upon a confluence: contained (cellular) and controlled energetic reactions (metabolism), coupled with a scripted code for reproduction (replication). The only way such processes could be initiated and sustained would be if Nature favored their occurrence. Otherwise, the odds of life ever emerging are exceedingly long indeed.

Life cannot be explained by our current laws of physics. ~ American astrobiologist Sara Imari Walker

❧ Origin on Earth ❦

The parameters surrounding the origin of life on Earth are known. The geological conditions have been discerned, the requisite chemistry understood, the timing apprehended. Yet life's onset retains mystery to its researchers. The origin of life on Earth bristles with puzzle and paradox.

What is not yet known is exactly how and where the ingredients that spawned life came together with the spark that started life's engine and kept it running, though the best possible spots for origination have been identified. The enigma that lingers lies in life's spontaneous generation: the force of coherence that led life to coalesce.*

What is known is that life on Earth began as soon as environmental conditions permitted. The mathematical probability of elements randomly assembling into a metabolizing,

---

* Nature commonly exhibits self-organization and hidden order within apparent disorder. Chaotically tumultuous fluids spontaneously create stripes of coherent flow alternating with turbulent regions. Liquids self-organize into crystalline structures: a phenomenon known as *disordered hyperuniformity*. Photons in laser light self-organize into fractal patterns. Viewed as particles in a system (instead of linearly), prime numbers exhibit an ordered structure.

self-replicating life is negligible. Yet it repeatedly happened, with nary a chance that it was happenstance.

All biological molecules used by living organisms are themselves synthesized by living organisms. The most important aspect of life's emergence was the process by which inheritable improvements were selected from a population of variants. This required molecules or molecular assemblies that can reproduce under certain kinetic constraints and resulted in the development of a specific kind of stability (known as *dynamic kinetic stability*) that is associated with the dynamics of reproduction.
~ French molecular biologist Robert Pascal

ဆ Geology ര

Continual bombardment by gigantic comets and asteroids abated 4.1 BYA. In the relative calm, life got its start; so marked the onset of the Archean eon.

Whether life initiated under even more hostile conditions remains an open question for lack of evidence. At most, 100 million years passed from the abating of constant bombardment to life emerging on Earth.

The cosmic assault seeded life on Earth. Meteorites commonly contain water, amino acids, and nucleobases, as well as chemicals that are rare or even nonexistent on Earth.

There have been numerous hypotheses as to where life arose. For a long time, the presumption was that life began on the surface, in primordial pools brimming with organic precursors. Hydrothermal environments on land existed early in Earth's history.

One hoary hypothesis is that lightning literally brought the spark of life; or, less dramatically, radiation. This notion was promoted by Russian biochemist Alexander Oparin in his influential book *The Origin of Life* (1936).

Simplifying assumptions scientists made led to thinking that life mechanistically came from a chemical brew. After the 2nd World War, Western biology moved away from thinking of cells in physicochemical (physical chemistry) terms, and toward a reductionist molecular biology approach, entranced by the nascent field of genetics. From this came a life origination hypothesis termed *RNA world*.

The steamy surface schema for life's origination is less likely than those scenarios that put life's start deep in the ocean or beneath it, near hydrothermal vents, which provide a constant source of both energy and warmth to catalyze reactions, relatively safe from bombardment. There, prebiotic evolution could have been sustained over long periods without disruption or loss of the chemical reducing power necessary for nonbiological synthesis of organic compounds. While temperatures and pressures would have been extreme, life exists there today.

It is certain that life began under inhospitable conditions, at least on the surface. Early Earth had a highly volatile atmosphere, without molecular oxygen, but with toxic gases from erupting volcanoes, methane-rich air, incredible electrical storms, and unscreened ultraviolet rays from the Sun. Acid rain was common.

By 3.5 BYA, Earth's seas were cool, with temperatures not unlike recent times. Areas of the deep oceans were heated by hydrothermal pipes that vented mantle heat.

The Moon orbited much closer to the Earth than it does now, raising huge tides.

Volcanoes spew pumice. As it is 90% porous, pumice floats. Pumice readily absorbs a variety of chemical compounds. In the unlikely event that life was born on the surface, its cradle may have been pumice.

♎ Cool Life ♎

Whereas most scenarios of the environment in which life got its start are heated, it is conceivable that life was first cradled in ice. Nooks and crevices within ice could have provided a cozy, safe place for originating lively organic molecules. As ice forms, pure water crystallizes, while salts and other bits of debris accumulate in watery pockets. These impurities lower the water's freezing point. Little pockets may remain unfrozen within an otherwise solid chunk.

RNA construction reactions can proceed under icy conditions, albeit slower than at ambient conditions. Reaction time is an insignificant factor compared to the environmental stability required for sustaining the necessary reactions.

Ice, a simple medium likely to have been widespread on the early Earth, can provide a propitious environment for RNA self-replication and evolution. Ice not only promotes the activity of an RNA polymerase ribozyme but also protects it from hydrolytic degradation, enabling the synthesis of exceptionally long replication products. ~ English molecular biologist Philipp Holliger *et al*

♎ Pond Spawn ♎

If you think the origin of life required fixed nitrogen, as many people do, then it's tough to have the origin of life happen in the ocean. It's much easier to have that happen in a pond. ~ Indian astrophysicist Sukrit Ranjan

With its tight binding, nitrogen provides an ideal chemical infrastructure for molecular architecture – but only in the right form, as nitrogenous oxides (NOx).

There are 2 ways that nitrogen may have got fixed to support abiogenesis. Nitrogen could have reacted with carbon dioxide bubbling out of hydrothermal vents in the deep ocean; otherwise, lightning.

Lightning is like a really intense bomb going off. It produces enough energy that it breaks that triple bond in atmospheric nitrogen gas, creating nitrogenous oxides that can then rain down into water bodies. ~ Sukrit Ranjan

Lightning crackling through the early atmosphere may have produced sufficient nitrogenous oxides to fuel abiogenesis in any body of water. The critical factor is getting NOx reacting with life-forming molecules.

Shallow ponds provide an ideal environment for prebiotic reactions. The shallower the pond, the less dilution of NOx, thereby giving life a better chance to take hold. By contrast, NOx falling from the sky into the ocean are more likely to be broken down by ultraviolet light near the surface or react with dissolved iron sloughed off from benthic oceanic rocks.

♂ Extremophiles ♌

Life likely began with *extremophiles*: single-celled organisms living under harsh conditions but in a somewhat stable environment. There are several forms of extremophiles extant today, each varying by tolerance to extreme conditions.

Extremophile-first augers for life beginning among rocks, sheltered deep underground, where planetary impacts had little to no effect. Life originating under Earth's surface or near deep-sea fumarolic vents would have meant that the earliest metabolisms were hydrogen and sulfur-based, and non-photoactive.

Hyperthermophilic microbes like it hot. The ribosomal RNA sequences in today's heat-loving extremophiles place them among the most ancient ancestral lineages.

Even now, an estimated 2/3rds of the microbes on Earth lurk in a subterranean world, down many kilometers deep; far from sunlight, frigid temperatures, and lethal radiation on the surface. Subsea gribblies traverse the vent system that belches steaming water and dissolved minerals into the surrounding ocean.

The winds that sweep over the seas affect deep ocean currents. At life's onset, the skies above may have had a long reach, touching life below, particularly life's prospects for spreading.

Bacteria play an integral role in liberating energy from rocks. Under mid-ocean ridges, the mere presence of bacteria coaxes surrounding minerals, such as quartz, to release hydrogen gas, which the microbes feast on.

৪৩ Chemistry ৪৩

The chemistry of life is distinguished by being both highly ordered and far from thermodynamic equilibrium. ~ Dutch physicochemist Rogier Braakman & American physicochemist Eric Smith

The ingredients of life were prescribed by chemistry, both by abundance and ability to well serve essential needs. It is likely that life everywhere in cosmos relies upon the same chemical players. Life is bound to be carbon based, as its flexibility is unparalleled, and its existence ubiquitous: the 4th most common element in the universe.

Life as we know it depends absolutely on solution chemistry in a highly unusual solvent (water), a complex range of chemicals based on a few simple small atoms, and relatively small values of free energy. ~ Andrew Clarke

Water is essential to life. Its unique properties ensure its necessity. Further, its constituents – hydrogen and oxygen – are readily reactive.

Conversely, the steadiness of nitrogen makes it ideal for structural stability. While carbon is the epitome of freewheeling, nitrogen provides a sturdy infrastructure.

♂ Phosphorylation ☿

Phosphorylation attaches a phosphoryl group to a molecule. A phosphoryl group is a randy radical of phosphorous and oxygen ($P^+O_3{}^{2-}$). Phosphorylation and its counterpart, dephosphorylation, are instrumental in numerous biological reactions essential to living.

A phosphorylation chemistry could have given rise, all in the same place, to oligonucleotides, oligopeptides, and the cell-like structures to enclose them.* That in turn would have allowed other chemistries that were not possible before, potentially leading to the first simple, cell-based living entities. ~ Indian organic chemist Ramanarayanan Krishnamurthy

A phosphorylating agent is necessary to getting crucial biological reactions going. Diamidophosphate (DAP) was likely the compound which helped chemically kickstart life.

With DAP and water, these 3 important classes of pre-biological molecules could come together and be transformed, creating the opportunity for them to interact together. It reminds me of the Fairy Godmother in Cinderella, who waves a wand and 'poof,' 'poof,' 'poof,' everything simple is transformed into something more complex and interesting. ~ Ramanarayanan Krishnamurthy

---

* Oligonucleotides are short strands of nucleotides, the chemical core ingredient for genetics. Oligopeptides are short chains of amino acids, which, when congregated coherently, comprise proteins.

Numerous phosphorus-nitrogen compounds have been found in interstellar gas and dust. DAP is a simple radical, and so is likely to have been tucked into early Earth minerals.

Functionally, there are 2 requirements for life: 1) elements requiring modest energy to cohere reactive substrates that can energetically perpetuate for metabolism, and 2) readily reproducible compounds for replication.

As the organic building blocks of life are readily built in space, it seems inevitable life would revolve around nucleobases and amino acids, as they demonstrate flexible complexity at a low energy cost.

The crust under the ocean is a rich reservoir of all possible ingredients for life, including rare earth elements. There exists the most extensive, concentrated library of chemical possibilities.

Hydrogen, a randy reductant, is a rich potential energy source. Hydrogen-consuming bacteria are known, though now largely confined to deep-sea hydrothermal vents where H_2 is still readily available.

Thermal vents on the ocean floor and subsea are places where life originating would have been most amenable. The structural similarity between the minerals precipitated at hydrothermal vents and the most ancient enzymes shows that only a petite push of coherence would be sufficient for life to take hold.

Geologic recycling of the Earth's crust leave scant evidence of atmospheric conditions during the first 650 million years of the planet's history, particularly regarding which form of carbon predominated early on: CO (carbon monoxide), CO_2 (carbon dioxide), or CH_4 (methane). By 4.1 BYA, when life originated on Earth, CO_2 was the predominant atmospheric gas.

Whereas reactions in carbon dioxide are low-yield and of limited variety, reactions in methane yield abundant diverse organic compounds. This difference in energy potential was a significant factor in where early life formed.

The presumed requisite conditions for the emergence of life include water and a sufficiently stable environment. Those conditions would have been largely met near Earth's surface by aerosols, volcanic and interplanetary dust particles, and organic films; and at depth by hydrothermal minerals, chemical precipitates, and vesicular structures of mineral and organic compounds.

Likewise, the necessary energy sources for organic compound production were available via sunlight on the surface, or in the ocean or underneath it by chemical disequilibria between hydrothermal fluids and seawater.

The reducing power of oceanic, dissolved iron was available in both environments. Iron plays a significant role in cellular respiration.

ᛩ Sugar ᛪ

The story of life is suffused with sugar, which readily provides energy release in reactions. Structurally, ribose, a simple monosaccharide, forms part of the backbone of DNA & RNA molecules.

Ribose rides on meteorites. The most elemental sugars of life, with 2 to 3 carbon units, drift through space in gaseous molecular clouds.

Sugars are superbly flexible compounds, capable of myriad configurations. The critical requirements for the sugars of life were ease of construction and stability.

ᛩ Chirality ᛪ

There is an asymmetry to life. Most biochemicals are chiral: having a handedness.

The weak nuclear interaction is the only known physical force to have a handedness preference; leaving slightly more left-handed electrons that find favor in life's little bits. Electron handedness is sufficient to cause chirality in organic molecules.

Homochirality is the geometry of a substance composed of chiral units. Sugars and amino acids are homochiral.

In spite of a century and a half of study, the origin of biochemical homochirality remains a central mystery in life's emergence. ~ American mineralogist Robert Hazen

Chirality is pervasive throughout Nature, including in its referential namesake: human hands.

✂ Abiogenesis ✃

Whenever a new planetary system is made, these kinds of things should go on. This potential to make organics and then dump them on the surfaces of any planet is probably a universal process. ~ American astrophysicist Scott Sandford

Abiogenesis is the study of how organic life arises from inorganic matter.

For life to begin and sustain, 3 means have to arise: 1) self-contained cells (containment); 2) usable energy to produce proteins (metabolism); and 3) replication (reproduction). Any decent hypothesis of life's origin must account for all these facets.

The requirements for life's emergence seem to present a chicken-or-egg problem of which came first: metabolism or replication. While scientific consensus has yet to comfortably square that circle, the answer is fairly certain.

Inorganic energy provided the impetus to put together the organic building blocks that resulted in biosynthesis and nucleobase production. The complexity of both are equivalent, and both are needed.

Sequence is not the issue. The trickiest aspect is not of molecular combinations. Instead, it is the synchrony required for all the ingredients to functionally cohere; for metabolic energy to consistently be applied within a cell where and how required to sustain life, and for RNA to become the basis for reliable memory in protein production.

To simply say that "life happens," or that energetic pathways are dictated by economy, misses the big picture as well as the myriad of details.

No matter how minute an organism may be, or how elementary it may appear at first glance, it is nevertheless infinitely more complex than a simple solution of organic substances. ~ Alexander Oparin

Aside from the biochemical substrates and processes, there must be a natural force of coherence that begets life. Finding favor in physics and chemistry is, by itself, inadequate.

> The underlying problem is complexity. Although we have no idea of the minimal complexity of a living organism, it is likely to be very high. It could be that some sort of complexifying principle operates in Nature, serving to drive a chaotic mix of chemicals on a fast track to a primitive microbe. ~ Paul Davies

Above all, the macromolecules that make for the principle players in cellular life must stably self-assemble yet have the ready flexibility for different conformations to act as information storage.* And all this must be achievable with a mere modicum of energy. These requirements highlight how even the simplest life itself possesses an inherent sophistication that places it well beyond random chance.

¤ ✧ ¤

> Just as many people have been speaking prose all their lives without realizing it, many organic chemists of the 19th and the first half of the 20th century were prebiotic chemists without realizing it. ~ English chemist Leslie Orgel

Modern organic chemistry began with German chemist Friedrich Wöhler accidentally synthesizing urea in 1828. Russian chemist Alexander Butlerov discovered the *formose reaction* in 1861, forming sugars from formaldehyde. This remains a cornerstone of prebiotic chemistry. These and other early experiments into synthesizing biochemicals were oriented toward practical applications, without interest in the origin of life.

---

* A misstatement by abbreviation is made here. Matter may store information, but matter cannot *use* information. Information is purely conceptual, and so immaterial. As the show called Nature is made of matter, essential concepts such as information are portrayed materially–whence brains and accoutrements which comprise an 'intelligence' physiology. Do not confuse appearance with actuality – the very mistake matterist scientists blithely make. Information implies mentation – a mind at work.

The first to have such an interest was American chemist Stanley Miller, who in 1953 reduced amino acids from a brew of heated CH_4, NH_3, H_2O, and H_2 subjected to an electrical discharge. In selecting his prebiotic compounds, Miller aimed to recreate the chemical conditions of early Earth.

Miller's assumptions were mistaken, as were many of the surmises that followed in his wake. Miller began what became a continuing quest by would-be Dr. Frankensteins to create life in a flask.

> Since we know very little about the availability of starting materials on the primitive Earth or about the physical conditions at the site where life began, it is often difficult to decide whether or not a synthesis is plausibly prebiotic. Not surprisingly, claims of the type "my synthesis is more prebiotic than yours" are common. ~ Leslie Orgel

¤ ✧ ¤

From the 1960s, the idea of abiogenesis via spontaneously assembling proteins was gradually displaced by hypotheses emphasizing replication: life's onset via RNA. This notion first developed when life was presumed to emerge from primordial pools, with RNA emerging by dint of fortuitous biochemical combination. Though the focus shifted, the axiom of life beginning via chemical elixir remained.

Another story of life's origin starts with fool's gold in the hydrothermal deep, arguing metabolism-first. There are similar scenarios with slightly different emphases.

Then there is the possibility that life on Earth came from outer space: a concept called *panspermia*. Demonstrating the natural force of coherence, organic molecules readily form where chemistry instructs they should not. One panspermia scenario has life on Mars coming here.

⚘ RNA World ⚘

> RNA is the Swiss army knife of molecules – it can have so many different functions. ~ American microbiologist Michael McManus

An early, formulaic school of thought to life's start was a gene-first, RNA origination, with ribozymes as the 1st actors on the stage. A *ribozyme* is an RNA-based enzyme.

The *RNA-world hypothesis* was first propounded by American biochemist Walter Gilbert in 1986, though the concept of RNA as the primordial molecule of life had been kicked around at least since 1962, by American biologist Alexander Rich. This scenario still has many researchers who are interested in discerning how RNA was able to self-assemble into a functional form and self-replicate.

> The idea that RNA was "invented" by a simpler genetic system is now a popular one, but no convincing precursor system has been described. The idea that some simpler genetic system preceded RNA opens Pandora's box. There is very little to constrain the type of molecule involved or the environment in which it first functioned. There are numerous double-stranded structures with backbones very different from that of RNA but held together by base pairing. ~ Leslie Orgel

¤ ✧ ¤

Catalysts function by lowering the energy necessary for chemical reactions, and thus increase the frequency of such reactions. Nucleobases, the basic building blocks for genetic storage (DNA and RNA), assemble spontaneously given the ingredients and proper setting.

In early RNA-world envisionings, a primordial soup that was half-sugar, half-base mixture was catalyzed with phosphate by ultraviolet light. The results of similar combinations were nucleotides of different types which zipped together to form RNA molecules. These RNA molecules, by their stable chemical structures, carried information that afforded replication. Thus, RNA acted as both a catalyst and template for self-replication.

> We are still too far removed from a comprehensive knowledge of the living organism to even dream of attempting their chemical synthesis. ~ Alexander Oparin

Construction of usable RNA is not simple. Researchers have had difficulty reproducing the lead actor.

> Basically, we took half a base, added that to half a sugar, added the other piece of base, and so on. The key turned out to

be the order that the ingredients are added and the way you put them together – like making a soufflé. ~ English biochemist John Sutherland

The chemistry on early Earth was different than today. Meteorites seeded the planet with soluble, reactive phosphorus that could be incorporated into prebiotic molecules. In its current form on the planet, phosphorus is relatively insoluble and nonreactive.

℘ Viral Analogy ଷ

RNA viruses may seem something of a model for the RNA-world scenario, but the analogy is inapt, because a virus can't replicate itself; the very thing that the RNA-world scenario aims to explain.

Retroviruses pack a tiny genome encoded in RNA. A retrovirus hijacks a host cell for replication, copying its RNA into the cell's DNA using a reverse transcriptase enzyme. The host then duplicates, with the retrovirus genetic instructions intact.

A virus works from DNA for copying itself but bundles itself up for inheritance using RNA to transmit the hereditary data. Thus, today's viruses are much too sophisticated, and yet not sufficiently self-sustaining, to aid in understanding how RNA replication arose.

A less obvious disadvantage to RNA life becomes apparent by comparing RNA to DNA. DNA is a richer storage medium. A retrovirus is about as complex as an RNA-encoded entity can be.

℘ RNA ଷ

The earliest forms of life may have arisen from a different set of nucleobases than those found in modern life. ~ Korean American chemist Seohyun Kim

RNA encoding produces proteins, which are the workhorse molecules of life. RNA represents the language of life.

Given decent thermal conditions, organic molecules naturally form. There is a physicochemical affinity: the physics of chemistry favor the coherence of compounds that form some of the basic building blocks of life.

RNA and its cognate proteins extend the harmony. The density profiles of different nucleobases in transcribing RNA closely resemble the profiles of amino-acid affinity for these same nucleobases in the proteins they code for.

The genetic code, as embodied in RNA, is an evolutionary consequence of the direct binding propensities of amino acids for appropriate nucleobases. Nucleobases and their resultant products – amino acids – are affine. There is a molecular conformity that runs between the amino acids that comprise proteins and the nucleobases that make up RNA.

♎ Precursors ♎

Making RNA requires both purine and pyrimidine nucleotides to be simultaneously available. ~ English organic chemist Matthew Powner

RNA and DNA are composed of 2 classes of organic compounds: purines and pyrimidines. Both can be assembled on the same sugar scaffold (aldehyde) to form the ribonucleotides used to construct RNA. Aldehyde is a simple sugar thought to have been present on early Earth.

Nucleobase construction on a preformed sugar moiety would provide the simplest strategy for divergent monomer synthesis. ~ English organic chemist Shaun Stairs

For RNA to form, both purines and pyrimidines must be present. This requires simultaneous synthesis of these 2 different compound classes.

A single chemical precursor – an 8-oxo-purine – *may* have been able to divergently generate both purine and pyrimidine ribonucleotides.

8-oxo-purine ribonucleotides may have played a key role in primordial nucleic acids. ~ Shaun Stairs

There is a stumbling block to this possibility. The resultant purine compounds have an oxygen atom bound to a carbon atom in the base, rather than a hydrogen atom as in the RNA purines today. There seems no simple way to exchange the wayward oxygen atom for hydrogen.

Once formed, purine and pyrimidine nucleotides bind to one another through specific molecular interactions that provide a mechanism which may copy and transfer information at the molecular level, given a guiding force of coherence for this process.

The unconventional oxygenated purines created via a single precursor might have been unable to form RNA analogs with the properties needed to spark life. This leave the precursor question unanswered, and as well how RNA evolved into its genic role.

RNA is far less stable than DNA: lasting, on average, only 2 minutes before degrading. But that is exactly why RNA still plays the key role in catalyzing biochemical reactions, leaving more stable DNA to heavy-lifting heredity. Unlike DNA, RNA can adopt many different molecular configurations which are readily rendered interactive.

An outstanding issue in the origination of RNA is how it became compartmentalized: packed tight enough to stay together and evolve into its functional role as an organic memory store. Various polymers are capable of compacting RNA. Which actor played that part in RNA's meaningful origination is unknown.

Once RNA is concentrated, its reaction rate ratchets up. RNA compacted into cellular form is 70 times more reactive than when unpackaged.

RNA nominally comes coiled, while DNA naturally forms a much more complex double helix. That is not the entire gospel of RNA vis-à-vis DNA.

RNA can also be folded onto itself to form complex secondary shapes that afford a variety of functions. Some viruses employ double-stranded RNA.

RNA does not approach the structural versatility of DNA. That DNA macromolecules can take a vast variety of shapes is a major factor in its sophistication for information storage.

For all that, swarms of RNAs mixing through interconnection could have led to successful cooperative ventures that sustained nascent life. Mixtures of RNA fragments do tend to self-assemble into self-replicating ribozymes, spontaneously forming cooperative catalytic cycles and networks. While the structural differences of DNA & RNA are considerable, there are only 2 tiny chemical discrepancies between the two.

1st, remove a single oxygen atom from ribonucleic acid (RNA) to render deoxyribonucleic acid (DNA). Reactive free-radical intermediaries found in hydrothermal vents present a ready catalyst for this transition.

The RNA to DNA transition requires a single enzyme: *reverse transcriptase*; the very enzyme that today's retroviruses pack in their tiny toolkit (HIV is exemplary).

While the chemical transition from RNA to DNA is easy enough, the structural transform is not so simple. As Watson and Crick observed in 1953, a double-helix structure from ribose sugar instead of deoxyribose would have been "probably impossible, as the extra oxygen atom would make too close a van der Waals contact."

The 2nd chemical difference comes in adding a methyl group (CH_3) to RNA uracil to get DNA's thymine. *Methyl groups* are reactive free-radical splinters of methane gas and are plentiful in hydrothermal vents. Methylation plays a significant role in *epigenetics*: regulating employment of DNA codes.

The relative simplicity and efficiency of RNA versus DNA suggests that RNA preceded DNA, but nothing substantiates this supposition.

A lot of questions lack answers in the RNA-world scenario, including the origination of metabolism.

RNA itself evolved from a humble start via self-assembly of organic compounds. The earliest ribozymes were structurally simple; a step away from peptides.

Varieties arose, but with similar structures. Complexity evolved later.

The structural dynamics of RNA are understood. But in considering replication, the large issues loom, still unanswered by scientific inquiry: how did RNA take on the role of first working biological memory, and by that rememberability for heredity?

It is a large leap from a variety of RNA able to store information to actually acting as a script for protein production.

No doubt a coherent force enabled this operational transition; one that science may only see by effect.

☿ Pyrite Life ☿

The iron-sulfur world experiments are aimed at long reaction cascades and catalytic feedback (metabolism) from the start. The maxim of the iron-sulfur world theory should therefore be "order out of order out of order." ~ Günter Wächtershäuser

German organic chemist Günter Wächtershäuser developed the *iron-sulfur world* theory in the late 1980s, arguing that metabolism arose as a prerequisite to replication. Wächtershäuser and others contend that organic compounds emerged on the surface of pyrite in seafloor hydrothermal vents.

Pyrite is a mineral comprising iron and sulfur (FeS_2), called *fool's gold* because prospectors sometimes mistook its glimmer for gold. Pyrite is both hoary and ubiquitous. The mineral is found everywhere, even in the oldest sedimentary rocks. The chimneys of hydrothermal vents largely consist of pyrite.

While its location may appear fortuitous, Wächtershäuser's construction was not only for geochemical reasons. On the bottom line is energy.

Pyrite is synthesized from hydrogen sulfide (H_2S) and an iron salt (FeS) – abundant ingredients on primordial Earth.

$$H_2S + FeS \rightarrow FeS_2 + H_2 + energy$$

Besides releasing chemical energy from which autotrophic life may have originated, hydrogen released in the production of pyrite provides the reducing power needed to synthesize organic compounds from carbon dioxide (CO_2).

Wächtershäuser developed a compelling story, but the devils in the details resulted in numerous objections by others in the field: complaints about unanswered questions concerning critical facets that are fundamental biochemical mechanisms.

Foremost, Wächtershäuser's model requires a bootstrap technique to get from fool's gold to life. That bootstrap is some chemical scaffolding acting as a protocell; a concept advanced by Scottish organic chemist Graham Cairns-Smith. Given a plausible scenario for a protocell, Wächtershäuser's pyrite-pulled chemoautotrophic model appears redeemed.

৶ Protocells ৶

> The origin of the cell is perhaps the most obscure point in the whole study of the evolution of organisms. ~ Alexander Oparin

Some form of containment – a *protocell* – is essential to any scenario explaining life's onset.

Cairns-Smith recognized that organic compounds were much too complex to be synthesized under prebiotic conditions. His proposed solution involved a 2-step to life: a literal scaffold of chemical reactions that afforded a more complex set of reactions that begat the origin of life's molecules. Once life was on its way, like a building constructed, the scaffold that acted as the original cellular container was removed from the scene, not leaving a trace.

To Cairn-Smith and other scientists inclined to metabolism-first, the scaffold 2-step solves the complexity problem.

Hydrothermal vents may have fostered mineral cells: the necessary encapsulation for relatively stable RNA to emerge. There are other possibilities.

van der Waals interaction, an intermolecular force, can assist binding macromolecules into membranes which lead to protocells. Water droplets are a simple example of such vesicles.

Fatty acids spontaneously form double-layered spheres, much like the double-layered membrane of living cells. These protocells incorporate additional fatty acids, and spontaneously divide.

The chemistry of RNA replication works in fatty acid vesicles only if citrate is present. Otherwise, the high concentration of magnesium required for RNA copying destroys such protocells by causing fatty acid precipitation: forcing the fatty acids into a lumpy, useless mass.

Citrate is found in many modern organisms, but its abundance 4 BYA is unknown. Similar molecules may have worked as well.

In 2016, physicists discovered that energetically active, chemical-laden droplets may grow through internal reactions and spontaneously divide as shape instabilities trigger division into 2 smaller daughters. Waste products may be discarded during division.

> Chemically active droplets can exhibit cycles of growth and division that resemble the proliferation of living cells. ~ American physicist David Zwicker *et al*

The study was theoretical, but illustrative. With the right metabolism in place, protocell reproduction looks plausible.

♎ L-Form ♎

Most bacteria have cell walls, but many can switch to a wall-free existence. This is termed an *L-form* structure, in reference to the chirality involved.

The most striking change associated with the L-form state is the way bacteria replicate. Instead of precise cell division, a bacterium simply bulges on one side and pinches off a daughter cell: a process termed *blebbing*.

Bacteria accomplish L-form by relying on fatty acids to hold their cells together, allowing them a shape fluidity they would lack with a cell wall.

In some ways, self-organized cellular containment seems the simplest facet in life's origination, as numerous mechanisms exist, each with their own plausibility, and a naturally occurring combination of techniques readily conceivable. As L-form bacteria demonstrate, a protocell using fatty acids may have sufficed.

The importance of selectively permeable membranes to cellular life cannot be understated. Despite seeming simplicity as a container, the cell membrane acts as both protection and conduit; a functional wile considering the antithetical natures of its twin purposes.

♐ Biomolecular Evolution ♌

All living organisms have a metabolism, a set of life-sustaining chemical transformations that provide the energy and matter needed for the functions of the cell. These metabolic transformations occurred very early in life. Organisms probably replaced chemical reactions already going on in the planet and internalized them into cells through development of enzymatic activities. ~ Argentinian biochemist Gustavo Caetano-Anollés

Using an extensive genomic database of life, Turkish geneticist Ibrahim Koç and Argentinian biochemist Gustavo Caetano-Anollés studied the genetic evolution of molecular functions for all realms of life.

The best way to understand an organism is through its functions. You can take an entire genome that represents an organism and visualize it through the collection of functionalities of its genes. The study of these 'functionomes' tells us what genes do, instead of focusing on their names and locations. ~ Gustavo Caetano-Anollés

Koç and Caetano-Anollés figured that the genes for the most ancient functions would be shared by all organisms and exist in relatively large numbers compared to later-evolved functionality. They found metabolism and cell cohesion (binding) to be positively primordial.

It is logical that these two functions started very early, because molecules first needed to generate energy through metabolism and had to interact with other molecules through binding. ~ Gustavo Caetano-Anollés

The next major advance involved functions that made macromolecule production possible, which is likely when RNA entered the picture. This coincided with the trend toward specialization. The first biomolecules were multi-purpose, becoming functionally tailored through evolution.

Ancient molecules served multiple functions and showed broad specificity. These molecular functions diversified into more specific and efficient counterparts during evolution, leading to the extraordinarily diverse and specific functions that exist in the modern biological world. That ancient enzymes were generalist multi-tasking proteins has been borne out thanks to protein resurrection experiments that use phylogenetic reconstruction to design ancestral sequences and synthesize the corresponding proteins. ~ Ibrahim Koç & Gustavo Caetano-Anollés

ᛞ Water World ᛉ

> Life takes advantage of unbalanced states on the planet, which may have been the case billions of years ago at the alkaline hydrothermal vents. ~ American geochemist Michael Russell

Derivative scenarios emerged after Wächtershäuser's work, relying upon similar chemistry and events. One shift was the venue. Scalding hot acidic vents – "black smokers" – were originally considered the place where pyrite life emerged, owing to the highly energetic environment.

> There is a lot of CO_2 dissolved in the water, which could provide the carbon that the chemistry of living organisms is based on. And there is plenty of energy, because the water is hot and turbulent. These vents also have the chemical properties that encourage molecules to recombine into those usually associated with living organisms. ~ English chemist Nora de Leeuw

More recent work has focused on hydrogen-saturated alkaline water meeting acidic, relatively CO_2-rich oceanic water at gentler, cooler underwater vents than black smokers.

In an inorganically-formed protocell pocket, this meeting – of CO_2 from the ocean and H_2 & CH_4 from a vent – would create a proton gradient and ion pump, providing the energy for biochemical synthesis.

> Life lives off proton gradients and the transfer of electrons. ~ American geochemist Laurie Barge

Naturally-occurring minerals in vents that would have provided a suitable substrate include pyrite (FeS_2), greigite (Fe_3S_4), and fougèrite ($Fe^{2+}_4 Fe^{3+}_2(OH)_{12}[CO_3] \cdot 3H_2O$).

The rare metal *molybdenum*, which is found in enzymes, is considered instrumental, as it transfers 2 electrons at a time rather than the usual 1; particularly useful in driving key chemical reactions.

Through erosion, rocks in deep-sea thermal vents contain labyrinths of tiny, thin-walled pores which could have acted as protocell containment, producing a proton gradient and easy electron transfer, with a concentration of organic ingredients from which complex proteins and RNA could emerge.

A cell membrane may have evolved via a simple protein that employs the influx of protons to pump sodium ions out of the protocell before largely sealing up, thus facilitating the requisite self-contained environment while sustaining limited permeability.

The earliest Ediacaran animals evolved in the deep ocean, which had a more stable environment than shallow waters. It seems likely that this scenario also applied to the origin of life. Extremophile archaea illustrate the possibility of life emerging in this watery world.

> Their biochemistry seems to emerge seamlessly from the conditions in vents. ~ English biochemist Nick Lane

ϐ Archaea ϙ

> Archaea weren't even discovered until 1977, and were thought to be rare and unimportant, but we are beginning to realize that they not only are abundant, but they have roles that have not fully been appreciated. ~ American oceanographer Andrew Thurber

Archaea are among the earliest life. Archaea are found most everywhere: in the seas and soil, the marshlands and the swamp known as the human colon. The methane of marsh gas, and of ruminant and human flatulence, are atmospheric contributions from resident methanogens, archaea all.

The archaea in oceanic plankton make them one of the most abundant organisms, comprising 20% of the Earth's biomass. As planetary movers and shakers, at least by numbers, archaea have long played important roles in the carbon and nitrogen cycles of the biosphere.

Both bacteria and archaea reproduce asexually, by binary fission, fragmentation, or budding; but, unlike bacteria and some eukaryotes, no known archaea produce spores. Spores are offspring that can ride out hard times.

Many archaea are extremophiles, with exotic chemical processes within their single cell that fend off disruption by the habitat. When the going gets too tough, archaea form tough protective shells, stop their life processes, and wait it out for better days. An archaean itself becomes like a spore. Of all life, archaea are the ultimate survivors.

✇ Panspermia ☡

The basic building blocks of life can be assembled anywhere in the solar system and perhaps beyond. The catch is that these building blocks need the right conditions in order for life to flourish. ~ English geologist Zita Martins

2,500 years ago, Greek philosopher and cosmologist Anaxagoras proposed *panspermia* (Greek for "all seeds"): life delivered to Earth from space. The notion persisted for a very long time despite scant evidence for it.

As it turns out, microbes riding in comets or meteorites could have seeded the solar system with life. Conversely, terrestrial microorganisms from Earth could have been launched into space via bolide impacts that ejected rocks as far away as Saturn.

The flux of organic matter to Earth via comets and asteroids during periods of heavy bombardment may have been as high as 10 trillion kilograms per year, delivering up to several orders of magnitude greater mass of organics than what likely pre-existed on the planet. ~ American chemist Nir Goldman

Organic material makes its way through the cosmos on a regular basis. DNA can withstand the rigors of space. As life originates whenever and wherever it can, panspermia is entirely possible.

An influx of dust has acted as a continuous rainfall of little reaction vessels containing both the water and organics needed for the eventual origin of life on Earth. Continuous, co-delivery of water and organics intimately intermixed. ~ American cosmochemist Hope Ishii

⌘ Life from Mars ☡

The evidence seems to be building that we are actually all Martians; that life started on Mars and came to Earth on a rock. ~ American chemist Steven Benner

In the early 20th century, American astronomer Percival Lowell, seeing what he termed "non-natural features" on

Mars, speculated that Martians had an advanced civilization, replete with crop irrigation via canals drawing water from the planet's poles.

The molecules that combine to create genetic material may have needed more than whatever primordial prebiotic soup might have been cooked up on Earth 4 BYA.

Adding energy to organic compounds does not generate RNA. It merely makes tar. Rendering RNA requires atoms to be coaxed into shape by templating atoms on the surfaces of crystalline minerals.

The most effective minerals for patterning RNA would have dissolved in the oceans of early Earth, at a time when the planet was probably enveloped by ocean. Mars still has extensive, deep reservoirs of water, little of which stays on the surface for long.

Further, while water is essential to life as we know it, it is also corrosive to biopolymers such as RNA. The long strands needed for information storage can't form in water.

The best RNA templating minerals are boron and molybdenum. Both are water-soluble.

Oxygenating boron births a borate (BO_3). Add oxygen to molybdenum to make a molybdate ($MoO_4{}^{2-}$ and variations).

Borate minerals prevent the organic building blocks of life from devolving into tar. Molybdate can bond to the carbohydrates that borate stabilizes and catalyze a rearrangement into ribose: the R in RNA.

Besides an uncongenial aquatic surface, for lack of free atmospheric oxygen, both borate and molybdate would have been practically nonexistent on early Earth at the time life arose. In contrast, 4 BYA, the atmosphere of Mars had much more oxygen than Earth.

> The early history of Mars seems to have been very similar to that of Earth, especially with respect to the ancient hydrosphere.
> ~ American geomicrobiologist Nora Noffke

Life may have evolved on Mars. From around 4.5–3.5 BYA, Mars was habitable, at least by hardy microorganisms. Even now, Earth methanogens could survive there.

While organic compounds were produced on Mars, there is no extant evidence that life ever emerged. Exploration of Mars has been slight; we know little.

The scenario of life coming to Earth from Mars seems a simple 2-step. A meteorite knocked a Martian rock with stowaway microbes aboard into space. The Martian transport becomes a meteorite that splashes down on Earth.

It is possible that a violent impact could eject material without generating so much heat that it would destroy a microbial passenger, especially if the traveler were shielded in the interior, not on the surface; likewise, in entering Earth's atmosphere. Life nestled inside a meteorite would have a better chance of surviving the searing heat in coming down.

If Martian microbes hitched a ride on a dust particle, blistering heat may have been avoided by a gentle deceleration. But then there is the issue of travel time.

Most earthbound bits spend a long time in space. One Martian meteorite traveled for 15 million years before landing on Earth. But 1 out of 10 million objects make the journey from Mars to Earth in less than a year. This would minimize exposure to ionizing interplanetary radiation.

That said, some microbes that exist now are highly resistant to radiation, as well as being able to handle the jostle involved in projectile space travel.

As with heat, the best place to not be radiated would be inside a sizable rock. Such comfortable snuggling would also help preserve a habitat.

But then, suspended animation is possible. And pebbles are more likely to make a quick trip than boulders.

The timing of life traveling from Mars to Earth would have to have been fortuitous. Mars had a relatively oxygen-rich atmosphere 4 BYA, but its magnetic field disappeared around that time, allowing the solar wind to strip the atmosphere away.

While life from Mars is literally far-fetched, abiogenesis anywhere is itself a fantastic story. All scenarios of life's emergence are stories of staggering complexity with intricately layered plotlines and distinct dependencies.

For one, the cellular containment issue in the pyrite-life scenario (scaffolding) also affects the plausibility of the RNA-world hypothesis, which relies heavily on catalytic peptides for protocells to form.

Like different sides of the same coin, protein-like enzymes play an analogous role in gene-first models to the ribozymes that play a central role in several metabolism-before-replication scenarios.

❧ Data at the Dawn of Life ❧

Life may be characterized by its distinctive and active use of information, thus providing a roadmap to identify rigorous criteria for the emergence of life. This is in sharp contrast to a century of thought in which the transition to life has been cast as a problem of chemistry, with the goal of identifying a plausible reaction pathway from chemical mixtures to a living entity. ~ Paul Davies

Just as physics has its information adherents, so too those interested in the origin of life. However insubstantial, it is a refreshing perspective in emphasizing functional processes over reactions.

Functionality is not a local property of a molecule. For example, the functionality of expressed RNA and protein sequences is clearly context-dependent— only an exceedingly small subset of these molecules is causally efficacious (i.e. meaningful) in the larger biochemical network of a cell. The most important features of biological information (i.e. functionality) are decisively nonlocal, subject to informational control and feedback, so that the dynamical rules will generally change with time in a manner that is both a function of the current state and the history of the organism. ~ American astrobiologist Sarah Imari Walker & Paul Davies

Characterizing life as an intelligent information processor sidesteps mechanics and focuses solely on the overarching issue of how life cohered. In this, the emphasis is decidedly mind over matter.

The key distinction between the origin of life and other 'emergent' transitions is the onset of distributed information control, enabling context-dependent causation, where an abstract and non-physical systemic entity (algorithmic information) effectively becomes a causal agent capable of manipulating its material substrate. ~ Sarah Imari Walker & Paul Davies

⊘ Life Cohering ⅋

Probably all the organic beings which have ever lived on this earth have descended from some one primordial form, into which life was first breathed. ~ Charles Darwin

The metabolism-1st → gene-2nd pyrite scenario places life's genesis deep in or under the ocean, relying on hydrothermal vent energy, while the gene-1st → ribozyme hypothesis starts life in a soup aboveground, dependent upon light energy to spark biogenic ignition, and enough environmental stability to sustain life.

Metabolism and replication must have come together in very short order for life to originate, as there is an interdependency between energy generation and reproduction. Life originating in the benthic zone is much more likely, as Earth still suffered continuing bombardment when life arose.

¤ ✧ ¤

There is nothing more complex than life. As abiogenic researchers have come to realize, the interdependent intricacies of even the simplest life form are staggering.

Traditional dogma has a creator breathing life into inanimate matter. Modern scientific consensus rejects that, touting a chemically deterministic "life happens." Yet life does not just happen. Experiments to recreate life by stimulating various soufflés have been most notable in their failure. The one thing experimentally proven so far is that there is something to life that exceeds mere combination of matter.

The inference from what is known is obvious: a natural force of coherence tilted the odds of life's emergence on Earth toward ineluctability. Further, life did not arise just once and then spread throughout the planet. The mathematical odds of that are even lower than life randomly coming into being. Instead, life blossomed innumerable times. Life forms being self-similar yet unimaginably diverse deepen the intriguing mystery.

How remarkable is life? The answer is: very. Those of us who deal in networks of chemical reactions know of nothing like it. ~ American chemist George Whitesides

ဩ Life Rising ၶ

The species we see today are but the smallest part of what blind destiny has produced. ~ French mathematician and philosopher Pierre Louis Maupertuis

The transition from RNA to DNA fortified life, affording survival and reproduction under exceedingly harsh conditions. DNA likely evolved independently many times by the wiliest mavens of genetic manipulation: viruses.

RNA viruses invented DNA to protect their genomes. ~ French molecular biologist Patrick Forterre

Archaea and bacteria are primitive DNA replicators which may have evolved under somewhat similar circumstances. They share the same DNA coding. Many details of their protein syntheses are the same. But the mechanisms of DNA replication differ greatly between archaea and bacteria. Further, viruses and plasmids have their own unique DNA replication systems.

Some viruses employ DNA; others RNA. While DNA is the universal coding schema of known life, replication systems are strikingly diverse.

The modern-type system for double-stranded DNA replication likely evolved independently in the bacterial and archaeal/eukaryotic lineages. ~ Russian biologist Eugene Koonin

DNA replication is not the only distinction between archaea and bacteria: their cell membranes and walls are quite distinct. These fundamental differences indicate independent origination.

♂ Microbes Everywhere ♀

There is a commonality of colonisation of the subsurface of the planet. ~ Canadian biochemist Barbara Sherwood Lollar

19 distinct microbes have been found that are distributed throughout the world, tucked deep within Earth's crust.

> It is easy to understand how birds and fish might be similar oceans apart. But it challenges the imagination to think of nearly identical microbes 16,000 kilometers apart in the cracks of hard rock. ~ American biogeologist Matthew Schrenk

Selfsame microorganisms live in a subterranean South African gold deposit and in frozen methane pockets beneath the Indonesian sea floor.

Evidence indicates that radically distinct species originated on Earth at different times in widely dispersed places. Yet they all share the same genetic schema; a most intriguing riddle of life's rising.

⚥ Common Ancestor ⚥

> Universal common ancestry is a central pillar of modern evolutionary theory. A universal common ancestor is at least $10^{2,860}$ times more probable than having multiple ancestors. ~ American biochemist Douglas Theobald

In the 1740s, Pierre Maupertuis made the first known suggestion that all of life had a common ancestor.

The common ancestor hypothesis necessitates one of 3 scenarios: 1) a prodigal life originated in 1 place, from which it spread to every location throughout the world; 2) the same life originated in multiple places; or 3) somehow life homogenized.

The notion of prodigal life (1) is logistically improbable, though conceivable. A population can widely diffuse geographically over hundreds of millions of years. Viruses, which form a worldwide community, are exemplary. But the differences between archaea and bacteria are not accounted for in this scenario.

Duplicative origination (2) is also conceivable but implies that life naturally coheres to a singular form; hence, the ready duplication of the same life. This contradicts Nature's proclivity for diversity: a fact prodigiously proven. More specifically, duplicative origination is belied by there being 2 quite different cell types – archaea and bacteria – in the most primordial life.

That leaves 1 possibility. American microbiologist Carl Woese proposed in 1998 that there was no universal common ancestor, but that homogenization occurred among ancient communities of cellular organisms via genetic transfer.

> The ancestor cannot have been a particular organism, a single organismal lineage. It was communal, a loosely knit, diverse conglomeration of primitive cells that evolved as a unit. ~ Carl Woese

Archaea and bacteria originated independently, but both ended up with DNA as their genetic coding regime.

> The very essence of the virus is its fundamental entanglement with the genetic and metabolic machinery of the host. ~ American molecular biologist Joshua Lederberg

Only 1 agent had the means for worldwide DNA delivery: viruses. Comprising a worldwide community, viruses either invented DNA or appreciated the innovation and then infected other archaic life with it. Viruses' ability to travel light and work their way into suitable hosts shows sufficient cleverness to bring off such a coup.

> Viruses are the creative front of biology, where things get figured out, and they always have been. ~ American virologist Luis Villarreal

Both archaea and bacteria sought to rid themselves of RNA viruses early on. Archaea migrated to warm biomes and then even took to extreme environments. Bacteria designed a thicker murein cell wall that likely blocked many ancient viral families. These developments spurred viral genic innovation: the more robust DNA, which it then spread.

> Viruses have contributed enormously to the communication between cells, and to the appearance of multicellular organisms on Earth. ~ French virologist Felix Rey

As a regular work practice, viruses insert genes into their hosts. 8% of the human genome has a viral origin.

Eukaryotic intercellular communication and organ development come courtesy of genetic information provided by viruses. Animals would have never evolved beyond blobs of cells without viral innovations. In putting all organisms on the same genetic program, viruses painted a facile picture of life having a universal common ancestor.

It was self-interest that drove viruses to unify life to a standard genetic organization. Modified genetic coding systems have independently evolved at least 34 times. Viruses have not figured out how to infect organisms with these peculiar regimes. It may be that these alternate genetic systems arose to evade viruses. Having a standard genic schema gives viruses an edge in maintaining the lifestyle to which they have become accustomed.

Viruses are embedded in the fabric of life. ~ Gustavo Caetano-Anollés

❧ Life Elsewhere ☙

It seems that the universe produces plentiful real estate for life that somehow resembles life on Earth. ~ American cosmologist Erik Petigura

Life requires molecular structures which can metabolize to obtain energy, and which can retain information and use it for replication. Once that occurs, given a permissible environment, life arises and evolves.

We used to think that the sort of chemistry that makes life could only happen on Earth. We were wrong. ~ German biochemist Leonie Mueck

The chemical compounds and natural reactions that led to life on Earth are in no way unique. Complex organic molecules grew on the icy dust grains that lived in the infant solar system, warmed by ultraviolet photons.

Methanol (CH_3OH) is a key substrate in the synthesis of organic molecules leading to life. In the right conditions, carbon monoxide (CO) on the surface of interstellar dust can react at low temperatures with hydrogen (H_2) to create methanol.

The cold clouds that give birth to stars are a cradle for methanol production. From there, a variety of circumstances conspire to build complex organic molecules in space.

Meteorites contain many organic compounds, some so complex as to appear as a preserved extraterrestrial life form, like cyanobacteria.

The traditional habitable zone is known as the Goldilocks zone. A planet needs to be not too close to its sun but also not too far away for liquid water to persist, rather than boiling or freezing, on the surface. But that theory fails to take into account life that can exist beneath a planet's surface. ~ British cosmologist Sean McMahon

The presumed habitable zone for planets in this solar system is between the orbits of Venus and Mars. Yet, as the life potential of Europa and Titan illustrate, a body's composition and dynamics are more important than its orbital band.

The Saturn moon Enceladus has a crust of ice 5–35 km thick. Underneath is a toasty, mineral-rich saltwater ocean capable of hosting life. The gaseous mixture in geysers near Enceladus' south pole suggests methane-releasing microbes.

The same may be said of Jupiter's moons Callisto and Ganymede; Triton, a moon of Neptune; and Pluto. Underneath icy crusts are oceans warmed by radioactive decay in the core and kept liquid via trace amounts of ammonia. There is a universal abundance of water in which life may brew.

Archaea on Earth survive in conditions like those in Enceladus' subsurface ocean and in other celestial bodies. There have been rich ecosystems of subsurface Earth microbes for billions of years.

So much of life is within the Earth rather than on top of it. ~ American microbiologist Karen Lloyd

There are over 640 billion planets in the Milky Way. Over 20% – 130 billion – may have a habitable surface. Many more likely harbor life within.

As the Milky Way illustrates, a single galaxy is vast: beyond imagination in scale. Earth illustrates the wide tolerance of conditions in which life can exist. Given the ease with which the molecular building blocks of life come together, it is inconceivable that Earth alone supports life.

The Catholic Church burned Italian Dominican friar Giordano Bruno at the stake in 1600 for suggesting the possibility of "a plurality of worlds." Today's clergy are more inclined to hedge their bets, figuring that "brother extraterrestrial would still be part of creation."

> If it's true that our species is alone in the universe, then I'd have to say the universe aimed rather low and settled for very little. ~ American comedian George Carlin

≈ Synopsis ≈

The Origin of Life

> A functioning cell must be entirely correct at once, in all its complexity. ~ German chemist Wilhelm Huck

➤ *Bombardment from space* seeded the ingredients for life on Earth. Meteorites commonly contain water, amino acids, and nucleobases.

> More than organic ingredients may have rained down. Life on Earth might have traveled from Mars. Alternately, life could have even come in from outside the solar system, traveling millions of years before finding a home.

➤ However mathematically improbable, *life originated* on Earth as soon as environmental conditions permitted.

> The complex combination of manifestations and properties so characteristic of life must have arisen in the process of the evolution of matter. ~ Alexander Oparin

➤ If life originated domestically, it probably began on or in the warm ocean bed, most likely the ancestor to archaea, with bacteria arising shortly thereafter. Archaea are archaic – a likely direct descendant of the earliest life.

➤ The force of *coherence* that composes Nature was crucial in begetting life, notably in the sophistication required for coordinated cellular organization, metabolic processes, and replication.

> Life requires chemistry, but the properties of the living state emerge from the dynamical properties of that chemistry, including the temporal and spatial organization of molecular networks and their information management. Life uses information to construct itself. The development of networks over

time may be more important than the specific chemical nature of their molecular components. A concept of information relevant to biological organization may be essential to identifying these networked processes. ~ English chemist Leroy Cronin & Sara Imari Walker

Life Rising

➢ Archaea and bacteria are primitive *prokaryotes* that both use DNA, but have distinctive differences indicating independent origination.

➢ By infecting primitive life with DNA, viruses created what appears as a *universal common ancestor*. This homogenization of life at the cellular level created the platform for the intimate interrelations among various life forms that defines Earth's biosphere.

Life Elsewhere

> We have no evidence whatsoever for any life beyond Earth.
> ~ Paul Davies

➢ With many trillions of planets in the universe, the probability of Earth being the only planet with life is negligible. *Life exists elsewhere.**

➢ Given that organic molecules are naturally occurring in space, and reflect energy economics prescribed by physics and chemistry, life elsewhere likely follows Earth's gyre.

➢ The most intriguing questions about extraterrestrial life go to early evolution. Do viruses universally evolve to nudge homogeneity as they did on Earth? If not, do fundamentally incompatible life forms evolve, creating an entirely different biospheric dynamic than on Earth? That would happen if gene sharing was not as ubiquitous among organisms as it has been on Earth.

---

* This conclusion is supported by the probability that life on Earth independently arose innumerable times in a variety of habitats.

❧ Cells ❦

Every cell is derived from another cell. ~ French chemist, naturalist and physiologist François-Vincent Raspail

The biological world is comprised of cells.* Though physically self-contained, much of the activity within a cell relates to what's outside. More than anything, a cell's life is defined by its ecology.

Inside a cell, the modus operandi is self-assembly. The components that make up a cell, from molecules on up, organize themselves. This continuing coherence is the miracle of life.

From a cellular perspective, life takes 2 forms: prokaryotic and eukaryotic. The terms for these cell types derive from ancient Greek: *eu* means "true," *karyon* "kernel." Hence, whereas a *eukaryote* has a compartmentalized nucleus, a *prokaryote* – "before the kernel" – does not.

ଞ Prokaryotes ଔ

Microbial single-celled organisms are prokaryotes. They include bacteria and archaea.

One of several differences between archaea and bacteria is archaeal tolerance for living in the most extreme environments, at temperatures and in chemical conditions that no other life can withstand. Bacteria are robust, but archaea are

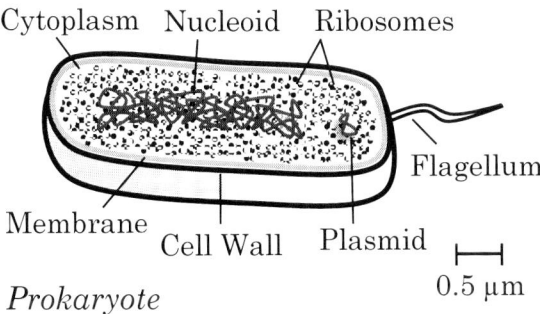

Cytoplasm   Nucleoid   Ribosomes

Flagellum

Membrane   Cell Wall   Plasmid

0.5 μm

Prokaryote

---

* The study of living cells is *cytology*.

tremendously tough. Despite distinctions between the two, their cell structure and functioning are largely selfsame.

Prokaryotes are everywhere: in the soil, water, and air. Their reach far exceeds all other life. Microbes are by far the most abundant and diverse life on the planet; some 25 times the total biomass of all animal life. There are well over a million different types of prokaryote.

Cells come in a variety of forms, shapes, and sizes. The smallest bacteria are 0.2 μm (micrometer) in diameter. At the other extreme are wiry neurons, which may extend for a meter or more. Such exceptions lie outside the rule that most cells fall into a rather narrow and predictable range. Bacteria are typically 1–5 μm, while the eukaryotic cells of plants and animals are 10 times that (10–50 μm).

The factors which constrain cell size are needing an adequate surface area/volume ratio and maintaining sufficient concentration of the substances involved in various cellular processes.

Surface area is important in allowing exchange between a cell and its environment. While cell volume limits the amount of nutrient intake and waste disposal, surface area constrains the membrane interface available for such uptake and excretion.

Given the tradeoff, cells stay within the range that optimizes material flow and spatial requirements for processing. That also includes the need to maintain an adequate concentration of essential compounds and enzymes to bestow reasonable reaction rates.

Functionally, all cells have much in common, as the requirements for living are identical. Compartmentalization, via membrane-enclosed *organelles*, is an evolutionary advance that afforded the greater size and complexity that eukaryotic cells possess.

Prokaryotic cells are also well-organized at the subcellular level. They just leave off the extensive employment of membranes. Small size has its advantages, notably stealth.

Other significant differences between prokaryotes and eukaryotes relate to the ways in which eukaryotes organize and utilize genetic material. While a prokaryote stashes its DNA in nucleoidal mass, a eukaryote has a highly organized categorization in chromosomes. Differences in gene expression reflect the more sophisticated system that eukaryotes employ.

ஐ Eukaryotes ଔ

The key events in the evolution of eukaryotes were the acquisition of the nucleus, the endomembrane system, and mitochondria. ~ American evolutionary biologist David Baum & American cytologist Buzz Baum

Eukaryotes arose from an evolutionary combination of prokaryotes. The partnership proved so winning that it established an environmentally dominant domain of life.

All multicellular organisms are eukaryotes. Plants and animals are eukaryotic.

While eukaryotic cells have a more elaborate structure, basic cell functioning of prokaryotes and eukaryotes is similar.

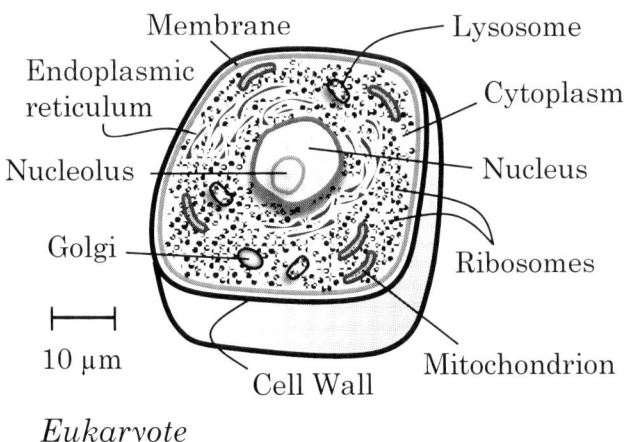

Eukaryote

ஒ Organelles ൞

Cellular organelles are simultaneously distinct from the rest of the cell and completely reliant on it for their identity and function. ~ American cytologists Suzanne Wolff & Andrew Dillin

A eukaryotic cell may have several thousand organelles which perform a vast variety of functions, including biochemical, regulatory, and motile processes. The composition of each organelle is tailored to its function and a cell's immediate need.

Cells distribute and position their organelles to optimize productivity; particularly, organelle-organelle interaction. Organelles communicate with each other and work together for the cell's benefit. Special proteins on organelles' outer membranes recognize their counterparts on other organelles and arrange physical contact to efficiently effect sharing. Cellular processes are kept running smoothly by an intricate orchestration of molecules that flow among organelles.

♂ Mitochondria ♀

Mitochondria are ancient relics of a purely single-celled world. ~ Suzanne Wolff & Andrew Dillin

Mitochondria are semi-autonomous organelles that contain their own genetic machinery. ~ American cytologist Eric Schon

The *mitochondrion* is a double-membraned organelle that acts as a cell's power plant: generating a supply of ATP, the universal cellular energy molecule. Digested nutrients enter a mitochondrion and are assimilated in a way that maximizes their energy potential.

The system for optimizing the extraction of energy from food molecules is versatile, and can be modulated in unexpected ways, in order to adjust to the dietary composition of nutrients, or to the specialized function of particular cell types. ~ Spanish biochemist José Antonio Enríquez

Mitochondrial ATP production is regulated by calcium flow into a mitochondrion from another organelle. That flow is controlled by various proteins which regulate the voltage across the inner mitochondrial membrane.

From physics and chemistry standpoints, there are numerous tradeoffs that affect the efficiency of ATP production.

Nature evolved a control structure finely tuned to effectively manage these tradeoffs with flexibility to adapt to changes in supply and demand, at the cost of higher enzyme complexity.
~ Indonesian biologist Fiona Chandra *et al*

Besides energy supply, mitochondria are also pivotal in the cell life cycle: production of cellular building blocks (e.g., amino acids, nucleotides), cell growth, cellular differentiation, intracellular communication, and cell death. There are multiple signaling pathways between various organelles and a mitochondrion.

Each mitochondrion has compartments dedicated to specific functions. It is one of the most complex organelles.

The proteins within a mitochondrion vary by tissue and organism species. The mitochondrial proteome is dynamically regulated.

The number of mitochondria in a cell varies widely by organism and tissue type. Some cells sport only a single mitochondrion, while others may have several thousand.

Producing power for cellular needs generates heat. Operating at 50 °C, mitochondria run 13 °C hotter than the rest of the cell.

Mitochondria do their best under stress to keep working. Contents of a damaged mitochondrion may be absorbed into a healthy one, thereby salvaging still-employable machinery. New mitochondria may be created; contributing to quality control by removing damaged ones.

Mitochondrial defects can lead to a wide variety of metabolic, muscular, and neurodegenerative diseases. Mitochondria are one basis for aging, and can incite age-associated diseases, including Alzheimer's, Parkinson's, and heart failure.

Hence, the health of mitochondria is crucial to cell wellbeing, and of the entire organism. The organelle plays an essential role in mind-body interactions, with long-term health consequence.

> Mitochondria contain their own genomes that, unlike nuclear genomes, are inherited only in the maternal line. ~ American cytologist Ruth Lehmann *et al*

Egg cells know that to get a good start in life they need reliable power plants. So, during their development (oogenesis), egg cells are choosy about the mitochondrial DNA they take with them, selecting the best available.

> The mitochondrion influences an organism's integrated response to psychological stress. ~ American pathologist Martin Picard *et al*

Entanglement of psychology and physiology is well documented. The importance of mitochondria to holistic health illustrates the integration of the mind-body at all scales: from organelle to organism. Such scale-independent unification suggests that organisms exist as synergistic life-energy gyres, with an emergent mind-body, in the physics sense of coming into being moment-by-moment, and with the vitality of the whole organism and constituents being reflective of each other.

With rare exception, organisms with sexual reproduction inherit only maternal mitochondrial genes. A prevailing hypothesis is that sperm mitochondria are exhausted after their fertilization competition, yielding an undesirable unreliability for genetic inheritance of this crucial capacity. Mitophagy, the process by which cells eat their own mitochondria, has a role in the selective elimination of paternal mitochondria.

℘ Nucleus ℞

> Links exist between the nucleus and the extracellular microenvironment that direct cell fate. ~ American biologists Valerie Weaver & Russell Bainer

The nucleus serves as the control center of a cell, maintaining the integrity of a cell's genes and their expression. The nucleus contains the *nuclear genome*, which holds most of a cell's genetic material, though the mitochondrion and cytoplasm also hold genomes relevant to their operations.

As with a mitochondrion, a double membrane envelops the cell nucleus. Nuclear envelope proteins – *nucleoporins* –

selectively allow molecules in and out of the nucleus. Proper functioning of these nuclear pore complexes is critical to maintaining cell health.

Amazingly, nucleoporins can manage over 1,000 molecules per second. Amino acid sequences within these proteins are optimized for efficiency in screening and transport. The physiochemical optimality of these sequences has been conserved evolutionarily from the early times of eukaryotes.

A nucleus has no membrane-bound sub-compartments but is organized for various operations. The best-known sub-nuclear body in the nucleus is the *nucleolus*, which is the site of ribosome assembly. A control center for cellular growth and health, the nucleolus takes up 25% of the nucleus. Malfunctioning nucleoli are a known cause of several human diseases.

⚐ Endoplasmic Reticulum ⚑

The endoplasmic reticulum (ER) is an organelle connected to the nuclear membrane; a membranous network of sac-like structures (*cisternae*) held together by the cytoskeleton.

The ER has a myriad of functions, including carbohydrate metabolism, lipid synthesis, glycoprotein production, and cell membrane manufacture.* The ER also plays a critical role in assisting mitochondrial division and replication. The ER and mitochondria have tightly coupled dynamics via extensive contacts.

When a cell is stressed, such as suffering nutrient deprivation or oxygen shortage, proteins degrade and become damaged. The ER coordinates response, invoking a variety of techniques. Several of the repair procedures are epigenetic in nature. Others include altering DNA sequences and facilitating first-aid functions to repair the damage.

¤ ✧ ¤

---

* A *glycoprotein* is a protein with an attached carbohydrate (glycan).

Once inside a cell, a wide variety of pathogens hijack the ER to do their dirty work. An endangered ER signals its dilemma by releasing inflammatory proteins which set off an immune response. To avoid triggering an alarm, bacterial pathogens often inhibit ER stress response. Others employ a cagier stratagem.

Brucella abortus actively encourages inflammation in pregnant animals that it infects. Its sepsis can provoke a spontaneous abortion, releasing the infected uterine contents. Animals nearby subsequently consume the infected tissues and themselves become infected. Thus *B. abortus* cannily spreads.

♂ Ribosomes ♀

Ribosomes translate sequences of nucleic acids into sequences of amino acids. ~ Israeli systems biologist Shlomi Reuveni *et al*

Although not membrane-bound, ribosomes are considered organelles. Ribosomes are protein factories; producing their products in an intricate, multi-step process. Nearly all the proteins required by cells are synthesized by ribosomes. Ribosomes receive their instructions from the cell nucleus and obtain their construction materials from the cytoplasm.

A ribosome assembles a particular protein by translating information encoded in *messenger RNA (mRNA)*. While many ribosomes churn out a wide variety of proteins, some specialize to manufacture only certain products. A single cell may have thousands of ribosomes.

Ribosomes are composed of 55 to 80 proteins, depending on organism type. Highly ordered, these structural proteins are unusually short and uniform in length. Ribosomes also have 2–3 strands of RNA, which account for up to 70% of the total mass of the ribosome.

The ribosomes of archaea, bacteria, and eukaryotes differ in structure and composition. But the subunits within are similar between prokaryotes and eukaryotes.

The ribosome is universal biology. ~ American biochemist Loren Williams

Prokaryotic ribosomes produce proteins using a slightly different process than eukaryotic ribosomes. But the core translation mechanics of the ribosome are essentially the same in all organisms. To meet production requirements, the outer regions of ribosomes expand and become more sophisticated as organisms become more complex.

Prokaryotic ribosomes are ~20 nm in diameter and contain 65% RNA and 35% ribosomal proteins. Eukaryotic ribosomes are 25–30 nm in diameter, with an RNA-to-protein ratio close to 1. The RNA in ribosomes account for ~85% of a cell's RNA pool. The number of ribosomes within a cell varies greatly, from less than 10,000 to up to 10 million.

Defective mRNAs result in aberrant, potentially harmful proteins; so special proteins surveil production and ensure quality control.

Ribosomal RNA itself plays a vital role in all stages of protein synthesis.

℘ Golgi Body ℞

Cells synthesize a sizable number of protein-based macromolecules. Some are used internally. Others are secreted for use by other cells (*exocytosis*).

The Golgi body is a stack of membranes that works in concert with an ER to package proteins before shipping the proteins off to their intended destination. Enzymes tune up newly synthesized proteins in the Golgi to improve their job performance potential.

In animal cells, the Golgi mysteriously breaks up and disappears at the onset of cell division (mitosis), reappearing during the telophase of mitosis: when 2 daughter nuclei form in the cell. In contrast, Golgi stacks stay intact through the cell cycle in yeast and plant cells.

ഌ Plasma Membrane രെ

The ability of organisms to form fluid structures of some rigidity in an aqueous environment is central to the existence of life on this planet. ~ American physicist Peter Collings

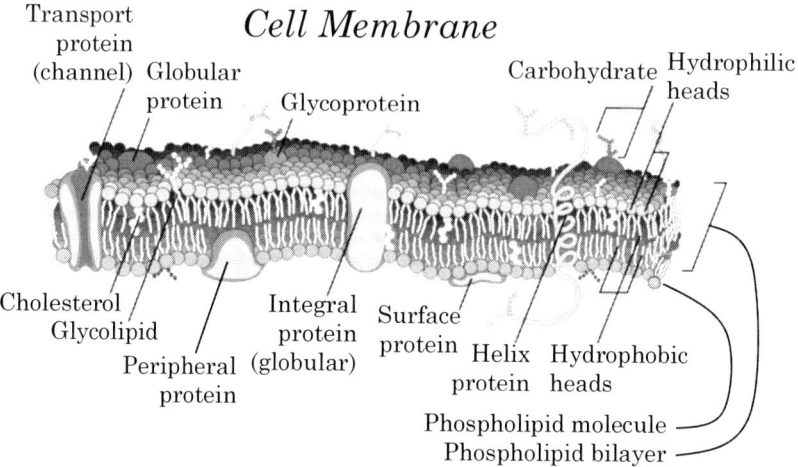

Cell Membrane

The contents of every cell are held within a *plasma membrane*, which is a flexible container of controlled permeability. The cell membrane supports its cell and helps maintain its shape. Cell membranes are living liquid crystals: able to maintain a necessary, delicate balance between fluidity and rigidity. The structure of the plasma membrane is similar in prokaryotic and eukaryotic cells.

Organelles within cells are also enclosed within a membrane. As an evolutionary artifact of endosymbiotic incorporation, the nucleus, mitochondria, and chloroplasts in plants have 2 membranes.

Cell membranes are a mix of proteins and lipids, with respective percentages varying by cell type. While lipids structure the cell membrane, proteins manage traffic through the membrane and keep the membrane in good working order. Cell membranes are organized into distinct domains, with different proteins performing a range of necessary functions.

The lipid composition of membranes can have profound effects on the behavior and activity of its resident macromolecules. ~ Eric Schon

3 lipids are employed in cell membranes: phospholipids, glycolipids, and sterols. While the amount of each varies by cell type, phospholipids are consistently the most abundant.

The cell membrane scaffold is a double layer of phospholipids. A *phospholipid* has a hydrophilic (water-loving) phosphate head and 2 hydrophobic (water-fearing) fatty-acid tails. Phospholipids spontaneously arrange themselves into a double-layered structure, with their hydrophobic tails pointing inward and their hydrophilic heads facing outward. Most biological membranes have this energetically favorable 2-layer structure, called a *phospholipid bilayer*.

A *glycolipid* is a lipid with sugar on top. In cell membranes, glycolipids self-assemble into highly organized domains. Glycolipids situated on a cell membrane surface act as signal markers, facilitating recognition by other cells. Their specific arrangement regulates numerous cellular processes which can cause disease if not properly done.

Lipids are hydrophobic. Their natural aversion to water causes them to clump together. That only partly explains the intricate integration of glycolipids on a cell membrane.

The sugar molecules in neighboring glycolipids interact to form an ordered crystalline structure, connected by hydrogen bonds. Thus, both lipid and carbohydrate molecules complement each other in forming well-organized glycolipid clusters.

> All sterols share a common property: the ability to regulate dynamics in order to maintain membranes in a microfluid state where they can convey important biological processes. ~ French biochemist Erick Dufourc

Sterols are essential in eukaryotic cells. In animal cells, cholesterol disperses between membrane phospholipids, keeping cell membranes from becoming too stiff, by preventing phospholipids from being too tightly packed together. Plant cell membranes lack cholesterol, opting instead for a sophisticated mix.

> In contrast to animal and fungal cells, which contain only one major sterol, plant cells synthesize a complex array of sterol mixtures. Sterols regulate membrane fluidity and permeability in a similar manner to cholesterol in mammalian cell membranes. Plant sterols can also modulate the activity of membrane-bound enzymes. ~ French molecular biologist Marie-Andrée Hartmann

A cell membrane protects the integrity of the cell, policing the molecules that go in or out. From without and within the cell, a cell membrane must process a wide variety of signals to initiate apposite responses to changing conditions.

> Membrane dynamics is essential for cellular life. ~ Erick Dufourc

Membranes are constantly changing, allowing migration of proteins and other products. This fluidity is essential for engulfing food, discharging waste, and secreting cellular products.

Membrane proteins act as border guards, membrane maintainers, cellular communicators (signalers and receivers), nutrient jitneys, and enzymatic actors playing various roles. Generally, there are 2 types of proteins associated with a membrane. *Integral* membrane proteins are inserted into a membrane and may pass through the membrane. Portions of these transmember proteins may be exposed on both sides of the membrane. *Peripheral* membrane proteins are on the exterior and connected to the membrane via interactions with other proteins.

Every cell is mostly water. Maintaining cell pressure, an essential function, means managing the flow of water.

Animal cells move about by managing water flow. Cell motility requires tightly regulated membrane dynamics and snappy cell shape change in the cytoskeleton. A moving cell is a symphonic exercise, conducted by certain proteins. *Morphogens* are signaling molecules that tell a cell where to go.

Aquaporins are proteins that act as channels for the flow of water across biological membranes in response to osmotic pressure changes. They provide the plumbing system for cells. Each aquaporin acts as a meticulous membrane pore.

All sorts of molecules pass through cell membranes, albeit selectively. Aquaporins permit the flow of small polar molecules, such as glycerol, while blocking others. As keeping electrical homeostasis is crucial, protons and certain ions are precluded passage.

ஐ Cytoplasm ൭

Within the membrane, a gel-like cytoplasm holds a cell's internals. The viscous liquid – *cytosol* – distributes needed substances throughout the cell. Proteins within cytosol intelligently manage the necessary processes.

Cytosol is 80% water. It is usually clear.

Most cellular activities happen within the cytoplasm. The flow of calcium ions in and out of the cytoplasm is how metabolic processes are signaled.

Calcium is a key to regulate many fundamental processes in cells. ~ Indian biochemist Muniswamy Madesh

Owing to its easy reactivity, calcium is a key bioelement. Calcium charges are essential to powering mitochondrial processing. Heartbeats happen as heart cells synchronously work calcium channels. The physiological correlate to thought processes within the glial cells of the human brain transpire via waves of calcium ions.

ഽ Cytoskeleton ಜ

Nestled amidst the cytoplasm is the *cytoskeleton*: tubular filaments of protein termed *microtubules*, which provide cellular scaffolding and form the internal structure of cell tails – both cilia and flagella. Cytoskeletal elements extensively and intimately interact with the cell membrane.

Tubulins are the family of proteins that comprise microtubules. Tubulins are the guides by which cells know their internal organization. The asymmetric configuration of cells, whether prokaryote or eukaryote, and even left-right organ placement in multicellular organisms, owes to tubulins driving patterning during development. How tubulins know and coherently provide their orientation guidance is a mystery.

The cellular patterns by which tissues and organs develop in a multicellular organism owe to the energetic database which materializes as genetics. But cell fate is also influenced by nutritional history.

What a cell takes in during its early development is a determinant in what it will become. Hence biases in development patterns of cell fates emerge from environmental signals.

ꙮ Cell Wall ꙮ

A tough but flexible cell wall surrounds the membrane, giving a cell structural (tensile) strength and protection from physical damage and attack by pathogens. Although it acts as a filter, a cell wall is also permeable.

A cell wall most importantly acts as a pressure vessel, preventing over-expansion when water flows into the cell. *Cytolysis* (*osmotic lysis*) happens from an osmotic imbalance, namely excess water getting inside a cell, causing the cell to burst.

Archaea and bacteria have cell walls. Bacterial cell walls are made of peptidoglycan: a polymer comprising sugars and amino acids, forming a mesh-like layer outside the plasma membrane. Archaea have various cell wall constructions.

Plants and fungi have somewhat similar cell walls, made of cellulose and chitin. As the concentration of solutes outside plant cells is typically less than inside, cell walls are critical in maintaining structural integrity.

Protozoa and animal cells, with rare exception, lack cell walls. Animals have connective tissues – bone and cartilage – to provide structural reinforcement. Animals regulate osmotic pressure by excreting excess water and salts, and by pumping ions across cell membranes.

Further, cell walls could be a handicap to animals by restricting movement. Muscle cells could not contract if they were encased in cell walls. Hence cell walls in animals are superfluous.

ꙮ Tails ꙮ

Once considered merely a vestige of evolution, cilia are in fact essential to many of the body's organs. ~ American medical science writer Mary Beth Gardiner

While a wall or membrane separates every cell from the outside, knowing what's outside is critical to knowing what to do, or one's role in the larger organismic scheme of things. And being able to get around may be just as essential.

Thus, almost all cells – whether prokaryote or eukaryote – have tails. Eukaryotes evolved from a tail-bearing prokaryote. Only seed-producing plants have cells without tails, though their genomes retain the knowledge of how to make tails.

Cell tails serve 2 purposes: motility and perception.

Tails may be a single whip-like flagellum or solitary sensory cilium, or a multitude of short hair-like cilia on the cell lining.

Mammalian sperm cells have a flagellum, which propels them toward a fertile destiny with an egg. Female eggs have rows of cilia, by which they make their way from the ovaries to the uterus.

Regardless of type, appearance, and motion, cell tails are structurally similar, and their employment selfsame. Each tail has fibers – *microtubules* – which rapidly shift positions among each other via a proton flux, generating movement. This set of fibers is enclosed in the membrane.

The electrochemical proton flux that propels tails is coordinated through membrane pores. Flux flow determines a tail's speed limit.

♎ Spirochetes ♎

Spirochetes are spiral-shaped, free-living, anaerobic bacteria. They are prokaryotes on the prowl. Having no need of oxygen indicates an ancient lineage.

Spirochetes are found in a wide variety of habitats, from swimming in mud to the guts of desert termites.

As a disease organism, a spirochetal bacterium, *Treponema pallidum,* causes syphilis in humans.

Spirochetes swim via a curious corkscrew motion, like the swiggles of sperm cells heading to an egg.

Some free-living spirochetes attach themselves to larger cells, providing propulsion. A taxi-driving spirochete gets its fare from leftovers of its cellular passenger.

Spirochete cilia are a ring of microtubules, like spun wires in an electrical cable. Some spirochete sport fine filaments spun from the same protein from which eukaryotes form microtubules. A characteristic animal pattern is 9 double tubules in a ring, with another double tubule holding down the middle, in a hub-and-spoke configuration.

Other bacteria prosper with different propulsion: flagella with helical filaments, driven by a rotary engine anchored on the inner cell membrane.

Flagella and cilia both work in coordinated fashion, albeit differently. Cilia move with a complex 3-dimensional swim, propelling with a power and recovery stroke.

Flagella rotate like a propeller rather than beating back and forth. These flagella are helical, and revolve 200–1,000 times per minute, propelling bacteria as fast as 60 cell-lengths per second. A cheetah, the fastest land animal, has a top speed of 25 body lengths per second.

Though similar structurally, the proteins that make up cell tails differs between bacteria, archaea, and eukaryotes. This suggests that tails are an instance of convergent evolution: coming to the same solution independently.

♎ Paramecia ♎

Paramecia have 2 sets of selfsame cilia. One puts a move on, while the other facilitates feeding, by sweeping nutrients into oral grooves.

Paramecia locomotion cilia are powered by a different motor than feeder cilia. Viscosity slows paramecia motion, but feeding cilia are unaffected. Getting around is less important than chowing down.

Tails are crucial cellular sensors. Depending upon type, cell cilia can sense fluid movement, chemicals, osmotic pressure, temperature, and/or gravity.

Mammalian adaptive immune system T-cells have no apparent cilium, but they do have the equivalent. The synapse that activates a T-cell acts like a cilium and is built much the same.

Cilia are critical to intercellular communication. In animals, they help maintain organ function via continuous feedback loops. In epithelial tissues, the cilia of host cells may actively recruit microbial symbionts.

The asymmetry of human bodies is established within a few hours after an embryo begins developing. This is done by the tails of cells sweeping clockwise, generating a net leftward flow, which tells left from right, and determines *situs solitus*: the position of organs.

Damaged tails often spell cell death. If the cilia of cells cannot function, disorders arise. Badly behaved or non-functioning cell tails are instrumental in many diseases.

ଈୠ Hygiene ଈଓ

The creative endeavors of ribosomes have clean-up counterparts. Worn-out organelles, waste materials, and cell debris must be swept up and recycled or disposed. This process is termed *mitophagy*. In contrast, *autophagy* is the process of removing a cell too worn-out to carry on anymore.

Mitophagy plays a critical role in cellular health. In humans, its disruption contributes to systemic degeneration, including intelligence system and cardiac diseases, diabetes, and cancer.

Lysosomes are a cell's waste disposal system. Plants have *lytic vacuoles* that serve the same function. These organelles contain acid hydrolase enzymes which break down cellular debris. Lysosomes digest the macromolecules from phagocytosis, endocytosis, and mitophagy: processes which ingest matter that needs recycling. Lysosomes also transport undigested material to the cell membrane for expulsion.

Cell organelles constantly monitor themselves. When an organelle no longer functions as it should, pro-mitophagy agents have the organelle mark itself on its membrane for delivery to the cell's recycler (lysosome or lytic vacuole). The same chemical marking is used for both organelle mitophagy and cellular autophagy.

❧ Apoptosis ❧

Aging seems to be the only way to live a long life. ~ French composer Daniel Auber

Programmed cell death (*apoptosis*) is part of life. During development, certain cells must surrender their lives to advance an organism to the next stage. Fingers emerge from paddle-shaped hands by apoptosis of the cells that form the webbing between digits-to-be.

Dying cells interact with their neighbors. They can send signals for other cells to proliferate and tell cells from afar that it is time for them to die too.

Good health depends on the strict regulation of cell division and cell death. Apoptosis -- a kind of suicide plan for cells -- is an important safety mechanism for the body to get rid of damaged, aged, or unneeded cells. ~ German cytologist Stephanie Bleicken

Cell death can run amok: killing cells in excess after a traumatic event, such as a heart attack or stroke, or in the course of degenerative diseases, such as Alzheimer's. Normally cell death is meticulously decided and controlled.

The apoptotic control network includes several positive feedback loops. Apoptosis spreads through trigger waves. ~ American systems biologist James Ferrell Jr. & Chinese cytologist Xianrui Cheng

Cell death begins with biochemical trigger waves. Once initiated, specific killer proteins in the cell, called *caspases*, activate.

When a cell is dying, it goes through a characteristic series of morphological alterations and, in the end, it is engulfed and digested by neighboring cells. ~ German cytologist Barbara Conradt

A cell undergoes a series of changes – altered protein production and gene expression – as it dies. Even during demise cells struggle to survive with remarkable resilience: able to bounce back from poisoning if the toxin is removed (*anastasis*, Greek for "rising to life"). The point of no return, when cell death is irreversible, is not yet known.

ᛒ The Evolution of Eukaryotes ᴈ

Everything you'll ever need to know is within you; the secrets of the universe are imprinted on the cells of your body. ~ American author Dan Millman

As the Earth and its atmosphere were transformed by cyanobacteria, viruses promoted prokaryotic evolution by providing genetic uploads.

~2.5 BYA, eukaryotes arose. The first step to eukaryotic life was through unification: one prokaryote incorporated another. The *host* was an archaeon.

An intracellular bacterial parasite gave rise to the mitochondria found in all eukaryotic cells. At some point, the bacterium that beget mitochondria became benign, then mutualistic. The mitochondrial bacterium went from stealing ATP to providing it.

Transition to accommodation is not unusual. Viruses go from devastating to their hosts to being tolerable or even beneficial, as they learn to prolong their residency by not inflicting untold damage, and thereby benefit from host longevity.

Mitochondrial incorporation came late in the evolution of eukaryotes. Already many eukaryotic hallmarks, including complex subcellular organization, were in place prior to added a dedicated power plant facility to the works.

♎ Colonial Choanoflagellates ♎

Bacteria are the important part of the multicellular story. ~ American cytologist Nicole King

Choanoflagellates are free-living unicellular and colonial flagellate eukaryotes. These plankton are the closest living relatives of animals.

In a colony, choanoflagellates are more than an aggregation. Instead, they are an interacting cluster, demonstrating the basic mechanics for multicellularity.

Colonial choanoflagellates only exhibit this behavior under the influence of resident bacteria. Bereft of bacteria, they do not inch toward acting in a multicellular manner.

Evolution is adaptation guided by energy economy. Hence the evolution of eukaryotes was a matter of task division according to relative performance of the different cells involved. A common genetic foundation afforded the necessary chemical communication network. This furthered adaptation geared to *efficiency*: the least energy expended to produce the desired result.

Over evolutionary time, most of the genes housed in the former bacterial endosymbiont, now mitochondrion-bound, migrated to the genome of the archaeal host. These genes became enclosed in a protective membrane, forming the cell nucleus.

The genes that stayed in the mitochondria were those needed for practicing its core business: coding for proteins which maintain redox balance, which must be synthesized locally to counteract the otherwise deadly effects of ATP-generating electron transport.

In modern eukaryotic cells, the mitochondrial genes resemble those of bacteria, while the original nuclear genome resembles a heritage of bacterial and archaeal origin. Eukaryotic DNA replication descends from archaea, not bacteria.

The acquisition of the mitochondrial power plant gifted eukaryotic cells with 200,000 times the energy available to the average prokaryotic cell. This enabled eukaryotic cells to expand their volume by up to 15,000 times that of the typical bacterium, and to support a genome 5,000 times larger.

After the 1st union of prokaryotes, one or more proto-eukaryote cell types had a 2nd round of endosymbiotic uptake.

Photosynthesis was acquired by endosymbiosis. The ancestor of light-fed algae and progenitor of plants incorporated a cyanobacterium, thus picking up chloroplasts. A *chloroplast* is the photosynthetic organelle found in algae and plant cells. It is a type of plastid.

Plastids is the catchall term for the major organelles found in the cells of plants and algae. Various plastids have different functions, such as storing starch or fat, or detecting gravity. Some plastids have several internal membrane layers, indicating an independent heritage of endosymbiont incorporation. Algae plastids typically differ from plant plastids.

Like the previous incorporation leading to proto-eukaryotes, many genes of the consumed cyanobacterium that became a chloroplast migrated to the host cell genome. Mechanisms evolved that allowed the proteins encoded by transferred genes to work for the chloroplast colonizer, so as to photosynthesize.

The plastid genome that remained kept its legacy of cyanobacterial origin. But the nuclear genome of plastid-containing eukaryotes is *chimeric*: containing both the proto-eukaryotic genome and genes derived from the cyanobacterial genome.

The genes to make the enzyme that manufactures the amino acid *phenylalanine* (*Phe*) was lifted from an ancient bacterium. Phe is used in many plant products, including *lignin*, which is critical for the strength of plant cell walls. For animals, Phe is an essential amino acid.

The incorporating evolution of eukaryotes was a milestone, but not as novel as once thought. Some bacteria have protein shell subunits which are functionally equivalent to eukaryotic organelles. Efficiency by division of labor, along with cooperative communication and coordination, existed even in early prokaryotes. There are numerous known examples of cooperative exchange and intimate relationships among prokaryotic microbes.

However impressive in effect, it is only an incremental step from prokaryotic cells aggregating, and sharing genetic material (plasmids), to cooperative envelopment involving different lineages of prokaryotes. Hence, multicellular life was an incremental adaptation from single-celled organisms. In aggregate forms, prokaryotic cells communicate, communally make decisions, and even differentiate; quite like organelles within eukaryotic cells, which communicate and coordinate; as do cells themselves within multicellular organisms.

Larger eukaryotic cells offer efficiencies as well as room for more sophistication. And so more complex organisms are eukaryotic. But there are tradeoffs.

Compared to eukaryotes, smaller size gives prokaryotes a greater surface-area-to-volume ratio, which translates to a higher metabolic rate and a faster growth rate, resulting in shorter generation times. This speed advantage, coupled with horizontal gene transfer, yields a formula for the rapid adaptive capabilities that microbes are known for. *Horizontal gene transfer* is the environmental depositing and pickup of genic packages which contain actionable intelligence. Bacteria are in the business of being genetic quick-change artists; as are viruses, which carry even less baggage than prokaryotes.

♎ Metamonads ♎

The presence of mitochondria and related organelles in every studied eukaryote supports the view that mitochondria are essential cellular components. This organism has evolved beyond the known limits that biologists circumscribed.
~ Polish molecular evolutionary biologist Anna Karnkowska

Metamonada is a large group of anaerobic flagellate protozoa. Most live as symbionts in the guts of animals, from insects to mammals. Protozoa are a diverse phylum of unicellular eukaryotes, partly classified together for their various abilities to live in harsh environments.

Metamonads get by without mitochondria, thanks to a cytosolic mobilization system that they acquired from bacteria. This system substitutes for essential mitochondrial functions.

In being an apparent step back toward prokaryotes, the loss of mitochondria is an instance of reversion evolution. Despite having no mitochondria, metamonads have otherwise sophisticated eukaryotic cells.

♐ Eukaryotic Uplift ♐

Geology played a part in eukaryotic evolution, notably in copious delivery of useful compounds.

Earth's landmasses collided and created the supercontinent *Nuna* 1.9 BYA. As Nuna was nudged into being, molten mantle material made its way to the crust and crystallized, forming a rare type of granite that began eroding quickly, distributing quantities of metal sulfides to coastal areas. This introduction enhanced the biogeochemical environment; a welcome gift to evolving biota. An ampler supply of zinc – a

trace element employed by microbes, plants, and animals –
was one such delivery.

Its geological enrichment accomplished, Nuna began to
fragment 1.5 BYA.

❧ Eukaryotic Cell Types ☙

Diversification of cell types afforded the evolution of multicellular organisms. The simplest are made of a few dozen
different cells, while humans are composed of over 200 kinds
of cells.

Somatic cells are the ordinary cells of a multicellular eukaryote. These cells are the basis of the 4 primary animal
tissue types: epithelium, muscle, connective, and intelligence
tissues.

Epithelial tissues line the surfaces and cavities of bodily
structures and form many glands. These cells secrete, selectively absorb, protect, and transport.

Muscle cells, being capable of contraction, provide for
movement. Muscle is the most abundant tissue in most animals.

Connective tissue supports, separates, or connects other
tissues. Connective tissue, immersed in body fluids, is composed of cells, fibers, and extracellular matrices.

In some animals, including vertebrates, the body physically centers its intelligence system in glial cells, which are
networked together via neurons. 85% of human brain cells
are glia.* Glia nurture and control neural growth and activity
within the brain, as well as receiving neutrally transmitted
information from external stimuli for further processing.

Stem cells, found in all multicellular life, are cells of flexible form, able to differentiate into diverse specialized cell
types, including somatic cells. Stem cells guide organism development. Stem cells can also self-renew: stir up more stem
cells.

---

* Whether cell or organism, an intelligence system is energetically
  driven by a mind.

> There is a fitness advantage to renewing your mitochondria. Stem cells know this and have figured out a way to discard their older components. ~ American biologist David Sabatini

Stem cells intelligently manage their resources: providing the highest-quality organelles to daughter stem cells. Such strategic thinking lessens cellular damage that can lead to stem cell exhaustion, thereby aging an organism from reduced tissue renewal.

> By dividing asymmetrically, stem cells can generate two daughter cells with distinct fates. Stem cells segregate their old mitochondria to the daughter cell that will differentiate, whereas a new stem cell will receive only young mitochondria. ~ Finnish cytologist Pekka Katajisto

If there is a shortage of stem cells, differentiated cells can take their place: generating various cell types, including more stem cells. Genes dormant in differentiated cells again become active when reverting back to stem cell status.

Germline cells are the special cells of sexual reproduction, producing *gametes*. In animals, the gametes are eggs and sperm. Plant germ cells produce ovules and pollen.

Animal germ cells develop in the embryotic stage. In flowering plants, germ cells come from somatic cells in adult floral *meristem*: the plant tissue where growth occurs.

ஐ Organized Activity ෬

Form and function in cytology are simply different perspectives, as the two are inexorably intertwined. In this, the precepts of physics and chemistry provide the foundation for life. That withstanding, living is a matter of intelligently applying energy.

ஐ Communication ෬

There is a constant flurry of communication and coordination in every cell. Chemical messages flit back and forth.

Cells maintain their integrity and reproduce via genetic material. Operationally, a cell's genome is a massive instruction manual for producing proteins and other bioproducts.

This manual expresses the language of life via coordinated conferencing.

Though the functions of prokaryotes and eukaryotes are comparable, the elaborate structure of even a single eukaryotic cell means that its communication needs are inherently more complex than those of a prokaryote.

⚭ Extracellular Matrix ⚮

The complexity of eukaryotes is massively multiplied with multicellularity. Most cells in a multicellular organism are in close, ongoing associations with their neighbors.

Tissues are cell masses joined by junctions, or by an *extracellular matrix* (ECM) of secreted glycosylated (carbohydrate-rich) proteins that create attachment bases for cells: holding tissue together without direct contact between neighboring cells.

Glycocalyx is a common glycoprotein ECM, produced by some bacteria as well as eukaryotic cells. Glycolcalyx forms a coating on the outside surface of a cell membrane. The slime on the outside of a fish is a glycolcalyx.

Glycocalyx plays various roles: cell recognition, cell adhesion, protection, and acting as a permeability barrier. Glycocalyx acts as one way for an organism to distinguish between its own healthy cells and those that are diseased, as well as assisting in recognizing invaders. Multicellular organisms expend tremendous time and energy building and maintaining molecular forests of glycocalyx.

ECM glycoproteins come in various shapes and sizes, but functionally they fall into 3 groups: transporters, ion channels, and receptors.

Transporters carry specific molecules, usually food, across cell membranes. Each type of molecule has its own jitney model.

Ion channels are gated communication pathways that form a selective signaling matrix. Communications across cell membranes are regulated by ion channels.

Receptors are the most diverse group of cell-surface glycoproteins. Each receptor is designed to respond to a specific signaling molecule. A signaling molecule binding to its receptor sets off a sequence of biochemical events which may regulate cell production, growth, or even death (apoptosis).

The various communications systems of multicellular organisms, such as the nervous and endocrine systems, are multifaceted matrices that use ECMs.

໕ Movement ଓ

> For a cell to move forward it must convert chemical energy into mechanical propulsion. ~ Swedish engineer Pontus Nordenfelt *et al*

The great advantage of multicellular organisms is *specialization*: differentiated cells that perform different tasks. Specialization requires coordination. Coordination involves movement; well, sometimes. Plant cells are immobile. The genetic and hormonal networks that control plant growth are completely different than for metazoa with their motile cells.*

Chemotaxis is cellular movement toward or away from a chemical stimulus. Among other uses, chemotaxis lets stem cells find, and reside in, proper locations.

During embryonic development (*embryogenesis*), chemotaxis is repeatedly employed to rearrange cells. This occurs during primordial germ cell migration, organ formation, and wiring the nervous system.

Embryogenesis involves undifferentiated cells arranging themselves into groups of functionally similar cells to form tissues. Developing embryos are awash in signals that guide cells and spur cell differentiation, thus forming organs.

> An individual in a collective consciously tries to align its movements with those of its neighbors, which involves orchestrated sensing and action. So it is with the collective migration of cells. ~ German cytologist Joachim Spatz

Many cells are migrant workers. Red blood cells flow to deliver oxygen. White blood cells roam the circulatory system. When one encounters a wound, it releases epidermal growth factor, a chemical transmission which summons other white blood cells to assist.

---

* A *metazoan* is a multicellular animal.

ℒ Healing ℒ

In recovering from a wound, new cells must locomote to the right location. To move into place, part of the cytoplasm temporarily rigidifies, affording a fluid creeping motion by stretching and retraction of the right parts at the right time – an elaborately coordinated exercise.

In an adult, chemotaxis mediates the trafficking of immune cells, and is crucial for inflammation. Chemotaxis also participates in wound healing, and in tissue maintenance.

Cells use slight differentials of molecular concentrations in fluid to tell which direction they should go.

Cells can detect differences in concentration as low as 2%. They're also versatile: detecting small differences whether the background concentration is very high, very low or somewhere in between. ~ American cytologist Peter Devreotes

Detecting gradients is a 2-step process. First, a cell tunes out background noise. Next, the side of the cell getting less of the chemical signal stops responding to it.

Then, the control center inside the cell ramps up its response to the message it's getting from the other side of the cell and starts the cell moving toward that signal. ~ Chinese cytologist Chuan-Hsiang Huang

As cells migrate, they transmit mechanical forces from their leading edge, creating a stress wave that propagates through tissue. This stress gradient creates a faint electrical field that is detectable by nearby cells.

These forces are generated backward, from cell to cell, through intercellular junctions. This build-up of stress gradients and voltage guides cells in the right direction.

Meanwhile, cell fragments move in the opposite direction of the electrical charge. Thus, wounded tissue is more easily cleared as replacements arrive.

Cells are not just sacks of jelly. They have a complex structure which includes scaffolding: a network of wiry molecules that maintain cell integrity, as well as microtubular piping acting as intracellular conduits for data transmission.

Around the edges of cells are networks of filaments. These filaments, composed of actin, are in constant flux, even when a cell is not moving.

For a vast variety of purposes, cells migrate collectively via intermittent bursts of activity. That takes planning.

Cells talk to nearby cells and compare notes before they make a move. ~ Russian American biophysicist Ilya Nemenman

A cell moves by converting chemical energy into mechanical force. 3 molecule families work together to enable cell migration: actins, integrins, and cadherins.

Actin is a family of proteins on the inside of the cell membrane which collectively form a cell's skeleton, and so help maintain cell shape as well as effect cell motility. Actin participates in many other cell processes, including communication, cell division (mitosis), and organelle organization.

Integrins are transmembrane receptor molecules on the cell surface that can attach themselves to other surfaces. Integrins act as communication bridges for cell-to-cell and cell-extracellular matrix interactions.

There are several types of integrins. They work alongside other cell receptors, including cadherins.

An adapter protein links actin and integrin together to effect cell movement.

Once the adaptor has connected the integrins and actin, the mechanical force from actin gives the 'green light.' The integrin molecule then binds itself to a nearby partner surface, through which the cell can move slightly forward.

Once the cell's migration is complete, the integrin and actin molecules separate from each other in this part of the cell, while another part of the cell becomes active. ~ Pontus Nordenfelt

Cell migration is governed by integrin molecules, which coordinate to become active in the necessary regions to propel movement in the intended direction. But the general mechanism for locomotion involves localized actin polymerization.

A cell moves by building filaments in the direction of movement, while tearing down filaments on the tail end. As this happens, the cell surfaces bulges, forming a lobe, and then uses molecular clamps to grab onto the underlying surface.

A cell swims and tugs its way. It is an intricate dance that takes a lot of energy.

Of course, a cell needs to know which way to go. Multiple cells often form a train, migrating as a group.

Cadherins are a class of cell membrane proteins. The term derives from "calcium-dependent adhesion," as cadherins operate via calcium ions (Ca^{2+}): a common cellular modus operandi.

Cadherins were long thought to merely act like mortar between bricks: holding cells together and preventing motility. But cadherins are not just glue.

> Cadherin is serving multiple purposes, all of which function together to coordinate the collective ability of these cells to sense direction. ~ American cytologist Denise Montell

Each of the different cadherins cluster together with others of their kind. Cadherins on one cell create a zipper-like connection with others on another cell.

Cells recognize one another and coordinate cell migration via cadherins and other cell adhesion molecules (CAMs) – tethering cells together and pulling them in the proper direction.

> Cells can sense not just the precise concentration of a chemical signal, but concentration differences. That's very important because in order to know which direction to move, a cell has to know in which direction the concentration of the chemical signal is higher. Cells sense this gradient and it gives them a reference for the direction in which to move and grow. ~ Ilya Nemenman

Another class of CAM provides adhesion without relying upon calcium. Some of these cell adhesion proteins are involved in forming nerve cell contacts and neural grouping.

When cells from 2 different tissues are dissociated mechanically, or by treatment with enzymes and then mixed in a culture medium, the cells re-aggregate into different groups according to their origin. The cell adhesion molecules on the cell surface provide markers identifying the different cells.

Cancer cells metastasize (spread) by altering CAM function to lose cell adhesion. Conversely, many pathogens and parasites use CAM proteins as signposts, and to colonize by binding firmly to mucosal surfaces.

Cells move differently, depending upon their situation within tissues: alternately using adhesive or electrostatic forces to propel themselves. To avoid detection, cancer cells mimic the way native cells move.

> If cancer is to spread from one part of the body to another, it will encounter a diverse range of tissue environments. Being able to switch movements is the most important factor in making a cancer cell dangerous. ~ English pathologist Erik Sahai

A *transmember protein* can travel through a cell membrane. Many transmember proteins act as border guards at membranes, denying or allowing specific molecules, either in or out. Transmembrane proteins recognize their clientele. They ferry some molecules across by particularly folding or bending to move the molecule through.

♀ Push & Shove ♀

> Force induces additional interactions at the atomic scale. Residues that had previously not been making contact are now interacting. These are force-induced interactions. ~ Chinese microbiologist Cheng Zhu

Much cell activity is lively chemistry. But jostling about is also important. Mechanical forces can regulate cellular chemical reactions and cell functions. Tensile pressures are applied to cells all the time.

A cell might rearrange its cytoskeleton to accommodate an applied force, or apply its own forces to do something, such as moving itself. Cells can vary their mechanical environment to affect their biochemical environment.

ಐ Self-Organization ಐ

Recognition of changing conditions is essential for proper cellular activity. The most sensitive possible perception would be for a cell to keep itself close to the edge of self-organized criticality. That is in fact how biology operates, from the macromolecular to the cellular level. Biochemical interaction networks form an operational state memory in multiple dimensions, provide the means for self-assembly for complex functioning, and facilitate organizational dissolution when a structure is no longer needed.

Successful development of multicellular animals requires cooperative intercellular interactions and coordination that ensure tissue integrity. Various mechanisms enforce these behaviors.

One such mechanism monitors genetic identity, preventing uncooperative cells from contributing. How genetic disparities are recognized is unknown but determining cell fitness via co-acting communications is a critical component.

> Cells within developing tissues that are recognized as mutant or compromised are competitively eliminated. ~ Swiss molecular biologist S.N. Meyer *et al*

Phase transitions are used to spatially organize and regulate cellular information storage and transfer. A small change triggering a single transition point can create a large response through cascading effect. An example of this butterfly effect is in phase transitions which organize cellular compartments.

> Cells have numerous examples of nonmembrane-bound compartments containing many proteins that perform complex biochemistry. These compartments form rapidly and are disassembled when not required. ~ Finnish biochemist Kai Simons & English cytologist Anthony Hyman

ಬಂ Longevity ಢ

How long a cell may live in a multicellular organism has a lot to do with its function and criticality.

Red blood cells have a 4-month lifespan. They are relatively long-lived service providers.

In contrast, white blood cells – warriors of the immune system – are lucky to last a month. The longest-lived white blood cells are lymphocytes, which retain the memories of past infections, to advise in future conflicts.

Turnover varies tremendously among proteins. Most are replaced several times during a cell's lifespan. The average protein half-life in a yeast cell is 90 minutes; in mammals, 1–2 days. Only a few proteins last the life of a cell.

There is an intricate accounting system for proteins within cells. Proteins are stochastically festooned with sequential age markers, indicating their service state and degradation to date.

An exception is DNA, which owes its long life to dedicated repair mechanisms that remedy damage. Ordinary proteins lack such support.

The histones that bind DNA are extraordinarily long-lived. These proteins are essential to DNA function, as they act as the spool around which the genetic codes are wound.

⚲ Nuclear Pore Complexes ⚳

The double-membraned eukaryotic cell nucleus has a selective permeability. Nuclear pores are complexes of large proteins – *nucleoporins* – which porter various molecules through the nuclear envelope.

A cell may have up to a couple thousand nuclear pore complexes (NPCs), depending upon cell type and life cycle stage. Each NPC has at least 456 nucleoporins, of which there are 30 distinct types. An NPC can conduct 1,000 translocations per complex per second.

This transport includes RNA and ribosomal proteins moving from the nucleus to the cytoplasm; and various proteins, carbohydrates, lipids, and signaling molecules moving into the nucleus. Although smaller molecules simply diffuse through, larger molecules must be recognized and ushered in or out of the nucleus with the help of nucleoporins.

While the turnover of nucleoporins is generally typical of hard-working proteins, scaffold components resist degradation, as the scaffold substructure may hold epigenetic depots critical to transmitting inheritance information. Otherwise, long-lived nucleoporins are those actively involved in translation, which is a later stage in gene expression. These too are mission-critical proteins.

Unlike other large protein complexes, such as ribosomes and proteasomes, an NPC is not replaced in toto. Instead, individual subcomplexes are exchanged at specific intervals. In other words, NPCs undergo regularly scheduled maintenance. Dismantling entire pore complexes might jeopardize the integrity of the nuclear envelope.

☙ Cell Division ❧

The *cell cycle* is the life cycle of cell, though the term is often used to refer to the *mitotic phase*, when cell division occurs. Between divisions, a cell is in *interphase*, which is its ordinary life.

Cell division is a daedal process of cell replication. There are a variety of regimes for cell division, depending upon the type of cell and organism.

Cell division is dynamically regulated, with much decision-making. Cells decide to divide based upon the environment they sense they are in, cell size, and the time of day. Changing cell fluid pressure helps drive cell division.

Prokaryotes asexually reproduce via *fission*: splitting their cells into 2 (binary fission) or more (multiple fission). Some organelles in eukaryotic cells also replicate via fission, as do some single-celled eukaryotes.

In eukaryotic cell division, a parent cell divides into daughter cells. The cell nucleus divides first (*mitosis*), followed by the cytoplasm splitting in 2 (*cytokinesis*). For unicellular eukaryotes, such as amoeba, cell division is equivalent to reproduction: a new organism is created.

Cell division in multicellular eukaryotes varies by cell type. Somatic and stem cells undergo *mitosis*. As part of sexual reproduction, germline cells divide via *meiosis*.

The primary concern of cell division is maintaining the parent cell's genome. Prior to division, the genomic encyclopedia stored in chromosomes (or prokaryotic genophores) must be precisely replicated, and the duplicated genome cleanly bifurcated between cells.

Considerable cellular infrastructure is involved in ensuring genomic consistency from one generation to the next. The intricacy of cell division illustrates the amazing intelligence in biological processes.

ഔ **Mitosis** ര

> The fact that mitosis is a universal eukaryotic property suggests that it arose at the base of the eukaryotic tree. ~ English biologists Adam Wilkins & Robin Holliday

Mitosis results in somatic or stem daughter cells, each with a full set of chromosomes. Daughter cells have the same genome as their mother cell. A crucial process within mitosis is chromosomal replication, which is regulated epigenetically.

A *chromosome* is a genetic package in the nucleus of a eukaryotic cell, comprised of DNA base pairs. A DNA *base pair* comprises 2 complementary nucleobases on opposite sides of a DNA double helix, linked by hydrogen bonds.

A chromosome is a single macromolecule, the largest in Nature. Human DNA has 46 chromosomes. Chromosome 1 in the human body, the biggest, has 10 billion atoms. The average human chromosome has ~140 billion base pairs.

Cell Division

Fission — Prokaryote — Eukaryote — Mitosis — Meiosis — DNA replication — Chromosome segregation — Cytokinesis — Genophore — Plasmid — Gametes — Zygote

Mitosis begins with chromosomal DNA base pairs being cleaved in half: the hydrogen bonds holding each ladder rung broken by enzymes. This happens in several sections simultaneously. To keep things tidy, proteins neatly stack the chromosomes in layers, forming multilaminar plates.

Then each half-ladder has its complement rebuilt. The result is 2 DNA ladders: half each of the original nucleotide string. The other side of each ladder is reconstructed.

Then these 2 sets of chromosomes are pulled apart from the center of the cell, in opposite directions, by protein ropes. These microtubules are part of the cytoskeleton, and play numerous roles in cell upkeep, mostly in moving bits about the cell.

Mitosis is an incredibly intricate process: an orchestration of many thousands of proteins moving through space, finding their exact positions at just the right time. While the proteins remember where they are supposed to go, and have an innate affinity to the right location, innumerable decisions must be made for cell division to succeed.

> For normal tissue structure and function, cells exert strict control over growth versus differentiation. The distribution of organelles influence the outcome of stem cell division. ~ American cytologist Elaine Fuchs

Epigenetic influences take effect during mitosis. Hence, life experiences are passed on during each cell replication.

Mitosis takes 15–30 minutes per chromosome. Replication of a human cell DNA takes up to 10 hours. Yet human cells may divide as often as every day.

When the whole DNA molecule has been duplicated, the chromosomes move to opposite ends of the cell. Then *cytokinesis* kicks in: the cytoplasm of a eukaryotic cell divides; whence 2 cells are constituted.

೮ Meiosis ೪

Meiosis is the special cell division for sexual reproduction, producing germline gametes (sperm or eggs). Meiosis also refers to the cell division process for making spores. Meiosis evolved 1.4 BYA as a derivative of mitosis.

In meiosis, the chromosomes replicate, but rather than pulling apart, the duplicate (homologous) chromosomes stay together. These doubled chromosomes appear as 4 lengths of ladder, side by side.

RNA mediates recombination of homologous DNA during meiosis. Ribosomal RNA molecules search for and sense homologous sequences.

The matched pairs then separate and move to opposite ends of the cell, as in mitosis. The chromosomes divide again, then replicate, thereby producing 4 daughter cells.

Because there have been 2 divisions with only 1 replication, these gamete cells have only half the complement of chromosomes, one of each matching pair. A human gamete has 23 chromosomes.

While meiosis produces gametes in animals, meiosis generates *spores* for fungi, bacteria, and other simpler life forms.

Spores are patient hibernators. Unlike seeds, which are stocked with food, spores have minimal energy supply. Still, spores keep their wits about them.

Once conditions become favorable, a spore comes to life by mitotic division, turning the haploid mitospore into a diploid living organism. Ferns, especially those adapted to dry habitats, produce diploid spores.

Sporulation is one process of *asexual reproduction*. Other asexual reproductions include: binary fission, where 1 parent becomes 2 daughters; budding (e.g. baker's yeast, hydra): a mother creating a smaller daughter; and fragmentation: a new organism grows from a fragment of the parent.

Vegetative reproduction is floral fragmentation. Herbaceous and woody perennial plants often practice vegetative propagation.

Single-celled organisms, such as the archaea, bacteria, and protists, reproduce asexually. Many multicellular plants reproduce asexually as well.

Sexual reproduction starts with fertilization. An egg recognizes whether the sperm before it is proper, beginning with it being species compatible. Unsuitable suitors are rejected.

A human sperm delivers its 23-chromosome package to an egg. 1 chromosome per pair is inherited from each parent.

The chromosome pairing process is not necessarily a neat, exact matchup of the genetic sequences from each parent. Genetic recombination of DNA segments occurs. This further increases genetic variability.

For humans, with 46 chromosomes per cell, there are over 8 million ($2^{23} = 8,388,608$) possibilities for the chromosomal pair in 1 gamete. Fertilization squares that to the neighborhood of 70 trillion.

But heredity is not a genetic numbers game. There is no direct relationship between chromosomal quantity and organism sophistication. For one thing, as a byproduct of evolution, the number of genes per chromosome varies among species.

Bacteria have 1 circular DNA molecule, but regularly complement their DNA with plasmids (gene packets) found in the ambient environment, often provided by other bacteria. This is how antibiotic resistance spreads among bacteria. Numerous strains of cave bacteria, cut off from the surface for millions of years, are resistant to several antibiotics. They remember the bacterial wars of old. These bacteria can easily pass antibiotic resistance to other bacteria.

Meiosis is not the only time that gene shuffling occurs. Diploid somatic cells of fungi, plants, and animals occasionally undergo *mitotic recombination* during mitosis. Recessive genes may be expressed through this genic jumble.

✤ Transport In & Out ✤

For nutrition, communication, and defense, the swapping of substances is essential to all cells. Trafficking materials in and out of cells is controlled by a variety of proteins found in cell membranes, termed *transporters*. This commerce comprises letting vital compounds in, provisioning data and cellular products, as well as disposing of wastes by trucking them out.

Prokaryotes and eukaryotes differ in their ingestion and secretion. As prokaryotes lack membrane-bound organelles, their transport mechanics are simpler.

Once a meal is found, an archaeon or bacterium internally produces enzymes, which it then exudes. The enzymes fracture the food down into digestible bits.

The prokaryote then sucks the substance in, either by osmosis or active transport. Via *osmosis*, food simply flows in through the cell membrane. If osmosis fails, a prokaryote practices *active transport*: actively gobbling by pulling savory juices through the cell wall.

Osmosis takes advantage of an auspicious *concentration gradient*, owing to an unequal distribution of ions across the cell membrane. When diffusion disfavors a hungry prokaryote, active transport must be employed.

By contrast, eukaryotic cells employ their cell membranes for both intake and output. In *endocytosis*, a eukaryotic cell internalizes material from outside the cell. The opposite process, *exocytosis*, secretes select contents, often proteins, out of the cell membrane, onto the cell surface, or into extracellular space.

Both endocytosis and exocytosis employ *vesicles* as containers. A vesicle is formed from the cell membrane for endocytosis and absorbed back into the cell membrane after exocytosis. Thus, these complementary processes continuously recycle the plasma membrane.

ॐ Endocytosis ॐ

In *endocytosis*, a eukaryotic cell engulfs material to absorb it. There are 2 main types of endocytosis, distinguished by vesicle size and cellular machinery involved.

Cells drink via *pinocytosis*. The vesicles are small: convex pits for fluid collection and modest molecular uptake.

Pinocytosis is indiscriminate. Whatever chemicals are in the water are taken in.

Via *phagocytosis*, cells gobble large particles, be they cell debris or entire microorganisms.

While all cells continually ingest via pinocytosis, phagocytosis is a specialty act. In multicellular organisms, only specialized cells wolf down large chunks. For instance, immune system macrophages defend their organisms by swallowing microbial invaders.

Single-celled eukaryotes eat by engulfing their prey. For amoebas and protozoa, *phagocytosis* is a fancy term for *dining*.

In multicellular eukaryotes, endocytosis can play a crucial role in conforming a cell; driving shape changes by remodeling the cell membrane. This is one of many ways that the environment can influence cell development.

ഔ Exocytosis ര

In *exocytosis*, a cell exposes cellular products to select organelles or outside the cell. Proteins sequestered within an intracellular vesicle are let loose as the vesicle membrane fuses with a cell's plasma membrane. The inner surface of the vesicle becomes the outer surface of the membrane.

Membrane fusion is an energetic exercise, requiring the coordinated interaction of adaptor molecules on both the vesicle and plasma membrane. The adaptors are highly selective, only allowing vesicles to fuse with membranes of certain organelles or the cell's plasma membrane.

Once the appropriate adaptors bind with each other in an elaborate docking maneuver, stored ATP energy is released, forming a fusion pore between the vesicle membrane and plasma membrane.

Vesicle contents are released as the pore widens. Ultimately, the vesicle is either absorbed in the plasma membrane or recycled to the cytoplasm.

Exocytosis can be constitutive or regulated. *Constitutive exocytosis* occurs all the time; placing proteins, such as receptors, on the outside of a plasma membrane.

Regulated exocytosis is triggered when a cell receives a signal from outside. Calcium ions are typically involved in triggering.

The received signal is a substance, such as a hormone or neurotransmitter, which binds to a specific receptor on the cell surface. This activates the synthesis or release of a 2nd messenger within the cell.

Many secreted cellular products are for the tissue type in which the cell resides. Otherwise, outputs are transmitted to a more distant part of the organism. Cholesterol and hormones are exemplary secreted products.

Most exocytic products are enzymes or other proteins which have undergone rigorous quality control in the endoplasmic reticulum and Golgi complex. At the downstream end of the Golgi network, cellular products are sorted and accumulated in exocytic vesicles.

Exocytosis is a common vehicle for intercellular communication. Immune systems are extensive employers of exocytosis.

A cell harboring a virus displays viral by-products on its surface; a danger sign that attracts immune cells. Upon arrival, an immune cell tells an infected cell to self-destruct, to save its neighbors. Failing that, a good cell gone bad may be engulfed via phagocytosis.

ଅ Metabolism ଔ

All cells have the ability to sense whether nutrients are scarce or abundant so that appropriate anabolic or catabolic programs can be initiated. ~ Norwegian cytologists Hilde Abrahamsen & Harald Stenmark

Metabolism comprises the cellular chemical reactions that provide energy to sustain life. The mitochondria in eukaryotic cells are the site of respiratory metabolism. Via metabolism, organisms grow, maintain themselves, ecologically interact, and reproduce.

Metabolism is sometimes defined more expansively, for an organism rather than at the cellular level, thus including digestion and the transport of substances between cells. By this broader definition, metabolic cellular processes are more specifically called *catabolism*.

Catabolism is the controlled cellular process of breaking down organic matter to harvest energy via cellular respiration. During catabolism, polymers are reduced to monomers. Polysaccharides are broken down into monosaccharides; lipids into fatty acids; nucleic acids into nucleotides; and proteins into amino acids. Cells use resultant monomers either for energy, by further breakdown, or to construct new polymers, via *anabolism*.

ഔ **Redox** ଓ

Cellular metabolism is driven by *redox reactions*: oxidation of one molecule and reduction of another. *Redox* is a portmanteau for *reduction-oxidation*: a chemical species having its oxidation state changed. *Reduction* is a gain of electrons, or a decrease in oxidation state by a molecule, atom, or ion. *Oxidation* is an increase in oxidation state via loss of electrons.

Reduction potential is the tendency of a chemical species to acquire electrons, and thereby be reduced. A *reductant* is a chemical species that donates an electron to another species. Hydrogen (H), with an oxidation state of +1, is the strongest reductant, as it freely gives its sole electron for chemical reactions.

Oxidation state, also termed *oxidation number*, refers to an element's bonding potential in a reaction when the element is in a molecule. Thus, oxidation state characterizes the charge potential of an atomic species within a compound. Oxidation number is expressed as an integer.

Carbon (C) has an oxidation state of –4. H = +1, while H_2 = 0, as H_2 is a stable monoatomic molecule comprising only hydrogen. Hence, H_2 has no oxidation potential.

Similarly, oxygen (O) = –2, while O_2 = 0. Hence, O_2 has no reduction potential.

Figuring oxidation state can be tricky. The oxidation number of oxygen in H_2O = –2, but in peroxides, such as hydrogen peroxide (H_2O_2), oxygen's oxidation state = –1.

As H_2O_2 is an electrically neutral compound, the sum of the oxidation states must = 0. Since each hydrogen atom has an oxidation number = +1, and hydrogen is more electropositive than oxygen is electronegative, to balance the oxidation number of H_2O_2 to zero, each O atom must have an oxidation state = –1.

An electrically neutral compound necessarily has a net oxidation state of zero. In figuring oxidation number, a more electronegative or electropositive element trumps a lesser one.

Fluorine (F), with an oxidation number = –1, is the most electronegative element, and so a strong oxidant. In fluorine monoxide (F_2O), the fluorine is more electronegative than oxygen, so balancing the oxidation state of this neutral compound requires that oxygen has an oxidation state = +2.

To summarize, *redox* is a change during a reaction that involves loss or gain of electrons, with *reduction* a gain and *oxidation* a loss.

৶ Electron Transport Chains ৷

Electron transfer is the elemental transaction in chemical reactions. An atom donates an electron to another atom in a process that takes a few quadrillionths (10^{-15}) of a second.

An *electron transport chain* comprises electron transfer between a series of electron donors and acceptors. Electron transport chains are formed by protein complexes embedded in a membrane that act concertedly during a sequence of redox reactions.

A complex is *reduced* by accepting electrons. Conversely, a complex is *oxidized* when it gives up electrons.

An electron transport chain works because each acceptor, the next in the chain, is more electronegative than the donor. For an electron transport chain to function – allowing electrons to pass through – an exogenous electron acceptor must be present at the end of the chain.

Some protein complexes use electron transport chains to transfer H^+ ions (protons) across a membrane. This is part of the *oxidative phosphorylation* process, which is a metabolic pathway to use energy released by the oxidation of nutrients to produce ATP.

৶ ATP ৷

The simple sugar *glucose* can be metabolized to release some energy and store the rest as ATP (adenosine triphosphate). ATP is the universal energy currency of life. All cells store surplus energy as ATP, which acts like a rechargeable battery for cellular energy.

Chemical energy transport within cells transpires by shuttling ATP. Cells use ATP not only for energy but also for

$$ATP$$

triphosphate

adenine

active region

ribose

communication within and between cells – a cellular coin of the realm.

In releasing energy to a cell, ATP turns into ADP (adenosine diphosphate). In storing energy for later use, ADP becomes ATP.

ATP is a nucleotide, comprising 3 main structures: a nitrogenous base (*adenine*); a sugar (*ribose*); and a chain of 3 phosphate groups (*triphosphate*), bound to the ribose.

The adenine ring and ribose sugar form *adenosine*, which is a purine nucleoside. In animals, adenosine acts as a neuromodulator, promoting sleep and suppressing arousal. On its own, *adenine* (A) is a nucleobase.

Nucleobases (nucleic acid bases) are nitrogen-based, ring-shaped molecules that comprise the cornerstones of *nucleotides*. Nucleobases comprise the individual units of the nucleic acids DNA and RNA.

The active part of ATP is the triphosphate. The 3 phosphorous groups are connected to each other by oxygen atoms, with side oxygens connected to the phosphorous atoms.

ATP is formed by adding a 3rd phosphate group to ADP, turning the diphosphate (ADP) into a triphosphate (ATP).

When extra energy is available – from food (in animals) or sunshine (in plants) – ADP is charged into ATP via *glycolysis*: an energy reservoir to provide a source of power when needed. Photosynthesizers power *photophosphorylation* from the energy of sunlight. Otherwise, via the internally powered redox of *cellular respiration*, cells turn ADP into ATP. Generally, ATP production has an energy efficiency of ~54%, well above that of the most efficient machines.

Energy can be gained from ATP by *dephosphorylation*: ditching at least 1 phosphate group via hydrolysis, turning ATP into ADP. The negatively charged side oxygen atoms in phosphate are uneasy in proximity and would like to escape the association; the escaping oxygen atom dragging its attached phosphate with it.

An ATP molecule would be content with just 2 phosphate groups instead of the 3 that it is saddled with. ATP (triphosphate) is just itching to become ADP (diphosphate).

$$ATP + H_2O \rightarrow ADP + P_i$$

Via *hydrolysis*, the easily triggered conversion of ATP into ADP releases energy that cells use to power themselves. Hydrolysis is the process of splitting H_2O into H^+ and OH^-; a redox reaction that creates usable energy. Conversely, when extra energy is available – from food (in animals) or sunshine (in plants) – ADP is charged into ATP via *glycolysis*, stocking a reserve to provide power when needed.

ଚ Respiration ଓ

Cellular respiration comprises the metabolic processes that convert nutrients into ATP. The many individual reactions of respiration are catabolic reactions involving redox. A cell gains useful energy via respiration.

Respiration relies upon an electron transport chain. Molecular oxygen is a highly oxidizing agent, and so is an excellent electron acceptor. Both aerobic and anaerobic respiration involve the transport of hydrogen ions (H+) or electrons to oxygen, which is then reduced.

Aerobic respiration has the advantage of oxygen as an input to the respiration process, thus affording the complete breakdown of a glucose molecule via *glycolysis*, a metabolic pathway. Hence aerobic respiration oxidizes glucose ($C_6H_{12}O_6$) to carbon dioxide (CO_2), and reduces oxygen (O_2) to water (H_2O).

$$C_6H_{12}O_6 + 6\ O_2 \rightarrow 6\ CO_2 + 6\ H_2O$$

Oxidizing 1 molecule of glucose via aerobic respiration produces 2.28 attojoules of energy.

Prokaryotes arose before atmospheric oxygen was readily available, and so had to make do with anaerobic respiration. Many anaerobic organisms are *obligate anaerobes*: they can respire only using anaerobic compounds. Oxygen is deadly to them.

Anaerobic respiration is respiration using a substitute electron acceptor rather than oxygen in the initial stages. Anaerobic glycolysis incompletely breaks down glucose, yielding ethyl alcohol (C_2H_5OH) and carbon dioxide.

$$C_6H_{12}O_6 \rightarrow 2\ C_2H_5OH + 2\ CO_2$$

Anaerobic respiration is much less energetically efficient than aerobic respiration. With smaller reduction potential, 9 times less energy is released per glucose molecule: 0.25 attojoules.

Eukaryotic tissues resort to anaerobic respiration when deprived of oxygen. Anaerobic glycolysis results in the lactate that appears in overworked muscles.

In being largely hydrocarbons, lipids are more complex in their metabolic redox than sugars. Energy from fat is not as easily released as it is from carbohydrates.

From a redox point of view, whereas plants are sun-powered, animal metabolism is driven by inhaled oxygen added to migrating food electrons.

ം Biosynthesis ൙

Biosynthesis is the cellular construction process: converting substrates into more complex products. In this context, a *substrate* is the material upon which enzymes act to create a bioproduct. Common biosynthetic products include carbohydrates, proteins, vitamins, and lipids.

¤ ✧ ¤

Biosynthesis and anabolism seem to be synonyms. Conceptually they are, but *anabolism* is most often used to refer to *metabolic pathways*: series of chemical reactions within a cell.

Metabolic Pathways

The organization of cooperating enzymes into macromolecular complexes is a central feature of cellular metabolism. A major advantage of such spatial organization is the transfer of biosynthetic intermediates between catalytic sites without diffusion into the bulk phase of the cell. ~ American biochemist Brenda Winkel

A *metabolic pathway* is a series of chemical reactions occurring within a cell, commonly with an intended biological end product, along with inevitable waste. *Metabolic rate* is the speed at which a pathway transpires.

A metabolic pathway modifies an initial molecule (*substrate*) into a product, which may be used in 1 of 3 ways: 1) immediately, as an end product; 2) intermediately: to initiate another metabolic pathway (a flux-generating step); or 3) stored by the cell for later.

Metabolic pathways are either productive (*anabolic*) or degradative (*catabolic*). Catabolism breaks matter down, anabolism builds it up. Anabolism is sometimes referred to as *constructive metabolism*; catabolism as *destructive metabolism*.

There are as many metabolic pathways as there are products, both catabolic and anabolic. Metabolic pathways are highly organized highways within cells, affording efficient channeling and construction of substrates into products. Efficacy in catabolic processing is abetted by limiting diffusion of intermediate products, thus keeping requisite chemical stocks in readily available pipelines.

Though there are several hundred pathways for any cell, only a couple dozen are critical to cell functioning. These critical pathways are identical in most forms of life – a package which has been conserved through evolutionary time.

Cell Intelligence

Living cells continuously measure, process, and store cellular and environmental information. ~ Swiss biologists Simon Ausländer & Martin Fussenegger

Cells constantly sense their own health and level of stress. Various proteins, such as filamentous actin, act as sensors. Signals are passed to the cytoskeleton and into the endoplasmic reticulum, which decide how to respond, and then let the rest of the cell know what to do.

Living cells are complex systems that are constantly making decisions in response to internal or external signals; like a table around which decision makers debate and respond collectively to information put to them. ~ French biologist Emmanuel Levy

Cells know and control the size and composition of their organelles to meet immediate needs, and to efficaciously allocate resources. There are both dedicated reporters of organelle condition and indirect functional readouts. These are used to optimize organelles during normal and stressful conditions.

Through the extracellular matrix, cells sense their external environment: both chemical and mechanical characteristics. A stem cell relies upon this information to determine what type of cell it should become.

Certain cell proteins have receptors which discern a variety of chemical attributes. Others sense external softness or stiffness based upon the mechanical stress applied to bonds that proteins put out.

It is critical for cells, whether microbial or part of a larger organism, to adaptively configure their biochemical operations to current conditions. Cells do so through a host of measures, including altering protein production and self-manipulating their DNA. All cells intelligently manage their lives at the molecular level.

If a cell is unable to make copies of its DNA, or if it overlooks mistakes in the DNA code, the result can be the production of cancerous cells or cell death. So, cells continuously monitor their DNA for damage.

Cells are flexible in managing their DNA. When the normal tools of DNA replication are damaged, cells adaptively try to work around the problem.

Cells communicate through protein pathways. Especially in the embryo, and during times of stress, quick decisions must be made. Through a variety of techniques, cell networks rapidly perform analog computations using optimal decision theory algorithms. Based upon inputs from cells, an intercellular network cooperatively decides a proper response to a situation and disseminates its decision to its member cells.

An astonishing example of cell intelligence occurs during multicellular organism development. *Morphogenesis* is the process of how a heritable body plan grows from a single-cell embryo, through various stages, into an adult. From a single cell emerges a diversity of cell types and tissues which appear in a strictly regulated sequence.

A mother stem cell can mint a daughter cell that is genetically identical yet given a different cell fate based upon immediate developmental need. This it knows by information from its external environment, combined with knowledge from its vast genomic encyclopedia within and the ability to distinguish slight chromosomal and epigenetic differences. The information processing capability of cells exceeds what the human mind can consciously comprehend.

A human embryo begins with a single cell that relentlessly divides, creating more cells. All of these are pluripotent stem cells, able to differentiate into any cell type.

7 days into embryogenesis, cells begin to specialize, crafting tissues and organs. Cues from 2 distinct vectors determine morphogenesis: one is chemical, the other geometrical.

Energetic/genetic developmental plans include geographic arrangement. Cells know what is expected of them by where they are physically situated.

Further, cells know how large they are, and the precise topography of neighboring cells, including those of different types. They carry in their minds detailed 3D maps of what is and what should be (the genetic "standard" plan). Such extensive memory is necessary because cells may act as service providers to cells of another type.

Using their knowledge, cells self-organize. This is how tissues are built. Extracellular networking and cell adhesion molecules play constitutive roles.

ଔ Healing ଓ

Wounds heal by regeneration. Regeneration occurs as cells recreate their lost brethren. Muscle cells create more muscle cells. Cartilage cells reproduce their own kind. All types of cells regenerate in the same intricate mixture as before the wound. Wound healing is only possible because cells know the schematic plan of the body in which they belong.

Until recently, cytologists wrongly thought that only pluripotent stem cells were able to facilitate wound repair. Instead, a somatic cell can differentiate into a different cell type if need be. Altering gene expression can turn one cell type into another. This process requires that cells recognize the situation and act accordingly.

In salamanders, dermal (skin) cells can regenerate either more skin or cartilage. Newts, but not salamanders, can regenerate lost eyes via somatic cells of different types.

Cell relationships are commonly symbiotic. Different types of cells actively cooperate to deal with a wound.

Scars occur because the regeneration system is less than perfect. In animals, ATP leaks from damaged cells. It is converted to adenosine, which promotes healing.

A scar forms when adenosine continues to be produced at a wound site after the injury has already healed, leading to a larger, thicker scar than what otherwise may have been.

Scarring is also an artifact of the cellular memory system, for remembrance as well as structural reinforcement. Scarring shows that cells absorb environmental influences.

Unless wound damage is severe, most scars appear cosmetic, and do not significantly affect function. To retain memory without impairing function, if possible, is a subtlety of healing.

Another reason for knowing the neighborhood is keeping invaders out. Cells sense when they are being fiddled with and respond to rid themselves of the nuisance.

Salmonella bacteria invade by injecting pathogenic proteins into a cell after breaking and entering using certain host enzymes. *Salmonella* prevent lysosomes, which are the first line of cellular defense, from getting the supply of toxic enzymes they need to do their jobs. Robbed of their ammunition, lysosomes become ineffective.

An assaulted cell recognizes the attack and sounds the alarm. This is an innate immune system response. Recognition and communication happen through various proteins working together in their various roles under cellular directive.

♎ Cancer ♎

As a defense, the human blood-brain barrier separates blood from brain extracellular fluid. Brain infections are rare because of it and the encasement of the brain in a protective membrane.

For every tumor that originates in the brain, 10 make their way from other organs. They enter in disguise.

Deception is a common ruse for pathogens. Cancer cells lie about who they are in order to spread. To enter the brain, breast cancer cells conceal themselves by impersonating neurons. Cancer cells know where they are and take the cagey steps necessary to gain illicit entry.

Via elaborate communications, cancer cells distribute tasks, share resources, differentiate, and decide how to proceed. Before sending cells to colonize tissues and organs (*metastasis*) spy cells are sent out reconnoiter the situation and report back. Only with hopeful prospects do metastatic cells leave a tumor and navigate to new posts. If they sense danger, cancer cells become dormant, then reactivate at will when conditions warrant.

Cancer cells change their environment to suit their needs, inducing genetic changes and enslaving surrounding cells. Cancers may fuse with other cells, or otherwise obtain the knowledge and equipment they need to metastasize. Having gone bad, cancer cells use wiles and determination to take down their host.

Cancer causes aging. Cancer cells drive the aging process in neighbouring non-cancer cells. ~ English cytologist Stuart Rushworth

≈ Synopsis ≈

The very existence of cells represents a notable element of standardization in Nature's design. ~ Steven Vogel

Prokaryotes & Eukaryotes

➢ The earliest living cells were *prokaryotes*. Most single-celled microbes are prokaryotes, including bacteria and archaea.

➢ *Eukaryotes* had a chimeric origin: formed from a host archaeon incorporating a bacterium. Eukaryotic evolution was an incremental step from the cooperative exchanges and intimate relationships common among prokaryotic microbes.

➢ Prokaryotes have compartmentalized functioning but lack the formality of the *organelles* which eukaryotes employ: specialized, membrane-enclosed cell structures. Though eukaryotic cells have a more elaborated structure, cellular functioning of prokaryotes and eukaryotes is similar.

➢ Cell form conveys a *speed-versus-sophistication tradeoff*. Prokaryotic cells have a greater surface-area-to-volume ratio than eukaryotes, which affords prokaryotes faster metabolism, higher growth rate, and faster generation time.

Eukaryotic Cell Types

➢ There are 3 basic *eukaryotic cell types*: germline, stem and somatic. The cells which comprise the body parts of a multicellular organism are *somatic*.

Germline cells are the special cells of sexual reproduction, producing *gametes*. In animals, the gametes are eggs and sperm cells. Plant germ cells produce ovules and pollen.

Stem cells are generic cells that differentiate into a specialized cell type, including somatic cells, or generate more stem cells. Somatic cells may sometimes act as stem cells.

Cell Division

➢ Prokaryotes replicate via *binary fission*, a form of asexual reproduction.

➢ Multicellular eukaryotic cell division varies by cell type. Stem cells and somatic cells undergo *mitosis* to create new cells. Germline cells divide during *meiosis*.

Metabolism

➢ *Metabolism* has 2 contexts: for an entire organism, and at the cellular level. For an organism, metabolism is the process of creating usable energy by digesting nutrients. Likewise, cellular metabolism – *catabolism* – is the controlled process of breaking down organic matter to harvest energy via cellular respiration.

➢ *Cellular respiration* converts nutrients into ATP. There are 2 cellular respiration techniques: *anaerobic* and *aerobic*. Early life did not have an ample supply of oxygen, and so had to employ anaerobic respiration, which is much less efficient; photosynthesis excepted. Aerobic respiration evolved in non-photosynthetic life as quickly as atmospheric oxygen became available.

➢ *ATP* is the chemical coin of metabolism, used to store and transport chemical energy within cells. ATP is also employed in communication within and between cells.

➢ *Biosynthesis* is the process of biological construction: conversion of substrates into more complex products. A *substrate* is the material upon which enzymes act to create a product. Common biosynthetic products include carbohydrates, proteins, vitamins, and lipids.

➢ A *metabolic pathway* is a series of chemical reactions occurring within a cell, typically with an intended bioproduct. *Metabolic rate* is the speed at which a pathway transpires.

➢ *Anabolism* and *catabolism* are complementary metabolic pathways: anabolism builds matter (biosynthesis), catabolism breaks it down.

➢ There are as many *metabolic pathways* as there are products, either catabolic or anabolic. A cell may employ several hundred pathways, but only a couple dozen are critical to cell functioning. The critical pathways are identical in most life forms.

Organization & Functioning

➢ From the molecular level on up, there is a natural tendency for biological systems to optimize self-assembly and responsiveness by operating near peak biological phase transition points. Such efficiency is exemplary of coherence in Nature.

Cell Intelligence

➢ Cells are *aware* of the condition of their organelles (cellular proprioception), and control organelle size and function as needed to meet cellular needs, given available resources.

➢ Cells retain *memories*, not only of themselves, but also their neighborhood.

➢ Based upon inputs from the cells they service, *intercellular networks* employ optimal algorithms to make quick decisions, which they then disseminate to their employers.

➢ When cells need to *migrate*, they discern the proper direction and proceed according to plan. Cell migration is commonly coordinated with other cells.

➢ *Healing* illustrates cell intelligence. Perhaps the most astonishing example of cell intelligence is *embryogenesis*.

➢ Ultimately, multicellular life relies upon cells *knowing* just what to do and when to do it, with intricate coordination among themselves.

❧ Genetics ❧

Genetics is about how information is stored and transmitted between generations. ~ English geneticist John Maynard Smith

Genetics is much more than an inheritance or evolution mechanism. It is instead about architectural plans for every cell and organism, and of keeping track of both ancient history and current events.[*] Genes provide both an encyclopedia of organismal knowledge and a ledger of current events, actively employed during a cell's life to survive, and to chart the future.

As the presumed porter of heredity, genes were long assumed to control cellular replication and organism reproduction on a trait-by-trait basis. This proved both an exuberant exaggeration and a gross simplification.

Science becomes dangerous only when it imagines that it has reached its goal. ~ Irish playwright George Bernard Shaw

The intertwining of genetic elements – chemically, geometrically, mechanically, contextually, and temporally – is so intricate that the paradigms and terms used in this discipline are largely obsolete, even as they remain in place.

The discontinuity between terminology and actuality makes genetics an especially difficult subject to impart. The sophistic constructs espoused within the discipline are so askew as to spell obfuscation rather than cogency. What nevertheless comes through is a miraculously sophisticated molecular information storage and processing system to adaptively sustain and propagate life. What remains unexplained, as it proceeds unseen, is the coherence guiding the genetics gyre: the intelligent energetic force behind the atomic manipulations.

¤ ✧ ¤

---

[*] As an empirical study, genetics is confined to physical artifacts. These nuclei acids and other molecules have no ipso facto import. Of significance are the energetics behind genetics (egenes), which show in organism traits and behaviors.

Recipes for biosynthetic products, such as proteins, are stored as polynucleotides and chemically coded in DNA.

> DNA is a history inside us – and yet one that is our master as well. We are indeed trapped in this particular vehicle of history as it is trapped in us. ~ Peter Ward and Joe Kirschvink

ಬಿ Nucleic Acids ಣ

> DNA was the first three-dimensional Xerox machine. ~ English economist and sociologist Kenneth Boulding

Swiss physician and biologist Friedrich Miescher isolated various phosphate-rich chemicals from the nuclei of white blood cells in 1869. He called them *nuclein* because they seemed to come from the nuclei of cells.

A few years later, nuclein were found to be slightly acidic, so they became known as *nucleic acids*. No one knew what the substance was or was for. As well as having different functions, nucleic acids are chemically different than amino acids.

Nucleic acids are biopolymers; specifically, RNA (ribonucleic acid) and DNA (deoxyribonucleic acid), both polynucleotides. A *biopolymer* is a polymer produced by a cell. A *polynucleotide* is a biopolymer of 13 or more nucleotide monomers chained together by covalent bonds.

ಬಿ Deciphering DNA

> It now seems certain that the amino acid sequence of any protein is determined by the sequence of bases in some region of a particular nucleic acid molecule. ~ Francis Crick

Continuous research of nucleic acids followed the typical path of scientific inquiry: mistaken hunches and endless experimentation to suss structure and function, until insight puts the pieces together and fills the evidentiary gaps.

By 1929, Russian American biochemist Phoebus Levene had identified the components of DNA & RNA and coined the term *nucleotide*, though his ideas about the structure of DNA were wrong.

English chemist and X-ray crystallographer Rosalind Franklin managed the first snapshots of DNA in 1951 via X-ray diffraction imagery. With her wealth of accumulated image data and analysis, Franklin significantly furthered understanding of DNA's intricate structure.

Franklin correctly surmised that DNA was a double-stranded helix, with each helix having 4 *nucleobases* (nucleic acid bases) attached to a phosphate-sugar backbone. The DNA bases are adenine (A), cytosine (C), guanine (G), and thymine (T).

These results she shared at a seminar attended by American biologist James Watson and English biologist Francis Crick, who were graduate students at Cambridge University. Franklin worked at nearby King's College.

Watson & Crick completed figuring out the 3-dimensional structure of DNA in 1953, after building innumerable physical models. Their first models were quite off: a triple helix, not double; and the bases on the outside, not inside where they belonged.

Watson & Crick got an advance publication copy of DNA work by American chemist Linus Pauling from Peter Pauling, Linus' son, who worked in the same lab as Watson & Crick. Pauling's paper wrongly had the triple helix, wrongly put the bases on the outside, and wrongly characterized the phosphate groups by not ionizing them. Pauling, then a world-renown chemist, made careless schoolboy blunders in his DNA work.

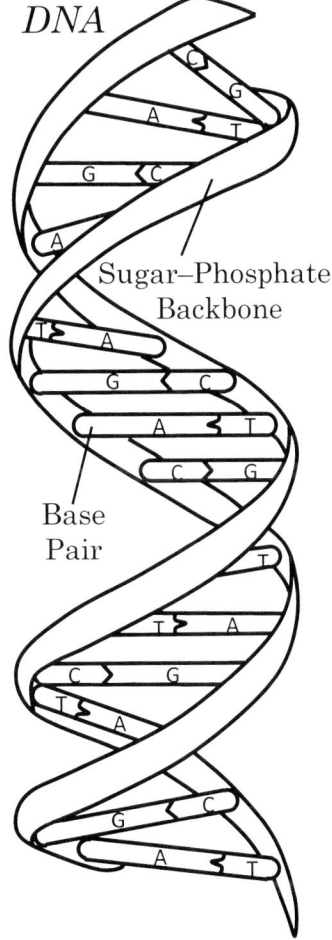

DNA

Sugar–Phosphate Backbone

Base Pair

Watson & Crick had previously made equally obtuse mistakes, including ignoring water's role in the DNA molecule. But, determined to secure their career prospects, Watson & Crick persevered post haste.

A crucial piece of the puzzle was handed to them by John Griffiths, a mathematics postgraduate student: that nucleobases A & T paired, as did C & G.

Linus Pauling had heard that years before, on a sea cruise where Griffiths and Pauling were on the same boat. But petulant Pauling, peeved at having his vacation interrupted, ignored Griffiths' input.

In the finale, Watson & Crick put the matching bases inside, and flipped the negative phosphorous ions so they wouldn't touch. The result was a tight-fitting double helix that checked out.

DNA comprises a double helix of paired molecules, joined by hydrogen bonds. The helix is like a twisting ladder, each rung a bonding of 2 complementary nucleobases.

Watson & Crick's triumph, shortly before others hot on the same trail, came solely from the chemical and physical observations done by others, without doing any original research themselves. Obsession paid off in a Nobel prize.

ಕ Chemical Composition ಚ

All of today's DNA, strung through all the cells of the Earth, is simply an extension and elaboration of the first molecule. ~ American physician Lewis Thomas

The polynucleotides of DNA and RNA form a long chain built from millions of nucleotide subunits. Nucleotides form the basic structural unit of both DNA and RNA. A *nucleotide* has a nucleobase and sugar and phosphate groups, all glued together by ester bonds. A *nucleobase* is a ring-shaped molecule with a nitrogen base.

Nucleobases pair up by chemical affinity. *Base pairs* come in a strictly limited variety of combinations, circumscribed by chemical bonding rules: A–T and C–G for DNA, and A–U and C–G for RNA.

Nucleobases

Adenine

Guanine

Purines

Cytosine

Thymine
(DNA only)

Uracil
(RNA only)

Pyrimidines

The larger nucleobases – adenine (A) and guanine (G) – belong to the *purine* chemical class. The smaller – cytosine (C) and thymine (T) and uracil (U) – are in the *pyrimidine* class. Whereas pyrimidines are simple ring molecules, purines have fused rings. Purines and pyrimidines are complementary by their sharing hydrogen bonds, which is particularly convenient for adaptable information storage because hydrogen bonds are easily broken.

The *backbone* of both DNA and RNA is provided by a *phosphodiester bond*: a group of strong covalent bonds between 2 5-carbon-ring sugars and a phosphate group, over 2 ester bonds.

Esters are ubiquitous organic compounds, formed by condensing an acid with an alcohol. Many lipids are fatty-acid esters of glycerol. Esters with low molecular weight are found in pheromones, the chemical compound used in scent-based communication.

⚘ DNA–RNA Differences ⚘

DNA and RNA differ in 3 main ways: 1) number of strands; 2) sugar composition; and 3) a single nucleobase difference. Each distinction indicates RNA as the precursor to DNA.

1st, whereas RNA is a single-stranded molecule, DNA is double-stranded. RNA has much shorter chains of nucleotides than DNA. Its functions are simpler.

2nd, RNA and DNA employ different sugars; a chemical signature distinction signified by R and D. RNA contains *ribose**, a slightly simpler monosaccharide than DNA's *deoxyribose†*, (actually, 2-deoxyribose). Deoxyribose is derived from ribose by the loss of an oxygen atom (reduction).

From an evolutionary viewpoint, deoxyribose is a chemical enhancement of ribose. RNA's ribose structure is less stable than DNA's deoxyribose because it is more prone to hydrolysis owing to the extra oxygen atom.

In this context, *hydrolysis* is a reaction that breaks a biopolymer down in the presence of water and an enzyme. Hydrolysis constantly occurs in cells as part of the basic reactions on sugars for metabolism and energy storage, both involving ATP.

3rd, RNA & DNA have 1 different nucleobase. RNA's uracil (U) is an unmethylated form of DNA's thymine (T). The other nucleobases are the same: adenine (A), cytosine (C), and guanine (G).

A *methyl group* is a hydrocarbon (CH_3) common in many organic compounds. In biological systems, *methylation* is a chemical process catalyzed by enzymes, where a methyl group substitutes for a hydrogen atom. Methylation plays a role in regulating *gene expression*: whether and how genetic information is used.

---

* ribose: $C_5H_{10}O_5$; H–(C=O)–(CHOH)$_4$–H
† deoxyribose: $C_5H_{10}O_4$; H–(C=O)–(CH$_2$)–(CHOH)$_3$–H

❦ Unlettered Nucleobases ❧

The letters that comprise DNA and RNA are not the only possible nucleobases. Other molecules make the grade in fitting into a double helix.

There is much more to the code than chemical composition. The nucleic acids are but the atomic nuclei of a larger scale system of information storage. Nature cohered to a form from a myriad of considerations, including element availability, spatial arrangement, and energy tradeoffs.

❦ Folding ❧

Chromatin structure stabilizes and compacts the genome to package it within the nucleus. This structure also serves as a dynamic regulator of gene expression, silencing or activating transcription depending on molecular signals impinging upon it. ~ American biochemist David Sweatt

DNA is always intricately folded: physically shortened by a factor of 10,000 or more to fit inside cells. In human chromosomes, the packing ratio can reach 10 million to 1. This is achieved via 10,000 nonoverlapping loops. Each cell distinctly packs its DNA to suit itself, following some ineffable schema.

The variability is truly astounding. In each being unique, chromosomes are like snowflakes. ~ American geneticist Brian Beliveau

A human cell has 1.8 meters of DNA, wound via histones to 90 micrometers (0.09 mm) of *chromatin*, which is the combined package of proteins and DNA that comprise genetic information, stored in the nucleus of a eukaryotic cell.

In packaging DNA in compact form, chromatin prevents damage as well as providing a ready means to regulate expression and DNA replication. Altering the efficiency of chromatin affects the employment of the genetic information contained within.

A chromatin has 120 μm of chromosomes after being freshly duplicated and condensed during mitosis. Thousands of proteins are involved in compacting DNA.

DNA is stored in a fractal Matryoshka pattern: nested self-similar globules of sequences, highly organized spatially, and systematically nested. The spatial architecture of chromosomes and other genetic structures is critical to their operation. This crucial aspect of genetics is barely understood.

In defiance of the topological complexity that characterizes folded DNA, identical sequences can recognize each other from a distance and even gather together. When bound in double-helix form, nucleobases are tucked away, hidden behind their nucleotide support structures of charged sugar and phosphate groups. A likely explanation for such sequence identification goes to conformity in shape and charge pattern, but this is an incomplete reckoning. The inherent intelligence involved in such recognition by these DNA strands remains a mystery.

♌ Recognizing Foreign DNA ♌

As part of the mammal innate immune system, IFI16 is a protein that recognizes foreign DNA by distinguishing it from native DNA.

DNA unfolds a bit to reveal a strand that needs to be read in order to make a bioproduct. In mammals, less than 60 base pairs are unzipped. Exposed pathogenic DNA strands are typically longer.

IFI16 inspects DNA exposures. 4 IFI16 can fit in 60 base pairs.

If longer exposed fragments are found, IFI16 congregate along the foreign DNA strands, chaining themselves together to envelop it. These filaments form a scaffold that temporarily thwarts the pathogen. IFI16 then signal for help from immune system enforcers.

DNA folding itself contains valuable information. How genetic information is packaged has a major impact on its functionality.

Mad cow disease is technically termed *bovine spongiform encephalopathy* in cattle and *Creutzfeldt–Jakob disease* (CJD) in humans. This disease is caused by a *prion*: a misfolded protein acting as an infectious agent. A prion is a template macromolecule that converts other proteins to its perversity. All known prion diseases, which are currently untreatable and universally fatal, affect the structure of brain cells or other intelligence tissue.

DNA begets RNA to extract relevant information content. This structured RNA naturally folds under the influence of several environmental factors. How RNA folds affects the processes – transcription and translation – in which it is employed.

DNA, RNA, and proteins are not the only molecules where folding is essential to proper functioning. Via folding, hemoglobin acts as molecular tongs in picking up a molecule of oxygen for transport.

✄ DNA Knots ✄

Linear DNA is formed from various isoforms – superhelical coils, knots, and catenanes – of closed circular DNA at thermodynamic equilibrium.

Organic molecular folding is understood by applying the mathematical concepts of topology, specifically knot theory. Every molecule has an *energy landscape*: a set of possible conformations, with each potential spatial configuration having an associated energy level.

Atomic interactions in molecules dictate molecular conformations that take energy to maintain. The energy level ultimately relies upon HD quantum mechanical properties.

Topologically, an energy landscape has hills and valleys of energy levels for different configurations. Applying knot theory to mathematically figure an optimal conformation for a complex macromolecule is beyond daunting because the range of possible shapes is gigantic.

The chains found in biologically significant proteins are a tiny subset of those possible. Biology cohered to optimal efficiency of all things that could be considered, given inherent tradeoffs in the dynamics of folding and unfolding.

Because folding is essential to storing and retrieving the complex coding for cellular work and reproduction, geneticists first hypothesized, wrongly, that evolution favored formations with relatively simple energy landscapes. Instead, unfolding a carbon chain requires energy, whereas a folded shape is in relative energetic repose. As DNA strands spend much more time in folded stasis, it would be wasteful to use energy to maintain the folded state.

ஐ Genes ☾

We do not know what most of our DNA does, nor how, or to what extent it governs traits. ~ English physicist and chemist Philip Ball

German zoologist Wilhelm Haacke came up with the concept of *genes* near the end of the 19th century. He imagined that hereditary traits were molecularly self-contained. A *gene* was originally idealized as a molecular unit of trait heredity.

English evolutionary biologist William Bateson coined the term *genetics* in 1905, from the Greek word *gennō*: "to give birth." Danish botanist Wilhelm Johannsen followed in 1909, using the term *genes* to define dollops of inheritance information.

These concepts were concocted in anticipation of pursuing the path trod by Gregor Mendel. Researchers set their sights on discovering what they wanted to see rather than exploring what was.

In the 1950s, as molecular biology progressed, a *gene* was partially redefined as a template for producing a protein. As proteins are the workhorse of cellular life, the fallacious assumption that proteins were associated with heredity lingered, by virtue of their terminological association as *genes*. Thus, genetics proceeded upon a dogma of woolly wishful thinking.

Simplism in genetics was part and parcel of the mechanistic mindset which has pervaded science since the 17th century, when Descartes formulated the reductionism that materialists have found so appealing.

The facile notion of genes as localized trait heredity units belies the incredibly sophisticated network of knowledge involved in a cell managing its life. Hewing to the simplistic *gene* paradigm impeded comprehension for decades, and has only loosened in the 21st century, as understanding of epigenetics has dissolved the formulaic concept of genes.

> Most of the assumptions that we operate on in molecular biology derive from the initial assumption that most genetic information is transacted by proteins. And while that's largely true in bacteria, it's not true for humans. ~ Australian molecular biologist John Mattick

Proteins are the prototypal *gene product*, but there are also innumerable DNA regions that do not code for protein, but which instead are employed to create useful RNA products. These different RNAs perform a diversity of tasks, including assisting protein synthesis and training, catalyzing biological reactions, cellular communication, and acting instrumentally in gene expression.

Genes are often described as if they are linear sequences, awaiting ready decoding as construction templates for useful products. Nothing could be further from the truth.

The DNA coding schema defies easy characterization because it defies topological comprehension. Further, workable regions vary dynamically, influenced by a variety of factors.

Hemoglobin is the iron-based oxygen transport protein in the red blood cells of vertebrates. Many different proteins go into hemoglobin, as well as other molecular constructions necessary to produce hemoglobin. The instructions for those different components lie on different chromosomes.

The example of hemoglobin highlights not only the distribution of DNA strands, but also the use of nested instructions. The formula for hemoglobin is a set of recipes for distinct components, where each component has its own template (coding sequence).

Genetic material is even more complex. The encoding represented by DNA is nothing more than the score of a symphony that is played by an orchestra, where every player contributes to the overall effect. Lacking a singular conductor, it is an interpretive exercise by a multitude. The way that template data are arrayed and move within cellular space profoundly affects their functioning.

Individual chromosomes occupy distinct territories in the cell nucleus. Where they reside, and what other chromosomes are in the neighborhood, can strongly influence whether the genetic material in a chromosome is active and how it functions.

> Operationally, the gene can be defined only as the smallest segment of the gene-string that can be shown to be consistently associated with the occurrence of a specific genetic effect. ~ American geneticist Lewis Stradler in 1954

The definition of *gene* is non-specific for good reason. A gene is conceptual, not an actual entity: a term for the information encoded within polynucleotides. Genes don't exist. They are only construals in the minds of geneticists.

20th-century biology was structured according to a linear Newtonian worldview. Molecular biologists were so set about linearity that when the gene came along, they took the gene to be the be-all and end-all of basic biology. That comes out of thinking in terms of particles and linear interactions. ~ Carl Woese

Mapping the notion of genes to reality has meant a constant revision of presumption. Definition of the term *gene* itself has been a moving target, and its meaning still varies widely. The concept itself is debased in understanding that, however the term is defined, a 'gene' is not functionally or structurally delimited. Theoretically, genetics is nothing more than sloppy sophistic philosophy.

The term *gene* is used as if the recipe and result were synonymous. They are not. Research into epigenetics has shown that genes as an adhered-to rulebook represents an inapt simplification.

However misrepresentative, the concept of genes is so ubiquitously doctrinal that it is the requisite context for introductory exposition. So we proceed.

ೞ Genetic Coding ೞ

> We may have totally misunderstood the nature of the genomic programming. ~ John Mattick

Canonically, the information encoded in a gene serves as a template for assembling a protein from the amino acid level on up. In other words, the construction of each protein is represented by a unique amino acid sequence, which is specified by the nucleotide sequence of the gene encoding the protein.

The relationship between a nucleotide sequence and the corresponding amino acid sequence represents the *genetic code*. The code is seldom simply translated, nor is it readily packaged in one place.

> The canonical genetic code is assumed to be deeply conserved across all domains of life with very few exceptions. ~ Russian geneticist Natalia Ivanova *et al*

Microbes pay no mind to the canonical dogma of geneticists. Viruses are especially prone to freely interpret genetic codes to suit themselves; a practice called *recoding*. The little parasites exploit the knowledge that their host typically follows standard scripting, thereby gaining leverage for their wily manipulations.

♂ Codons & Cistrons ♀

DNA and RNA each have 4 nucleobases. Combining base pairs comes up short in expressing enough variations to encode 20 different amino acids: $4^2=16$. But $4^3=64$. So, encoding 20 amino acids requires an arrangement capable of more combinations than base pairs alone can provide.

Hence, genetic coding is not as simple as matched base pairs on the rung-by-rung DNA ladder. Instead, amino acid templates were originally presumed as specified by nucleotide triplets, termed *codons*, which run along the length of a DNA ladder; not base pairs.

It was once thought that genes would be codon sequences regularly arranged in some discernible order, and that a single gene was a unit of heredity for a trait. Instead, as little as 10% of human DNA has a known coding function. Genetic instructions turned out to be enormously more complicated than codon sequences.

The term *codon*, while facilely descriptive of the way that genetic data is stored, became considered as insufficient. Codons are not the physical equivalent to genetic function once thought.

Genetics turned out to be infinitely more intricate than early optimism justified. Sometimes the definition of a word is clarified with further understanding, as it was with *nucleic acid*. Other times, as in the case of *codon*, where the original definition turns out to be partly inaccurate, a new word is concocted to cover the deficiency.

Though *codon* is still commonly used, presumed precision of coding sequence to gene is now termed a *cistron*. A cistron is a hypothetical localized segment of DNA with all the template information required for producing a single protein. The term *cistron* emphasizes that a gene provides for a specific trait.

The terms *gene* and *cistron* are synonymous; 2 names for a baroque complex for producing bioproducts – particularly proteins – needed by cells. These words speciously congeal what Nature adroitly conceals.

✃ Genetic Variations ✄

A poet can survive everything but a misprint. ~ Irish writer Oscar Wilde

Altering a sequence of nucleotides may change the corresponding amino acid sequence, which in turn may affect the structure or function of a protein encoded by DNA. This is the basis of genetic *mutation*.

A gene's *locus* is its position in a genophore or chromosome: the highest level of DNA packaging. Determining the locus for a certain biological trait is termed *gene mapping*. Genes occasionally move their locus.

Sometimes a gene has a single form. More often there are *alleles*: genetic variants at the same locus. If the alleles at a locus are the same, they are *homozygous*; if different, *heterozygous*.

Different alleles may result in distinct traits, but not necessarily. Alleles may have different levels of influence on genetic expression: equal, or unequal, and unequal to varying degrees.

One allele may be *dominant*, the other *recessive*. The dominance hierarchy of alleles is controlled by small RNA (sRNA) molecules that form a regulatory network.

A dominant allele may completely mask a recessive allele. A recessive trait often appears only with homozygous alleles that are recessive.

There are more than 70 known alleles at the gene locus for determining blood type, the ABO locus. This creates a plethora of blood types, some of which are compatible for transfusions between types, others not.

In humans, ~250 different forms of each gene exist. Over half the genes in an individual are unique to that person. Albeit often subtle, this spells a tremendous diversity in personalized proteins and other genetic products.

Our genomes possess an intrinsic level of instability, resulting from the misincorporation of RNA, the chemical sister of DNA. ~ English biochemist Keith Caldecott

While DNA is the universal storage medium for genetic information, cells are awash in RNA, as RNA is employed in decoding the instructions locked away in DNA. There is an operational tension between the two. RNA may mistakenly be incorporated during DNA synthesis. So, there are several quality control and repair processes that ensure DNA integrity.

The notion that the purity of DNA is an intrinsic property of its synthesis is wrong. Effort is required by cells to ensure that genetic material retains its DNA identity. ~ Keith Caldecott

≋ Gregor Mendel ≋

Gregor Mendel (1822–1884) was an Austrian monk with an inquisitive mind and a love of gardening. Between 1856 and 1863, Mendel cultivated and examined 29,000 pea plants, from

which he derived laws of genetic heredity, which he exposited in 1866. His paper was understood as being about hybridization, not heredity, and was ignored for over 3 decades, when it was rediscovered.

Mendel hypothesized hereditary units, as well as speculating about how inheritance manifested, hybridization, and expression of dominant or recessive characteristics. Mendel's heredity laws were: 1) the law of segregation, and 2) the law of independent assortment.

While genes are paired in normal cells, they are segregated in sex cells (eggs or sperm), which unite to form a gene pair. The pair express either a dominant or recessive characteristic.

A dominant gene trumps a recessive gene, so it takes 2 recessive genes for a recessive characteristic to be expressed. Whence Mendel's *law of segregation*.

Mendel's law of segregation is nominal, and subject to violation. Plants are known to sometimes employ ancestral alleles, not parental genes. This *paramutation* may occur by inheritance via double-stranded RNA, not DNA.

Mendel's 2nd law – *independent assortment* – was that the expression of any one genetic characteristic is not influenced by another. It may have seemed that way for pea plants at first blush, but inheritance is generally much more complex than that.

Genes are often linked. For example, biorhythms determined by an organism's circadian clock are the product of a gene complex. Many processes are under biorhythmic sway: from organs to tissues to cells. Even the production of ATP in mitochondria oscillates by a molecular clock.

☿ Supergenes ☿

Adaptation is commonly a multidimensional problem, with changes in multiple traits required to match a complex environment. ~ English zoologists M.J. Thompson & C.D. Jiggins

The original quest of geneticists was to comprehend the units of heredity which determined traits – hence the concept of a *gene* defining a *trait* with a one-to-one correspondence.

Nature is not so neat. A single gene may influence multiple traits (*pleiotropy*). Conversely, groups of genes may be inherited together because of close genetic linkage and being functionally related in an evolutionary sense. Such an inherited genetic group is termed a *supergene*.

Supergenes commonly express variations in traits among a population; an effect achieved by *cis-regulatory elements*, which are noncoding DNA that regulate transcription of nearby genes. Trait variety may promote population survival when environmental stresses take a toll.

Supergenes and the processes that create them have diverse and far-reaching roles in adaptation and evolution across many groups of organisms. ~ Swedish geneticist Andreas Wallberg

⚙ Expression & Regulation ⚖

Each cell expresses only a limited amount of its full genetic potential. ~ American biochemist Gordon Tomkins *et al*

The value of a gene is in its *expression*: the process of using the genetic information to synthesize a functional protein or other bioproduct.

The conventional view has been that traits manifest on a gene-by-gene basis. Instead, many genes are expressed in groups, and their expressions are affected by innumerable interactions, even contact among DNA strands.

DNA is coiled and tangled like spaghetti inside the cell. So there are many places where the DNA touches and intersects. These interactions could be crucial to how the information in the DNA is read and interpreted by the cell. ~ South African geneticist Marc Weinberg

Regulation is the control of gene expression. There are many ways in which expression may be modified or silenced.

Conventional wisdom holds that modifying a gene to make the encoded protein inactive — 'knocking out' the gene — will have more severe effects than merely reducing the gene's expression level. However, there are many cases in which the opposite occurs. In fact, the knockout of a gene sometimes has no discernible impact, whereas the reduction of expression (knockdown) of the same gene causes major defects. ~ American obstetrician Miles Wilkinson

♎ Sea Lampreys ♎

Sea lampreys are an ancient jawless fish with over a half-billion years in lineage. Their approach to gene-driven development is extremely conservative.

The genic instructions that guide embryonic development produce pluripotent stem cells, which can differentiate into any cell type. To prevent untoward problems, these potent genes are laid aside in lampreys after early development; sealed away so as not to risk their being misexpressed.

> Lampreys experience rampant programmed genome rearrangement and losses during early development. The genes are restricted to the germline compartment suggesting a deeper biological strategy to regulate the genome for highly precise, normal functioning. The strategy removes the possibility that the genes will be expressed in deleterious ways. Humans, on the other hand, must contain these genes through other epigenetic mechanisms that are not foolproof. ~ American biologist Chris Amemiya

A genetic script is not always expressed as it was coded. Gene expression is a multiple-step process, with numerous agents that act with some degree of independence in their performance, albeit under chemical guidance from past events. The protein produced by a gene may itself regulate its natal genetic expression.

Gene expression is a more complex process in eukaryotes than earlier-evolved prokaryotes; an evolutionary elaboration granting further adaptive flexibilities, and perhaps something of a compensation for the ready ability of prokaryotes to share and selectively absorb new genetic information via horizontal gene transfer.

Extracting information from DNA and employing it to produce a biological product is an elaborate process, albeit with 2 principal steps: transcription and translation.

℘ Transcription ℘

The transcription of a gene is fundamental to the parsing of the genetic code and is highly regulated by the cell. ~ American geneticist Gautham Nair & Indian American molecular biologist Arjun Raj

Transcription is the process of producing an RNA copy from a DNA sequence. An *RNA polymerase* enzyme unwinds a specific strand of DNA determined by a *promoter*: a special nucleotide sequence that provides a secure binding site for the RNA polymerase.

The RNA polymerase breaks the hydrogen bonds between the complementary nucleotides, separating the 2 strands of DNA. It then adds matching RNA nucleotides that pair with the complementary DNA bases.

An RNA sugar-phosphate backbone is formed, with assistance from the RNA polymerase. The RNA copy – a *transcription unit* – is complete. The hydrogen bonds of the untwisted RNA + DNA helices break, freeing the transcription unit.

¤ ✧ ¤

A transcription unit encodes at least 1 gene. If a transcribed gene encodes a protein, the transcription unit is termed *messenger RNA* (*mRNA*). Otherwise, the transcription unit may encode various other products: a regulatory RNA (e.g., microRNA), ribosomal RNA, a component used in protein assembly, or a ribozyme.

A central dogma of genetics has long been that the RNA transcription unit is a faithful copy of the DNA master. Geneticists came to this axiom from studying *E. coli* bacteria, a common gut microbe. They presumed that what worked for a popular prokaryote was a universal genetic truth. Instead, RNA transcription units are often subject to tampering.

Transcription is a highly regulated process, with many decisions made. Various actors in the process communicate in a regulatory network that extends across the *transcriptome*: all the RNA molecules in a cell or population of cells. Alterations emerge from networked decision-making.

Most genes in the cell are regulated by several transcription factors in a combinatorial fashion, as parts of a complex network. There is also a layer of time-based regulation. ~ Chinese chemist Long Cai

RNA misspellings are common, and they are not random. On average, 20% of RNA copies of human genes contain misspellings. The most common transposition is changing DNA–A to RNA–G. How the misspellings occur is not yet known, nor is their effect understood.

Transcription has other tricks as well. RNA may copy from the strand opposite the one that codes a protein. This too is mysterious.

Protein coding sequences are strongly conserved over evolutionary time. In contrast, changes in transcription binding often factor in speciation. Altering the regulation of transcription is a common avenue for evolutionary adaptation.

¤ ✧ ¤

RNA polymers are compacted and organized in cells to allow protein synthesis. ~ Canadian geneticist Daniel Zenklusen

In prokaryotes, the mRNA created by transcription is ready for translation. The eukaryotic path is more complex.

DNA normally never leaves a eukaryotic cell nucleus. But mRNA can. Messenger RNA carries its amino acid codes out of the nucleus and into the cytoplasm, to a nearby ribosome, which synthesizes proteins from peptide pieces.

In eukaryotes, the mRNA transcription product undergoes a series of modifications prior to translation. Part of this may be an editing process termed *RNA splicing*, with cutting and pasting segments of exons and introns.

❀ ❀ ❀

Genes in prokaryotes are continuous DNA strands. In contrast, eukaryotic genes have coding regions (*exons*) interspersed with noncoding segments (*introns*).

Introns are nucleotide segments in either DNA or RNA, some of which may be *self-splicing*: able to extract and insert themselves into gene products. Some introns encode specific proteins; others, functional RNA.

Introns exist in the genomes of bacteria and eukaryotes. Their capabilities are not well understood, but they are known to enhance gene expression. Introns in yeast cells have been found to promote resistance to starvation and promote growth.

DNA coding regions typically comprise several separated exons (coding sequences) that are joined as an RNA transcript. Exons are formed from precursor RNA segments (introns) that are removed from a gene by RNA splicing.

Simultaneous encoding of amino acid and regulatory information within exons is a major functional feature of complex genomes. The information architecture of the received genetic code is optimized for superimposition of additional information, and this intrinsic flexibility has been extensively exploited in evolution. ~ American geneticist Andrew Stergachis *et al*

When proteins are made from intron-containing genes, RNA splicing is part of the RNA processing pathway that follows transcription and precedes translation.

✂ Translation ✂

Translation is the process by which the genetic information contained in mRNA is used to determine the sequential order of amino acids in a protein. ~ English biochemist Michael Ibba & German biochemist Dieter Söll

During *translation*, messenger RNA is decoded by a ribosome to produce the intended *polypeptide*: an interim product that will later be folded into an active protein. Just prior to initiating translation is a key point for regulating gene expression – before a ribosome has committed to the energy-intensive process of synthesizing a polypeptide.

Gene expression during translation may be regulated by affecting the stability of mRNA, or by altering whether and how translation transpires. Besides edits, stifling gene expression, either altogether or to a certain degree, plays a significant role.

Proteins may be modified after translation. Some proteins require modification before becoming active.

♎ Actin ♎

> Closely related protein isoforms can exhibit functional differences. ~ Russian biochemist Anna Kashina *et al*

Actin is a family of proteins essential to cell functioning. Actin is found in all eukaryotic cells except roundworm sperm. Actin has equivalent cousins (homologs) in prokaryotes.

Distinct actin variants – *isoforms* – perform a wide variety of different roles, including maintaining cell shape, cell motility, cell division, cell signaling, and intercellular communication. These isoforms are created by mRNA selecting specific exons or through post-translation modifications.

Humans have 6 actin isoforms. 2 in particular – ß-actin and γ-actin – are nearly identical in their structure.* Yet these near-twin proteins carry out distinct roles. In mammals, ß-actin is critical to embryogenesis, whereas γ-actin plays a regulatory role in managing the proteins in a cell's cytoskeleton.

Via epigenetic activity, a single gene can produce multiple proteins. Very minor physical changes can have a profound impact on proteome diversity and the behaviors of proteins.

Conversely, different genes may encode selfsame bioproducts. Despite being nearly identical proteins, ß-actin and γ-actin are encoded by separate genes.† The epigenetic activities in producing ß-actin and γ-actin are significantly dissimilar yet yield physical self-similarity.

> The parts of genes that we think of as being silent actually encode very key functional information. ~ Anna Kashina

Actin illustrates how labyrinthine genetics is, but also shows that there is an energetic component essential to life at the molecular level. The study of genetics has been confined to physical artifacts – DNA sequences – and associated

---

* ß-actin and γ-actin exhibit only minor differences in 4 amino acids in just 1 region of these proteins. Actin altogether comprises 376 amino acids, folded into a labyrinthian arrangement that defines the functional potentialities of this protein family.

† 6 actin genes are expressed in birds and mammals.

processes upon those artifacts. Neither genetics nor biochemistry can explain how a slight physical difference in proteins affords very distinct behavioral profiles, as with ß-actin and γ-actin. More generally, these sciences have no explanation for how DNA sequences can encode the divergent behavioral paradigms which proteins exhibit.[*] Knowing about epigenetic tweaks does not demystify how protein personalities exist.

Matter transformations cannot explain coherent energy patterns. The issue becomes completely perplexing when considering how molecules such as proteins can behave intelligently through their production via genetics: how matter can inform knowledge and decision-making ability, which are clearly traits of a mind, not a molecular body.

Cephalopods provide a vivid illustration of how genetics is so much more convoluted than geneticists ever imagined, and that so little is understood.

♎ Coleoids ♎

Octopus, squid, and cuttlefish – the coleoid cephalopods – are surprisingly savvy creatures. Scientists have long wondered how soft-bodied coleoids are so much cleverer than their hard-shelled cousin: the nautilus.

One evolutionary hypothesis is that in losing their protective shell, these short-lived creatures compensated with superior acumen; a hypothesis not far removed from hominins losing physical power and gaining abstract reasoning as recompense. This is a big-picture view with molecular implications.

After transcription (transcribing DNA into RNA), coleoids extensively edit directions for making proteins, particularly those involved in making the cells of intelligence: glia and neurons. All told, ~12% of the protein-building instructions related to brain cells are selectively edited. Coleoids

---

[*] Geneticists cannot even explain how genes encode the patterns of folding which practically define the behavioral potentialities of proteins.

also edit RNA related to other tissues, but not nearly as extensively.

> It introduces immense complexity and diversity. ~ Israeli geneticist Eli Eisenberg

Coleoid RNA editing has evolutionary significance. Limiting DNA alterations in favor of RNA editing has slowed evolution in coleoids. 10–26% fewer DNA mutations are found in RNA-edited genes than others.

While gene manipulation in these marine mavens seems correlated with intelligence, there is insufficient evidence to infer causality. A mystery lingers.

✄ Quality Control ✄

> The cell places a high priority on ensuring that translation produces proteins that accurately reflect the corresponding genetic information. To this end, quality control can be seen at every step in translation where errors might accumulate. ~ Michael Ibba & Dieter Söll

In complex organisms, hundreds of thousands of different proteins are constantly being produced to replace degraded ones. A lot can go wrong in producing proteins, and regularly does. Preventing putting defective proteins on the job can be critical to health.

Protein production quality control is termed *nonsense-mediated mRNA decay* (NMD). As suggested by its name, NMD focuses on recognizing defective messenger RNA, and efficiently degrading them so that pathetic proteins are not produced.

> Messenger RNAs exist in many different configurations in cells, including a stable closed-loop conformation. ~ Indian geneticist Srivathsan Adivarahan

For quality control, mRNA carry a specific protein, termed up-frameshift1 (UPF1). UPF1 is normally removed from the messenger RNA (mRNA) by the ribosome that processes the protein formula carried by the mRNA. But if a ribosome finds the mRNA suspicious, it lets UPF1 stick, thus tagging the mRNA as defective. The ribosome then recruits enzymes to break the bad mRNA down.

Quality control is also applied to ribosomes fresh off the assembly line in the cell nucleolus, before they are exported to the cytoplasm for production work. To ensure that a ribosome has been successfully assembled, a protein border guard does not let the ribosome pass until an enzyme acting as export inspector gives the go-ahead.

⚹ Prokaryotic Adaptive Immunity ⚹

In 1987, Japanese molecular biologist Yoshizumi Ishino noticed an oddity in an *E. coli* gene he was studying. It had short, repeating sequences of nucleotides, with 2 repeaters having unique sequences (now known as *spacers*) between them.

It took 2 decades for geneticists to figure out what Ishino's discovery meant. In 2007, researchers showed that that the genic repeaters and spacers served as part of an adaptive immune system, herein called *pais* (an acronym for *prokaryotic adaptive immune system*).[*] Microbes evolved innumerable such immune systems which work in slightly various ways.

> Overall, prokaryotes appear to have evolved a nucleic acid-based "immunity" system. ~ French American geneticist Rodolphe Barrangou *et al* in 2007

Prokaryotes have ever been plagued by viruses. To remember the experience (if they live through it), they preserve the remnants of encountered viral villains within a DNA profile (a spacer bookended by 2 repeater caps).

> Prokaryotes can store information in specific loci in their DNA to remember encounters with invaders (such as bacteriophages – viruses that infect bacteria). ~ Israeli microbiologists Rea Globus & Udi Qimron

Spacers are read by specific enzymes that then cut out any exogenous matching DNA they find, which left untouched would spell an infection.

---

[*] The *prokaryotic adaptive immune system*, encapsulated as *pais,* has hitherto been awkwardly known as CRISPR/Cas. The gene editing tool called CRISPR/Cas9 is covered at the end of the chapter.

Pais is powerful, but not all microbes have them. 90% of archaea have a pais, but only 35% of bacteria do.

Pais is useful when microbes encounter enough variety of viruses to make adaptive memory worthwhile. But if there is too much viral variety, or viruses are rapidly adapting, pais won't help, because a microbe might never encounter the same virus again.

All known microbes that live in super-hot environments have pais, as the environment is a fairly stable ecosystem, with a middling viral diversity, which means pais might help.

> No immunity comes without a cost. ~ Israeli microbial ge-
> neticist Rotem Sorek

Pais has downsides. Microbes may accidentally make spacers from bits of their own DNA, creating an auto-destruct sequence. This rarely happens, as there are built-in preventative mechanisms against it.*

Viruses can fight back against pais, morphing into unrecognizable forms. Alternately, they may develop counter weaponry.

The bacterium *Pseudomonas aeruginosa*, which resides in soil and water, and can cause dangerous infections in macrobes, has a vigorous pais. Some viruses are not in the least fazed by it. That's because those viruses have wily proteins that gum up *P. aeruginosa's* pais.

Viral anti-pais measures are so common that it leaves geneticists wondering how many pais systems are truly active. There is a tremendous diversity in how vigorously microbes employ their pais as an immune response.

Some *E. coli* carry a pais that they leave switched off. Why bother? Microbes decide what cellular baggage they keep. They could pitch their pais if it made no difference – thus it must, even if seemingly inactive.

There are many mysteries about pais. For one, spacers should reflect the individual story of the viruses a microbe

---

* Incoming viral DNA is linear, facilitating its recognition as foreign. A microbe's genophore is protected because of its circular form. But should a sequence break off and become linear for too long, such as during a stalled replication process, there is a risk of the DNA being taken as alien and encapsulated as a spacer.

has encountered. Some do, but most seem generic, and the contents of many remain a conundrum.

> Is it the case that there is a huge, unknown amount of viral dark matter in the world? ~ Eugene Koonin

One bacteriophage (a virus that infects bacteria) carries its own pais with it, using it to fight the bacterial defense system that the virus encounters upon infection. The viral pais smartly chops up the segment of bacterial DNA that normally inhibits phage infection.

Beyond the problematic fight against viruses, it's not always smart for a prokaryote to keep out foreign DNA, which may contain the makings of a useful trait.

Microbes that lack pais are not helpless. Far from it. As much as 10% of the genome of a pais-poor prokaryote may be dedicated to other hawkish defense systems.

Plus, a prokaryote may acquire a pais as conditions warrant. Prokaryotes are prodigious acquirers of environmentally available gene packages, through horizontal gene transfer (HGT). As a form of community altruism, bacteria commonly produce and exude helpful HGT packages for others, as well as picking up on such actionable intel when seeking a solution to their own problems.

Pais may serve as more than just an immune system. Spacers sometimes act to silence genetic expression. By selectively silencing genes, a bacterium may stop making molecules on its surface that are readily detected by a macrobe that the bacterium is intent on infecting. Without a pais in place, the bacterium would blow its cover and be killed.

> This is a fairly versatile system that can be used for different things. ~ Russian geneticist Konstantin Severinov

❦ Genomes ❧

> The genome is a highly sensitive organ of the cell that monitors its activities and corrects common errors, senses unusual and unexpected events, and responds to them, often by restructuring. ~ American cytogeneticist Barbara McClintock

A *genome* is (the idea of) the total complement of genes in a cell or organism. If a gene is a recipe, a genome is a recipe book.

> Different cell types express different portions of their genome.
> ~ Gordon Tomkins *et al* in 1969

It was long supposed that all cells in an organism had the same genome, as the above quote suggests. But that is not so. Multicellular organisms comprise a population of cells, each with its own personal genome (*pergenome*). Even cells of the same type have their own pergenome.

Prokaryotes have a flexible genome that can change during a single life cycle. This can happen because prokaryotes can readily pick up new genetic material.

Chromosomal mosaicism – genetic variation among cells – can occur by a variety of means, including errors during chromosome segregation or DNA replication, copying variations, gene rearrangement, single-nucleotide variation, or other instabilities.

Such mutations can occur at any stage of development: in stem cells, differentiating cells, and in somatic cells (which are nominally terminally differentiated). The genetic makeup of a multicellular organism is multifarious.

Over evolutionary time, all organisms selectively incorporate alien genetic material. Human DNA includes gene packages from at least 8 retroviruses. Some of these viral genes are essential to human reproduction.

¤ ✧ ¤

A genome comes in no particular order. While genome structure is surmised as significant, it is more likely to have been preserved simply by inertia.

> Intuitively, you wouldn't believe that just by chance things would be conserved for 500 million years. ~ French molecular biologist Daniel Chourrout

The number of genes in an organism is a meaningless statistic, especially in comparing organisms in the same kingdom. Some prokaryotes have thousands of genome copies (polyploidy) .

For multicellular eukaryotes, only a fraction of a genome is actively employed. Most of a genome is kept as a historical reference: a database of possibilities for the future from the experiences of the past. This legacy information is accessed as needed.

Plants commonly experiment genetically. They may duplicate their genome, with the original serving as a reference, and the copy as a testbed. For instance, 70 million years ago, the tomato triplicated its genome: keeping a preserved master copy and generating 2 spare copies to adaptively mutate. One result was the birth of the potato, a tuber-producing evolutionary offspring.

> Replication is like a mirror that reflects the evolutionary history of living beings: the first genes to be replicated are the oldest, while those that replicate later on are the youngest. ~ Spanish biologist Alfonso Valencia

In replicating a genome, the most valued, conserved genes are copied first. Newer genes, in evolutionarily active regions, are copied afterwards.

> The regions that replicate late also have a compact and inaccessible structure; they are hidden zones in the genome that act as evolutionary laboratories, where these genes can acquire new functions without affecting essential processes in the organism. ~ Spanish biologist David de Juan

☙ Genophores & Chromosomes ❧

Cell central holds the principal genetic material, but not all of a cell's genome. For a prokaryote, cell central is the *nucleoid*, with the primary genetic package in a *genophore*. For a eukaryote, cell central is the *nucleus*, containing *chromosomes*.

☙ Genophores ❧

A prokaryote's nucleoid holds a cell's genome in a single *genophore*: a large double-stranded DNA molecule, generally circular in shape. A nucleoid is not enclosed in a membrane. Having only a single copy of each gene makes a prokaryote haploid.

A typical prokaryote has 2,000 to 4,000 genes; a housefly, mouse, or human: ~20,000. An ocean bacterium, *Pelagibacter ubique*, has the most efficient genome known: 1,354 genes; no clutter, no noncoding sequences, no duplicate entries, no viral genes, nor any introns.

220 human genes come courtesy of horizontal gene transfer from a prokaryotic pal: bacteria. This was a direct transfer, not a product of ancestral lineage.

> A fundamental concept in biology is that heritable material, DNA, is passed from parent to offspring, a process called vertical gene transfer. An alternative mechanism of gene acquisition is through horizontal gene transfer (HGT), which involves movement of genetic material between different species. HGT is well known in single-celled organisms such as bacteria. HGT has contributed to the evolution of many, perhaps all, animals and that the process is ongoing. ~ English biochemist Alastair Crisp *et al*

⚕ Genetic Exchange ⚕

Hereditary genetic transmission – from one cell generation to the next – is *vertical gene transfer*. In contrast, gene sharing among cells or organisms is *horizontal gene transfer*.

Prokaryotes, particularly bacteria, pick up genetic material in 3 main ways. 1st, they may scavenge gene-bearing snippets from dead cells in the vicinity. 2nd, viruses inject genes into infected cells. 3rd, bacteria often voluntarily release genes for others, in packets called *plasmids*.

A plasmid is a tightly-folded ball of DNA which can replicate independently of the cellular DNA. A plasmid is considerably smaller than a nucleoid.

Prokaryotes are prolific plasmid exchangers. Horizontal gene transfer is rampant among microbes. As a social service, to preserve a community under attack, bacteria practice HGT to provide antibiotic resistance to others.

The ecology of animals creates a network of constant gene exchange for the microbiome within. The microbiome is a major determinant of an animal's health. Genetic exchange is a critical component of that.

HGT is also employed by eukaryotes. HGT is not as common among animals as it is plants, which frequently exchange genes, even entire genomes.

Plants that grow in close vicinity may incorporate a neighbor's genome via chloroplast capture: obtaining the genome of another plant by uptake of an organelle. Parasitic plants steal the genes of their host to better understand and adapt to their roost.

Viruses have the simplest genophores. Their RNA or DNA lack structural protein templates. Viruses get away with this sketchy setup by hijacking other cells for their replication.

Viruses are fervent traders, acquirers, and manipulators of genic matter. They know what they need, and how to get and use genes to attain their objectives.

๛ Chromosomes ๏

> Every complete set of chromosomes contains the full code. The chromosome structures are instrumental in bringing about the development they foreshadow. They are architect's plan and builder's craft in one. ~ Erwin Schrödinger

In contrast to prokaryotes, a eukaryotic cell has a membrane-enclosed *nucleus*, with proteins – *histones* – that fold and pack DNA into highly organized chromosomes.

A *chromosome* is an elaborately coiled and knotted package of genetic material in a eukaryotic cell, comprising DNA genes, regulatory elements such as histones, and other nucleotide sequences.

Chromosomes are not simply compacted gene chains, evenly spaced. A chromosome is a complex structure, with distinct spatial features that play important roles in replication, transcription, and regulation of gene expression.

> Chromosome folding follows a pattern, which is important for ensuring their proper function. ~ American geneticist Elphege Nora

Within a chromosome are regions densely packed with working DNA, while other areas are genetic deserts. The genome is organized in a fractal fashion: a self-similar Matryoshka of nested genetic information, efficiently organized by usage.

Nuclear chromosomes are packaged by proteins into organic origami. This compact packaging lets long DNA molecules fit into a cell nucleus.

While the sizes of chromosomes vary considerably by species, all are elongated cylinders. This self-organized superstructure is ubiquitous because it is the most efficient shape to access the layered information within.

Each chromosome has 2 arms: one short and one long. Chromosomes are tucked into cell nuclei; hence the origin of the term *nucleic acid*.

A DNA molecule may be linear or circular, extending from 100,000 to 10 billion nucleotides in a long chain. Stretched out, a single DNA molecule may be several centimeters long, but it would take a stack 50,000 deep to be as thick as a human hair.

A *nucleosome* is the basic nuclear DNA package in eukaryotes: a DNA segment wound around a core of 8 histones, like a thread wrapped around a spool.

Nucleosomes are folded in a successive hierarchical series of structures, eventuating in a chromosome. This both highly compacts DNA, and, by virtue of histones, creates a layer of regulatory control, which ensures proper gene expression, at least in healthy cells.

Histones do not merely act as a central hub and spooler for DNA. They also act as an antibacterial agent. Histones hang out on lipid droplets: ubiquitous cellular fat-storage organelles. If a bacterium is detected, a histone troop heads out to kill it.

As a higher level of organization, chromosomes are spun into *yarns*. A chromosomal yarn is a related group of genes and the regulatory elements necessary for gene activity.

> The DNA of individual genes is wrapped around nucleosomes to form a 'beads-on-a-string' structure. These beads-on-a-string subsequently fold up to form 'yarns-on-a-string,' where each yarn is a group of genes. This domainal organization of chromosomes is a fundamental organizing principle of genomes. ~ American geneticist Job Dekker

Via yarns, chromosomes are spatially folded so that genes are functionally organized into isolated domains. The folding pattern brings together genes and their regulatory elements into a spatial cluster, so that the activity of the genes in a yarn is easily coordinated, without interference from other genes.

The three-dimensional organization of chromosomes allows distal genomic elements to be brought together and to functionally interact with each other. At certain points during development it is thus possible to precisely orchestrate the activity of genes that are far away from each other on the linear chromosome thread, but that are actually in contact physically, within a chromosome yarn. The downside of this type of organization is that a single mutation altering the organization of such a 'chromosome yarn' can affect a whole group of genes. ~ Elphege Nora

⚘ Life Cycle ⚘

The contents of chromatin and its chromosomes undergo structural changes during the cell cycle.

Interphase is the 90% of a cell's life cycle when it lives its everyday life: eating, producing bioproducts, doing cellular business, and growing. Interphase also includes preparation for cell division.

During interphase, nucleosomes with active genes are more loosely packaged than inactive genes. In preparing to divide, during *prophase*, chromatin packages tighten up. The nucleolus disappears. DNA has already been replicated prior to prophase.

Cell Cycle

Interphase

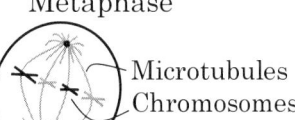

Metaphase

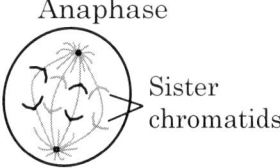

Anaphase

Chromatin compaction is a dynamic process, full of decisions which can foreclose access to the genetic codes within, thus thwarting transcription. This suppression happens via epigenetic regulation.

Metaphase is the stage of cell division where chromosomes migrate to opposite poles of a cell. During *anaphase,* 2 identical daughter chromosomes form.

Next comes *telophase*, which starts with 2 daughter nuclei forming. *Cytokinesis* follows, with the cytoplasm bifurcated. The outcome of telophase is 2 daughter cells, each with a selfsame set of chromosomes.

◊ Ploidy ◊

Ploidy is the number of chromosome copies a eukaryote has. Ploidy varies. With 2 sets of chromosomes, humans are diploid, as are nearly all mammals.

Plants switch between haploid and diploid states during their life cycle, as do algae, fungi, and slime molds. This is termed *alternation of generations.*

While female eusocial insects, such as bees, wasps, and ants, are diploid, males are haploid because they develop from unfertilized, haploid egg cells. Hence these eusocial insects are referred to as *haplodiploid.*

◊ Telomeres ◊

Telomeres ensure genome stability. ~ American microbiologist Janelle Vultaggio

A *telomere* is a region of repetitive nucleotide sequences at each end of a *chromatid* (a freshly copied chromosome).

Telomeres protect the end of the chromatid from deterioration, or from fusion with neighboring chromosomes. Telomeres have been likened to the plastic caps at the end of shoelaces, as they keep the ends from fraying or becoming entangled.[*]

Each time a human cell divides, its telomeres shorten a bit. A telomere is refurbished by its telomerase enzyme, but cell division takes a toll.

---

[*] The end of a telomere tidily loops back into the main body of the telomere.

A cell reaches decrepitude and no longer divides when its telomeres become too short. Such cellular senescence is the natural aging process.

As telomeres shorten during normal aging, they activate a DNA damage response to arrest cell growth, which protects DNA from harm. The pathway controlling growth arrest, however, is commonly altered in cancer cells, allowing malignant cells to divide despite shortened telomeres. ~ Austrian cytologist Jan Karlseder

Telomere dysfunction triggers autophagy. Activation of autophagy is critical for cell death. Loss of autophagy function is required for the initiation of cancer. ~ Jan Karlseder *et al*

ꝸ G-Quadruplexes ꝷ

Guanine-rich DNA sequences of a particular form have the ability to fold into 4-stranded structures called G-quadruplexes. ~ English chemist Julian Huppert & Indian chemist Shankar Balasubramanian

DNA is typically a double-stranded helix, coiling into densely packed chromosomes. But a strand may double up, particularly at telomeres, which are rich in guanine.

Via hydrogen bonds, G-rich strands naturally self-associate into G-quadruplexes. These squarish 4-strand DNA structures act in various cellular pathways, including gene expression, DNA replication, and telomere maintenance. As an aberration, G-quadruplexes are instrumental in cancer.

ꝷ Comparative Chromosomes ꝸ

The number of chromosomes varies widely between species. Humans have 46.

Most organisms carry 10 to 50 chromosomes. A salamander has 20 times more DNA than a human. A mosquito has 6 chromosomes, but a silkworm has 56. A mouse has 40, a duck 60, a goldfish 94, and a toucan 106. One species of fern has 630 chromosome pairs per cell.

Humans are most closely related to chimpanzees and bonobos, which have 48 chromosomes: 2 more than people. But 1 human chromosome has the information stored in 2 chimp chromosomes.

Genetic drift of humans from chimps and bonobos began 5 MYA. Yet the DNA sequences differ by less than 1%. Chimp blood can substitute for human in transfusions.

Chimps and bonobos diverged after humans left the lineage. Chimp-bonobo genetic drift started 2 MYA.

Humans lack 510 DNA sequences that chimps, macaques, and mice share. Most of those sequences are thought to be genetically unimportant, if not entirely vacant of genes.

Though genetic differences between humans and chimpanzees may be statistically slight, they are phenotypically significant. One lost sequence allows expansion of certain brain regions in humans during development. Another controls production of sensory facial whiskers and penile spines, which humans lack.

Penile spines help males ejaculate quickly during intercourse. Quick impregnation increases the immediate prospect for reproduction. Lacking penile spines results in longer copulation times, affording emotional bonding between mating partners; something quite instrumental in human evolutionary success.

✥ Genic Content ✥

Of the 6 billion base pairs in every human cell, only 120 million code for proteins. Over 98% of the human genome is *noncoding DNA*: genes that do not encode protein sequences. Humans are not alone in this. Most codons in higher eukaryotes are deemed noncoding.

The "extra" genes in eukaryotes exist as introns or repetitive sequences. The enzymes that duplicate DNA sometime slip extra copies of a gene into a chromosome. These genetic replicas, which often have a slight variation, comprise ~5% of the human genome.

Selfsame regions are repeated hundreds or even thousands of times. This is a genetic legacy of evolution, beginning with the combination of genomes from single-celled prokaryotes that joined together in a eukaryotic endeavor.

The massive expansion of genetic code in later-evolved organisms likely came from invasive elements. Although this proliferation may represent something of a burden for coordinated gene expression programs, it also affords genomic plasticity and a data-oriented path for evolution, as well as some degree of stochastic gene regulation.

Noncoding DNA often plays some role in biochemical functions, such as during transcription, in promoting and regulating conversion of DNA into RNA. Near-duplicate genes in the human genome may have been responsible for brain enlargement in early hominids.

Eukaryotic microorganisms have fewer introns. 70% of the genes in the yeast *Sacchromyces cerevisiae* encode protein.

The coding genes of many prokaryotes exceed 90%. Sequences may be repeated in prokaryotes, but usually only a few copies.

There is a 300-fold difference between the genome sizes of yeast and mammals, but only a modest 4- to 5-fold increase in gene number.

The ratio of coding to noncoding and repetitive sequences is somewhat indicative of the complexity of the genome. Unicellular fungi have sparse noncoding DNA compared to any multicellular organism.

♎ Bladderwort ♎

Some evolved species have no truck with noncoding DNA. The carnivorous bladderwort plant is one.

An unusual and highly specialized plant, the bladderwort lives in fresh water and wet soil, and is endemic to every continent except Antarctica.

The bladderwort's vegetative organs are not clearly distinguished into roots, stems, and leaves, as in most other angiosperms (flowering plants). But its bladder trap is one of the most sophisticated structures in the plant kingdom.

While many later-evolved species are biased toward archiving noncoding DNA, the bladderwort keeps its genome trim. The bladderwort has 28,500 coding genes: comparable to its relatives, the grape and the tomato. Whereas a grape has 590 million DNA base pairs, and a tomato has 780 million, one bladderwort species (*Utricularia gibba*) carries only

80 million. This is especially surprising considering that the bladderwort underwent 3 complete genome doublings since it split from the tomato lineage.

Replication Accuracy

In multicellular organisms, somatic cell replication is essential to replacing worn-out cells with fresh copies. While plants can grow themselves past genetic defects to a limited extent, animals must have good working replacements for proper functioning.

Nevertheless, from one cell generation to the next genomes are transmitted with many mistakes: somatically acquired deletions, duplications, and other mutations. This is even true of nerve cells in the human brain, which continue to function properly despite large numbers of genetic errors.

Error-prone replication is selective. Cells replicate the transcriptionally active portions of their genomes with care, then rush through the silent sections. Cells are often careless about replicating the unused parts of their genomes.

The brain may be particularly well-suited to coping with scattered genomic errors at the cellular level. During development, an overabundance of neurons are connectively networked. Then, during maturation, those cells that don't sufficiently contribute are eliminated – a process analogous to plants pruning leaves which don't photosynthetically pony up.

❁ ❁ ❁

It may even be that genomic eccentricities have a benefit yet unknown. Knowledge of genetics is in its infancy.

♂ Sex Chromosomes ♀

The test for whether or not you can hold a job should not be the arrangement of your chromosomes. ~ American politician Bella Abzug

One animal chromosome is of especial significance: X. In his investigation of chromosomes in 1890–1891, baffled German cytologist Hermann Henking came upon a stand-out. This outlier he named X. No one knows what he meant by it;

perhaps merely a failure of imagination. Anyway, the designation stuck.

In many animal species, including mammals, X is one of 2 chromosomes determining sex. The other is Y, named as the next letter in the alphabet.

A female mammal has 2 X chromosomes; a male carries XY. If an egg gets Xs from both parents, a female is in the offing. A Y from dad means a male will be had.

The difference between X and Y is enormous. X is the largest, most gene-rich chromosome: more than 153 million base pairs, with around 10% (2,000) of the 20,000–25,000 human genes. By contrast Y is puny: 1/3rd the size of X, with but 78 genes, and many repetitive sequences.

At one point in evolutionary history, X and Y were equal in size and gene count. 300 million years wrought a mismatch in evolution of sexual specialization, with Y slacking off and X picking up the slack.

XY does more than determine sex. The combination is a risk in of itself.

Many genes vital to brain development reside in X. The X chromosome is also instrumental in human sperm production. This is an evolutionary advance in the past 80 million years, when mice and men diverged from a common ancestor.

Y lacks the complement of many X genes. Because of that, any recessive mutation on the maternally derived X becomes dominant in males, making males genetically the weaker sex. And a flaw in a male's single X spells trouble.

Having 2 copies of a gene is handy. If one is defective, the other becomes the production template.

In females, one of the X chromosomes is largely deactivated, in a process termed *X inactivation*.

❧ Alternation of Generations ☙

Alternation of generations (AoG) refers to alternate asexual and sexual reproductive modes during a multicellular organism's life cycle. For algae, plants, fungi, and slime molds, AoG also involves different genetic forms at different stages of life: alternately haploid (1 set of chromosomes) and diploid (2 sets of chromosomes).

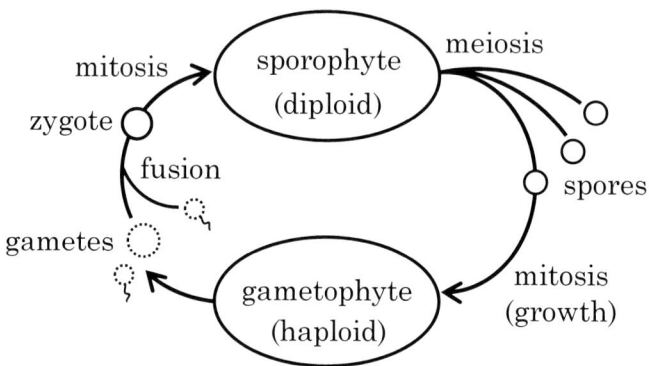

Alternation of Generations

One generation is predominant during life. The other form is typically part of early development.

ഹ Animals ର

Many invertebrates alternate between sexual and asexual generations; notably protozoa, jellyfish, and flatworms.

Alternation of generations for animals refers solely to reproductive mode: sexual (heterogamic) or asexual (parthenogenic) reproduction. Parthenogenesis may happen in nematodes, parasitic wasps, some bees, some scorpions, and a few vertebrates, including fish, amphibians, and reptiles.

In some species, switching between heterogamy and parthenogenesis may be triggered by the season (aphids, some gall wasps), by a lack of males, or by conditions that favor rapid population growth (e.g., rotifers and water fleas).

Asexual reproduction occurs either in summer (aphids), or as long as conditions are favorable. Asexual reproduction allows a successful genotype to spread quickly without the fuss of sex or wasting resources on males who can't reproduce.

Animals are always diploid. As such, animal AoG has a limited context when contrasted against the genomic gyrations of plants.

ꍰ Plants ꍰ

All land plants have a genetically complex life cycle that alternates between sporophytes (spore-producing organisms) and gametophytes (gamete-producing organisms). Whereas sporophytes are diploid, gametophytes are haploid.

The earliest plants – liverworts, mosses, and hornworts – have the gametophyte generation predominant in their life cycle. At the onset of life, a minute and nutritionally dependent sporophyte grows upon the body of a gametophyte.

In ferns, a (diploid) sporophyte produces (haploid) spores via meiosis (cell division which cuts the number of chromosomes in half). A spore grows into a gametophyte by mitosis (still haploid, but bigger).

The fern gametophyte is a *prothallus*: a small, flat, free-living organism bearing reproductive organs and feeding itself via photosynthesis. The gametophyte produces gametes via mitosis, often fabricating both sperm and eggs on the same prothallus.

A motile flagellate sperm fertilizes an egg that remains attached to the prothallus. The fertilized egg is a diploid zygote that grows (via mitosis) into a fern plant. Thus, the sporophytic phase predominates the fern's life cycle.

Gymnosperms (e.g., conifers) and angiosperms (flowering plants) generally follow the fern plan. The major difference for seed plants (from ferns) is that the gametophyte is even more reduced: sometimes only 3 cells, and the gameotpyhte is entirely dependent upon the sporophyte for all its nutrition. There are a diversity of variations to the overall scheme, including the sizes of spores and gametes and the means by which sperm and egg meet in fertilization.

ꍰ Fungi, Slime Molds, Algae ꍰ

Land plants descended from algae, many of which exhibit alternation of generations (AoG). Fungi and slime molds also practice AoG.

Fungal mycelia are typically haploid. Each mycelium has a sexual orientation. When mycelia that may mate meet, they produce a pair of multinucleate cells which fuse to form diploid nuclei, in a process termed *karyogamy*.

Karyogamy produces a diploid zygote. This short-lived sporophyte soon undergoes meiosis, forming haploid spores. The spores germinate, developing into new mycelia.

Slime mold AoG is similar to fungal mode. Haploid spores germinate into *myxamoebae*: swarm cells that find each other and create a colony.

Swarm cells fuse to form a diploid zygote, which develops into a *plasmodium*. A plasmodium matures, whereupon producing 1 or more fruiting bodies (depending upon species) which shed haploid spores.

✲ Benefits ✲

Genetic alternation of generations is a hedge against adverse environmental conditions, as well as a nod to the benefit of genetic diversity via sexual reproduction. Spores can outlast inhospitable habitats, waiting until their time has come. AoG practitioners can gauge growth prospects and time their *ontogeny* (course of development) accordingly.

✲ Genetic Complexity ✲

Especially for plants, growth and tissue differentiation in each generation is governed by different genetic programs, which are initiated either by fertilization (haploid to diploid) or meiosis (diploid to haploid).

Having 2 distinct ontologies demands a rich genome: far beyond the needs of animals, with a single ontogeny that is, however mind-boggling, relatively simple. Yet development is triggered by a few key genes capable of jiggering regulatory networks.

To take full advantage of AoG, its practitioners rely upon environmental cues, from which genetic connection is made via existential epigenetics. One might otherwise call it "life experience," but spores are not exactly living it up.

❧ Epigenetics ☙

Genes are genetic recipes. Edits that alter the recipe are epigenetic.

The term *epigenetics* is a portmanteau of *genetics* and *epigenesis*; coined by English geneticist Conrad Waddington in 1942 before details of genetics were known. Waddington meant epigenetics as a notion of how genes might interact with their environment to alter a *phenotype*: the visible traits of a cell or organism. While Waddington's gist was generally correct, the definition of epigenetics has become more gene specific.

That gene expression may be suppressed by DNA methylation was discovered in 1975. It was not until the 1990s that the term *epigenetics* became common in research circles.

¤ ✧ ¤

Epigenetic mechanisms provide the key to understanding the size and organization of eukaryotic genomes. ~ American biochemist Nina Fedoroff

Epigenetics involve factors outside DNA sequencing by which life experiences are encoded into cells, and which may be passed on to cell offspring when a cell divides. Epigenetics provides for cellular memory that lives on in descendant cells.

Single-celled paramecia have 2 distinct mating types, called even (E) and odd (O). Sex between E and O occurs by *conjugation*: reversible cell fusion, during which partners exchange genes before separating into progeny cells.

Although offspring all start with identical, mixed genomes, each cell retains the memory of the mating type of its parent. It does so via epigenetics. Small RNA molecules communicate mating type between parent and progeny cells.

Epigenetics refers to regulating, modifying, or suppressing gene expression without altering DNA sequence, which instead would be a genetic mutation. Epigenetic effects also include changes to the chromatin proteins associated with DNA, whereby expression may be silenced or engendered.

> Chromatin marks are highly specific and localized. They are induced by the signals cells receive during embryological development or in response to changed environmental conditions. Once induced, the information about cellular activities that is carried in a chromatin mark can often he transmitted in the cell lineage long after the inducing stimulus has disappeared. The chromatin-marking systems are therefore part of a cell's physiological response system, but they are also part of its heredity system. ~ Israeli geneticist Eva Jablonka & English evolutionary biologist Marion Lamb

Epigenetics also encompasses modulating epigenetic effects. As such, epigenetics involves an interleaved context-dependent network that tempers gene expression. All epigenetic mechanisms are interrelated.

Genetically identical cells living in the same environment can display markedly different traits. Both extracellular triggers and internal influences can drive a cell to change its lifestyle or fate.

> Genes do not automatically stay "on" or "off" once activated or repressed. Rather, those states of gene expression require the continual activities of the specific regulators to maintain that state of expression. ~ American molecular biologist Mark Ptashne

Epigenetic marks are an ongoing dynamic. Lifestyle and mental well-being have an epigenetic effect which is inherited by offspring.

> We inherit more than just genes from our parents. Acquired environmental adaptations are passed to our offspring. ~ Italian geneticist Nicola Iovino

Memories are mental products, but they have physical correlates. Memories are epigenetically encoded via de novo protein synthesis. This holds for all organisms. These gene expression programs – memories – are readily transmitted to offspring.

> Epigenetic processes mediate long-term memory formation.
> ~ German molecular biologist André Fischer

Stress of any sort makes a deleterious mark that long out-lasts its source. Conversely, meditation and a calm mind pro-mote healthier genetic expression and boost the immune system.

Biologically based survival lessons are inherited. For in-stance, smells that signal danger are epigenetically encoded and transmitted to future generations.

Epigenetic changes are crucial for the development and dif-ferentiation of the various cell types in an organism, as well as for normal cellular processes. ~ English cytologist Alex Ec-cleston *et al*

Epigenetic activity guides organism development. This provides adaptive flexibility. The onset of mammalian pu-berty is epigenetically triggered. Epigenetic changes con-tinue throughout an individual's life.

Families tend to have similar epigenetic patterns. Hand-edness and sexual orientation are both familial epigenetic phenomena. Homosexual men have a different epigenome than heterosexuals.

Sexual orientation is implanted via epigenetic changes related to testosterone exposure during fetal development. These epigenetic marks are passed from one generation to the next.

Epigenetics explores how genetically identical entities, whether cells or whole organisms, display different characteris-tics, and how these are inherited. ~ French geneticist Jonathan Weitzman

¤ ✧ ¤

The cells themselves must be influenced ultimately by that mysterious force which we call *life*. ~ American physician Duncan Bulkley

Epigenetics is an integral part of the cell life cycle. All somatic cells retain the full complement of DNA present in germline cells and stem cells.

A stem cell generates a different cell type by invoking ep-igenetic alterations. The specialized functions of somatic cells are programmed epigenetically. That a somatic cell may differentiate into a different cell type in exigent circumstance is an epigenetic exercise of intelligence.

> Various cell types respond differently to the environment by using distinct circuits of genomic reprogramming. ~ Italian biochemist Paolo Sassone-Corsi

Cells regularly adapt to changes in their environment by regulating gene expression, thus modulating cellular behaviors. Cell memory and intelligence are expressed by epigenetic responses.

Organisms have an array of strategies to recognize and restrain invasive foreign DNA, such as those introduced by viruses. One way is by remembering previous gene expression and tracking expression changes. This epigenetic memory silences foreign genes from one generation to the next.

> Novel protein-coding genes can arise either through re-organization of preexisting genes or *de novo*. ~ French geneticist Anne-Ruxandra Carvunis *et al*

Occasionally, new genes are expressed. This activation is passed on as an epigenetic memory. Hence, a eukaryote may adopt a foreign gene, by a decision process not yet understood.

Plants and animals both produce many thousands of RNA molecules that do not code for proteins. Instead, these molecules may communicate the present state, and memories, thereby selectively silencing or promoting gene expression.

These RNA epigenetic messengers may travel from cell to cell, stifling or activating genes as they go. Hence, an epigenetic response to a stimulus may be carried far and wide from its point of origin.

For good or ill, epigenetic effects store an organism's life experiences in chromatin. Behavior patterns, depravations, addictions, and illnesses, both physical and mental, are encoded epigenetically. Autism stems from epigenetic inheritance.

An organism lives an integrated experience. Hence, epigenetic marks are not isolated occurrences. Epigenetic changes may be different for different cell types, even as they emanate from the same experience.

Brain and nerve cells are affected by epigenetics. But the brain is not the origin of many behaviors, including those commonly considered as conscious choices. Impulses of all sorts are epigenetically encoded cellular imperatives from regions distant from the brain.

For better or worse, epigenetics provides the mechanism for lessons learned, neglected, or ignored. Being a creature of habit is embedded in cells.[*]

The external world, inclusive of food, toxins, carcinogens, and many other day-to-day factors, has a significant impact on cellular regulation. ~ American geneticist and cytologist Amber Willbanks *et al*

¤ ✧ ¤

Genes influence the behaviors of a cell and traits of an organism through the proteins and other molecules constructed from them. Getting from nucleotide code to bioproduct is a convoluted process where much can come out differently than DNA dictates.

Transcription comes first: making an RNA copy from DNA. Most epigenetic regulation happens by inhibiting transcription in very specific ways.

During *translation*, the RNA template is used to create a polypeptide, which is then folded up into an active protein during *post-translational processing*. These processes can render a functional protein quite different than what would be expected from reading the DNA recipe.

Hence, the genetic code alone provides only a partial picture of inheritance. Due to epigenetic influences, actions during protein synthesis and genomic activity during cell differentiation are as much effect as cause.

Germline cells erase epigenetic memory at critical points during development, thereby resetting the epigenome. But

---

[*] As the mind-body is an entangled energetic gyre, trying to suss cause-and-effect between mind and body is like trying to untangle a gnarled enigma with imaginary tweezers.

the erasure is commonly incomplete. Certain epigenetic indications escape reprogramming and are transmitted to offspring. By this, parental life experiences are passed to progeny.

> Genes have adapted to allow for the correct balance of memory versus flexibility. ~ Indian epigeneticist Sandip De & American epigeneticist Judith Kassis

¤ ✧ ¤

Epigenetics is the story of a cell's life experiences, of markings that are passed from one generation of cell to the next. Epigenetically, the offspring of organisms are simply an extension of cellular memory.

Most traits are the product of multiple genes, or even epigenetic tweaks to gene complexes. Few traits are traceable to a single gene. Even then, epigenetics factors in.

ɞ Genetic Evolution ଓ

A long-standing presumption about evolution is that increased complexity in organisms owed to a greater number of genes that encode information about development and growth. In other words, evolution was partly an outcome of quantitative genetic growth. Evidence has shown otherwise.

> Most animals have a similar number of genes encoded in their DNA. ~ geneticist Miloš Tanurdžić

The number of genes in an organism is merely a historical artifact, not a telltale indicator of evolutionary 'progress', whatever that may be intended to mean. Instead, complexity and diversity were largely achieved via various epigenetic mechanisms.

> Gene regulation is responsible for the evolution of diversity. ~ Miloš Tanurdžić

A related misconception has been that genetics itself evolved – particularly, a greater sophistication in gene complexes or in regulatory mechanics. Not at all. Primordial sponges that lived 800 million years ago had all the genetic savvy that humans have today.

> Gene regulatory complexity was fundamental for the evolution of multicellularity and diverse forms and functions. ~ Miloš Tanurdžić

> The regulatory landscape used by complex bilaterians was already in place at the dawn of animal multicellularity. ~ evolutionary and molecular biologist Federico Gaiti *et al*

The upshot is that genes themselves say little about an organism. It is their employment – involving epigenetics – that matter.

Further, whatever evolution genetics itself underwent has, so far, eluded detection. As with the origin of life, the inception of genetics is lost in the mists of time.

℘ Junk Debunked ℘

> So much "junk" DNA in our genome. ~ Japanese American geneticist Susumu Ohno in 1972

The academic discipline known as genetics arose from an interest in comprehending heredity. Hence the presumption that the molecular knowledge base of life was merely a currency for inheriting traits. That genetics is a fountain of ongoing cellular activity is a recent idea.

At the beginning of the 20th century Wilhelm Johannsen created the terms *phenotype* and *genotype* to correlate traits to genes on a one-to-one correspondence. Research proceeded on that basis, to expanding dismay. Bewildered by the apparent disorder of DNA in humans, geneticists declared much of it useless, as it did not fit their preconceptions.

> We were using the idea of "junk" in the genome in the sixties in Cambridge. ~ South African biologist Sydney Brenner

This simpleminded misattribution continued as dogma for decades. Relentless exploration of DNA begat more questions than answers. As complications piled up, the tidy notion of genes unraveled.

Beginning in the 1st decade of the 21st century, DNA sequences previously considered *junk* – those that do not code for proteins – were found to regulate access and employment of genetic coding.

> What was dismissed as junk because it was not understood
> may well turn out to hold the secrets to human complexity and
> a guide to the programming of complex systems in general.
> ~ John Mattick

Only 1.5% of the human genome consists of protein-coding genes. The rest is now called *intergenic*: DNA sequences between genes. All told, 8.2% of the human genome is presently presumed functionally employed.

Intergenic DNA and RNA play critical roles in regulating genetic expression; the development of cells and organisms, and their intelligence system; enhancing biomolecular performance; and ensuring biological propriety and health. Among other diseases, autism correlates with anomalies in intergenic DNA.

Among other employments, intergenic DNA is read to produce small certain RNA molecules which inhibit protein production by shutting down a ribosome's protein assembly line. This technique is used when cells are stressed and need to instigate quick responses. RNA molecules can be manufactured much faster than proteins.

Genes are the coding of heritable traits in only the vaguest sense. Such vagary does not serve scientific inquiry.

It makes little sense to have distinctions without meaningful difference: genetic versus intergenic versus epigenetic. All are essential facets of using the artifactual knowledge base by which cells manage their affairs.

Genetics terminology is an instance of historical continuity obscuring understanding, as umbrella terms have been reinterpreted through time to mean different things to different people.

Meanwhile, new terms are piled on to an obsolete paradigm. This jungle of jargon is ill-serving.

The commonly bandied term *noncoding* is even vaguer than *gene*, as it encompasses a far more amorphous realm of greater magnitude than protein templates. Further, the notion is inconsonant, as so-called noncoding DNA/RNA functionally codes for regulation rather than production.

In many instances, functionality was simply overlooked. If researchers discovered a sequence that did not correspond to preconception, it was labeled "noncoding" and dismissed.

Such disregard was blithe: a product of ignorant assumption of how DNA works. A 2013 survey of protein production uncovered 193 proteins produced by supposed noncoding sequences.

> The fact that proteins came from DNA sequences predicted to be noncoding means that we don't fully understand how cells read DNA, because clearly those sequences do code for proteins. ~ Indian molecular biologist Akhilesh Pandey

In essence, *noncoding* is merely a replacement for nucleotides previously dismissed as *junk*; a belated turn of muck into brass, albeit with equally vacuous terminology.

In conventional parlance, *long noncoding RNA* (lncRNA) differs from *small noncoding RNA* by the arbitrary distinction of being an identifiable sequence greater than 200 nucleotides. The same goes for *microRNA* (22 nucleotides), and *circular RNA* (circRNA), a recently discovered enigma. Naming solely by how something looks under a microscope is hardly helpful, especially when looks bely a plethora of functions among cast members.

> The vast majority of trait-associated DNA variations occur in regions of the genome that were once labeled as "junk DNA" because they do not code for proteins. We now know that these regions harbor genetic elements that control where, when, and to what extent specific genes are expressed to make functional RNA and protein products. Therefore, most trait-associated DNA variants are thought to alter not the gene itself, but rather, the regulatory elements that control the process of gene expression. ~ American geneticist Terrence Furey & Indian geneticist Praveen Sethupathy

When cells divide, chromosomes are distributed to daughter cells. Centromeres – specialized chromosome regions – ensure that the chromosomes correctly segregate.

The human DNA for centromeres transcribes to a long noncoding RNA. Instead of producing a protein itself, the RNA product recruits and binds 2 proteins so that a centromere functions properly.

Research into erstwhile "junk" is proving to be an exploration of an incredible web of molecular knowledge with staggering intricacy – the font of life's unicity.

The Encyclopedia of DNA Elements (ENCODE) project has systematically mapped regions of transcription, transcription factor association, chromatin structure and histone modification. These data enabled us to assign biochemical functions for 80% of the genome, in particular outside of the well-studied protein-coding regions. Many discovered candidate regulatory elements are physically associated with one another and with expressed genes, providing new insights into the mechanisms of gene regulation. ~ The ENCODE Project Consortium

℘ Epialleles ☙

Epialleles are heritable, nongenetic (epigenetic) differences in DNA methylation. ~ American geneticist Laura Zahn

Inheritance of most characteristics in complex organisms involves a confluence of factors, part genetic and part environmental: nature and nurture. While alleles provide a genetic basis for trait variation, *epialleles* are analogous epigenetic factors that afford divergence from straightforward gene expression.

Epialleles are alleles but include the characteristic marks that may affect expression of an allele. But even epialleles leave the inheritance picture incomplete, as many, if not most, epigenetic effects are from chemical attachments outside of the gene being epigenetically regulated.

☙ Mechanisms ❧

Processes in free-living cells are modulated to fit the environmental conditions. The possible programs a given cell can execute are defined by the cell's genome, and the optimal program is selected based on the level of environmental signals sensed by the cell. Regulation of gene expression is a main component in the selection of the optimal program. Gene regulatory networks are based on simple building blocks such as promoters, transcription factors and their binding sites on DNA. Simple elements of transcription regulation form a highly flexible toolbox that can generate diverse functions. ~ Hungarian geneticist Alexander Hunziker *et al*

There are an epic number of epigenetic options: transcription codon choice, gene silencing, histone and chromatin remodeling, RNA regulation, X chromosome inactivation, paramutation, bookmarking, imprinting, reprogramming, translocation and transvection, among others. Their application to alter gene expression is interrelated, forming a tensor network of relations and effects.

This brief survey introduces a few of the mechanisms which illustrate the intelligent, intricate, and delicate nature of cell life which affect an organism in ways both subtle and profound.

✂ Methylation ✂

DNA methylation is one of a number of epigenetic mechanisms that can determine which proteins are made in different cell types without changing the underlying DNA sequence. ~ English geneticist Hannah Long *et al*

Epigenetically regulated regions have characteristic markings from specific chemical attachments. One such marking is *methylation*, which is a frequently employed method for epigenetic inheritance. A methyl group (CH_3) attaches to either cytosine or adenine, stifling the associated gene so that the it will not properly express in a cell or its offspring.

DNA methylation is not an on-off switch; instead, strands may be methylated to varying degrees.

Methylation may serve as an agent of evolution. ~ American epigeneticist Gregory Hannon

Methylation can drive changes in DNA sequence. In such a scenario, methylation acts as prototyping for genic reprogramming. If the methylated change is productive, DNA may be recoded to produce the new desired output.

Methylation patterns vary depending upon cell type and function. Methylation marks are important in determining how a cell develops and how it behaves when mature.

DNA methylation is essential for the survival of the embryo, and its occurrence is dynamically regulated during development. ~ French geneticists Sylvain Guiber & Michael Weber

Mammalian cells are wiped clean of their methylation marks and reprogrammed in 2 instances: for germline and embryonic stem cells. The reason for this erasure and reinscription remains mysterious – perhaps to anticipatorily clarify signal quality. Germline cells go on to serve a reproductive role. Embryonic stem cells are the basic cell type for differentiation into distinct duties during development.

These 2 cell types represent the output of the 2 reprogramming waves, and as such are the basis or 'ground state' for all that will follow, over the life of the individual cell and the organism itself. ~ French geneticist Antoine Molaro

♎ Breeding Urges ♎

Many animals breed seasonally. Hormonal changes invoke and extinguish breeding urges.

Breeding timing is tied to *melatonin*, a nocturnally produced hormone that accounts for day length, and so serves as a biological clock. Melatonin changes alter the level of methylation of a regulatory DNA region in the hypothalamus, a brain region active during periodic biological imperatives, including hunger and sleep, as well as mating and parenting behaviors.

For mammals, the more methylated a DNA strand is, the less active its expression. Methylation of the sequence which encodes glucocorticoid receptors is exemplary.

Glucocorticoid receptors relay signals from stress hormones in the blood into the portions of the brain that control behavior. Methylation of the glucocorticoid receptor code affects anxiety level and handling of stress.

Twice in the lives of mammal cells, methylation marks are wiped clean and then reprogrammed. The purpose of this erasure and reinscription is not known.

DNA methylation patterns alter as a person ages. These changes can contribute to age-related diseases, including cancer.

Methylated regions can cumulatively lose methyl groups, turning strands back on that increase the risk of infection and diabetes. Osteoarthritis results from reduced methylation of a destructive enzyme. Demethylation in an aging brain causes cognitive decline.

Methylation can be reversed by demethylation at any age. Plants appear particularly adept at employing methylation and demethylation to suit their expression needs.

Methylation is only one facet of a complex epigenetic regulatory network which includes many mechanisms.

> There is a huge amount of flexibility in what can be done to reach different endpoints from the same DNA blueprint. Hormones play a key role in shaping the genetic basis of traits.
> ~ American evolutionary biologist Robert Cox

♎ Honeybee Memories ♎

> The development of different bees from the same DNA in the larvae is one of the clearest examples of epigenetics in action.
> ~ English biochemist Mark Dickman

Honeybees in a colony are caste-bound. The queen prolifically produces offspring sisters. A few are fated to follow as queens, while a vast multitude of workers collect and store food, tend to the young, and maintain the hive.

Royal jelly is fed to larvae via secretions by worker bees. How much royal jelly a larva receives determines her future role in life. Royal jelly consumption affects hormone signaling, which alters gene expression patterns via methylation and histone modification.

A worker bee finds an ample food supply that is communicated to fellow foragers. She must quickly learn that route, then later just as easily forget it to better retain new routes. Honeybees have a methylation system that is instrumental in storing and erasing memories.

So too mammals. Neural DNA methylation promotes associative learning.

Other insect species, including ants, have active methylation systems; but not fruit flies, which have been by far the genetics study subject of choice. Hence, researchers long overlooked the significance of epigenetics.

♎ Seeds & Embryos ♎

Procreation in higher mammals begins with a *zygote*; the initial cell formed from the fertilizing union of 2 gamete cells during sexual reproduction. A zygote is the source of both the embryo and the placenta. The placenta nourishes a developing embryo but is not part of the growing offspring.

Seeds are similar. An endosperm nourishes the embryo.

Flowering plants and placental mammals are the only known organisms that employ *genetic imprinting*: identifying regulating genetic expression by the sex of the parent.

When horses and donkeys mate, they produce sterile mules. This owes to imprinting bias toward paternal genes.

In plants, the paternally derived gene copy carries the methylation, but this copy is not always the one that is inactivated. Hence, in plant imprinting, methylation tells how a gene was inherited, not whether the gene should be expressed.

♎ Methylation Variation ♎

Methylation technique and effect varies. Both plants and higher mammals silence the genes of repetitive elements, albeit using different techniques.

Duplicate plant genes are individually methylated. This is more precise than in mammals, where repetitive elements are silenced by methylating chromatin, the gene packaging. In contrast, honeybees do not methylate repetitive elements.

Methylating the body of a gene which encodes amino acids variably regulates expression. In contrast, methylating the promoter site stifles gene expression.

Plant and insect genes are often methylated in a way that modulates, but does not deter, expression. Regulating but not silencing expression is used for long-term mammal memory, particularly during brain development. Otherwise, methylation that stifles gene expression is common in mammals.

Methylation illustrates how evolution reuses the same materials and mechanisms in different ways. With coherent intelligence applied in creating combinatorial regulatory networks, organic chemistry has provided an incredibly flexible toolkit for variety and adaptability.

The key players orchestrating DNA methylation all work together in an elegant way. ~ American epigeneticist Scott Rothbart

✄ Histone Alteration ✄

Histone modifications are important markers of function and chromatin state, yet the DNA sequence elements that direct them to specific genomic locations are poorly understood. ~ American geneticist Graham McVicker *et al*

Histones are the proteins that package DNA in an orderly way into nucleosomes. Histones tightly bind to inactive DNA sequences. The binding is loosened where sequences are actively engaged in protein synthesis.

The strength of binding between a histone and DNA may be epigenetically impacted. Histones may be biochemically altered by removing an acetyl group (deacetylation), or via methylation, which blocks gene expression by preventing transcription.

Impacting a histone may have a knock-on effect, as histone behaviors are often coordinated. Histones of related DNA sequences communicate. How a histone complex carries on a conversation depends on how its constituents are chemically modified.

Like methylation, a histone epigenetic effect is passed on to descendant cells. Methylation and histone modification are often harmonized.

Histone regulation plays a major role in controlling organism development.

♎ Rite of Spring ♎

Plants commonly must endure a prolonged cold spell to provoke flowering. *Vernalization* is the term for an angiosperm requiring a cold winter to flower in spring.

Vernalization is a form of state memory, via epigenetics, particularly histone alteration. The flow of epigenetic activity that ultimately regulates flowering varies depending upon the season and the developmental stage that a plant is in.

A seed has no need to know of flowering. The epigenetics associated with vernalization have been reset.

As a plant grows, memories of its development are preserved from one cell generation to the next. Those memories are stored epigenetically.

Having grown strong during summer into fall in its 1st year, and then survived the winter, a plant is prepared to bring forth the next generation. The epigenetic marks in its cells tell it so.

✂ RNA Regulation ✂

RNA has become widely suspected as the culprit behind almost every case of epigenetic regulation. There continues to be a shift in how we conceptualize this remarkably versatile macromolecule, once regarded primarily as mere intermediary of the "central dogma" stating that information moves unidirectionally from DNA to RNA to protein. ~ Chinese American geneticist Jeannie Lee

DNA is used to make RNA. RNA is used to make proteins. Proteins are the principal actors of biological functions.

That is the classical script for RNA. But many types of RNA have other functions besides protein coding. Those functions involve RNA communiqués that alter protein production or gene expression. There is an ever-growing compendium of regulatory agents involved. 2 worth noting are RNAi and microRNA.

RNA interference (RNAi) affects which genes are active, and how active genes are. RNAi limits expression, sometimes to the point of silencing a gene, by cleaving its target: messenger RNA (mRNA).

✂ microRNA ✂

microRNA (miRNA) are a diverse class of short noncoding regulatory RNA molecules that inhibit expression by binding to *microRNA response elements* (MREs), thereby decreasing the stability of messenger RNA (mRNA) or limiting the efficacy of protein translation.

microRNAs work in middle management: regulating protein manufacture. They help a cell maintain balance by not making unnecessary proteins and help prevent build-up of potentially harmful proteins.

An evolutionarily ancient avenue of genetic regulation, microRNA pathways are well conserved in eukaryotes.

The repertoires of plant and animal microRNAs evolved independently, with different ways of working. Animal microRNAs are specific in the binding, while plant microRNAs may bind at both coding and noncoding regions.

microRNA offers combinatorial regulation. A microRNA may have different mRNA targets, and any given site subject to regulation may be targeted by multiple microRNAs.

Modest alterations by microRNAs can have a butterfly effect, including changing the appearance of an organism.

microRNA activity is essential to learning. Associative memories which impart survival skills can be epigenetically passed on to offspring. microRNA is the likely physical mechanism for inheriting primal memories.[*]

microRNA plays a role in numerous diseases, including cancer. Some protect against cancer, while others promote it.

Other RNAs may compete in binding to microRNAs. By this, MREs mediate relevant communication, allowing different types of RNA to converse and build regulatory networks which act epigenetically.

ℬ X Inactivation ℬ

X inactivation (aka *lyonization*) silences gene expression for 1 of the 2 X chromosomes that mammalian females possess.

There are many cell divisions prior to X inactivation, but it happens early in embryonic development. The timing of X inactivation affects development.

---

[*] A physical molecule cannot encapsulate a meaningful memory, but its HD energy gyre can. Many mental attributes, such as memory and stress, have physical counterparts. Here is an instance.

For many animal species, after some fetal development, X inactivation varies cell by cell. Some work the mom X, while others the dad X. This confers genetic variety. The timing and specifics of X inactivation is one contributor to identical female twins not being entirely identical.

Cells retain their lineage whether X-inactivated or not. In maternal tissue sustaining a fetus, paternal X chromosomes are inactivated, as with marsupials.

Calico and tortoiseshell cats, which are always female, show their X linkage by their patchwork coats: the light, dark, and orange areas detail the pseudo-random X inactivation of hair cell lineages.

Not all double-X genes are inactivated: 15–25% escape inactivation. These are *housekeeping genes*: sequences for the basic cellular processes required by all cells.

X inactivation initiates at the *X Inactivation Center* (XIC), a specific spot on the X chromosome. Several actors play roles on the XIC stage. The X inactivation script has an intricate plot.

A leading X-inactivation actor in placental mammals is *Xist*: a X-inactive-specific transcript. Xist is not a protein production template. There is no Xist protein. Instead, Xist is an odd messenger RNA: while processed like other mRNAs, Xist stays untranslated.

> Xist evolved from a protein-coding gene. The loss of protein-coding function of the proto-Xist coincides with the four flanking protein genes becoming pseudogenes. This event occurred after the divergence between eutherians and marsupials, which suggests that mechanisms of dosage compensation evolved independently in both lineages. ~ French geneticist Laurent Duret *et al*

¤ ✧ ¤

> Xist is not sufficient. There have to be other factors, on the X chromosome itself, that activate Xist and then cooperate with Xist RNA to silence the X chromosome. ~ American geneticist Sundeep Kalantry

Xist acts to lyonize the X chromosome to which it is attached, in a multiple-stage process of smothering. Xist identifies its target regions by recognizing their folded shape. As Xist copies are made, they plaster the target chromosome.

Then Xist RNA attracts histones and methylating factors. Finally, the Xistified X chromosome is crunched into a compact blob: a *Barr body*, named after its discoverer, Canadian physician Murray Barr.

Duplication of the Xist gene on another chromosome inactivates that chromosome.

Sometimes *translocation* occurs: chromosomal bits get dislodged, such as a genetic bit of X with Xist on it; an abnormal rearrangement happens, and another gene gets inactivated, at least partially.

Translocations happen in around 1 in 500 human newborns, and also occur with genes other than Xist-covered bits. Some translocations are harmless. But unfortunate translocations can cause Down syndrome, infertility, or cancer.

♎ Marsupial X Inactivation ♎

Marsupials are 334 extant species of mammal that carry their young in a pouch. Well-known marsupials include kangaroos, koalas, and opossums.

Marsupial X inactivation is a more ancient evolutionary state than the more modern mammal lineage – eutherians (placental mammals) – that diverged from marsupials.

Marsupials work the maternal X. The paternal X is largely inactive.

Marsupials lack Xist. Because of that, marsupials lack the benefit of genetic variety from the selective X inactivation that transpires in placental mammals.

Xist is a most significant difference between marsupials and placentals.

♋ Protein Post-Production ♋

Proteins interact with each other, work together, and perform individual steps in chain reactions, sometimes collaboratively. Affecting protein production and/or interactions can have profound effects on cellular activity.

As macromolecules, proteins can be influenced at multiple interaction sites. A protein's active (enzymatic) site is a small fraction of its surface. This means that most of a protein's surface is available for binding to other proteins, and for changing the shape or activity of a protein. Binding at a place other than an active site is *allostery*.

℘ Allostery ঌ

Allosteric site binding among proteins results in numerous interactions, not only between those proteins, but also with others not physically connected. These interactions may affect protein functioning and may be affected by allosteric regulation.

> Allostery is the process by which biological macromolecules (mostly proteins) transmit the effect of binding at one site to another, often distal, functional site, allowing for regulation of activity. ~ American molecular biophysicist Hesam Motlagh *et al*

Allostery functions as a dynamic interrelated network, creating ensemble behavior in affected proteins. Allostery effectively entangles proteins into a regulated web, behaving like the protein equivalent of quantum nonlocality.

The biomechanics of allostery are only partly understood; but allostery works via conformational changes in the proteins involved, and thermodynamics within the effective domain of the allosteric network.

Drugs often function via allostery. Further, allostery provides for adaptive evolution outside any changes in genetic code.

From archaic bacteria to humans, practically all cells can tweak proteins by changing their chemical properties after production. This capability provides ready, adaptive flexibility, enabling cells to react quickly to changing conditions or needs.

2 post-translational modifications – phosphorylation and lysine acetylation – are intracellular communication signals that can have epigenetic effect. *Lysine acetylation* affects histone employment with a downstream epigenetic effect. Phosphorylation influences expression of protein-building genes by adding a phosphate group. This can alter proteins involved in building other proteins.

ೋ Epigenome ಌ

The epigenome (the constellation of all epigenetic modifications in the nucleus) constitutes a primary interface between environmental factors and the genome. ~ American molecular biologist James Shapiro

The somatic cells in a multicellular organism collaborate in building a body and keeping it going. Most are specialized, each with slightly different genomes.

During the continual replication that occurs throughout the human body during a lifetime, the 200 different kinds of cells are reproduced by reading different scripts written in DNA. There is a 2nd layer of instructions embedded in the special proteins that package the DNA of a genome. This 2nd layer – the *epigenome* – controls access to genes; allowing each cell type to activate (express) its own special genes, while blocking access to much of the rest of the genome, because the particular cell type does not need that knowledge.

A genome comprises regulatory genes whose protein products – *transcription factors* – control the activity of other genes. There is also a subset of master regulatory genes that control the lower-level regulators. Gene regulation is hierarchically networked and interrelated.

¤ ✧ ¤

Transcription factors have to commute to work. Finding the work site is nontrivial. These proteins amble about chromosomes, attaching to specific DNA sequences along the way, until they hit their target. Other proteins bound to the chromosome act like roadblocks, slowing the search. Once the binding site is found, a transcription factor slides over it several times, checking the target out before binding to it.

The master transcription factors act on each other's genes in a way that sets up a circuitry. The output of this circuitry shapes the initial cascade of epigenomes spun from a fertilized egg.

Though epigenetics forms an interactive entangled network, the organization of epigenomes emanate from information inherent in the genome. Systemic genetic entanglement is a fact of life.

The field long has been focused on identifying genes that manufacture proteins. The epigenome is just as important. ~ Chinese geneticist Ting Wang

♎ Wood Work ♎

In trees, transcription factors take control of a cascade of genes that controls the production of wood by differentiating cells into the needed components in the proper proportions. The primary controller protein for wood production regulates gene expression on multiple levels, ensuring proper growth.

What is unusual about the controller in poplars and rockcress is its residence in the cytoplasm, outside the cell nucleus. This is odd, as transcription factor proteins are otherwise always in the nucleus. In this instance, a nucleus-based transcription factor comes and ushers the cytoplasmic controller protein into the nucleus to begin wood work.

Multicellular organisms have many epigenomes. Epigenomes define not only which genes are accessible in each type of cell, but also control the timing of when accessible genes may be expressed.

DNA code is only a recipe. The epigenetic proteins that selectively turn a recipe into a gene product are as essential as a cook is to making dinner.

While the genome is fairly stable, the epigenome is constantly altered by life experiences.

The epigenome is dynamically regulated over the lifetime of a person, perhaps in response to environmental signals, life experiences, and as part of the normal aging process. ~ David Sweatt

֍ Lifestyle Epigenetics ֎

Gene expression is modulated by lifestyle and environmental factors. ~ Mexican toxicologist Jorge Alejandro Alegría-Torres

A man contributes to its offspring's genetic inheritance, but a fetus develops within a woman's womb, which strongly influences fetal gene regulation and expression.

A fetus is not the only one genetically affected. Fetal DNA can persist in its mother for the rest of her life. That DNA may benefit a mother's health or stir adversity.

All mammal mothers undergo a range of hormonal change prior to and after birthing. Stimulating the hypothalamus, oxytocin promotes affection. In healthy animals, these and other changes combine to engender maternal behaviors.

Most furred mammal mothers, including rodents and dogs, lick their pups. Pups mothered by generous lickers fare better under stress than those stingily succored. Neglected rat pups who don't get loving licks become neglectful mothers.

The psychological effects of parenting can be profound and lifelong. The quality of parenting, especially mothering, creates a perpetuating generational cycle. This has been repeatedly observed in rodents and primates, including people.

A foster pup, going from a poor licker of a mother to a good one, develops a better stress response; one more like that of its foster upbringing than its biological mother.

While epigenetic effects tend to persist, the permanence of methylation is in flux during early development. The earlier methylation goes unabated, the more pronounced and pervasive its impact. Epigenetic alterations account for differences in stress response in identical twins.

Social interactions of every sort affect gene regulation, as they are a form of stress. From fish to humans, competitive interactions influence testosterone levels with consequential impact on gene activity.

Altered genetic expression from stress is passed to the next generation. Chronically stressed pregnant women bear children with greater proclivity to physical, psychological, and behavioral disorders owing to greater sensitivity to stress.

Fathers as well as mothers pass the effects of their diet, temperament, and lifestyle to their offspring.

Epigenetically inherited stress increases the risk of depression, obesity, and autoimmune diseases. Dampened glucocorticoid-receptor-gene activity renders people more aggressive and impulsive. This makes men particularly inclined to abusiveness that perpetuates through generations.

It makes no difference the source of stress – physical or psychological – for parent or offspring; such distinction is clinical anyway. Existence is holistic; so too health and illness.

Methylation patterns vary with diet. Early malnutrition can create a host of problems, such as hyperactive stress response, with wide-ranging effects that last throughout life.

Exposure to pollutants, including alcohol and tobacco, can have lasting epigenetic effects which may be passed to offspring who are never exposed to the triggering pollutant. Such effects can last for generations.

Many chronic diseases are epigenetically endowed as a culmination of lifestyle. Autoimmune diseases, such as rheumatoid arthritis, are exemplary.

> Common diseases are due to many changes with small effects on a handful of genes. ~ American geneticist Peter Scacher

The speed at which the epigenome changes relates to lifespan, both in individuals and across species. For animals, eating less slows the rate of epigenomic change.

⚥ Cancer ⚥

> Cancer is due to errors in the mode of living. ~ Duncan Bulkley

Cancer is an umbrella term for diseases characterized by uncontrolled cell growth. Cancer essentially emanates from environmental stimulus, though genetic expression tilts the risk for the disease.

Several epigenetic alterations characterize cancer, including lowering histone levels, which help activate the growth and ensure the survival of cancer cells. Cancer is a product of defective gene regulation, though the progression of cancer often has genetic repercussions as well.

ಬ Comparative Epigenetics ಆ

While significant in all life forms, epigenetic inheritance plays a stronger adaptive role in other life forms than it does in mammals. Insects, plants, and yeast transfer extensive adaptations epigenetically.

Cold numbs even nerve cells. Signaling slows down. Amazingly, octopi in the frigid waters near Antarctica adapted without any genetic change from those in tropical seas. An enzyme that specializes in editing mRNA alters the blueprints for octopus nerve cells by adjusting nerve cell timing for ambient water temperature. This epigenetic trait has been conserved through evolutionary time. Cold-water survivability is so handy that it independently evolved in several octopus lineages (convergent evolution).

ಬ Plant Epigenetics ಆ

Plants have a more complex and redundant array of epigenetic mechanisms than animals. ~ Nina Fedoroff

Plants and fungi lack the early segregation of germline that is characteristic of multicellular animals. Epigenetic inheritance offers an avenue of adaptability via an ample assortment of actions. RNA sequences carry guidelines for epigenetic activity from one generation to the next. Plants use epigenetics to regulate various developmental processes, including vernalization, flowering, stem cell maintenance, and in response to hormonal and environmental stresses.

A plant's reproductive success depends critically on the precise timing of flowering in the springtime. ~ American epigeneticist Karissa Sanbonmatsu

Plants only flower after a certain number of cold weather days. They plan their blooms by remembering the number of days since winter set in. Noncoding RNAs are the physical mechanism for this memory.

Other functions are accomplished by chromatin remodeling, including histone modification and replacement of canonical histones with variants.

Paramutation alters gene expression. In paramutation, one allele induces a heritable change in a homologous allele at the same locus. This is a common floral epigenetic technique, but rare in animals. Paramutation is meiotically inheritable, and so violates Mendel's law of segregation.

Plant parts select which genes to express. A certain part may revert to an ancestral gene rather than the parent version adopted in the rest of the plant.

There is a tradeoff between growth and defense in many organisms, but this is especially pronounced in plants. The more resources devoted to defense against pathogens the slower the growth rate.

Plants in a pathogen-rich habitat tend to be stunted. This environmental adaptation is passed on epigenetically: no gene mutation occurs for this tradeoff to be carried by offspring. By conveying critical information to seedlings, this epigenetic head start improves the odds of survival for next-generation plants.

Plants assiduously avoid inbreeding. When closely related individuals do mate, offspring tend to be less fit.

This *inbreeding depression* has long been attributed to genetics: reduction in gene variability. In plants, epigenetics also plays an active part in reducing fitness from inbreeding. This lessens the likelihood that inbred plants will further propagate and weaken a population.

Epigenetic inheritance is both more common in plants than animals and more stable: lasting for hundreds of generations. Epigenetic inheritance in plants is as stable as genetic inheritance. Plant epigenetic reprogramming is much less pervasive and thorough than in animals, so epigenetic marks linger unscathed. Nonetheless, for flexible adaptation, marks remain reversible.

❧ Genomic Protection ☙

There are several strategies employed to protect DNA from damage. The spacing between introns is one such structural technique.

Numerous proteins keep a close watch on genomic stability. The class of enzymes known as *helicases* is exemplary. Helicases are vital to all organisms. Their main role is unpackaging nucleic strands for employment. They also help ensure integrity for their objects of attention.

Not all genomic threats come from without. Some lurk within.

✄ Transposons ✄

Transposable genetic elements (TEs) comprise a vast array of DNA sequences, all having the ability to move to new sites in genomes either directly by a cut-and-paste mechanism (transposons) or indirectly through an RNA intermediate (retrotransposons). ~ Nina Fedoroff

A *transposon* is a DNA sequence which can change its position within a genome; informally called a "jumping gene." Transposable elements (TE) were discovered by American cytogeneticist Barbara McClintock in the early 1940s while studying maize.

Several major types of TE are recognizable in the genomes of a wide range of organisms; these differ in their transposition mechanisms. ~ English evolutionary biologists Deborah Charlesworth & Brian Charlesworth

In prokaryotes, transposons are essential in cataloging viral encounters, thereupon creating their adaptive immune system (pais) library. Transposons also comprise a large fraction of most eukaryotes' genomes, as transposition often results in TE duplication. 67% of the human genome comprises transposable elements.

Most TEs are found in intergenic DNA or (to a lesser extent) in introns. In some regions of the genome, TEs can be very densely packed, with jmulitple elements inserted within one another. ~ Deborah Charlesworth & Brian Charlesworth

Transposons are generally considered junk DNA. They have also been characterized as "selfish." Such blithe dismissal belies biological reality. Depending upon context and instigation, transposons may be beneficial or a bane.

Transposon insertions can have beneficial effects for their respective host organisms. ~ Thomas Eulgem

Transposons help cells adapt to stress and serve as cellular defense against viruses. Insects can quickly become resistant to pesticides thanks to transposons.

Transposable elements can drive evolution by creating genetic and epigenetic variation. ~ Japanese plant cytologist Tokuji Tsuchiya & American plant cytologist Thomas Eulgem

Conversely, these genetic gypsies can disable genes where they impose themselves, even triggering cancer, and contributing to neurodegenerative disorders such as schizophrenia and Alzheimer's.

Transposons do not just jump. Instead, they usually leave a copy behind at their original location.

If the copy and paste were left unchecked, TEs could explode the genome. But the process is regulated.

After a certain number of copies are made, *transposase* – the enzyme that catalyzes jumping – reaches a critical threshold, and transposition ceases.

Transposons are not rogue genetic elements. Instead, they are often part of an intricate complex of epigenetic functioning. Transposons associate in families, and the jumps they make transpire through that affiliation. In plants, transposons play a role reprogramming the germline.

The activity of transposons does not only depend on themselves, but also on factors which the host cells produce. ~ Serbian geneticist Ana Marija Jakšic

In a specific adaptation which repeatedly occurred, fish adapted to freshwater from saltwater via transposons.

A single adaptive genetic innovation repeatedly allowed marine fish to colonize and diversify in freshwater. Transposons were responsible. ~ American biologist Jesse Weber *et al*

Transposons orchestrate the genetic expressions responsible for the prolonged pregnancy of placental mammals. This dramatic evolutionary divergence from marsupials transpired ~90 million years ago.

> The evolution of pregnancy was associated with a large-scale rewiring of the gene regulatory network. Transposable elements are potent agents of gene regulatory network evolution. ~ American evolutionary biologist Vincent Lynch *et al*

Though the specifics are not well understood, transposons appear to have played a major role in evolution via cross-species jumps.

> Jumping genes introduce themselves into other genomes. ~ Australian geneticist David Adelson

ꙮ Piwi-interacting RNA ꙮ

Discovered in 2001, p*iwi-interacting RNA* (*piRNA*) constitutes a huge group of RNA regulatory molecules that keep transposons in check. In humans, piRNA variety may number into the millions. piRNA operate in conjunction with *piwi proteins* (*piwip*).

On their own, piwi proteins work to bind or cleave RNA. Piwips are present in both plants and animals.

piRNA-piwi protein complexes (*piRNA+piwip*) act as a genomic molecular defense system, analogous to an organism's immune system. Like an adaptive immune system, piRNA+piwip can tell friend from foe, mobilize a response to a jumping gene, and adapt to new TEs. These genomic guardians have a memory of past threats and actions taken, stored epigenetically.

piRNA police track their quarry: they genomically jump just like TEs.

When a transposon migrates, it stands a chance of landing in a piRNA+piwip cluster. When that happens, the TE is captured. piRNA+piwip can recognize a transposon by it never being expressed.

Once entrapped, complementary DNA sequences are produced to thwart the genetic interloper elsewhere, thus gaining protection from that particular TE. In distinguishing between *self* and *uninvited*, piRNA+piwip become transposon specific.

piRNA+piwip are prolific. Their variety in mammals may number in the millions.

piRNA are not all-powerful. While they do tackle TEs on their own, piRNA+piwip often enlist help from other RNA management specialists. Genomic protection is a team endeavor.

Some piRNA+piwip do not have transposon targets. Instead, these molecules facilitate cellular learning.

Retrotransposons are a subclass of transposon. They amplify themselves using RNA intermediates, including mRNA. Using RNA intermediates allows rapid copying.

Retrotransposons are ubiquitous in eukaryotes and are particularly abundant in plants, where they may comprise a majority of nuclear DNA. At least 42% of the human genome comprises retrotransposons.

About half of the human genome consists of highly repetitive DNA, what was once considered "junk." These repetitions are essential to repress retrotransposons and thereby protect the genomic integrity of stem cells.

Methylation is the primary silencing mechanism for retrotransposons in somatic cells. In contrast, stem cell expression is not suppressed by methylation. Chromatin repetitions safeguard stem cells.

☙ Endogenous Retroviruses ❧

We suspect that these viruses are forced to make a choice: either to keep their 'viral' essence and spread between animals and species, or to commit to one genome and then spread massively within it. ~ English zoologist Robert Belshaw

Another type of transposon is endogenous retroviruses (ERVs). ERVs are endogenous viral elements that closely resemble retroviruses. Some evolved from retroviruses.

ERVs are a unique combination of pathogen and selfish genetic element. ERVs may replicate either as a transposable element or a virus.

Some ERVs proliferate by infecting germline cells, as typical retroviruses do. Others lack the gene required for virions to enter cells. These behave like retrotransposons.

ERV lineage in eukaryotic organisms is primordial. In evolutionary time, they played an active role in shaping genomes.

ERVs are abundant in jawed vertebrates. ERVs occupy 8–10% of the human genome.

ERVs independently evolved into retrotransposons multiple times. This explains their surfeit in mammal genomes.

The majority of vertebrate ERVs are so ancient as to be inactivated by mutation. Hence, many are merely genetic artifacts in their host.

∞ The Genetics of Behavior ∞

Almost all aspects of life are engineered at the molecular level. ~ Francis Crick

Nature versus nurture is a long-standing debate about the degree to which behaviors are biologically bound as opposed to learned. That autonomic functions and reflexes are genetically encoded has never been controversial. The debate centers on behaviors from conscious decisions which may involve planning.

The crux is of cause versus correlation. Does behavior shape biology, or vice versa (causality)? Otherwise, do behavior patterns and biochemistry coincide (correlation)?

The epigenetic answer is "yes" to both cause and correlation. Behaviors invoke epigenetic changes. Genetic activity prompts behavioral modification. And behavioral and genetic dynamics go together as self-reinforcing.

An animal's diet has epigenetic effects. Conversely, epigenetic changes may occur that shift dietary preferences. One may reinforce the other. This applies to a wide spectrum of seemingly complex behaviors, from foraging to mating and parenting.

Behaviors arise from a suite of traits. All traits are biologically based.

The relations among genes, the brain, and social behavior have complex entanglements across several different time scales, ranging from organismal development and physiology all the way to evolutionary time. Genes do not specify behavior directly, but rather encode molecular products that build and govern the functioning of the brain through which behavior is expressed. ~ American biologists Gene Robinson, Russell Fernald, & English psychologist David Clayton

Life experiences impress cells epigenetically. They also do so genetically.

Immediate early genes are certain genes which are instantly activated in response to cellular stimuli. Immediate early genes play important roles in behaviors, learning, memory, immune system activity, and many other bodily functions.

The emotional impacts of life are encoded genetically. Conversely, routine patterns of behavior and thought are genetic expressions.

Experiences and behaviors form a self-reinforcing cycle. Biological programming parameterizes future behaviors into predictable patterns; what is called one's *character*.

Character is simply habit long continued. ~ Greek essayist Plutarch

As epigenetic effects are transferred generationally, and strongly set during early development, nurture is nature and vice versa. In that animals are creatures of habit, breaking a behavior pattern seems to be going against the genetic grain. But then, strength of will is also in one's nature.

♎ Mice Digs ♎

Rodent burrows are a product of planning, which involves choice of location based upon soil type, and design, including number of tunnels, width, length, and usage. Oldfield mice and deer mice both dig tunnels, but their burrows are different.

Oldfield mice consistently dig a burrow with a long entrance tunnel, along with a 2nd tunnel that stops short of the surface. The 2nd tunnel allows the mice to escape predators.

In contrast, deer mice burrows are shallow and lack an escape route.

Crossbreeding between these mice resulted in a variety of different tunneling patterns and revealed the genes behind the burrows. Also unearthed was the unsurprising revelation that complex burrowing in mice evolved incrementally and piecemeal, by combining genic modules responsible for simpler digging behaviors.

Complex behaviors may be encoded by a just a few genetic changes. ~ American biologist Hopi Hoekstra

ꙮ Gene Editing ꙮ

Cells evolved 2 strategies to search their genome for specific information. Transcription factors and restriction enzymes recognize a specific DNA sequence through interactions in double-stranded DNA grooves, whereas other proteins are dynamically programmed by an RNA or single-stranded DNA to recognize complementary nucleic acid sequences through base pairing. ~ Swedish molecular biologist Johan Elf *et al*

In his accidental 1987 discovery of pais (prokaryotic adaptive immune system), Yoshizumi Ishino also inadvertently uncovered what would become the standard gene editing tool, called CRISPR/Cas9.

CRISPR is an acronym for "Clustered Regularly Interspaced Short Palindromic Repeats." CRISPR applies to segments of prokaryotic DNA that have short, repetitive base sequences. The term *palindromic* refers to a sequence of nucleotides that are the same in both directions.

After each repetition in a CRISPR is a short DNA segment, called a *spacer*, that came from exposure to foreign DNA, such as a virus or plasmid. This foreign bit serves as a memory of a survived attack, and is instrumental in dealing with a subsequent, similar infection. For gene editing, spacers serve as the search sequence.

Situated next to CRISPR sequences are small clusters of genes known as *Cas* (CRISPR-associated system): a gene that encodes a particular enzyme which acts upon CRISPR to effect an immune response in the prokaryote in which it resides. The 9 in Cas9 refers to a certain enzyme that can search and cleave a specific DNA sequence given a target RNA sequence as a guide. Cas9 came from a common *Streptococcus* bacterium and was chosen because geneticists could figure out how to fiddle with it.

> An intracellular search requires Cas9 to unwind the DNA double helix to test for correct base pairing to the guide RNA. ~ Johan Elf *et al*

A Cas9 is bound to each potential target for less than 30 milliseconds. A single search typically takes 6 hours.

Altogether, CRISPR/Cas9 provides a generic gene-editing tool. Like cut-and-paste in a word processor, a specific genic sequence can be edited out and a substitute sequence inserted – at least theoretically, though not easily in practice.

> To cut DNA with CRISPR, it's like trying to remove 1 specific word on a particular page in a novel. ~ geneticist Bruce Conklin

A cell having its genome disrupted by CRISPR/Cas9 naturally attempts to repair any DNA damage. If the attempted insertion is successful, the artificially suggested substitute is employed in the rehabilitation.

> When CRISPR makes a cut, the DNA is broken. To survive, the cell recruits many different DNA repair factors to that particular site in the genome to fix the break and join the cut ends back together. ~ Australian geneticist Beeke Wienert

Gene editing is problematic in at least 4 ways. 1st, the guidance system may go awry, with the CRISPR molecules leading the search enzyme to parts of the genome that are similar, but not selfsame to the intended target.

The 2nd problem has proven most vexing to geneticists: quality of repair. Cells take 2 general approaches to repairing DNA damage. A cell may stitch severed strands back without much regard to accurate reproduction of what was there before: a simple patch job. The other way more carefully repairs a break: with guidance from what the cell considers a reliable facsimile, usually DNA inherited from its mother.

Cells prefer the quick-and-easy method of patching. They only bother with precision repairs a minuscule percentage of the time. Error-free repair is more likely during cell cycle phases when sister chromatids are present, thus providing a ready corrective-instruction guide. A tiny protein is instrumental in making the repair decision. How the protein intelligently does so is a mystery.

The 3rd difficulty is that more than a DNA sequence is involved in incorporating introduced genic material. Subtle physical geometrics play a critical role; something which gene editors cannot control with their insertions.

> Cells rely heavily on active-site positioning and structural features of the DNA, rather than direct sequence recognition, to localize DNA integration to the CRISPR locus. ~ American molecular biologist Addison Wright *et al*

The 4th issue is that any desired edit must reach every cell. If an embryo is being edited, even partial failure leads to genetic *mosaicism*, where only some of the cells are edited. If the aim is to eliminate a genetic disease, mosaicism risks nullifying the intended effect.

The answer to the risk of mosaicism is to use a powerful genetic driver. This creates a 5th dilemma: genetic proliferation. Engineered DNA may race through a population so easily that a small number of rigged organisms may spread the mutation.

Gene editing can have profound, unforeseeable consequences to an ecosystem. This has already occurred with the industrial agriculture which relies upon genetically modified organisms.

New problems with gene editing keep being found. Geneticists manipulate genetic codes despite knowing next to nothing about the processes underlying genic employment and repair by cells. The genetic modifications publicized in mass media are only of success, which are a minuscule fraction of attempts to play with the essential ingredients of life.

Next to nothing is known about the ongoing consequences of artificial gene editing, either for the organism involved or for the environment in which the organism lives. The monitoring required is extensive, and there has been almost none of it. Gene editing is in its infancy, and geneticists as responsible as infants in what they unleash upon the world.

> Gene editing is super-powerful, but so far is a lot of trial and error. ~ American geneticist Jacob Corn

Given the environmental track record of humanity, the idea of this primate playing God is rightfully frightening. Just because you can do something does not mean that you should.

≈ Synopsis ≈

➢ Organisms are largely built of, and operate, via *proteins*. The artifactual rules for the construction and activities of proteins are coded in *polynucleotides*: DNA & RNA.

➢ DNA became the preferred form of storage for genetic data because it is stabler than RNA. Yet, in the process of extracting information content, DNA is first transcribed to RNA (*transcription*). RNA is then read (via *translation*) to render biological products, notably proteins.

➢ The macromolecules of life are characterized by their chemistry, information content, and spatial conformity. The *origami* of proteins and polynucleotides is critical to their performance. The intricacies of these compounds are astounding, as is the intelligence involved in their fabrication and employment.

History

➢ *Heredity* has long fascinated natural philosophers. Ancient Greek physician Hippocrates, Aristotle, and ancient Indian physicians writing in *Charaka Samhita* posited early hypotheses about biological inheritance.

➢ English naturalist Charles Darwin reprised Hippocrates' hypothesis of *pangenesis* as the mechanism for heredity, whereupon offspring inherit holistic attributes of their parents through some atomic contrivance.

➢ Studying pea plants, Austrian monk Gregor Mendel observed genetic variations (*alleles*) in the early 1860s. He hypothesized them as *heredity units*.

➢ In imagining traits as embedded in biological dollops of heredity, Danish botanist Wilhelm Johannsen coined the term *gene* in 1909, well before the basics were understood.

➢ The dominant paradigm of *genes* as encoded units of heredity impeded apprehension of the molecular encyclopedia of life for nearly a century. Researchers continued thinking that spatially cogent genes and bodily traits had a strong correspondence long after evidence showed otherwise; whence the ignorant notion of *"junk" DNA*.

Rather than adopt a more appropriate conceptual framework, geneticists inappropriately kept redefining old terms and adding new ones. Hence, the most intricate study of life at the level of molecular intimacy has become obtuse. The study of genetics continues to be full of surprises only owing to the illusion of knowledge.

Genetic Codes

➢ A *gene* is a conceptual entity, not a physical chemical complex. Genes are just ideas in the minds of geneticists.

➢ Genes were first conceived as molecular packages of trait heredity. They were later redefined as recipes for making proteins. Once other biological products were discovered to be important, and their templates stored in DNA, the definition of *gene* was extended to include them as non-protein coding genes. Now genes are conceived as molecular packages of trait heredity.

➢ Historically, a gene was assumed to be a certain strand of DNA that cogently carried a coding sequence for synthesizing a protein. Such facilely imagined localized genes are relatively rare, and their codes are not necessarily faithfully transcribed.

1st, physical trait data may be in multiple places. 2nd, translating genetic instructions into the intended product is a tortuous path, where much material goes unused or is waylaid in the expression process.

➢ DNA sequences act as instructions for specific manufactures. *Gene expression* is the process of using the information in a gene to synthesize a functional bioproduct, which is typically a protein.

➢ Prodigious research efforts have been expended toward *gene mapping*: pinpointing the loci of coding sequences for traits. The quest is mostly quixotic. Inherited traits are commonly the product of a network of genes, and gene expression is far more than just transcription at a gene's locus. A trait may be the outcome of many genes, either activated or suppressed, in a long sequence of reactions leading to a result, with innumerable other factors in play.

➢ An *allele* is a gene variation at the same locus. Different alleles may result in different traits, but not necessarily. One allele may be *dominant*, while another *recessive*. A dominant allele may completely mask a recessive allele.

➢ A *genome* is the total complement of genes in a cell or organism. Individual cells have their own genomes. How that comes to be is not known.

➢ A prokaryotic cell has a single *genophore* stuffed with genetic material. Prokaryotes readily pick up new genetic material in non-genophore *plasmid* packages, which are exchanged with other microbes via *horizontal gene transfer* (HGT). Eukaryotes also employ HGT, though not nearly as frequently. A eukaryotic cell keeps its genome packed in *chromosomes*.

  The genetic makeup of a multicellular organism is multifarious. Each cell has its own individualized genomic package (*pergenome*). Genetic variation among cells is termed *chromosomal mosaicism*.

➢ Evolutionary descent frequently involves carrying more *genetic baggage*. While over 90% of the genome of prokaryotes is employed, less than 2% of human genes are used for protein production.

DNA not directly recognized as used in protein production is termed *noncoding*. Such genetic material is often repetitive, sometimes with slight variations. This unobvious genetic material is employed in subtle ways that are just beginning to be appreciated.

➤ Algae, slime molds, fungi, and plants practice *alternation of generations*: alternate reproductive and genetic modes during their life cycle: asexual spore production while haploid, and sexual reproduction while diploid. A few animals may alternate reproductive modes – sexual and asexual – but all animals are always diploid.

➤ There is a *genomic protection* system that maintains the integrity of a cell's genome, especially against damage that may be caused by self-serving jumping genes that otherwise proliferate.

Epigenetics

➤ Genes as the porters or heredity is an incomplete picture. Heredity also occurs epigenetically: outside genetic coding per se. Epigenetics is a cellular inheritance mechanism via gene regulation, without changing the structure of the DNA involved. Even for identical twins, epigenetic inheritance creates variations, thus rendering every organism genetically unique.

➤ An integral aspect of a cell's life cycle, epigenetics is one way *a cell adapts* to its environment without changing genetic code. Through epigenetics, cellular memory may be passed to descendant cells. Epigenetic inheritance can be as stable as genetic inheritance.

➤ *Gene regulation* modulates the complicated process of gene expression. A DNA sequence is often not expressed as it was coded.

➤ There are many *epigenetic mechanisms*. Gene regulation comprises an interrelated network of information exchange among numerous molecular participants in the various processes of genic employment.

➤ *Epigenetic marks* are sometimes referred to as "switches," implying that genes are either on or off. Instead, there are degrees of expression/suppression.

➤ An *epigenome* is the idea of a summated cellular set of instructions affecting access and expression of genes, including the timing of expression. A multicellular organism may have as many epigenomes as it has cells.

➤ *Stress* can cause epigenetic changes that are passed to the next generation. Exposure to pollutants results in epigenetic changes that can be inherited, even if offspring are never exposed to the pollutant.

➤ *Homosexuality* is epigenetically implanted during the development of a fetus. This epigenetic effect can last for generations.

➤ *Meditation* and a calm mind have a positive epigenetic effect and boost the immune system as well. There is no physiological explanation for how this is possible.

> DNA is not the be-all and end-all of heredity. Information is transferred from one generation to the next by many interacting inheritance systems. ~ Eva Jablonka & Marion Lamb

❧ Conclusion ❦

Two dangers threaten the universe: order and disorder.
~ French philosopher Paul Valéry

This universe is one of many in an eternal cycle of cosmic creation. As the natural sciences show, its fabric is ordained.

⚸ Physics ⚹

Physics begets chemistry. Chemistry defines biology. Biology circumscribes life. By this cascade of energy to animate matter, all that manifests is intertwined.

As energy encompasses existence, physics focuses on characterizing energy. Following constructal law, one modus operandus of energy is economy. Matter is a slow-motion product of energetic relations.

Nothing happens until something moves. ~ Albert Einstein

That matter is made of energy is indisputable. Energy is nothing more than an abstraction which may be measured by its effect on matter – a tidy circularity. Energy is the cogent immateriality by which the fabric of materiality is woven.

Physics has proven that Nature is a façade for a noumenal reality. From there are the strange mechanics which generate existence.

Attempting to understand the roots of phenomena has demonstrated that our 3-dimensional (3D) vista provides only a partial picture. All facets of modern physics expose a universe with extra dimensions (ED) beyond spatial 3D.

Nonlocality is a fundamental facet of actuality. Subatomic particles are often entangled, responsively changing synchronously. This too indicates a holistic dimensionality (HD) beyond the experiential 4D of spacetime. It also shows something even more fundamental about Nature as a ruse.

Nature functions by integration. ~ François Jacob

Entanglement demonstrates that spacetime itself is an emergent property: an outcome of interconnected interactivity at every level of existence. Phenomena appear via bottom-up, moment-by-moment fabrication into a tight weave of manifestation.

The long-sought unified theory of physics – from the quantum to the cosmological scale – can only evolve from the realization that everything is entangled. The geometry of spacetime, and all the energy that flows through it, are defined by this web.

Physics has already progressed to the point of realizing that its formulas yield only approximations. The elemental remains empirically elusive because it lies beyond phenomena.

Ultimately, science is an interpretive exercise of observed events. Intuition is as essential as mathematics to propelling physics forward.

> Try and penetrate with our limited means the secrets of Nature and you will find that, behind all the discernible concatenations, there remains something subtle, intangible and inexplicable. ~ Albert Einstein

Self-organization from non-equilibrium driving forces has been observed at both the molecular and cosmological levels. *Coherence* is the only possible explanation for the order in Nature – the natural, fundamental force from which phenomena arise. This is most poignantly true of *life*, which, in its statistical impossibility, clearly lies beyond random chance.

☿ Chemistry ☋

Chemistry susses the relations between species of matter. Grounded in the relative solidity of atoms and molecules, chemistry – at first blush – appears to validate empiricism; but chemical reactions readily happen where they should not, as the tentacles of quantum mechanics touch all phenomena. The classical laws of chemistry render interstellar space too cold for organic molecules to form. Yet they do, prodigiously so.

The exercise of matter involves vibrancy within an energy economy. The stability of chemical compounds relies on discrete tolerances to energetic disturbances.

This energy reactivity/stability yin-yang is manifest in the elements of royalty in organic chemistry: carbon and nitrogen. Carbon is the most flexible element for molecular combination, and so is copiously employed as a manifest information conduit. Contrastingly, nitrogen's inherent stability renders it a cornerstone for cellular structures.

๙ Life ๖

Chemistry seemingly defines biology, but the gap between inorganic and organic chemistry is more than carbon combinations and nitrogen backbones. Life is infused with vital sparks of consciousness.

Operationally, organisms need energy (metabolism), animation (movement), and the ability to perpetuate (reproduction). Metabolism may be understood within the confines of classical chemistry. Animation and perpetuation require something more.

Movement involves a plan to achieve its aim of translocation: a fluid interaction between the instant state of structural composition and the energy for an incremental change to achieve the intended next state. Movement illustrates how life relies upon memory. Reproduction is a penultimate product of remembrance: replication as a means for perpetuation.

In contrast to physics, life is about using information well: intelligence. Energy is essential yet incidental.

The phenotypes and behaviors of organisms are statements of memory in form and function. Even the simplest organisms are organized with specialized structures and pathways attuned to specific tasks.

While these structures and pathways ultimately exist as chemical compounds, they are organized according to a plan. The plan itself is an embedded memory, as is the pattern that comprises the life cycle.

DNA and its accoutrements provide a physical manifest of life, both as a memory and oracle: of what may be along with what has been. Every cell carries the memory of its evolution, going back toward its historical origination. Life experiences are encoded epigenetically: a remembrance passed to offspring.

Knowing the steps behind affords ready adaptation: an evolutionary plan of action based upon the experiences of the past. The biochemical database of life – genetics – provides a practically unlimited vocabulary of expression; hence the unimaginable variety of life on Earth.

Plants are particularly adept at manipulating their own genes, and at understanding the genomic processes of other organisms. They do so by comprehending the codes of coherence – a feat matched only by viruses in nuanced understanding.

Genetics is essentially selfsame for all organisms. Its homogeneity afforded cellular interdependence among the vast diversity of life forms, which is exactly how life evolved: interdependently.

Without genic compatibility, the coordination necessary for multicellular life would be lacking. Incompatible genetics would have resulted in a much different world – one with evolution stymied.

By homogenizing the language of life, viruses were the agent of organic harmony: the simplest life acting as the most profound.

Spokes of the Wheel continues with *Book 2: The Web of Life.*

❧ Glossary ❦

~ : approximately.

#

0th law of thermodynamics: the hypothesis of comparative thermodynamics among systems: that if 2 thermodynamic systems each are in thermal equilibrium with a 3rd system, then they are in thermal equilibrium with each other.

1st law of thermodynamics: the hypothesis that the total energy in an isolated system is immutable: that energy can be neither created nor destroyed in a closed system.

2nd law of thermodynamics: the hypothesis of there being a tendency over time toward entropy in an isolated physical system. The 2nd law outlaws perpetual motion machines.

3-center 2-electron bond: a covalent bond between 3 atoms sharing 2 electrons.

3rd law of thermodynamics: the hypothesis that the entropy of a system approaches a constant value as its temperature approaches absolute zero.

4D (aka *spacetime*): the 4 dimensions of everyday experience: 3 of space (3D) + 1 of time. See *HD* and *ED*.

A

abiogenesis: the study of how life arose.

ΛCDM (Lambda cold dark matter) model: the current standard cosmological model, positing a 13.82 BYA Big Bang based upon the 1st observable light for cosmic origination; physics-defying cosmic inflation; a cosmological constant, denoted by lambda (the Greek letter Λ); the cosmological hypothesis of a homogeneous and isotropic universe; with cosmic expansion presently accelerating. ΛCDM has been disproven on multiple fronts.

acid (chemistry): a molecule capable of donating a hydron. Acids react with bases. Contrast *base*.

acidophile: an organism that lives in a highly acidic habitat.

actin: a family of globular multifunctional proteins that form microfilaments. Actin is found in all eukaryotic cells except roundworm sperm. Actin has equivalents (homologs) in prokaryotes.

activator (chemistry): an enzyme that increases reaction rate. Contrast *inhibitor*.

active site (organic chemistry): the position on a protein where substrates bind and undergo a chemical reaction.

active transport: ingestion of molecules across a cell membrane.

actuality: the world experienced sensorially. Contrast *reality*.

adaptation (evolutionary biology): the teleological process of adjusting to ecological circumstance.

adaptive immune system (aka *acquired immune system*): the portion of the immune system that learns to recognize specific pathogens. Contrast *innate immune system*.

adenine (A) ($C_5H_5N_5$): a nucleobase of DNA and RNA, complementary to *thymine* in DNA or *uracil* in RNA.

adenosine: a nucleoside of adenine.

ADP (*adenosine diphosphate* ($C_{10}H_{15}N_5O_{10}P_2$)): the product of ATP dephosphorylation, which provides energy for a cell. See *ATP*.

adsorption: the process of a gas, liquid, or solution gathering on a surface in a condensed layer.

aerobic: living with oxygen. Contrast *anaerobic*.

aerobic respiration: cellular respiration which employs oxygen. Contrast *anaerobic respiration*.

aerosol: a suspension of fine liquid or solid particles in gas. Aerosol particles are less than 1 micrometer in diameter.

aether (aka *ether, quintessence*): a long-presumed ethereal substance that pervades empty space. The assumption was eventually abandoned by physicists in the early 20th century after a futile search.

aka: "also known as."

alchemy: the study of matter transmutation, which evolved into *chemistry*.

aldehyde: a common organic compound comprising a carbonyl center with a hydrogen sidekick, connected to a *side chain*

(R): R-C=O-H. An aldehyde group without the side chain is termed an *aldehyde group* or *formyl group*. Formaldehyde (CH_2O) is the simplest aldehyde. Aldehydes are aromatic. Many fragrances are aldehydes. Compare *ketone*.

alga (plural: *algae*): a eukaryotic protist, usually unicellular or colonial, that photosynthesize via chloroplasts.

algorithm: a step-by-step procedure, often employed for mathematical problems. Compare *heuristic*.

aliphatic compound: the group of hydrocarbons that do not link together to form a ring.

alkali metal: a group of shiny, soft, highly reactive metals, owing to having an affable outermost electron (i.e., an outermost electron in an s-orbital that renders it readily sharable). The 6 alkali metals are: lithium (Li), sodium (Na), potassium (K), rubidium (Rb), cesium (Cs), and francium (Fr).

alkaliphile: an organism that lives in a highly alkaline habitat.

alkane: a hydrocarbon bonded exclusively by single bonds.

alkene: a hydrocarbon with double bonds between carbon atoms.

alkyl: a univalent, aliphatic radical C_nH_{2n+1} (e.g., methyl, ethyl) derived from an alkane by removal of 1 hydrogen atom.

alkyl group: a chemical functional group, usually designated as *R*, comprising alkanes. A *methyl group* is an alkyl derived from methane (CH_4). An *ethyl group* is an alkyl derived from ethane (C_2H_6).

alkyne: a hydrocarbon with triple bonds between carbon atoms.

allele (aka *allelomorph*): one of multiple forms of a gene; a variation of a gene at the same locus. Selfsame alleles at a locus are *homozygous*; if different, *heterozygous*.

allometry: growth of a body part relative to the entire organism; also, the study thereof.

allopatric speciation: evolution into distinct species owing to populations being isolated from each other.

allosteric activator: an enzyme that enhances activity at an allosteric site. Contrast *allosteric inhibitor*.

allosteric inhibitor: an enzyme that lessens activity at an allosteric site. Contrast *allosteric activator*.

allosteric site: a site on a protein other than its *active site*.

allostery: regulation of an enzyme or other protein by binding an effector molecule at the protein's allosteric site.

allotrope: a molecular structure of the same atomic species that may take various forms; that is, where element atoms may be bonded together in different ways.

allotropy: the property of an element existing as an allotrope (structural variations).

alpha particle: 2 protons and 2 neutrons bound together into a particle identical to a helium nucleus. Comprising the equivalent of doubly charged helium atoms (stripped of 2 electrons), alpha particles are a relatively slow-moving particulate radiation (*alpha decay*).

alternation of generations (AoG): alternate asexual and sexual reproductive modes during a multicellular organism's life cycle. For algae, plants, fungi, and slime molds, AoG also involves different genetic forms at different stages of life: haploid and diploid.

altruism: unselfish behavior.

aluminum (Al): the element with the atomic number 13; a soft, ductile, silvery-white, nonmagnetic metal; the 3rd-most abundant element in Earth's crust (after oxygen and silicon (silica)), and the most abundant metal. For a metal, aluminum has remarkably low density.

Alzheimer's disease: an incurable degenerative disease leading to dementia. Symptoms advance to confusion, irritability, mood swings, trouble with language, and memory loss.

amine: a derivative of ammonia.

amino acid: an organic molecule comprising a carboxylic acid group, an amine group, and a side chain specific to the specific amino acid. The key elements in amino acids are carbon, hydrogen, oxygen, and nitrogen, with other elements found in the side chain.

ammonia (NH_3): a colorless gas that figures in biology because of its nitrogen content. In certain microbes, atmospheric nitrogen is converted into ammonia by enzymes termed *nitrogenases*, in a process called *nitrogen fixation*.

amphiphilic: a chemical compound with both hydrophilic and lipophilic properties.

amplify (genetics): copy.

amplitude: the height of a wave.

anabolism: the metabolic pathways for constructing biopolymers. See *biosynthesis*. Contrast *catabolism*.

anaerobe: an organism that does not require oxygen.

anaerobic: living without oxygen. Contrast *aerobic*.

anaerobic respiration: cellular respiration without oxygen. Anaerobic respiration is less efficient than *aerobic respiration*.

analyze: to ascertain and separate an entity (material or abstract) into constituent parts or elements; to determine essential features. Contrast *synthesize*.

anaphase: the stage of cell division where replicated chromosomes split and 2 daughter chromatids migrate to opposite poles of a cell. See *interphase, telophase*.

anapole: a toroidal dipole: a solenoid field bent into a torus.

anastasis: the process of a cell recovering from dying.

Andromeda (aka *Messier 31, NGC 224*): a spiral galaxy 780 kiloparsecs (2.5 million light-years) from Earth; the closest major galaxy to the Milky Way.

angiosperm: a flowering plant, descended from gymnosperms. Angiosperms arose 245 MYA, incorporating several innovations, including leaves, pollen, flowers, and fruit. Angiosperm proliferation began 144 MYA. Over 254,000 species are extant.

angular frequency: the rate of change in the phase of a sinusoidal waveform.

animal: a kingdom of eukaryotic heterotrophs. Most animals are motile. The other kingdoms of eukaryotes are *fungi, plants*, and *protists*.

anion: a negatively charged ion (indicating a surplus of electrons). Contrast *cation*.

annual (botany): an angiosperm that lives 1 year. Compare *biennial, perennial*. See *herbaceous*.

anoxic: oxygen depleted.

antiferromagnetism: the material state where the magnetic moments of atoms or molecules align in a regular pattern of neighboring electron spins pointing in opposite directions. Compare *ferromagnetism*.

antimatter: antiparticle matter. Matter encountering antimatter results in their mutual annihilation.

antioxidant: a molecule that inhibits oxidation of other molecules.

antiparticle: the electromagnetically opposite partner to a subatomic matter particle. For instance, the *positron* is the antimatter equivalent of the *electron*.

ape: a tailless primate; not a monkey.

apeiron: an eternal coherence that creates phenomena; a concept proposed by Anaximander. See *coherence*.

aphid (aka *plant lice*): an extraordinarily successful insect herbivore comprising 4,400 species in 10 families. Aphids exist worldwide but are most populous in temperate zones. Aphids can migrate great distances by riding the winds. Their success has labeled them as one of the most destructive crop pests in temperate climes. Many aphid species are *monophagous*: feeding on only 1 plant species. Others forage on hundreds of plant species across many families.

Apollo (technology) (1966–1972): the NASA human spaceflight project that culminated with landing humans on the Moon (20 July 1969) and returning to Earth.

apoptosis: programmed cell death in multicellular organisms.

aquaporin: a cell membrane protein that forms a selective pore in the membrane of a cell.

arborescent (botany): a plant with wood; a treelike plant. See *herbaceous*.

archaea (singular: *archaeon*): the group of prokaryotes from which eukaryotes arose; taxonomically a domain of life, alongside bacteria and viruses. Archaea are an extremely robust and versatile life form. Archaea are plentiful in the oceans. The archaea in plankton make them among the most abundant organisms on the planet. Archaea play roles in the carbon cycle and nitrogen cycle. Social to a fault, archaea are commonly mutualists or commensals. No archaeal pathogens or parasites are known.

Archean (3.9–2.5 BYA): the eon when life first appeared on Earth.

arsenic (*As*): the element with atomic number 33; a metalloid; notoriously poisonous to multicellular life, albeit an essential dietary element to some animals in minute amounts; in humans, a carcinogen that severely damages the intelligence system, causing dementia.

asexual reproduction: biological reproduction from a single parent. Contrast *sexual reproduction*.

asteroid: a small rocky body orbiting the Sun. Most asteroids emerge from the asteroid belt between Mars and Jupiter.

atom: the smallest particle of a chemical element, comprising at the simplest a proton and an electron (hydrogen).

atomic clock: an electronic clock kept by microwave emissions from atoms cooled to near 0 K.

atomic decay: particulate radiation by subatomic particles from atomic nuclei. Compare *beta decay*.

atomic number: the number of protons an atom has.

atomic species: atoms of the same type (same number of protons).

atomic spectral line: a spectral measurement of an electron changing energy level.

atmosphere: the layer of gases that surround a body with enough mass to keep the gas layer. The atmosphere is held in place by the gravity of the body.

atomism: the philosophy that Nature consists of 2 fundamental aspects: atom and void. Atomism developed in both ancient Indian and Greek traditions.

ATP (*adenosine triphosphate* ($C_{10}H_{16}N_5O_{13}P_3$)): the universal cellular energy storage and source molecule. ATP acts like a battery for cellular power. See *ADP*.

attojoule: a unit of energy equal to 10^{-18} joules.

attosecond: 10^{-18} seconds.

audition: sound perception.

aurora: a luminous plasma region of charged particles that appears in a planet's atmosphere. Auroras sporadically occur in Earth's upper atmosphere, primarily at the high latitudes of both hemispheres: the *aurora borealis* (northern lights) and *aurora australis* (southern lights).

autism: a developmental mental disorder characterized by impaired communication and social interaction, and restricted and repetitive behavior.

autophagy: the catabolic process of recycling and waste disposal in cells. Compare *mitophagy*. See *lysosome, vacuole*.

autopoiesis: a dynamic of self-sustaining activity; a system capable of maintaining and reproducing itself. A biological cell maintaining itself is an example of autopoiesis. Compare *homeostasis*.

autotroph: an organism that makes its own food. Autotrophs are lithotrophs or photoautotrophs. Lithotrophs consume electrons from inorganic chemicals for energy. Phototrophs take light as their primary energy source. Contrast *heterotroph*.

B

B-mode: a curly light polarization pattern.

background extinction: extinction limited to relatively few species. Contrast *mass extinction*.

bacteria (singular: *bacterium*): a taxonomic domain of single-celled prokaryotes, abundant in most ecosystems. Bacteria play vital roles in various facets of the biosphere.

bacteriophage: a virus that preys on bacteria.

bad metal: a metal in which electrical conductivity does not lessen with heat.

Barr body: an inactivated X chromosome.

baryon: a composite particle of ordinary matter: protons and neutrons, which each consists of 3 quarks.

base (chemistry): a molecule capable of accepting a hydron. Bases react with acids. Contrast *acid*.

base pair (genetics): 2 complementary nucleobases on opposite DNA (or certain RNA) strands, linked by hydrogen bonds.

base sequence (genetics): an order of nucleotide bases (1 of a base pair) in a DNA molecule.

basement (rock): a rock below a sedimentary platform. Basement rock is igneous or metamorphic in origin.

bat: a mammal with forelimbs forming webbed wings. Bats are the only mammal capable of sustained flight. 1,240 bat species are known; 70% are insectivores.

bee: a flying insect of 20,000 species. Bees, like ants, are a specialized form of wasp. Bees are best known for their product from pollinating flowering plants: honey.

Bell's theorem: a 1964 theorem by John Stewart Bell that quantum mechanics must necessarily violate either the principle of *locality* or *counterfactual definiteness*. Bell held that locality is violated and counterfactual definiteness applies.

benthic zone: the ecological region at the bottom of the ocean or other water body, including the sediment surface and subsurface layers.

benzene (C_6H_6): an aromatic hydrocarbon; a natural constituent of crude oil; a human carcinogen.

berkelium (*Bk*): the element with atomic number 97; a soft, silvery-white radioactive metal with a half-life of 330 days.

beryllium (*Be*): the element with atomic number 4; a rare, toxic, insoluble metal which only occurs naturally in combination with other mineral elements. Beryllium was first isolated in 1828.

beta decay (*β-decay*): radioactive decay of atomic nuclei or particle transmutation, emitting beta particles (electrons or positrons), mediated by the weak force. Compare *atomic decay*.

beta particle: an electron or positron on a mission as part of beta decay.

Big Bang: the hypothesis that the universe began with an initial energetic cosmic explosion from a dense, hot state of singularity. That this universe started with a Big Bang ~14 BYA is a myth. The universe is much older. See *cosmic inflation.*

binary fission: a form of asexual reproduction where a single parent becomes 2 daughters.

biochemistry: the chemistry of organisms.

bioelement: a planetary ecological element. The bioelements include the atmosphere, lithosphere, hydrosphere, and biota.

biennial (botany): an angiosperm that takes 2 years to complete its life cycle. Contrast *annual*, *perennial*. See *herbaceous.*

biofilm: a colony of prokaryotes encased in a stabilizing polymer matrix, commonly known as *slime.*

biogenesis: biological origin (genesis).

biological pump: the ocean's biologically driven sequestration of carbon and other essential nutrients into the deep ocean.

biology: the science of life.

biome: an area where organisms live with similar conditions, both geographically and climatically.

biopolymer: a polymer produced by a cell.

bioproduct: a biologically synthesized chemical compound.

biosphere: the global summation of Earth's ecosystems.

biosynthesis: a cellular construction process; conversion of substrates into more complex products. See *anabolism*.

biota: the organisms in an environment.

biotrophic: dependent upon another organism as a nutrient source.

bioturbation: displacement and mixing of sediment by fauna or flora.

black body: an idealized opaque/non-reflective object which absorbs all incident electromagnetic radiation. The term was coined by Gustav Kirchhoff in 1862.

black-body radiation: an electromagnetic radiation about a *black body*. Black-body radiation has a specific spectrum and intensity that depends only on the temperature of the body.

black hole: an infinitely dense celestial void that draws in matter and light, rendering the singularity black. Albert Einstein knew of the idea of black holes as a side effect of general relativity but did not think they could exist, writing in 1939 that the idea was "not convincing." Tom Bolton discovered the first evidence of a black hole in 1971.

black hole evaporation: an alternate term for *Hawking radiation*.

bladderwort: a freshwater, carnivorous, flowering plant in the genus *Utricularia*, with 233 species; found in wet soil or in the water; extant worldwide except Antarctica.

bleb: reproduction by breaking off a daughter cell in bacteria that lack cell walls (L-form state).

blood: an animal body fluid employed to transport nutrients to and waste products from cells.

blood-brain barrier: an animal defense mechanism to protect the brain from infection by separating circulating blood from brain extracellular fluid.

bolide: a brighter-than-usual meteor; officially defined from a perspective on Earth as a fireball brighter than any of the planets.

bond (chemistry): a shared electron pair between 2 atoms.

bond energy: a measure of the strength of a chemical bond.

bond order: the number of chemical bonds (bonding electron pairs) between a pair of atoms.

bonobo (*Pan paniscus*): a peaceable ape, closely related to the chimpanzee and human species. Bonobos have a matriarchal society. Bonobos are notably fond of sexual behaviors.

bookmarking (genetics): an epigenetic mechanism of cellular memory by marking genes during mitosis in a way that persists. Bookmarking is vital for maintaining a lineage of cell specialization, so that one cell type does not become another.

boron (*B*): the element with atomic number 5; a water-soluble metalloid concentrated on Earth in borate mineral compounds. Because boron is produced entirely by cosmic ray spallation (cosmic rays bombarding objects) and not by stellar nucleosynthesis (stellar fusion debris), there is little of it in the solar system, including Earth's crust.

Bose-Einstein condensate (*BEC*): a coherent state of matter for a dilute gas of weakly-interacting bosons cooled near 0 Kelvin. BEC exhibits extraordinary quantum mechanical properties at a macroscopic scale. Named after Satyendra Bose and Albert Einstein, who predicted this matter state in 1924.

boson: a quantum that carries a fundamental force, according to quantum physics' Standard Model. Named after Satyendra Bose. Contrast *fermion*.

bow-tie (paradigm): a processing structure capable of handling a diversity of inputs (fan-in) and producing divergent outputs (fan out).

Bragg peak: the apex of ionizing radiation; named after its 1903 discoverer, William Henry Bragg.

brane: a string theory construct of an HD membrane.

braneworld: a physical model using branes. Braneworld models are extensions from earlier *M-theory* and *D-brane* models.

bridgmanite ((Mg,Fe)SiO₃) (formerly *perovskite*): a ferromagnesian silicate mineral; the predominant mineral (38%) in Earth's lower mantle.

brown dwarf: a substellar body too low in mass to sustain fusion reactions in its core, unlike stars, which do.

Brownian motion: the seemingly random movement of particles suspended in a fluid (gas or liquid). Named after Robert Brown.

Brucella: a genus of bacteria named after David Bruce. *Brucella* cause *brucellosis*: a zoonosis transmitted by direct contact with an infected animal or ingesting contaminated food.

budding: a mother creating a smaller daughter. Baker's yeast reproduces by budding.

butterfly effect: a sensitive dependence on initial conditions, where an incremental change at one place in a nonlinear system creates a cascade which results in large changes.

BYA: billions of years ago. BY as an acronym for "billion years" is deprecated in modern geophysics in favor of *Ga*, shorthand for *gigaannum*. The author prefers a sensible acronym to one which is a head-scratcher.

C

cadherin: a calcium-dependent cell adhesion (CAM) protein.

calcium (Ca): the element with atomic number 20. Calcium is a soft, gray, alkaline, earth metal. See *calcium channel*.

calculus: the mathematical study of change.

Callisto: the 2nd-largest moon of Jupiter, after Ganymede, and the 3rd-largest moon in the solar system (Titan is larger).

(large) calorie: the amount of heat needed to raise the temperature of 1 kilogram of water 1 °C.

Cambrian (542–485 MYA): the 1st period of the Palaeozoic era, when the fossil record evidences a vast proliferation of complex life. The name derives from Latin for the area in Wales where the best Cambrian rocks in Britain are exposed.

cancer: a disease characterized by uncontrolled cell growth.

capacitance: the ratio of change in electric charge to change in electric potential.

capillary: a tiny tube in a multicellular organism, typically to facilitate fluid flow to cells.

capillary action: the ability of a liquid to readily flow when narrowly confined in a solid tube, essentially ignoring gravity.

carbene: a molecule comprising a neutral carbon atom with a valence of 2, and 2 unshared valence electrons; also used to refer to *methylene*.

carbohydrate: a macromolecule containing carbon, hydrogen, and oxygen. Carbohydrates are sugars of varying complexities.

carbon (*C*): the element with atomic number 6; an extremely friendly element, with 4 electrons available to form covalent bonds. Life is based upon molecules made with a carbon backbone.

carbon cycle: the gaseous cycling of carbon exchange among the geosphere, pedosphere, hydrosphere, atmosphere, and biosphere.

carbon dioxide (*CO_2*): a colorless gas that has fluctuated in concentration in Earth's atmosphere through geologic time. Plants breathe CO_2; animals exhale it. CO_2 is a greenhouse gas.

carbon–nitrogen–oxygen (*CNO*) cycle: a catalytic fusion reaction cycle by which stars combust. See *proton–proton chain reaction*.

carbonyl: a chemical functional group of a carbon atom double-bonded to an oxygen atom (C=O).

carboxylic acid: a polar molecule (–CO_2H) connected to a hydrocarbon. A carboxylic acid completes itself with a *side chain*.

Carnot cycle: a 19th-century theory by French engineer Nicolas Carnot about efficiently converting heat into work.

carotenoid: a tetraterpenoid organic pigment occurring in photosynthetic organelles of plants (e.g., chloroplasts).

Cas (*CRISPR-associated system*): a gene associated with a CRISPR.

cascade event: an event which results in related follow-on events (*cascade effect*).

Casimir effect: a facet of quantum field theory about physical forces arising from a quantized field. Named after Hendrik Casimir.

caspase (an acronym for *cysteine-aspartic proteases*): a family of protease enzymes that play a critical role in inflammation and programmed cell death (including apoptosis, pyroptosis, and necroptosis).

catabolism: the controlled cellular process (metabolic pathway) of breaking down organic matter to harvest energy via cellular respiration. Compare *anabolism*.

catalysis: an increase in the rate of chemical reaction due to a catalyst.

catalyst: a molecule that causes a change in rate of a chemical reaction by lowering the energy necessary to effect a reaction.

catenane: a molecular compound containing multiple interlocked rings without being chemically bonded.

cation: a positively charged ion (indicating a deficit of electrons). Contrast *anion*.

causality (aka (noun) *cause and effect*, (adjective) *cause-and-effect*): the idea that one phenomenon provokes a succeeding phenomenon. Contrast *correlation*.

cause (verb) (physics): to effect; to bring about.

cell (biology): the basic physical unit of living organisms.

cell cycle (aka *cell-division cycle*): the cellular life cycle, descriptively emphasizing cell division/replication. A cell lives ~90% of its life in *interphase*. Cell division begins with *prophase*, as cells tighten their genetic package in preparation for segregation and division. Plant cells have a preliminary step to prophase, termed *preprophase*, in which the nucleus migrates to the center of the cell. Following prophase, eukaryotic somatic cells enter *prometaphase*, in which the nuclear membrane breaks apart, and the chromosomes inside form protein structures called *kinetochores*. Prometaphase is sometimes considered part of the end of prophase, and early metaphase. During *metaphase*, chromosomes are pulled toward opposite ends of the cell. In *anaphase*, 2 identical daughter chromosomes form. In the 1st step of *telophase*, 2 daughter nuclei form. The cell is bifurcated in the process called *cytokinesis*, whereupon telophase ends with 2 daughter cells.

cell division: eukaryotic cell replication. See *cell cycle*.

cell wall: the flexible membrane holding the contents of a cell and providing an interface to the external environment.

cellular respiration: a set of metabolic reactions within a cell to convert biochemical energy from nutrients into ATP and then release waste products.

Celsius (aka *centigrade*): a commonly used temperature scale; named after Anders Celsius, who devised the inverse of an otherwise similar scale in 1742. In 1954, following the 1743 suggestion of Jean-Pierre Christin, the scale was revised to its current form, a more scientific standard related to the Kelvin scale, with the triple point of purified water as a key reference point. Celsius and Kelvin have the same magnitude of degrees. The difference is that the two are at an offset: 0°C = 273.15 K; −273.15°C = 0 K. See *Kelvin*.

Centaurus: a bright constellation in the southern sky; known to Ptolemy in the 2nd century. Named after the *centaur*, which is an ancient Greek mythological creature that is a human upper torso on a horse's body.

centripetal force: a force that makes a body follow a curved path. The mathematical description of centripetal force was derived by Christiaan Huygens in 1659.

centromere: the part of a chromosome that links sister chromatids, which are the identical copies (chromatids) formed by replication of a single chromosome.

centrosome: an organelle in cells that serves as the main organizing center of microtubules.

cephalopod: a class of marine animals in the mollusk phylum. Squid, octopuses, cuttlefish, and nautilus are among the over 800 extant species of cephalopods.

cesium (*Cs*) (aka *cæsium*): the element with the atomic number 55; a soft, silvery-gold alkali metal; with a melting point of 28.5 °C, cesium is 1 of only 5 elements that is liquid at ambient temperature.

chaos theory: the study of dynamic systems highly sensitive to initial conditions, yielding widely divergent outcomes depending upon incremental differences early on. See *butterfly effect*.

chaperonin: a protein that provides a scaffold for initial protein folding.

Charaka Samhita (aka *Compendium of Charaka*) (3rd century CE): a compendium of 8 books on traditional Indian medicine (Ayurveda).

charge (electric): the force of electromagnetism per unit of time, measured in *coulombs*. Electrochemical charge is measured in *faraday*.

charge conjugation: a transform of a quantum particle into its antiparticle.

charge order: the orderliness in arrangement of electrons and holes with the same spin and momenta.

charge separation: the process of an atomic electron being excited to a higher energy level by absorbing a photon, and thereby by leaving home to join a nearby electron acceptor.

cheetah (*Acinonyx jubatus*): a large feline indigenous to Africa and part of the Middle East. The cheetah is the absolute fastest land animal: able to accelerate from 0 to 100 km/h in 3 seconds and sustain 115 km/h for short distances (500 m). The cheetah's agility and ability to anticipate the escape maneuvers of its specific quarry gives it the hunting edge it needs.

chemical species: atoms or molecules that are energetically equivalent.

chemistry: the study of matter, especially chemical reactions.

chemosynthesis: employing chemical reactions to generate usable energy.

chemotaxis: cell or organism orientation or movement toward or away from a chemical stimulus.

Cherenkov radiation: an electromagnetic radiation caused by charged particles polarizing molecules in a medium, resulting in radiation during the medium's return to its ground state. The characteristic blue glow of nuclear reactors owes to Cherenkov radiation. Named after Pavel Cherenkov, Cherenkov radiation had been predicted by Oliver Heaviside in 1888.

Chicxulub: site of a 66 million-year-old impact crater underneath the Yucatán Peninsula in Mexico.

chimeric: an organism of diverse genomic constitution.

chimeric gene: a gene formed from a combination of different coding sequences to produce a new gene.

chimpanzee (*Pan troglodytes*): a medium-sized ape, closely related to bonobos and humans.

chirality: handedness that demonstrates asymmetry. In organic chemistry, chirality is most often caused by an asymmetric carbon atom within the molecule.

chlorine (*Cl*): chemical element with atomic number 17. Chlorine is in the halogen group of elements. Chlorine is typically a yellow-green gas of diatomic molecules. Chlorine readily combines with other elements. Chlorine has the highest electron affinity, and the 3rd-highest electronegativity of all elements. Chlorine is a strong oxidizing agent.

chlorofluorocarbon (*CFC*): an organic compound comprising carbon, chlorine, fluorine, and hydrogen, produced as a volatile derivative of ethane and methane.

chlorophyll: the green biomolecule in plants that absorbs light for photosynthesis.

chloroplast: the photosynthetic organelle (plastid) found in algae and plant cells.

chloroplast capture: obtaining the genome of another plant by uptake of an organelle.

choanoflagellate: a unicellular, flagellate, planktonic eukaryote.

chondrite: a stony (nonmetallic) meteorite; formed from dust and small grains in the early solar system by accretion into primitive asteroids.

chromatid: a copy of a newly copied chromosome which is still joined to the original copy by a single centromere.

chromatin: the combined package of proteins and DNA that comprise physical genetic information storage in the nucleus of a eukaryotic cell.

chromophore: the moiety that causes a conformational change of a photosensitive molecule when hit by light.

chromosomal mosaicism: the condition of individual cells in a multicellular organism each having their own genome (pergenome).

chromosome: an elaborately coiled molecular package of genetic material in a eukaryotic cell, comprising DNA, regulatory elements such as histones, and other nucleotide sequences. *Chromosome* is sometimes used to refer to *genophore* as well, thus fuzzing its definition. For clarity, the distinction between the two is maintained in *Spokes*. Compare *genophore*.

chromosphere: the 2nd of 3 main layers in the Sun's atmosphere. The Sun's corona lies outside the chromosphere.

chron: the duration of consistency in Earth's magnetic field before reversing. A chron may last 0.1–1 million years, with an average of 450,000 years.

ciliate: a group of protozoans characterized by cilia.

cilium (plural: *cilia*): a hair-like protuberance from a cell, employed for sensory perception and/or locomotion (motile cilia). Flagella and motile cilia comprise a group of organelles termed *undulipodia*. Compare *flagellum*.

circadian rhythm: a biological process entrained to an endogenous oscillation of ~24 hours.

cis-regulatory element: a region of noncoding DNA that regulates transcription of nearby genes.

cisterna (plural: *cisternae*): a flattened membrane that is part of the *Golgi body*.

cistron: a segment of DNA with all the template information required for producing a genetic product; a synonym for *gene*.

citrate: a derivative of citric acid ($C_6H_8O_7$), which is a weak organic acid.

classical information theory: a branch of applied mathematics and electrical engineering concerned with quantifying information. Computers are working examples of classical information theory.

classical physics (*mechanics*): the Newtonian model of physics, notably gravity as a force of attraction; epitomized by Newton's 3 laws of motion and the laws of thermodynamics. Compare *modern physics*.

clutch: a group of laid eggs.

CMB (*cosmic microwave background*): thermal radiation permeating the observable universe.

CNO cycle: see *carbon–nitrogen–oxygen cycle*.

coding DNA (or *strand* or *region*): a DNA sequence, composed of *exons*, that codes for a protein.

codon: a nucleotide triplet which runs along the length of a DNA ladder. Codons were once though descriptive of the way that genetic information is stored but were found to be only a partial picture. See *cistron*.

cofactor: a molecule that binds to a protein to have the protein perform a task. Enzymes are typically activated by cofactors, which act as helper molecules. A cofactor molecule may either be an inorganic ion or organic (coenzyme).

coherence: the intelligent interaction behind Nature.

Cold War (1947–1991): the political and military tension after World War 2 between the United States (and its allies) and the Soviet Union (and its minion nations).

collective excitation: a subatomic particle not recognized in the Standard Model which behaves like a boson. Phonons and plasmons are collective excitations. Contrast *quasiparticle*.

coleoid: a soft-bodied cephalopod. Squid, octopuses, and cuttlefish are coleoids.

color charge: an abstracted indication of a particle's strong interaction according to quantum chromodynamics theory. Color charge is a property of a subatomic particle's field interaction with the strong nuclear force.

comet: a ball of ice and dust originating in the Oort cloud, in the outer reaches of the solar system.

common ancestor: the hypothesis that all life somehow arose from a single life form.

compactification (astrophysics): a hypothesis that any extra spatial dimensions that may exist do so at less than Planck length.

complex conjugate: a complex-number pair where the real components are identical but the imaginary parts, though of equal magnitude, have opposite signs. *1 + 2i* and *1–2i* are exemplary complex conjugates.

complex number: a number in the form of *a* + *bi*, where *a* and *b* are real numbers, and *i* is an imaginary number ($\sqrt{-1}$). Complex numbers extend a conceptual 1-dimensional number line (of real numbers) to a 2-dimensional complex plane (of real and imaginary numbers).

complex system: a nested hierarchical network.

compound (chemistry): a combination of elements bonded into a molecule.

concentration gradient (biology): an unequal distribution of ions across a cell membrane, causing a solute flow. Such movement is termed *diffusion*.

concept (aka *idea*): an abstract construct involving discriminatory categorization.

conceptualize, conceptualization: mentally resolving perceptions into a concept.

condensate: a condensed medium, typically a gas or liquid.

condensation reaction: a chemical reaction combining 2 moieties or molecules that results in a larger molecule, albeit at the loss of a small molecule.

conduction: (atomic) thermal transfer. See *convection*.

conductor (chemistry): a material amenable to transmitting electric charges and/or heat. Contrast *insulator, resistor*.

conformation (chemistry): a spatial configuration of elements.

conjugation (microbes, particularly bacteria): a term used for horizontal gene transfer (HGT) by researchers in 1946, who analogized HGT process to sex.

connective (tissue): 1 of the 4 primary animal tissue types. Connective tissue supports, separates, or connects other tissues. Immersed in body fluids, connective tissue is composed of cells, fibers, and extracellular matrix. See also *epithelium, muscle,* and *intelligence* (tissue).

consciousness: the platform for awareness in an individual life constituent, such as a protein, cell, or organism. Compare *Consciousness*.

Ĉonsciousness: the unified field of consciousness. Ĉonsciousness naturally localizes into individualized consciousnesses. Compare *consciousness*.

conservation (evolutionary genetics): preservation of a trait through generations (of cells or offspring).

conservation of energy: the unproven hypothesis that the energy of an isolated system is constant; that energy can be neither created nor destroyed in a closed system. Related to the *1st law of thermodynamics*.

conservation of momentum: the theory, implied by Newton's laws of motion, that the total momentum of a closed system is a constant.

constructal law: the tenet that the design and evolution of all forms aim to facilitate flow; postulated by Adrian Bejan in 1996.

continent: a gigantic landmass, 7 of which are currently extant on Earth: Africa, Antarctica, Asia, Australia Europe, North America, and South America.

continental drift: the movement of tectonic plates that causes continental masses to move about.

continuum mechanics: the study of matter as a process (rather than their particulate appearance).

convection: the concerted, collective movement of fluids (liquids, gases) and rheids (a solid deformed by viscous flow).

convection zone: the outermost layer of a star's interior, where turbulent energetic convection occurs.

convergent evolution (aka *parallel evolution*): the independent evolution of similar traits in organisms of separate species which are usually not closely related.

Cooper pair: 2 fermions, typically electrons, entangled via a phonon. Named after Leon Cooper, who first described the phenomenon in 1956.

Copenhagen interpretation: a conclusion formed in the late 1920s that wave/particle duality is merely computational, not actual.

Copernican principle: the hypothesis that the Earth is not the center of the universe. Named after Nicolaus Copernicus.

coral: a colonial marine invertebrate comprising numerous identical polyps.

core-accretion theory: a simplistic cosmological model of planetary development in star systems.

corona: an extremely hot plasma layer toward the outer edge of a star's atmosphere.

coronal mass ejection: a prominent release of plasma and magnetic field energy from the solar corona. Compare *solar flare*.

correlation: the fact that multiple phenomena coincide. Contrast *causality*.

cortisol: a glucocorticoid released when blood sugar is low or in response to stress.

corticosteroid: a class of steroid hormones involved in many vertebrate physiological responses, including stress, immunity, and regulation of carbohydrate and protein metabolism, inflammation, blood electrolyte level, and behavior.

cosmic inflation: a myth about the early cosmos, claiming that the universe had a near-instantaneous massive inflation 3×10^{-36} seconds after the onset of the Big Bang, which abruptly stopped. Cosmic inflation outrageously violates physics as understood.

cosmic microwave background (CMB): thermal radiation permeating the observable universe.

cosmic ray: radiation from outer space.

cosmogony: a conjecture about the origin of the universe.

cosmological constant: as an adjunct to general relativity, a construct first coined by Einstein to create a stationary universe.

cosmological principle: the false axiom that the distribution of matter in the universe is homogeneous and isotropic when viewed on a large-enough scale.

cosmology: the study of the universe.

cosmotrophic: an organism that can survive in space.

Coulomb force (aka *Coulomb's law*): the physics of static electricity, proposed by Charles-Augustin de Coulomb in 1784. More properly, an inverse-square law characterizing the electrostatic interaction between electrically charged particles. Coulomb's law was crucial in the development of the theory of electromagnetism, most prominently advanced by James Clerk Maxwell.

counterfactual: contrary to facts.

counterfactual (physics): values which could have been measured but were not. This is distinct from normal usage of the term.

counterfactual definiteness (*CFD*): a theory that phenomena are probabilistically consistent in repeatability. CFD is related to quantum mechanics and the uncertainty principle, regarding locality and superluminal interaction (entanglement). The validity of CFD was under consideration in *Bell's theorem*.

covalent bond: a stable chemical bond by sharing 1 or more pairs of electrons between atoms of a molecule.

CP (physics): an acronym for 2 hypothetical particle symmetries: charge conjugation (C) and parity (P).

CP violation: violation of charge conjugation (C) and/or parity (P) in a CP symmetrical system.

CRISPR (*Clustered Regularly Interspaced Short Palindromic Repeats*): a cluster of short, repeated DNA sequences found in prokaryotes from encounters with foreign DNA.

critical point: the point (in temperature, pressure, or composition) at which no phase boundaries exist for a substance.

critical temperature: the temperature below which a material becomes super-conducting.

crust (baking): the hard exterior of a bread loaf; the pastry portion of a pie.

crust (geology): the outermost solid slab of a rocky planet.

crystal: a solid characterized by an orderly, repeating pattern. A *lattice* is a typical crystalline pattern.

cryptochrome: a photosensitive protein, particularly blue light.

current (electrical): a flow of electric charge through a medium; alternately, a measure of charge passing through a point every time unit; measured in *amps*.

cyanobacteria: photosynthetic eubacteria; often called blue-green algae, though they are not in the same group as algae.

cyclic cosmology: a model that posits universes eternally coming and going. The cyclic model supposes a *multiverse*.

cycloalkane: a hydrocarbon with 1 or more rings of carbon atoms.

cytogenetics: the branch of genetics studying the structure and functions of eukaryotic cells, especially chromosomes.

cytokine: a group of small proteins critical to cell signaling.

cytokinesis: the process by which the cytoplasm of a eukaryotic cell divides.

cytology: the study of living cells.

cytolysis (aka *osmotic lysis*): an osmotic (water) imbalance from excess water inside a cell, causing the cell to burst.

cytoplasm: the watery gel that holds a cell's organelles within a plasma membrane.

cytosine (*C*) ($C_4H_5N_3O$): a nucleobase of DNA and RNA. Cytosine is complementary to *guanine*. Cytosine is inherently unstable and can spontaneously change into *uracil* (spontaneous deamination). If not repaired, spontaneous deamination results in a *point mutation*.

cytoskeleton: filaments of protein within a cell, providing cellular scaffolding.

cytosol (aka *cytoplasmic matrix* or *intracellular fluid*): cytoplasmic fluid (the liquid within cells), comprising mostly water, along with dissolved ions and various molecules, including proteins.

D

D-brane: a higher dimensional (HD) object; related to *M-theory*.

Dada (aka *Dadaism*): an early 20th-century art movement in Europe and North America which rejected reason and the aestheticism of modern capitalism for nonsense and an anti-bourgeois sentiment.

dark energy: an aberration in ΛCDM of a hypothetical energy that permeates 3D space, exerting negative pressure, thus tending to accelerate the expansion of the universe.

dark matter: a discredited hypothetical matter that supposedly exists extra-dimensionally (ED), lending only gravitational distortion to experiential 3D space. Despite extensive search, no evidence of dark matter has been found. Contrast *baryon, light matter*.

data: factual information.

daughter cell: a cell formed from a *parent cell*.

de novo: anew.

decay event: an event initiating radiation.

decision theory: (statistics): quantitative methods for reaching optimal decisions for defined problems. Decision theory is related to *game theory*.

decoherence: loss of quantum coherence (*superposition*) via environmental interactions.0

deduction (logic): the method of inferring a conclusion about particulars from general principles. Contrast *induction*.

degree of freedom: 1 of a limited number of ways in which a dynamic system may change.

delocalize: to free from locality.

density wave: an oscillation in the galactic gravitational field that influences star motion.

dephosphorylation: removing at least 1 phosphate group from an organic compound via hydrolysis. Energy is gained from ATP by dephosphorylation; ATP is turned into ADP. Contrast *phosphorylation*.

determinism: belief in cause and effect, from which emanates the doctrine that all facts and events exemplify natural laws.

deuterium (aka *heavy hydrogen*): a stable isotope of hydrogen, comprising a nucleus of a proton and a neutron. Contrast *protium*.

dialectic: logical argumentation based upon the interaction of juxtaposed ideas.

diamidophosphate ($PO_2(NH_2)_2-$): a simple ion of phosphorous, nitrogen, and hydrogen.

diatom: an alga; one of the most common phytoplankton.

diatomaceous earth: a soft, siliceous sedimentary rock, easily crumbled into a fine whitish powder. Diatomaceous earth comprises fossilized *diatoms*.

diatomic: 2 nuclides of the same *atomic species*.

diffraction: the bending of energy waves around obstacles.

diffusion (chemistry): the passage of molecules between chemical species.

dihydrofolate reductase (*DHFR*): an enzyme critical to producing DNA precursors.

dinosaur: a diverse clade of largely extinct reptiles, excepting birds; an arbitrary exclusion, as birds descended from dinosaurs.

diploid: a cell having 2 sets of chromosomes. Most eukaryotes, and almost all mammals, are diploid: 2 sets, 1 from each parent, typically through sexual reproduction. Compare *haploid*.

dipole (physical chemistry): a polar molecule.

dipole (physics): a pair of equal and opposing electric or magnetic poles, separated by an infinitesimal distance.

Dirac equation: a relativistic quantum mechanical wave equation that characterizes the spin of fermions; created by Paul Dirac in 1928.

Dirac fermion: a fermion with mass and charge; named after Paul Dirac. Ordinary matter is made of Dirac fermions. Compare *Weyl fermion, Majorana fermion*.

diradical: a molecular species with 2 electrons occupying 2 degenerate (equal energy) orbits. O_2 and CH_2 (methylene and carbene) are exemplary diradicals.

dirt: see *soil*.

disordered hyperuniformity: coherent patterning within an apparently disordered system.

dispersion relation: the effect of dispersion on waves in a medium. Dispersion occurs when pure plane waves of distinct wavelengths have their own propagation velocities, so that a wave packet of mixed wavelengths tends to spread out in space.

dissolved organic matter (*DOM*): the slowly sinking remains of oceanic life.

distributed causality: multiple agents in a nonlinear dynamic system that render initial causality uncertain.

divergent (mathematics): in context, an integral that sums to infinity.

DNA (*deoxyribonucleic acid* ($C_5H_{10}O_4$)): a long, double-stranded molecular chain employed as a physical template for biochemical production. DNA is physically heritable. There is no reasonable explanation based upon known facts that the information essential for trait inheritance is portered by DNA; quite the contrary: DNA itself cannot possibly be the energetic agent of heredity. See *RNA*.

domain (biological classification) (aka *empire*): the 2nd highest taxon (below *life*), with 3 classes: archaea, bacteria, and viruses.

dominant (trait): a genetic trait that masks a *recessive trait*.

DON: dissolved organic nitrogen.

Doppler shift (aka *Doppler effect*): a change in observed frequency relative to the source of a generated wave; proposed by Christian Doppler in 1842.

double bond: a chemical (covalent) bond of sharing 2 pairs of electrons. Compare *single bond* and *triple bond*.

doublet (chemistry): a diradical with a spin of 1/2. Contrast *singlet* and *triplet*.

Down's syndrome (aka *Down syndrome, trisomy 21*): a human genetic developmental disorder that causes physical growth delays and intellectual disability.

dwarf galaxy: a relatively small galaxy, with up to a few billion stars. The term *dwarf* is relative to the Milky Way galaxy, which has 200–400 billion stars.

dynamic kinetic stability: the ability of a dynamic system to maintain homeostasis.

E

$E = mc^2$: an equivalence of energy and mass, embodying the concept that the mass of an object is a measure of its energy content; formulated by Albert Einstein in 1905.

E. (*Escherichia*) *coli*: a rod-shaped enterobacteria commonly found in the lower intestine of endothermic organisms. *E. coli* normally colonize an infant's gut within 40 hours of birth, delivered by food, water, or mere handling. *E. coli* has long been a model organism in microbiology studies; one of the first organisms to have its entire genome sequenced, in 1977.

Earth: the 3rd planet from the Sun; the densest and 5th largest.

ecology: an interactive interface; patterns of relations among entities; as a subdiscipline of biology, patterns of interrelations between life forms (e.g., cells, organisms) and their environment (including other organisms); more broadly, the relations between bioelements.

ecosystem: the community of biota in a biome, and the abiotic (non-living) elements within the area.

ectotherm: an animal species without internal means to maintain thermal homeostasis. Ectothermic species, such as reptiles, practice behaviors to regulate body temperature, like lying in the Sun to warm oneself. Commonly misnamed *cold-blooded*, ectotherms' blood is just as warm as endotherms. Compare *endotherm*.

ED: extra dimensions (or extra dimensionality); the dimensions of existence beyond those that are perceptible and measurable. See *4D* and *HD*.

effector molecule: a regulatory molecule that binds to a protein and alters the protein's activity.

egene: (the idea of) an energetic hereditary unit which conveys all the information needed to create a trait or biological effect. Nucleic acids alone cannot explain heredity. Compare *gene*.

egg: an organic vessel in which an embryo first begins development. See *sperm*.

eigenstate: a measured state of an object with quantifiable characteristics, such as position and momentum. The state must be measurable and have a definite value (*eigenvalue*).

electric dipole moment (*EDM*): a measure of electrical polarity by measuring the separation between negative and positive charges.

electric potential (aka *electric field potential*, *electrostatic potential*): the amount of work needed to move a positive charge inside a field without creating acceleration.

electrical resistance: a measure of opposition to flow of an electric current. See *conductor* and *resistor*.

electrodynamics: the branch of classical physics studying the interactions between electric charges and currents.

electrolyte: a substance that releases ions when dissolved in water. Salts, acids, and bases are electrolytes.

electromagnetic radiation (*EMR*): energy emitted and absorbed by charged particles. EMR exhibits wavelike behavior as it traverses space.

electromagnetic spectrum: a continuum of increasing energy intensity, from longer wavelengths to shorter.

electromagnetism: one of the fundamental physics forces, affecting particles that are electrically charged. Except for gravity, electromagnetism is the ambient physical interaction responsible for practically all phenomena encountered in everyday life.

electron: a negatively charged fermion. An electron hypothetically has 1/1836 the mass of a proton when at rest, but an electron is never at rest.

electron acceptor: an atom or molecule that accepts electrons.

electron cloud: the cumulative electron shells of an atom.

electron diffraction: reference to the wave nature of electrons.

electron orbital: the orbit of an electron about an atomic nucleus. See *shell*.

electron transfer: the donation of an electron from one atom to another.

electron transport chain: an electron transfer by coupling an electron donor and electron acceptor, with a transfer of hydrons across a membrane. For an electron transport chain to function, allowing electrons to pass through, an exogenous electron acceptor must be present at the end of the chain. Cell respiration requires an electron transport chain.

electron volt (*eV*): an energy measurement unit. 1 eV is the energy that an electron gains in passing through an electric field with a potential difference of 1 volt.

electronegativity: the measure of a chemical species to take electrons; *electroaffinity* would be a more accurate term. Contrast *electropositivity*.

electropositivity: the measure of a chemical species to donate electrons; *electrocharity* would be better. Contrast *electronegativity*.

electrostatics: the branch of physics studying electric charges at rest.

electroweak force: a quantum field theory uniting the electromagnetic and weak forces.

elegance (mathematics): an attribution to a physical model that is relatively simple and mathematically cogent.

element (chemistry): a species of atoms with the same number of protons in their nuclei.

elementary particle: a subatomic particle that has supposedly no constituents, even though they do: *virtual particles*. Elementary particles are the supposed bottom-up building blocks of the cosmos, and, by their continuous 4D/ED interaction, comprise an interface between observable (4D) manifestation and actual (HD) existence. See *HD*.

embryogenesis: the process by which an embryo develops.

emergence: the way that complexity arises from a multiplicity of simple interactions. The idea of emergence has been around at least since Aristotle, who expressed that the totality of reality is greater than the sum of its parts; a non-reductionist sentiment. In physics, emergence refers to existence coming into being on an infinitesimal moment-by-moment basis.

empirical: based upon fact.

empiricism (epistemology): the presumption that knowledge derives solely from sensory experience.

empiricism (philosophy of science): the belief that Nature may be entirely explained by physical forces.

Enceladus: the 6th largest of Saturn's 62 moons.

endocytosis: the cellular process of absorbing macromolecules, such as proteins, by engulfing them. All cells employ endocytosis. Contrast *exocytosis*.

endogenous (biology): originating within an organism. Contrast *exogenous*.

endogenous retrovirus: a transposable element that resembles a retrovirus.

endoplasmic reticulum (ER): an organelle connected to the nuclear membrane; a membranous network of sac-like structures (cisternae) held together by the cytoskeleton. ER serves various functions, including carbohydrate metabolism, lipid synthesis, glycoprotein production, and cell membrane manufacture. ER also plays a critical role in assisting mitochondrial division and replication.

endosperm: the nutritious tissue inside flowering plant seeds.

endosymbiont: an organism living within the body or cells of another organism, forming a symbiotic relationship.

endosymbiosis: the evolutionary incorporation of an organism by another.

endotherm: an animal species with internal means to maintain thermal homeostasis. Birds and mammals are endotherms. Endothermy raises an animal's metabolic needs compared to ectothermic animals. Compare *ectotherm*.

endothermic reaction: a chemical reaction that absorbs thermal energy. Contrast *exothermic reaction*.

energy (physics): the idea of an immaterial force acting upon or producing matter. Energy is characterized relatively and by type (how it affects matter). Energy manifests only through its effect on matter. Though the foundational construct of existence, energy itself does *not* exist. As matter is made of energy, this fact tidily proves *energyism*.

energy landscape: a set of possible conformations, with each potential spatial position (conformation) having an associated energy level.

energyism (aka (*philosophical*) *immaterialism*): the monistic doctrine that Nature is a figment of the mind. Energyism differentiates between actuality and reality. Whereas actuality is phenomenal, reality has a noumenal substrate, emergently spawning a shared actuality (showtivity) via a unified Ĉonsciousness. Contrast *matterism*.

entanglement (physics): distinct phenomena behaving synchronously. Entanglement defies *locality*.

entropy (physics, particularly thermodynamics): the tendency of energy to dissipate and equilibrate; a measure of thermal energy unavailable for work; introduced by Rudolf Clausius in 1865. An entropic interaction is one where energy is locally lost. Gravity is entropic.

environment: a designated spatial region or conceptual realm.

envirotype: the ecological influences on an organism and typical organism interactions with the environment.

enzymatic: (an) enzyme catalyzed or inhibited (reaction).

enzyme: a protein that facilitates the activities of other proteins or substrates. Enzymes typically act as catalysts.

epiallele: the idea of an allele loaded with epigenetic information, affording divergence from straightforward gene expression.

epigenetics: (the study of) gene regulation and physical heredity mechanisms without changing the structure of the DNA involved – that is, without genetic mutation.

epigenome: the conceptual sum of instructions in a cell affecting access and expression of genes.

epipubic bone: a pair of bones projecting forward from the pelvic bones of modern marsupials and most non-placental fossil mammals.

epithelium (plural: *epithelia*): 1 of the 4 primary animal tissue types. Epithelial tissues line the surfaces and cavities of bodily structures and form many glands. Epithelial cells secrete, selectively absorb, protect, and transport. See also *muscle*, *connective tissue*, and *intelligence* (tissue).

equivalence principle: following Galileo's conception, Albert Einstein's proposition regarding apparent acceleration: that there is no way to distinguish the effects of acceleration (inertial mass) from the effects of gravity (gravitational mass).

erythrocyte: red blood cell.

Escherichia coli: see *E. coli*.

essential amino acid: an amino acid necessary for health that cannot be synthesized by the human body and so must be obtained via diet.

ester: an organic compound comprising a *carbonyl* adjacent to an *ether*.

ether: a class of organic compounds characterized by an oxygen atom bonded to 2 carbon atoms (C–O–C).

Euclidian geometry: a mathematical system limited to 3D, attributed to Euclid. Euclidian geometry has a small set of axioms from which theorems can be deduced. The 5th axiom (the parallel postulate) was found independent of the first 4 in the 19th century. Its breakage led to non-Euclidian geometry.

eugenics: beliefs and practices aimed at improving the genetic quality of humans.

eukaryote: an organism with cell structures (organelles) separated by membranes. Multicellular life is eukaryotic. Compare *prokaryote*.

Euler Beta function: an equation used to characterize scattering amplitude; employed in *string theory*.

eusocial: an animal species that has: 1) overlapping generations; 2) cooperative care of the young; and 3) reproductive division of labor. Contrast *presocial*.

Eutheria: the placental mammal clade that diverged ~160 MYA. Eutherians lack epipubic bones, allowing for an expanding abdomen during pregnancy.

eutrophication: the process by which a body of water becomes enriched with dissolved nutrients that stimulate the growth of microbial aquatic life, which typically results in depleting the oxygen dissolved in the water.

event: a perceived process with an outcome.

event horizon: a boundary in spacetime beyond which events cannot affect an outside observer. An event horizon is typically portrayed as the "point of no return" into a *black hole*.

evolution (evolutionary biology): the process of adaptation, most apparently seen as a distinctive change across successive generations of a population.

existence: corporeality, including both matter and energy. See *actuality*, *manifestation*, *Nature*, and *phenomenon*.

exocytosis: the cellular process of secreting proteins outside the cell. Contrast *endocytosis*.

exogenous (biology): originating outside an organism. Contrast *endogenous*.

exon: a polynucleotide sequence in a nucleic acid that codes for protein synthesis. An exon is copied and spliced together with other such sequences to form *messenger RNA*. Compare *intron*.

exothermic reaction: a chemical reaction that releases thermal energy. Contrast *endothermic reaction*.

extensive property: a physical property of a system that depends upon system size or materiality. Examples include mass and volume. Contrast *intensive property*.

extinction: the demise of a species.

extinction event: a period of mass extinction.

extracellular matrix (ECM): a biological matrix composed of different glycosylated proteins that create attachment bases for cells, holding tissue together without direct contact between neighboring cells. Glycocalyx is a common ECM.

extra dimensions: see ED.

extremophile: an organism that thrives in an environment adverse to most life.

F

fact: recall of an experienced event.

> Facts are of not much use, considered as facts. They bewilder by their number and their apparent incoherency. Let them be digested into theory, however, and brought into mutual harmony, and it is another matter. ~ Oliver Heaviside

Fahrenheit: an obsolete temperature scale, named after Daniel Fahrenheit, who suggested it in 1724. Fahrenheit is used only in Belize, the Bahamas, the Cayman Islands, Palau, and the United States. The Fahrenheit scale was set upon 3 references: 1) a frozen mixture of water, ice, and salt (0°); 2) where water nominally freezes (32°); and 3) typical human body temperature in the mouth or under the armpit (96°). Water boils at 212° F. Conversion to Fahrenheit: $[°F] = [K] \times \frac{9}{5} - 459.67$. Room temperature of 296 K is 73° F (23° C). See *Celsius, Kelvin*.

falsifiability (aka *refutability*): a statement (hypothesis or theory) which may be tested for validity through observation. The concept was introduced by Karl Popper in 1994 as a cornerstone of scientific epistemology. Statements which are not supported by falsifiability are *pseudoscience*.

fat (chemistry): a broad group of compounds comprising carbon, hydrogen, and oxygen; a subgroup of *lipids*. See *saturated fat* and *unsaturated fat*.

fatty acid: a carboxylic acid with a long aliphatic tail (chain).

fauna (plural: *faunas* or *faunae*): animals (metazoa). Compare *flora*.

femtometer (*fm*) (aka *fmeometre*): 10^{-15} of a meter.

Fermat's last theorem (aka *Fermat's conjecture*): a 1637 number theory by Pierre de Fermat that no 3 positive integers (a, b, c) satisfy the equation $a^n + b^n = c^n$ for any integer value of n greater than 2. The first successful proof was by Andrew Wiles in 1994.

Fermat's principle (aka *principle of least time*): a 1658 optics principle by Pierre de Fermat that light always travels most efficiently: from one point to another in the least time.

fermion: a quantum of matter under quantum physics' Standard Model; named after Enrico Fermi. Contrast *boson*.

fern (aka *Pteridophyta*): the first vascular plant.

ferromagnetism: the ability of a material to become a permanent magnet.

field: an energy associated with a spacetime point or region.

first law of thermodynamics: see *1st law of thermodynamics*.

fission (cytology): cell division into 2 (binary fission) or more (multiple fission) cells.

flagellum (plural: *flagella*): a whip-like appendage protruding from a cell, employed for locomotion and sensory perception. Compare *cilium*.

flatworm: a relatively simple unsegmented, bilateral (head and tail), soft-bodied invertebrate.

flavor (quantum mechanics): generic term for the qualities that distinguish the various quarks and leptons.

flora (plural: *florae* or *floras*): plants. Compare *fauna*.

fluid: a substance that deforms (flows) under an applied shear stress. Gases, plasmas, and liquids are fluids. Contrast *solid*.

fluorine (*F*): the element with atomic number 9; molecularly diatomic (F_2). At standard pressure, fluorine is a pale, yellow gas. With a −1 oxidation state, fluorine is the most electronegative element, and so a strong oxidant. Fluorine is the 13th most common element in Earth's crust, naturally occurring as a fluoride ion. Fluorine is not essential biologically. The few organisms that employ fluorine in their biochemistry do so to make poisons.

food web: the energy production and consumption interrelations between biota in an ecosystem.

force (physics) (aka *interaction*): an influence that causes a change in Nature. There are 5 known forces: coherence, strong (nuclear), weak (nuclear), electromagnetism, and gravity.

formaldehyde (($CH_2O(H\text{-}CHO)$) aka *methanal*): a naturally occurring organic compound that is a precursor to many other chemical compounds.

formose reaction: the formation of a sugar from formaldehyde; a portmanteau of *form*aldehyde and ald*ose*.

fossil fuel: a fuel formed from dead organisms. Coal, natural gas, and petroleum are fossil fuels.

fougèrite (aka *green rust* ($Fe^{2+}_4 Fe^{3+}_2(OH)_{12}[CO_3]\cdot 3H_2O$)): a naturally-occurring mineral.

fractal: a set of scale-invariant self-similar iterative patterns.

fractional particle: a subatomic particle (e.g., electron) exhibiting dichotomous or incongruent properties.

fragmentation (biology): a form of asexual reproduction, where a new organism grows from a fragment of the parent.

free electron: an electron not bound to an atom.

freezing: the physical process of a liquid turning into a solid.

frequency: the number of repetitious occurrences per time unit.

fumarolic (*vent*): a hole in a volcanic region from which hot gases and vapors issue.

functional group (chemistry): the specific group of atoms within a molecule responsible for the molecule's characteristic chemical reactions.

fungus (plural: *fungi*): a kingdom of eukaryotes that includes microorganisms such as yeast and molds, as well as macroscopic mushrooms.

fusion (physics): the energetic process of multiple atomic nuclei fusing.

G

G-quadruplex: a guanine-rich 4-stranded DNA structure, squarish in shape.

galactic web: the interconnection of galaxies via gravitational and energetic filaments.

galaxy: via a massive black hole, a gravitationally bound cluster of star systems and stellar remnants, swirling in an interstellar mixture of gas and dust.

Galilean relativity (aka *Galilean invariance*): a 1632 hypothesis by Galileo Galilei that the laws of motion are the same in all inertial frames.

gall wasp (aka *gallfly*): a small wasp of 1,300 species. The larvae of most gall wasps develop in plant galls which they induce. Oak is the wood of choice for many gall wasps.

game theory: theorization of outcomes and dynamics in situations involving parties with conflicting interests.

gamete: a cell or cell nucleus that undergoes sexual fusion to form a zygote. In animals, gametes are eggs and sperm cells. Plant germ cells produce ovules and pollen.

gametophyte: the haploid, gamete-producing phase of plants and algae that undergo alternation of generations; the prothallus in ferns, and the embryo sac in angiosperms. Compare *sporophyte*.

gamma ray: electromagnetic radiation above 10 exahertz (>10^{19} Hz); extremely high energy/frequency radiation.

Ganymede: Jupiter's largest moon, and the largest in the solar system. Ganymede is the 9th-largest body in the solar system, and the largest without an atmosphere to speak of.

gas: a fluid that may be airborne.

gauge boson: a quantum force carrier.

gene: the idea that a set of nucleic acids provide instructions for producing an organic molecule, typically a protein. Genes do *not* exist; they are merely a construal. The actuality of genetics is more intricate than supposed by matterist geneticists, as heritable bioproduct information is stored energetically, with organic molecules as illusory material substrates.

gene expression: employment of a gene; the conceptual process by which genetic information is used to synthesize a bioproduct.

gene mapping: the process of determining the locus for a specific biological trait.

gene product: the biochemical material resulting from gene expression. A protein is the typical gene product, though RNA is also a gene product.

gene regulation: control of gene expression, including stifling gene expression.

general relativity: a geometric physical theory that treats gravity as a property of spacetime, based upon the mass of objects; proposed by Albert Einstein in 1915. Gravity distorts 4D spacetime extra-dimensionally under general relativity. See *relativity, special relativity*.

generation (physics): a division for fermions, based on mass. Only 1st-generation fermions make up everyday matter. 2nd- and 3rd-generation fermions rapidly decay.

genetic code: the conceptual rulebook by which information is encoded in genetic material.

genetic drift (aka *allelic drift*): a difference in genome between species in a hereditary lineage.

genetic mutation: a change in a DNA sequence.

genetics: the study of heredity and variation in life forms at the molecular level. The 4 major subdisciplines of genetics are transmission genetics (heredity), molecular genetics (chemistry), population genetics (traits in populations), and epigenetics (influences of living on inheritance).

genome: the (idea of the) entire set of genes within an organism. Like genes, a genome is merely a concept, not phenomenal.

genophore: a package of DNA in a prokaryote's nucleoid. Compare *chromosome*.

genotype: the energetic constitution of an organism, as artifactually represented by genome. The *gen* in genotype refers to genesis (not genetics).

genus (plural: *genera*): a category of organisms, more generic than *species*.

geodetic effect: the curvature of spacetime caused by an orbiting body, such as a planet around a star.

geodynamic: relating to dynamic processes or forces within Earth.

geology: the science of the solid matter that comprises Earth, especially in the crust.

geosphere: within Earth, including the crust and mantle. Compare *pedosphere*.

germline cell: the line (sequence) of cells that may be passed to offspring. Contrast *soma*.

GeV (giga-electron volt): a unit of energy equal to a thousand million (10^9) electron volts (eV).

ghost field: a field that affects the mass of a boson via interrelations with other bosons and fermions. Ghost fields are necessary to maintain mathematical consistency in quantum physics' Standard Model. Ghost fields are conventionally presumed to be solely a mathematical device, and not exist, despite their being the origin of *virtual particles*, which supposedly manifest.

glacial period (aka *glaciation*): a period of glaciers, typically thousands of years, within an ice age, marked by colder temperatures and glacial advances. By contrast, interglacials are periods of warmer climate within an ice age. The last glacial period ended 15,000 years ago. The present epoch, the Holocene, is the current interglacial.

glia: the predominant cell type in animal brains. Neurons (nerve cells) support glial cells via their interfaces outside the brain.

glass: an amorphous (non-crystalline) solid.

glass transition: a temperature associated with phase transition from glass to liquid. The glass transition temperature is always lower than the melting temperature.

glucocorticoid: a corticosteroid that regulates glucose metabolism. The most important human glucocorticoid is *cortisol*.

glucose ($C_6H_{12}O_6$): a simple sugar used in glycolysis to form ATP.

gluon: the boson that porters the strong force.

glycan (aka *glycosyl group*): ostensibly a synonym for *polysaccharide*, but commonly used to refer to the carbohydrate bonded to a protein or other glycoconjugate.

glycerol: a simple alcohol compound, comprising 3 hydroxyl groups (3 molecules of hydrogen and oxygen).

glycocalyx: extracellular polymeric material comprised of glycoproteins. See *extracellular matrix*.

glycoconjugate: a carbohydrate covalently bonded to another chemical species, including peptides, proteins, and lipids.

glycolipid: a lipid with an attached carbohydrate. Glycolipids provide energy and act as markers for cellular recognition.

glycolysis: a metabolic pathway of 10 reactions that results in free energy; often used to form ATP.

glycoprotein: a protein containing a carbohydrate (glycan) attached to a polypeptide side chain.

gold (*Au*): the element with the atomic number 79; a dense, malleable, and ductile metal that is a bright reddish-yellow (golden) in hue. Gold is one of the least reactive elements.

Goldilocks (aka *The Story of the Three Bears*): a fairy tale in which an intrusive little girl pilfers porridge from homebody bears.

Golgi body (aka *Golgi complex*, *Golgi*): an organelle comprising a stack of membranes that works in concert with the endoplasmic reticulum to package proteins inside a cell before shipping the proteins off to their intended destination. Discovered by Camillo Golgi in 1898 while investigating the human nervous system.

GPa (*gigapascal*): *pascal* (*Pa*) is a standard unit of pressure. Geophysicists use *gigapascal* (*GPa*) for tectonic stresses with the Earth. Herein, *GPa* is used for intense pressures related to superconductivity.

GPS (*global positioning system*): a satellite-based navigation system.

granite: a coarse-grained igneous rock, at least 20% quartz by volume.

graphite: a crystalline, semimetal form (allotrope) of carbon. Graphite is a native element mineral, and a form of coal.

graviton: the hypothetical boson of gravity.

gravity: an entropic spacetime distortion caused by mass. Generally considered one of the 4 fundamental forces, though that is something of a misconception, as the other 3 interactions – strong, weak, and electromagnetism – are significant to subatomic particles, whereas gravity is not.

greenhouse gas: a gas in the atmosphere that absorbs and emits radiation within the infrared range. The primary greenhouse gases in Earth's atmosphere are carbon dioxide, methane, nitrous oxide, and ozone. Water vapor acts as a greenhouse gas.

greigite (Fe_3S_4): an iron sulfide mineral, found in clays and hydrothermal veins. One commonly found impurity in greigite is nitrogen. (FeNi)S clusters are somewhat common in enzymes, while the cubic Fe_4S_4 unit of greigite is employed by proteins for metabolism.

Grotthuss mechanism: the process of a proton moving through the hydrogen bond network of water molecules or other hydrogen-bonded liquids via the formation and concomitant cleavage of covalent bonds of neighboring molecules. Proposed by Theodor Grotthuss in 1806; an astonishing theory at the time, as the water molecule was thought to be HO, not H_2O, and ions were not understood.

ground state: the lowest energy state of a quantum mechanical system.

guanine (*G*) ($C_5H_5N_5O$): a nucleobase of DNA and RNA. Guanine is complementary to *cytosine*.

gymnosperm: a group of seed-producing plants, including conifers, cycads, ginkgo, and gnetophytes.

gyre: a conceptual framework treating a physical system as a dynamic vortex. A gyre is characterized by its structure, qualities, thermodynamics, and interactions. See *tensor*.

H

Haber process: a process for synthesizing ammonia, involving the nitrogen fixation reaction via hydrogen gas and nitrogen gas, catalyzed by enriched iron or ruthenium; named after Fritz Haber, its inventor.

habitat: the environment in which a species population lives.

Hadean (4.55–3.9 BYA): the 1st geologic eon, originally thought to be before life originated on Earth (but life started 4.1 BYA).

hadron: a composite subatomic particle made of a variety of quarks. Matter is comprised of *baryons*: hadrons composed of 3 quarks.

half-life: the duration required for a material to decay to half of its initial value. The probabilistic term is commonly used in nuclear physics to state the radioactive decay rate of atoms. Medical sciences use half-life to refer to the biological breakdown of chemical substances in the body.

halogen: a group of chemically related elements, so named because they all produce sodium salts with similar properties (*hal* being Greek for *salt*, and *gen* for *generate*). The 4 natural halogen elements are fluorine (F), chlorine (Cl), bromine (Br) and iodine (I). Astatine (At) is a halogen that exists as a short-lived radioactive isotope, as is the artificially conceived element 117 (ununseptium (Uus)).

halophile: an organism that lives in a salty habitat.

Hamilton's principle: the principle that the dynamics of a physical system are determined via variation in the Lagrangian function, which contains all information about the system and the forces acting upon the system. Originally formulated for classical mechanics by William Rowan Hamilton in 1833. Hamilton's principle also applies to classical fields (e.g., electromagnetism, gravity), and is relevant to quantum mechanics, quantum field theory, and criticality theories.

Hamiltonian mechanics: a reformulation of classical mechanics used in characterizing quantum mechanical systems. The Hamiltonian refers to the total energy of a system. A Hamiltonian system is a dynamic system governed by Hamilton's equations, which were derived by William Rowan Hamilton in 1833.

haplodiploid: a sex-determination system where the sex of offspring is determined by the number of sets of chromosomes received. Female eusocial insects, such as bees, wasps, and ants, are diploid, but males are haploid because they develop from unfertilized (haploid) egg cells.

haploid: a cell having 1 set of chromosomes. Compare *diploid*.

Hawking radiation: black-body radiation emitted by black holes, predicted by Stephen Hawking in 1974.

HD (*holistic dimensionality*): the totality of cosmic dimensions. HD refers to the universe having more than 4 dimensions (4D = 3 spatial dimensions and 1 time vector). HD = 4D + ED, where ED = extra (spatial) dimensions.

heat capacity (aka *thermal capacity*): the amount of heat needed to raise the temperature of a substance. See *specific heat capacity*.

heavy water (*deuterium oxide* (D_2O)): water with a higher hydrogen content (deuterium) than typical (light) water.

Heisenberg's uncertainty principle: see *uncertainty principle*.

helicase: a class of enzymes that unpackage nucleic strands (DNA, RNA). Helicases are vital to all organisms.

heliocentrism: the theory that the Sun is the center of the solar system around which planets orbit, including Earth.

heliosphere: a plasma bubble of charged particles in space blown by the solar wind.

helium (He): the element with atomic number 2; a colorless, odorless, tasteless, non-toxic, inert, monatomic gas. Helium is the 2nd lightest and 2nd most abundant element, behind hydrogen.

hemoglobin: the iron-based oxygen transport protein in the red blood cells of vertebrates.

herbaceous: an angiosperm that has leaves and stems which die down to the ground at the end of the growing season. Herbaceous plants have no persistent woody stem above ground. Herbaceous plants may be *annuals*, *biennials,* or *perennials*. Contrast *arborescent*.

heredity (genetics): inheritance of traits from one generation of life form to the next.

heterogamy (reproductive biology): sexual reproduction, as contrasted to parthenogenetic generation; in the context of *alternation of generations*. Contrast *parthenogenesis*.

heterotroph: an organism that cannot make its own food. All animals are heterotrophs. Compare *autotroph*.

heterozygous: different alleles at the same locus. Contrast *homozygous*.

heuristic (psychology): a simple, efficient rule employed to form judgments, solve problems, or make decisions. Compare *algorithm*.

hierarchy problem: the inability to explain why a theorized physical parameter value vastly differs from its effective (measured) value.

Higgs boson: a massive but elusive subatomic particle. Finding the Higgs put a finishing touch on the Standard Model of particle physics, by providing a means for fermions to have mass, while bosons supposedly don't, though at least 2 actually do (W & Z).

Higgs field: according quantum physics' Standard Model, the universal field that imparts mass. Quanta hypothetically swim in the Higgs field, interacting at different strengths, and so maintain distinct masses, or are massless if the Higgs field fails to impress. The quantum representing the Higgs field is the *Higgs boson*. See *Higgs mechanism*.

Higgs mechanism: the continuous process whereby gauge (W & Z) bosons acquire mass via spontaneous symmetry breaking (SSB). The Higgs mechanism exemplifies the basic mechanism by which Nature is composed: universal fields localizing, with local fields quantizing into particulate form. The exposition of Ĉonsciousness works similarly: from universal to localized field (individual consciousnesses).

Hilbert space: a geometry capable of characterizing any number of dimensions. Named after David Hilbert by John von Neumann.

Hinduism: an indigenous religion of the Indian subcontinent, dating to the 7th century BCE.

histone: a highly alkaline protein in a eukaryotic cell nucleus that packages DNA into a nucleosome. Histones also act intracellularly as an antibacterial agent.

HIV (human immunodeficiency virus): an enveloped RNA retrovirus, termed for the immune system deterioration it causes, leading to AIDS (acquired immunity deficiency syndrome).

hole (physics): a conceptual absence of an electron in an environment where electrons are abundant. An electron excited into a higher state leaves a hole in former, less energetic state. Contrast *positron*.

holism: the idea that systems and their properties should be viewed holistically (from the perspective of being a whole), not just as a collection of components. Contrast *reductionism*. See *synergy*.

holograph (aka *hologram*): a recording made via storage of interference patterns.

holographic principle: a conjecture, derived from string theories, that the universe is an information structure painted on a cosmological canvas, with energy and matter as incidentals.

homochirality: the geometry of something made of chiral units.

homeopathy: a pseudo-medicinal treatment of drinking water that has a specific substance diluted beyond measurement. Homeopathy can be effective via the *placebo effect*.

homeostasis (biology): a regulatory process by which an organism strives for holistic health. Compare *autopoiesis*.

homeostasis (physics): a tendency toward stability within a system.

homogeneous: the same at all locations. Compare *isotropic*.

homolog (biology): a shared evolutionary ancestor.

homologous (chromosomes): duplicate chromosomes (having the same allelic genes). See *homozygous*.

homozygous: selfsame alleles at the same locus on homologous chromosomes. Contrast *heterozygous*.

horizon problem (aka *homogeneity problem*): the conundrum that the cosmic microwave background exhibits a uniformity which cannot be explained by known physics.

horizontal gene transfer (*HGT*): sharing of genetic material between organisms. Contrast *vertical gene transfer*.

hormone: an organic compound intended for long-distance intercellular communication; from the Greek word for *impetus*.

host (biology): a cell, virus, or organism in/on/to which another organism has an interest or relationship.

host cell: a cell hosting an *endosymbiont*. Eukaryotes arose from an archaeon hosting one or more bacterial endosymbionts.

host range: the cell type(s) that a virus infects by recognizing cell surface receptors.

housekeeping gene: a coding sequence for a basic cellular function, expressed in all cells of an organism.

Hubble's law: a cosmological observation that deep space objects are observed via a *Doppler shift* relative to Earth, owing to their receding (moving away) from Earth.

Hubble sequence: a classification of galaxies by their appearance (visual morphology), devised by American astronomer Edwin Hubble in 1926.

Hubble Space Telescope: a 2.4-meter aperture telescope carried into Earth orbit by a US space shuttle in 1990.

human: a bipedal, largely furless mammal in the *Homo* genus.

Huntington's disease: a degenerative disease affecting muscle coordination, leading to cognitive decline and mental problems.

Huygens–Fresnel principle: a verified mathematical characterization of wave propagation by Christiaan Huygens (1678) and Augustin Fresnel (1818). See *principle of least action*.

hydrocarbon: a molecule comprising only hydrogen and carbon.

hydrogen (*H*): the element with atomic number 1, constituting in its simplest form a single proton and solitary electron (protium, ^{1}H); the lightest element, and the most abundant chemical in the universe, comprising 75% of cosmic baryonic mass. Hydrogen plays an important role in acid-base chemistry. Hydrogen is a proton donor in many reactions between soluble molecules.

hydrogen bond: a chemical bond between a hydrogen atom and either an oxygen, nitrogen, or fluorine atom in a molecule. Water is exemplary of hydrogen bonding.

hydrogenation: the process of turning an unsaturated fat into a saturated one via high-temperature heating.

hydrological cycle (aka *water cycle*): the cycling of water in the biosphere.

hydrolysis: (in context) a reaction that breaks a biopolymer down in the presence of water and an enzyme. Broadly, a chemical reaction in which water molecules (H_2O) are split into hydrons (H^+) and hydroxyls (*OH^-*).

hydron: a hydrogen cation (H^+).

hydronium (*H_3O^+*): an ion that is essentially water (H_2O) with a hydrogen hanger-on.

hydrophilic: having a *high* affinity for water. Contrast *hydrophobic*.

hydrophobic: having a *low* affinity for water. Contrast *hydrophilic*.

hydrosphere: the bioelement of water, including the participants in the water cycle.

hydrostatic pressure: the pressure exerted by a fluid at equilibrium because of gravity.

hydrothermal vent: a fissure, usually on the seabed at a volcanically active location, from which geothermally heated water issues.

hydroxide: a chemical compound with a *hydroxyl* group.

hydroxyl (OH⁻): a functional group comprising an oxygen atom covalently bonded to a single hydrogen atom. Compare *water* (H_2O).

hypernova: an exceptionally large supernova: at least 140–200 solar masses, which entirely explodes, leaving no core material.

hyperon: a 3-quark particle comprising up, down, and strange quarks; formed within a neutron star turning into a quark star.

hyperthermophile: an organism that can survive at 80°C or greater.

hypothalamus: a brain region found in all vertebrates. The hypothalamus controls body temperature and regulates episodic biological imperatives (circadian rhythms), such as thirst, hunger, fatigue, sleep, and mating and parenting behaviors.

hypothesis: a guess gussied up in scientific garb. Under the scientific method, hypotheses are ripe for falsifiability testing. Compare *theory*. See *falsifiability*.

I

ice: frozen water.

igneous (rock): rock formed by cooling and solidification of magma or lava. Compare *sedimentary* and *metamorphic*. See *basement*.

iguana: a genus of herbivorous tropical lizards.

illusion of knowledge: someone thinking that they know more than they do.

imaginary number: the square root of a negative number.

immediate early gene: genes which are instantly activated in response to cellular stimuli.

immune system: a biological system that wards against disease, especially infection. For macrobes, an immune system acts as a microbiome management system. See *innate immune system*, *adaptive immune system*.

imprinting (genetics): an epigenetic inheritance mechanism, where the gene expression of specific alleles is silenced based upon the sex of the parent gene set. Imprinting involves methylation and histone modifications.

in toto: entirely; as a whole.

induction (logic): the method of inferring a generalized conclusion from particulars. Contrast *deduction*.

inductivism: the traditional scientific method of evolutionary theory formation via fact accumulation; stated by Francis Bacon in 1620, who proposed incrementally (in terms of scale) proposing natural laws to generalize observed patterns. Disconfirmed laws are discarded.

In 1740, David Hume noted limitations in using experience to infer causality. 1st is the illogic of *enumerative induction*: unrestricted generalization from specific instances to all such events. 2nd is the presumptiveness of conclusively stating a universal law, since observation is only of a sequence of perceived events, not cause-and-effect. Nonetheless, Hume accepted the empirical sciences as inevitably inductive.

Alarmed by Hume, Immanuel Kant posited *rationalism* as favored by Descartes and by Spinoza. Kant noted that the mind serves to bridge the human experience with the actual world, with the mind creating space, time, and substance. With this, Kant trashed the *naïve realism* of science as only tracing appearances (phenomena), not unveiling reality (noumena). Compare *falsifiability*.

inertia: indisposition to a change in motion.

inertial reference frame (aka *inertial frame*, *Galilean reference frame*, *inertial space*) (under classical physics and special relativity): a frame of reference in which bodies are at rest or at constant velocity in a straight line; more generally, a frame of reference that describes time and space uniformly (homogeneously and isotopically), and in a time-independent manner. Conceptually, the physics of a system in an inertial frame that is self-contained, with no external causes.

infinity (∞): something without limit. Mathematics often treats ∞ as a special number, but that is a conceptual error. Infinity is beyond numerics.

inflaton (astrophysics): a hypothetical quantum particle (scalar field) of inflationary energy. No scalar fields have been observed in Nature. There is no evidence for the existence of inflatons.

inflationary energy: a hypothetical energy force of dense, intense negative pressure that allowed *cosmic inflation*.

influence (noun): the act of producing an effect indirectly.

influence (verb): to affect or alter, typically by indirect or intangible means.

information: an esteemed apprehension of an order among concepts.

information theory: a theory related to mathematical content quality in communication.

infrared (*IR*): electromagnetic radiation between 1–400 THz (terahertz). Most thermal radiation at room temperature is infrared. Infrared is emitted or absorbed by molecules when they change their rotational or vibrational mode.

inhibitor (chemistry): an enzyme that decreases reaction rate. Contrast *activator*.

innate immune system: the non-learning portion of the immune system. Contrast *adaptive immune system*.

insulator (chemistry): a medium that resists the flow of electrical current. Contrast *conductor, resistor*.

intelligence: an attribution for behaving appropriately; the process of gathering and analyzing information.

intelligence (tissue): 1 of the 4 primary animal tissue types. Glia and neurons are the primary cell types of intelligence tissue. See also *epithelium, muscle,* and *connective tissue*.

intensive property (aka *bulk property*): a physical property of a system that does not depend upon system size or materiality. Examples include temperature, density, hardness, and refractive index. Richard Tolman introduced the terms *intensive property* and *extensive property* in 1917. Contrast *extensive property*.

interaction (physics): see *force*.

interconnection: mutual connection.

interdependence: a system where one feature dynamic may affect another.

interface: the boundary between phases or systems.

interferometry: a measurement technique for fields via superimposing one wave upon another to extract information about the target wave.

intergenic: a DNA sequence located between genes.

interglacial: a period of warmer climate within an ice age. Compare *glacial period*.

interphase: the period of the cell cycle during which a cell lives its everyday existence. Interphase is 90% of a cell's life cycle. See *anaphase, telophase*.

interplanetary magnetic field: the solar magnetic field, carried with the solar wind out into the solar system.

intron: a polynucleotide sequence in a nucleic acid that does *not* code for protein synthesis. Introns are removed before translation of messenger RNA. Compare *exon*.

inverse-square law: Isaac Newton's formulation of gravity as a force: that the gravitational attraction between 2 objects is directly proportional to the product of their masses, and inversely proportional to the square of the distance between them.

invertebrate: an animal that is not a vertebrate.

ion: an electrically charged subatomic particle, atom, or molecule. See *anion* and *cation*.

ion channel: a chemical communication pathway comprised of pore-forming proteins that establish and control voltage gradients across the plasma membranes of cells by allowing the flow of ions down electrochemical gradients.

ionic bond: an electrostatic attraction resulting in 2 oppositely charged ions coupling. An *anion* and a *cation* join in an ionic bond.

ionic lattice: a lattice-like structure conducive to electrical conductivity.

ionization: the energetic process of converting an atom or molecule into an ion.

ionization energy (*potential*): the energy required to remove an electron from a gaseous atom or ion.

ionosphere: the ionized portion of Earth's upper atmosphere, at 85–600 km altitude.

iron (*Fe*): the element with atomic number 26; a metal. Iron is the most common element (by mass) in Earth, forming much of its core.

iron-sulfur world (theory): a theory developed by Günter Wächtershäuser that life originated in seabed hydrothermal vents, nestled in pyrite.

isoform: functionally similar proteins with a similar (but not identical) amino acid sequence which had been encoded by different genetic instruction sets.

isomer: a compound in one of various molecular structures (shapes). Isomers with the same chemical formula may have quite different properties.

isotope: a variant of a chemical species. Isotopes vary by number of neutrons in the nucleus.

isotropic: the same in all directions. Compare *homogeneous*.

J

joule: the energy equivalent of passing a 1-amp current through 1-ohm resistance for 1 second. Named after James Prescott Joule, who studied energetic relationships.

junk DNA: a DNA sequence that does not directly code for producing a protein.

Jupiter: the 5th planet from the Sun; a gas giant 2.5 times the mass of all other planets in the solar system. Jupiter has 63 sizable moons, 1 more than Saturn.

K

karyogamy: the final step in the process of fusing together 2 haploid eukaryotic cells. Karyogamy specifically refers to the fusion of the 2 nuclei.

Kelvin (*K*): an absolute temperature scale. Kelvin is the primary measurement unit in the physical sciences. From a perspective of classical thermodynamics, 0 K is the temperature at which all thermal motion ceases. Kelvin has the same magnitude as Celsius, albeit at a different offset. Absolute zero (0 K) is −273.15 °C. The Kelvin scale is named after Lord Kelvin, who expressed the need for an "absolute thermometric scale."

ketone: an organic compound comprising a carbonyl center connected to 2 side chains (R): R-C=O-R'. Many sugars are ketones. Compare *aldehyde*.

kinase: an enzyme that promotes reversible phosphorylation. More generally, kinases act on and modify the activities of specific proteins.

kinematics: often referred to as *the geometry of motion*, kinematics is a branch of classical mechanics that describes the motions of bodies and systems without considering the forces that cause movement.

kinetic energy: energy associated with motion.

kinetics: the branch of mechanics concerning forces which act upon matter.

kingdom (biological classification): the taxon above *phylum* and below *domain*. There are 4 eukaryotic kingdoms: protists, plants, fungi, and animals.

knot theory: the study of mathematical knots. Mathematical knots differ from physical ones in that the ends are joined so that the knot cannot be undone.

knowledge: cognition of facts or principles about Nature. Compare *omniscience*.

Kuiper belt: the region of the solar system extending past the orbit of Neptune to 50 AU (astronomical units) from the Sun, populated by cosmic debris.

L

L-form (state): a mode of existence for a bacterium of *not* having a cell wall (which most bacteria have).

Lacerta (astronomy) (aka *Lizard*): one of the 88 modern constellations; Latin for *lizard*; conceived in 1687 by Johannes Hevelius.

Lagrangian: the mathematical function of Lagrangian mechanics.

Lagrangian mechanics: a 1788 reformulation of classical mechanics by Joseph-Louis Lagrange.

Lamb shift: an energy difference between 2 energy levels of the hydrogen atom, according to quantum electrodynamics. Named after Willis Lamb.

Lambda cold dark matter model: see ΛCDM (under A because Λ looks like an ersatz A).

lambda point: the triple-point temperature below which fluid helium turns into superfluid helium: 2.172 K at 1 atmosphere (101,325 Pa).

Landau–Fermi liquid theory: a theoretical model of fermion interactions for most metals at low temperatures. The theory explains why some properties of an interacting fermion system are selfsame to those of the Fermi gas (i.e., non-interacting fermions), and why other properties differ.

Lanikea: the galactic supercluster in which the Milky Way galaxy resides.

lanthanum (La): the element with the atomic number 57; a soft, ductile, silvery-white metal which rapidly tarnishes when exposed to air and is so soft as to be easily cut with a knife.

Laplace's demon: the idea that existence would be utterly predictable (deterministic) to an intellect (the demon) that was omniscient; posited by Pierre-Simon Laplace in 1814.

Large Hadron Collider (LHC): the most powerful particle collider and the largest machine in the world; built 1998–2008. The LHC lies in a tunnel 27 km in circumference and as deep as 175 meters beneath the France-Switzerland border near Geneva.

latent heat: how much thermal energy (heat) can be absorbed or released by a body without changing the body's temperature.

lattice (chemistry): a repetitive arrangement of atoms.

lattice (mathematic): symmetrical order within a set.

lattice (physics): a repetitively arranged (lattice-like) physical model.

lattice constant (aka *lattice parameter*): the physical dimensions of unit cells in a crystal lattice.

law: a conclusion about a universal tendency in Nature.

law of independent assortment: a hypothesis by Gregor Mendel that the expression of any 1 genetic trait is not influenced by another. This so-called law is bogus.

laws of motion (Newton's): 1) a body has constant velocity unless acted upon by an external force; 2) acceleration is proportional to force and inversely proportional to mass; and 3) the mutual forces of action and reaction between 2 bodies are equal, opposite, and collinear.

law of segregation: an observation by Gregor Mendel that an *allele* in a diploid organism may express as *dominant*, masking a *recessive* allele that would express a different trait.

laws of thermodynamics: classical physics laws related to heat energy and entropy. The laws of thermodynamics all assume a universe that is an energetically closed system 4D. This presumption renders the laws fictional, because the cosmos has extra spatial dimensions (ED), with a constant energy exchange 4D and ED. Nonetheless, physicists still take these laws seriously, as they are taught as being cardinal. The laws of thermodynamics do provide proximate results at the ambient scale where they are typically applied.

lecithin: a yellow-brownish amphiphilic fat found in plant and animal tissues.

length contraction: a moving ruler that appears at rest to an observer will measure shorter than otherwise. Unnoticeable except at a frame of reference approaching the speed of light.

lepton: a subatomic particle not subject to the strong force. Electrons, muons, and neutrinos are leptons.

life: anything capable of perceiving its environment.

light: electromagnetic radiation visible to the human eye, at a wavelength of 380–740 nanometers.

light matter: ordinary matter. Contrast *dark matter*.

light-year: how far light travels in a year at light speed (as fast as light can travel); the standard unit used to express astronomical distances. A light-year is ~9.461 trillion kilometers.

lignin: an amorphous polymer related to cellulose. Lignin is an integral part of the cell walls of plants and some algae.

linear (chemistry): a molecular shape created by a central atom surrounded by 2 electron groups having bonding angles of 180°.

lipid: a broad group of relatively complex nonpolar carbon-based compounds, used for energy storage and a wide variety of biological functions.

lipid droplet: a ubiquitous cellular fat storage organelle for energy production and as a biosynthetic precursor.

lipophilic: having a high affinity for lipids.

liquid: a fluid that flows freely. Water is a liquid at room temperature.

liquid crystal: matter in a state with properties of both liquids and crystals.

lithosphere: the outermost shell of a rocky planet. Earth's lithosphere comprises its crust and upper mantle: the portions that behave elastically over vast expanses of time.

localization (biochemistry): control of allosteric regulation at a specific position on a protein via specific molecular binding configuration (sequence).

locality (physics): the idea that an object can only be influenced by its immediate surroundings. See *entanglement*. Contrast *nonlocality*.

locus (genetics): a gene's position in a genophore or chromosome.

London dispersion forces: a weak intermolecular force, arising from emergent quantum polarization multipoles in molecules; named after Fritz London.

lone pair: an electron pair not shared with other atoms.

loop quantum gravity: a quantum theory that quantizes all geometry, including space and gravity. Loop quantum gravity attempts to reconcile quantum mechanics with relativity.

Lorentz symmetry: the idea that all physical laws are the same for all observers; named after Hendrik Lorentz.

lunar cycle: the periodicity of the Moon's orbit about Earth.

lunar mare: a large dark basaltic plain on the Moon, formed by ancient volcanic eruptions.

lyonization (aka *X inactivation*): the process in which 1 of 2 copies of the X chromosome in female mammals is inactivated.

lysine acetylation: an epigenetic mechanism that affects histones by introducing an acetyl functional group.

lysis: viral reproductive release by cell wall rupture: killing the host cell in a violent outburst that releases a multitude of offspring. Contrast *lysogeny*.

lymphocyte: a type of white blood cell in the vertebrate adaptive immune system. The *innate* immune system operates through genetically programmed responses. In contrast, the *adaptive* immune system remembers past foes, to better dispatch nefarious invaders upon arrival.

lysogeny: a virus integrating itself into its host cell and replicating with the cell, secreting progeny viruses. Contrast *lysis*.

lysosome: the membrane-bound organelle in animal cells responsible for *autophagy*.

M

M-theory (physics): a physical theory that extends string theory into HD branes, postulating 11-dimensional spacetime.

macrobe: non-microbial life; any life form not requiring a microscope to be seen. Contrast *microbe*.

macromolecule: a large compound molecule, commonly created by polymerization of smaller subunits into polymer chains or 3D shapes. Nucleic acids, proteins, carbohydrates, and lipids are macromolecules.

macrophage (derived from the Greek for "large eater"): a type of phagocyte employed in vertebrate immune system defense.

Magellanic Clouds: irregular dwarf galaxies orbiting the Milky Way.

magic number (nuclear physics): the number of nucleons (either protons or neutrons, separately) forming a complete nuclear shell for an element. The most common magic numbers are 2 (helium), 8 (oxygen), 20 (calcium), 28 (nickel), 50 (tin), 82 (lead), and 126 (for neutrons). The term came from Eugene Wigner in the mid-1940s.

magma: molten rock made underground. Igneous rocks come from cooled magma.

magnesium (*Mg*): the element with atomic number 12; an alkaline metal. Magnesium is the 4th most common element in the Earth, behind iron, oxygen, and silicon. Most magnesium is in the mantle.

magnet: a metal, such as iron, that sports an external magnetic field.

magnetic dipole moment: the potential exertion force of magnetism upon a particle.

magnetic resonance imaging (*MRI*): a medical imaging technique used in radiology to form pictures of anatomy and bodily physiological processes.

magnetic reconnection: in conductive plasmas, rearrangement of magnetic topography and conversion of magnetic energy to heat, kinetic energy, and particle acceleration.

magnetism: a class of physical phenomena where atoms or molecules react from the influence of a magnetic field, which causes attraction or repulsion to nearby matter that is magnetically charged. Magnetism is a facet of *electromagnetism*. See *ferromagnetism*, *antiferromagnetism,* and *quantum spin liquids.*

magnetoreception: sensory reception of the Earth's magnetic fields by biochemical means.

magnetosphere: the area of astrological space where charged particles are controlled by a heavenly body's magnetic field. The Earth's magnetosphere is an outer layer of the ionosphere.

magnon: a collective excitation of spin waves with magnetic effect.

Majorana equation: a physics wave function using only real numbers. Most subatomic particles are defined by the *Dirac equation*, which necessitates complex numbers, with wave functions that result in complex conjugates. The Majorana equation characterizes *Majorana fermions.*

Majorana fermion: a fermion that is massless and chargeless; named after Ettore Majorana. The as-yet undiscovered Majorana is not included in the Standard Model. The Majorana is unique in being its own antiparticle. Compare *Dirac fermion*, *Weyl fermion.*

maltose ($C_{12}H_{12}O_{11}$; aka *maltobiose* or *malt sugar*): a disaccharide formed from 2 bonded units of glucose, formed via a condensation reaction.

mammal: a class of air-breathing, vertebrate animals, characterized by endothermy, hair, and females with functional mammary glands.

manifest (adjective): a) capable or readily and instantly perceived by the senses; b) capable of being easily understood or recognized at once by the mind.

manifestation: an outward, perceptible expression of Nature. Compare *phenomenon*.

mantle: the layer of Earth above the core and below the crust.

mantle plume: the rising of hot rock from the core-mantle boundary through the mantle to become a diapir (intrusion) in the Earth's crust.

many-body problem: a set of equations that characterize a system comprising many interacting components. Attempting to solve a many-body problem is computationally intensive. Approximations are often relied upon.

> It would indeed be remarkable if Nature fortified herself against further advances in knowledge behind the analytical difficulties of the many-body problem. ~ Max Born in 1960

many-body theory: a physics theory which models a system characterized by many interacting particles.

many-worlds interpretation (aka *parallel universes*): a fanciful extension of wave/particle duality (Schrödinger's equation) which posits that quantum waveforms represent an infinity of actual parallel universes. First suggested by Erwin Schrödinger in 1952, then formally proposed by Hugh Everett III in 1956. The many-worlds interpretation discards Heisenberg's uncertainty principle, which posits that figurative quantum waveforms collapse into actual quanta via observation; substituting an observer-free interpretation, in insisting that all the potentialities of quantum waveforms are actualized; whence many worlds.

Mars: the 4th planet from the Sun, with 2 small irregularly shaped moons.

marsupial: a clade of mammals, characterized by giving birth to relatively undeveloped live young. An infant marsupial (joey) develops within its mother's pouch.

mascon (*mass concentration*): a sizable gravity anomaly in a terrestrial body, often caused by compression from meteorite impacts.

mass (classical physics): a measure of inertia. Contrast *weight*.

mass (quantum mechanics): the energy level at which an elemental quantum may make an observable appearance.

mass extinction: the indiscriminate extinction of many species during an extinction event. Contrast *background extinction*.

mathematics: the systematic treatment of relations between symbolic entities.

Matryoshka doll: a set of hollow wooden dolls of decreasing size which can be placed one inside another.

matter (physics): something with mass, constructed of fermions. See *energy*.

matterism (aka (philosophical) *materialism*): the monistic belief that reality is made of matter. Matterism ignores that matter of made of energy and supposes that the mind is a figment of something substantial. Contrast *energyism*.

meditation: a practice intended to achieve a transcendental state of consciousness.

meiosis: the special cell division for sexual reproduction, producing germline gametes (sperm or eggs). Meiosis also refers to the cell division process for making spores. Compare *mitosis*.

Meissner effect: the complete expulsion of magnetic field lines from inside a superconductor as it transitions to a superconducting state. Named after its discoverer: Walther Meissner.

melanin: a group of pigments found in most organisms.

melatonin: a hormone found in microbes, plants, and animals. In animals, melatonin levels cyclically vary every day, affording entrainment of circadian rhythms.

mellitene ($C_{12}H_{18}$, in the structural formula $C_6(CH_3)_6$; aka *hexamethylbenzene*): an aromatic hydrocarbon derivative of benzene, composed of 6 methyl groups where carbon atoms have 6 bonds (not the usual 4).

membrane (cytology): a lipid bilayer surrounding a cell, providing a barrier between the cell and the outside world.

meniscus: the characteristic curve in the upper surface of a liquid at the top of the surface of a narrow container. A meniscus is caused by surface tension of the confined liquid. Water has a convex meniscus, while mercury has a concave meniscus.

mentotype: the psychological constitution of an organism, including cognitive orientations and capacities, awareness loci, and innate worldview. Compare *phenotype*.

Mercury: the planet closest to the Sun, and the smallest. A Mercury year is equivalent to 88 Earth days.

mercury (*Hg*): the element with atomic number 80. Mercury is the only metal that is liquid at room temperature and pressure. Mercury has a melting point of –38.83 °C and a boiling point of 356.73 °C, making it a metal with one of the narrowest ranges for its liquid state.

meristem: plant tissue where growth occurs.

meson: a hadronic subatomic particle comprising 1 quark and 1 antiquark.

messenger RNA (*mRNA*): an RNA molecule with the physical blueprint for a protein product.

metabolic pathway: a series of chemical reactions within a cell, typically with an intended biological end product.

metabolic rate: the speed at which a metabolic pathway transpires.

metabolism: cellular chemical reactions which provide energy for vital processes. See *anabolism, catabolism*.

metabolite: a product of metabolism.

metal: an element that readily conducts heat and electricity. 91 of the 118 elements are metals. Some elements have both metallic and nonmetallic phases. Compare *metalloid, nonmetal*.

metallicity (astronomy): the proportion of matter in a heavenly body other than hydrogen and helium.

metalloid (aka *semimetal*): a chemical element with properties of metals and nonmetals. There is no standard definition of a metalloid, but the term is common in chemistry.

metamorphic (rock): a rock arising from transformation via heat and pressure. The original rock (*protolith*) may be igneous, sedimentary, or a previous incarnate metamorphic. Compare *igneous* and *sedimentary*. See *basement*.

metamorphism (geology): the recrystallization of a rock owing to heat, pressure, or chemically active fluids.

metamonada: a group of anaerobic flagellate protozoa, most of which live as gut flora symbionts.

metaphase: the stage of cell division where chromosomes migrate to opposite poles of a cell.

metastasis: change of position, state, or form; commonly used to indicate spread of a disease within a body.

metazoan (plural: *metazoa, metazoans)*: a multicellular animal.

meteorite: a sizable rock from space that managed to smack a terrestrial body's surface. A meteorite might be a comet or an asteroid.

methane (CH_4): a flammable, explosive gas, which is colorless, odorless, and tasteless to humans. Methane forms in marshes and swamps, from decaying organic matter.

methanogen: anoxic archaea that produce methane as a metabolic byproduct.

methanol (CH_3OH): a simple alcohol; a polar liquid; a byproduct of anaerobic metabolism; a key substrate in the synthesis of organic molecules leading to life.

methoxy: a methyl group bound to oxygen.

methyl group: an alkyl derived from methane (CH_4).

methylation: an epigenetic mechanism that stifles or inactivates a gene by attaching methyl groups to nucleobases.

methylene (H_2C; aka *carbene*, $\lambda 2$-*methane*): a colorless gas that is the simplest carbene.

Michelson-Morley experiment: the 1887 experiment by Albert Michelson and Edward Morley to demonstrate the existence of the cosmic aether. The experiment failed and is widely considered the most famous failed experiment in physics history.

microbe: a microorganism, typically single-celled. Microbes include archaea, bacteria, fungi, and protists. Contrast *macrobe*.

microbial loop: recovery of otherwise lost organic energy by bacteria.

micrometer (*aka micron*) (*μm*): 1 millionth of a meter (1×10^{-6}).

microRNA (*miRNA*): a class of post-transcriptional regulators which bind to *microRNA response elements* (MREs), thereby decreasing the stability of protein-coding *messenger RNAs* (mRNA) or limiting their protein translation. The result is typically stifling or silencing gene expression. See *RNAi*.

microtubule: a rope-like macromolecule of protein; part of the cytoskeleton. Macrotubules are employed in cell structural maintenance, intracellular transport, forming the spindle during mitosis, and other cellular functions. Microtubules are comprised of *tubulins*.

microwave: a radio wave with a wavelength ranging between 1 meter and 1 millimeter; equivalently, between 300 MHz (0.3 GHz) and 300 GHz frequency.

Milankovitch cycle: a 1920 hypothesis by Milutin Milanković relating changes in sunlight, and thereby climate, to variations in Earth's orbit about the Sun. Earth has an elliptical orbit, with eccentricities in that orbit, as well in its axial tilt and precession (rotational orientation). Milankovitch cycles are now used extensively to explain the timing of glacial-interglacial cycles in Earth's evolution.

Milky Way: the spiral galaxy containing the solar system, formed 13.2 BYA; 120,000 light-years in diameter, containing up 200–400 billion stars, and at least 640 billion planets.

mind: an intangible organ for symbolic processing.

mineral: a solid homogeneous crystal.

mineralogy: the study of minerals.

miscibility: the capability of being mixed.

Mister Dog: a children's book by Margaret Wise Brown. Belying a neatness fetish, Mister Dog smoked a pipe and wore a straw hat.

mitochondrion (plural: *mitochondria*): an organelle that acts as a cell's power plant, generating a supply of ATP. Mitochondria play other important roles in the cell life cycle, including growth and aging. Mitochondria maintain their own genome, independent of the cell nucleus. Some eukaryotic cells have multiple mitochondria, others none. Whereas human red blood cells have no mitochondria, liver cells may have over 2,000.

mitophagy: cell organelle recycling. Compare *autophagy*.

mitosis: the eukaryotic cell division process. Compare *meiosis*.

mitotic recombination: a relatively rare genetic recombination that occurs in somatic cells during mitosis.

model (mathematics): a mathematical construct. See *physical model*.

modern physics: post-Newtonian conceptions of physics, including Einstein's relativity theories and models related to matter at the subatomic scale. Compare *classical physics*.

moiety: a small molecule of a chemical functional group.

mole (chemistry): a standard molecular weight unit, with the unit symbol *mol* (because keeping that last letter in would make it too damn obvious). 1 mole equals 12 grams of carbon–12 (^{12}C), the standard isotope of carbon.

molecule: a combination of atoms.

molecular geometry: the study of molecular shapes: the spatial arrangement of atoms in molecules.

molybdenum (*Mo*): the element with atomic number 42. The metal molybdenum naturally occurs in various oxidation states in minerals; never as a free metal itself.

momentum (physics): mass times velocity. Momentum is a vector quantity, with both direction and magnitude.

monatomic: a molecule comprising a single atomic species. Helium is monatomic.

monkey: a primate, excluding *apes*.

monomer: a molecule that may bind with other molecules to form a *polymer*.

monopole: a magnetic pole considered (theorized) in isolation.

monosaccharide (aka *simple sugar*): a simple carbohydrate with the formula $(CH_2O)_n$, where $n = 3$ (triose), 5 (pentose), or 6 (hexose). Glucose, fructose, and ribose are exemplary monosaccharides. See *disaccharide*.

Moon: Earth's solitary satellite; the 5th largest satellite in the solar system.

morphogen: a signaling molecule that directs cell movement and guides tissue development (morphogenesis).

morphogenesis: biological development of form.

morphology: form and structure (ostensibly of organisms, such as plants and animals).

multipole: a form of monopole with no pole strength or net charge.

multiverse: the idea that a multitude of universes come into and go out of existence on a vast HD canvas of endless time. The multiverse hypothesis is posited upon an ED spatial envelope, with individual universes appearing as membrane manifestations. See *cyclic cosmology*. There are various distinct multiverse concepts, some far-fetched. These arise from the assumption that simplistic physical models mirror Nature. See *many-worlds interpretation*.

muon: an unstable lepton, similar to an electron.

murein (aka *peptidoglycan*): a polymer comprising sugars and amino acids that forms a mesh-like layer outside the plasma membrane of most bacteria.

muscle (tissue): 1 of the 4 primary animal tissue types. Muscle cells are capable of contraction, and so provide for movement. See also *epithelium*, *connective* (tissue), and *intelligence* (tissue).

muscular dystrophy: a group of muscle diseases that eventuates in weakening and breakdown of muscles.

mutation: a change in a DNA sequence.

mutualism: a regular interaction between 2 organisms that provides mutual benefits.

MYA: millions of years ago. MY as an acronym for "million years" is deprecated in modern geophysics, in favor of *Ma*, shorthand for *megaannum*.

mycelium (plural: *mycelia*): a thread-like filament of mesh-like mass of fungal filaments (*hyphae*).

myxamoeba (plural: *myxamoebae*): a swarm cell that finds others of its kind and creates a slime mold colony.

N

naïve empiricism: the belief that knowledge can only be gained through empirical examination of Nature. See *naïve realism*.

naïve realism (aka *direct realism, commonsense realism, scientific realism*): the belief that actuality as perceived is reality.

NASA (*National Aeronautics and Space Administration*): the US government civilian space agency.

nautilus: a genus of hard-shelled cephalopods that emerged during the late Cambrian period.

natural genetic engineering: the process of altering cell functioning based upon genetic information.

natural genetic engineering toolkit: the set of biochemical capabilities a cell has to restructure its genome by cleaving, splicing, and synthesizing DNA chains.

natural philosophy: the study of Nature from a holistic perspective; the common methodology of comprehending Nature until the 17th century, before modern science barged in with its strictly empirical *scientific method*. See *natural science*. Contrast *science*.

natural science: natural philosophy coupled to the scientific method.

Nature: the exhibition of existence. See *coherence*.

nebular hypothesis: a hypothesis by Emanuel Swedenborg, that the solar system formed by swirling accretions of matter.

necroptosis: programmed inflammatory cell death (necrosis).

negative pressure: an expansive (inflationary) pressure.

neon (*Ne*): the element with atomic number 10; a colorless, inert gas. Neon is the least reactive element, and the 2nd-lightest gas, behind helium. Although the 5th most common in the universe (by mass), neon is rare on Earth. Commercial neon, which glows reddish-orange as a plasma in a vacuum discharge tube, is extracted from air, which contains trace amounts.

Neptune: the 8th and farthest planet from the Sun in the solar system. Neptune's orbit is ~165 Earth years.

nerve (cell): see *neuron*.

neuron (aka *nerve cell*): an electrically excitable intercellular signaling cell as part of the nervous system, employed for sensory or motor communication. Functionally, neurons are managed by glia.

neutralization: the reaction of an acid with a base by proton transfer, forming a salt.

neutrino: an electrically neutral, weakly-interacting subatomic particle.

neutron: a subatomic particle at home in the nucleus of an atom. Lacking an electromagnetic charge, neutrons act as a glue to hold feisty protons together in an atomic nucleus, which naturally repel each other.

neutron star: a stellar remnant from the gravitational collapse of a massive star (supernova). Neutron stars are made mostly of neutrons condensed to the utmost extent.

newton (*N*): the standard unit of force that produces an acceleration of 1 meter per second per second on a 1-kilogram mass. Used as a measurement of weight. Named after Isaac Newton.

nitrogen (*N*): the element with atomic number 7; a colorless, tasteless, odorless element that, as a diatomic gas (N_2), is relatively inert.

nitrogen cycle: the cycling of nitrogen in the biosphere.

nitrogen fixation: fixing atmospheric nitrogen gas into a biologically employable form. Only certain microbes have mastered the craft of fixing nitrogen.

nitrogen oxides (NO_x): the generic term for nitric oxide (NO) and nitrogen dioxide (NO_2).

nitrogenase: an enzyme employed by microbes to fix atmospheric nitrogen into biologically usable form.

noble (chemistry): an element that is chemically inert (inactive), thus not given to molecular combinations.

nociception (aka *nocioception*, *nociperception*): detection of stimuli which are hazardous. In animals, nociception usually causes pain.

nodulation: the process of forming a nodule where rhizobia can perform nitrogen fixation for legumes.

non-Euclidian geometry: a geometrical system that postulates curved, higher-dimensional (HD) space. Non-Euclidian geometry diverges from Euclidian geometry in relaxing the parallel postulate.

noncoding (*DNA/RNA*): a polynucleotide strand that does not encode for protein production. At 200 nucleotides, an arbitrary distinction is made between noncoding sequences deemed *small* and those called *long* (the RNA form abbreviated as *lncRNA*). Examples of noncoding RNA include ribosomal RNA, transfer RNA, piwi-interacting RNA, and microRNA.

nonlocality (physics): entanglement of objects at some distance from each other. Contrast *locality*.

nonmetal: a chemical element lacking metallic attributes. Nonmetals tend to be highly volatile (easily vaporized), good insulators of heat and electricity, have low elasticity, and tend to have high ionization energy (gaining or sharing electrons when reacting). 17 (of 118) elements are nonmetals; 11 are gases, 5 solids, 1 liquid (bromine).

nonpolar: an electrically neutral molecule, owing to its constituents sharing electrons equally.

nonsense-mediated mRNA decay (*NMD*): the quality control process in cellular protein production, via recognizing defective mRNA and efficiently degrading them.

noumenon: outside of existence. A *noumenon* is beyond perception, as contrasted to *phenomena*.

NP hard: nondeterministic polynomial-time hard. In computational complexity theory, NP hard comprises a class of problems that can make computers break down and cry.

nuclear cluster: a cluster of nucleons with relative stability based upon the bosonic character of the nucleons in an atomic nucleus.

nuclear genome: the genetic contents of a cell nucleus.

nuclear pore complex (*NPC*): a protein complex that porters molecules across a cell nuclear envelope.

nucleation: the 1st step in a transition to a new thermodynamic phase or structure via self-organization. The term is commonly used to describe ice crystal formation (ice nucleation).

nucleic acid: an acidic biomolecule comprising a nucleotide, discovered by Friedrich Miescher in 1869. DNA and RNA are nucleic acids.

nucleobase: a nucleic acid base; a nitrogen-based, ring-shaped molecule that comprises the basic building block of nucleotides.

nucleoid: an irregularly shaped region within a prokaryotic cell containing a single genophore.

nucleolus: site of ribosomal RNA synthesis within a eukaryotic cell nucleus.

nucleon: a subatomic particle in an atomic nucleus. Each atomic nucleus has 1 or more nucleons. Protons and neutrons are the 2 known nucleons.

neucleoporin: a protein which is part of a nuclear pore complex.

nucleoside: a nucleobase bound to a sugar (ribose or deoxyribose).

nucleosome: the basic nuclear DNA package in eukaryotes: a DNA segment wound around a core of 8 histones, like a thread wrapped around a spool.

nucleosynthesis: the process of creating atomic nuclei from preexisting nucleons (protons and neutrons).

nucleotide: an individual structural (monomer) unit of nucleic acid (DNA, RNA); a nucleobase packaged with sugar and phosphate groups, held together by ester bonds.

nucleus (cytology): an organelle in eukaryotic cells that acts as a cellular control center. The nucleus contains most of a cell's genome (the *nuclear genome*).

nucleus (physics) (plural: *nuclei*): the central core of an atom, comprising protons and neutrons (nucleons).

Nuna (aka *Columbia*): a supercontinent created 1.9 BYA. Nuna began breaking up 1.5 BYA.

O

objectivity: the idea that Nature and reality are independent of consciousness. Contrast *showtivity*.

obligate: obligatory.

obliquity (astronomy) (aka *axial tilt*): the angle between an object's rotational axis and its orbital axis; equivalently, the angle between an equatorial plane and an orbital plane.

Occam's razor: a principle of parsimony in logic, courtesy of William of Ockham. In explaining a system, Occam's razor states that the hypothesis with the fewest assumptions and simplest reasoning is most logically appealing. Science employ's

Occam's razor as a heuristic in developing theories and models – whence the failing of science through untoward simplification, as Nature never adheres to Occam's razor.

ocean: a large, deep body of saltwater.

octopus (plural: *octopuses*, *octopi*, or *octopodes*): a highly intelligent cephalopod.

ohm: a unit of electrical resistance, named after Georg Ohm.

Onion, The (1988–): American satirical news organization.

ontology (biology): an organism's course of development.

oogenesis (aka *ovogenesis*): the differentiation of an ovum (egg cell) into a cell which may become a zygote.

Oort cloud: a hypothesized cloud of comets nearly a light-year from the Sun. The outer edge of the Oort cloud defines the cosmographical boundary of the solar system, where Sun's gravity holds sway. See *Kuiper belt*.

orbit (physics): the gravitationally curved trajectory of an object.

organelle: a subunit within a eukaryotic cell that has a specialized function. Organelles are membrane-bound. Cell organelles evolved through endosymbiotic union with an archaeon host cell and a bacterial endosymbiont.

organic: related to living organisms; from a chemistry viewpoint, a complex molecular structure based upon a carbon backbone.

organism: a life form; an animated organic structure.

organitype: the paradigms which constitute an organism: the combination of phenotype, mentotype, and genotype.

origami: the traditional Japanese art of paper folding, originating in the 17th century.

Orion–Cygnus Arm: a minor spiral arm of the Milky Way galaxy, 3,000 light-years across and 10,000 light-years long. The solar system swirls in the Orion–Cygnus Arm.

ortho-water: an isomer of water with symmetric wavefunctions and atomic nuclear spins summing to 1. Contrast *para-water*.

orthogenesis (aka *orthogenetic evolution, autogenesis*): a hypothesis that organisms have a goal-directed (teleological) vector of evolution; introduced by Wilhelm Haacke in 1893; now considered moribund.

osmophile: an organism capable of growing in a sugary habitat.

osmosis: movement through a semipermeable membrane.

osmotic pressure: the pressure required to prevent inward flow of water across a semipermeable membrane, such as a cell membrane.

osteoarthritis: a joint disease from breakdown of joint cartilage and underlying bone.

owl: a bird among 200 species of mostly solitary and nocturnal birds of prey; typified by an upright stance, a large, broad head, sharp vision and hearing, and feathers that provide silent flight. Owls are found in all biomes except the coldest (Antarctica, most of Greenland).

oxidant: a compound capable of oxidizing other compounds that it encounters. Oxidation often damages cells.

oxidation: an increase in oxidation state by loss of electrons. Contrast *reduction*.

oxidation state (aka *oxidation number*): a characterization of the charge potential of an atom within a chemical species. An electrically neutral compound necessarily has net oxidation state of zero. The more electronegative or electropositive atoms in a compound are considered 1st in calculating the oxidation state of molecular atoms.

oxidative phosphorylation: a metabolic pathway that uses energy released by the oxidation of nutrients to produce ATP. Almost all aerobic organisms carry out oxidative phosphorylation to synthesize ATP. See *respiration*.

oxygen (*O*): the element with atomic number 8; a highly reactive nonmetallic element that readily forms compounds (notably oxides) with almost all other elements. Oxygen is the 3rd most common element in the universe.

oxytocin ($C_{43}H_{66}N_{12}O_{12}S_2$): a neurohypophysial hormone that acts in the brain as a sensation modulator. Oxytocin has various effects in different animal species. In primates, oxytocin is instrumental in facilitating social bonding.

ozone (*O_3*, aka *trioxygen*): a triatomic molecule comprising 3 oxygen atoms. O_3 is less stable than O_2 (dioxygen). Ozone is formed by ultraviolet radiation of dioxygen.

P

pais (*prokaryotic acquired immune system*): an adaptive immune system used by prokaryotes, commonly known as a *CRISPR/Cas system*.

paleoatmosphere: the atmosphere before life arose.

pangenesis: an ancient hypothesis of holistic heredity via an atomic biological mechanism.

panspermia: life delivered to Earth from space.

para-water: an isomer of water with asymmetric wavefunctions and atomic nuclear spins summing to 0. Contrast *ortho-water*.

paradigm: a construed pattern, often used as a framework for perception.

parallel postulate: Euclid's geometric 5th postulate, which states (for 2D geometry): if a line segment intersects 2 straight lines forming 2 interior angles on the 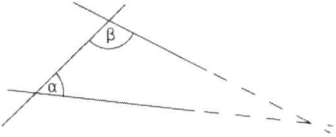 same side that sum to less than 2 right angles, then the 2 lines, if extended indefinitely, meet on that side on which the angles sum to less than 2 right angles. Unlike Euclid's other 4 postulates, the 5th postulate was not self-evident, as attested by efforts through the centuries to prove it.

parallel universes: see *many-worlds interpretation*.

paradox: a statement which appears self-contradictory or absurd, but which may express an insight.

paramecium (plural: *paramecia*): a unicellular ciliate, widespread in all watery habitats, including brackish water. Paramecia were among the first ciliates seen by early microscopists in the late 17th century. Their easy cultivation led to being widely studied.

paramutation: an allele causing a heritable change in expression of a homologous allele. Paramutation results in an epigenetic state that is inherited meiotically as well as mitotically. Paramutation is common in plants but rare in animals.

parasite: an organism living in/on/with another organism, obtaining benefits that usually reduces the fitness of its host.

parent cell: a cell dividing into 2 *daughter cells* as the result of cellular division (replication).

parity: a representation of a physical system capable of spatial transformation, transforming the system into its mirror image (*parity inversion*); a property of a symmetrical physical model.

parity transformation: an inversion of a spatial coordinate system. Also termed *parity inversion*.

Parkinson's disease: a degenerative disease affecting the intelligence system. The most obvious early symptoms affect movement: shaking, rigidity, slowness, and difficulty walking. Later symptoms include cognitive and behavioral problems.

parsec: an astronomical length unit; about 3.26 light-years, just under 31 trillion (3.1 x 10^{13}) kilometers.

parthenogenesis: asexual reproduction without fertilization. From the Greek for "virgin birth." Contrast *heterogamy*.

particle (physics): a point in spacetime, typically used to ascribe a quantum-sized field. Contrast *wave*.

particulate radiation: radiation comprising high-speed particles.

pascal (Pa): the SI unit of pressure, stress, and tensile strength; a measure of force per unit area; named after Blaise Pascal.

Pauli exclusion principle: a theoretical requirement that 2 fermions cannot occupy the same space simultaneously; formulated by Wolfgang Pauli in 1925.

pedoscope: a shoe-fitting X-ray fluoroscope.

pedosphere: the outermost terrestrial layer of Earth, comprising soil. Compare *geosphere*.

pemphigus: a rare autoimmune disease that affects the skin and mucus membranes, creating horrendous ulcers.

peptide: a short chain of amino acids: 2 to 50 or so. A longer chain is properly termed a *protein*.

peptidoglycan (aka *murein*): a polymer comprising sugars and amino acids, forming a mesh-like layer outside a cell's plasma membrane.

perceive, perception: mentally integrating sensory input (*sensation*) using memory. Perception is a 3-stage process: 1) turn a sensation into a symbolic representation, 2) identify sensed symbols using memory and categorization, then 3) derive the meaning of the identified symbols, especially regarding affinity or avoidance. See *conceptualization*.

perennial (botany): a plant that is present aboveground throughout the year, and which lives for more than 2 years. Woody plants, such as shrubs and trees, are perennials. Compare *annual*, *biennial*. See *herbaceous*.

pergenome: the personal genome of a cell in a multicellular eukaryote, as contrasted to the genome of the organism.

perihelion: the closest point to a star of an orbiting body.

periodic table of elements: a tabular display of atomic species (chemical elements), presented in increasing order of their *atomic number* (number of protons), with columns (groups) and rows (periods) based upon electron configuration.

permeable: a membrane that has pores through which molecules may pass.

permittivity (electromagnetism) (aka *absolute permittivity*): the measure of charge (capacitance) when forming an electric field in a certain medium.

perovskite: see *bridgmanite*.

perturbation theory: mathematical methods to squeeze an approximate answer from equations that refuse to resolve to an exact solution.

pH: a measure of acidity which ultimately relates to the number of protons in a solution. 7 = neutral; < = acidic; > = base (alkaline).

phagocyte: an animal cell which protects it host body by ingesting harmful foreign particles, select microbes, and dying or dead cells.

phagocytosis: the process of engulfing and ingesting cellular material; a form of endocytosis.

phase (physics, chemistry): a physically distinctive form of matter. Common phases, corresponding to temperature/energy levels, are gas, liquid, plasma, and solid.

phase transition: change from one operational state, or state of matter, to another.

phenomenal: known through perception.

phenomenon (plural: *phenomena*): a perceptible event. See *actuality*. Contrast *noumenon*.

phenotype: the composite visible traits of an organism: physical, physiological, and behavioral. Compare *mentotype*.

phenylalanine (*Phe*; $C_6H_5CH_2CH(NH_2)COOH$): an electrically-neutral amino acid used to form proteins.

philosopher's stone: legendary alchemical substance capable of turning base metals into precious metals (gold or silver).

philosophy: a set of consistent definitions pertaining to a system which yields a hierarchical construal.

phlogiston theory: the notion of fire as an element.

phonon: a collective excitation of interacting quanta characterized by a vibrational mode.

phospholipid: a class of lipids that are a major component of cell membranes, as they can form bilayers which afford regulated communication flow. The first phospholipid identified was lecithin, found in the egg yolk of chickens by Theodore Gobley in 1847.

phosphorus (P): the element with atomic number 15; as a mineral, always maximally oxidized. A component of RNA, DNA, ATP, and other biocompounds, phosphorus is essential to life.

phosphoryl group ($P^+O_3^{2-}$): a radical of phosphorous of oxygen.

phosphorylation: attaching a phosphoryl group to a molecule. Phosphorylation and dephosphorylation are extensively employed in cellular processes. In eukaryotes, protein phosphorylation is an extremely common genetic post-translational modification. The addition of a phosphate group to a protein that can alter gene expression by altering the proteins involved in building other proteins. Contrast *dephosphorylation*.

photochemistry: the branch of chemistry about the chemical effects of light.

photolysis (aka *photodissociation*, *photodecomposition*): chemical decomposition via radiant energy.

photon: a hypothetical bosonic particle of light; more properly, a *packet* of light energy, as light exhibits both particulate and wave appearances. Though photons supposedly do not interact with each other, they somehow porter the force of *electromagnetism*.

photophosphorylation: the phosphorylation of ADP to form ATP during photosynthesis.

photosphere: the visible surface of the Sun; the layer of the Sun's atmosphere that emanates light; the lowest (1st) of 3 main layers of the Sun's atmosphere. The *chromosphere* is above the photosphere.

photosynthesis: (an organism) converting sunlight into energy.

phylum (plural: *phyla*) (biological classification): the taxon above *class* and below *kingdom*. Phylum typically refers to a uniquely identifiable body plan.

physical chemistry (aka *physiochemistry*): the study of particulate phenomena in chemical systems; in other words, the study of physics in chemistry.

physical model: a typically geometric or algebraic mathematical model yielding a mathematical characterization of the embodied phenomena.

physical property: any measurable property of a physical system.

physical quantity: a measure of a physical property.

physical system: a portion of a physical universe chosen for examination. Everything outside the system is its environment.

physical theory: an explanation of relationships between various measurable phenomena. A physical theory may include a model of physical events (i.e., a physical model).

physics: the natural science of matter and its patterns of motion, with the intent of understanding how the universe behaves.

physiochemistry: see *physical chemistry*.

pi bond: a *covalent bond* formed by overlapping atomic orbital lobes. Compare *sigma bond*.

picosecond: 1-trillionth (10^{-12}) of a second.

piezophile: an organism that lives at a high hydrostatic pressure, such as in an ocean trench.

pilot wave theory: the deterministic theory that there is an inherent wave/particle duality for every elementary particle; proposed by Louis de Broglie in 1927. Contrast *uncertainty principle*.

piwi-interacting RNA (piRNA): a special group of noncoding RNA molecules that combine with certain proteins to protect the integrity of a genome.

placebo: a simulated medical treatment intended to inspire the recipient, thereby provoking the *placebo effect* of working to relieve or even cure the targeted affliction. The placebo effect illustrates the powerful sway that the mind has over health.

placebo effect: a rejuvenation owing solely to mental invigoration via belief in a placebo (totemic treatment).

placenta: an organ that connects a developing fetus to the uterine wall of its mother. Placentas are found in certain mammals, including humans, and some snakes and lizards.

Planck constant (aka *Planck's constant, Plank's action quantum*): a physical constant reflecting the size of energy quanta in quantum field theory. Planck's constant states the proportionality between the momentum and quantum wavelength of every subatomic particle. The relation between the energy and frequency of quanta is the *Planck relation*.

Planck length: the minimal theoretical limit to spatial distance; a measure derived from Newton's gravitational constant, the speed of light in a vacuum (c), and Planck's constant. Planck length is 1.616199 x 10^{-35} meters.

Planck mass: the theoretical amount of mass in a sphere with a radius Planck length, with a density of 1093 g/cm^3.

Planck satellite: a space probe launched in 2009 by a European consortium to measure cosmic radiative energy.

Planck time: the theoretical limit of temporal measurement; the time required for light in a vacuum to travel a single Planck length. At 5.391 x 10^{-44} seconds, Planck time is the shortest sprint imaginable.

Planck unit: a system of natural units used in physics, particularly *Planck length* and *Planck time*.

planetary nebula: a cloud of ionized gas emitted by a star toward the end of its life.

plankton: a minute organism living in a water column (freshwater or salt) that is incapable of swimming against a current. The term *plankton* is both singular and plural (they're just too damn tiny to count).

plant: a kingdom of eukaryotic autotrophs, including mosses, ferns, conifers, and flowering plants (angiosperms).

plasma: an ionized gas; one of the 4 fundamental states of matter; the others being gas, liquid, and solid.

plasma membrane: the membrane holding a cell's cytoplasm and other contents within.

plasmid: a DNA globule, useful to microbes for *horizontal gene transfer* (swapping genetic material).

Plasmodium: a genus of parasitic protozoa. Infection of plasmodia is *malaria*.

plasmon: a quantum of plasma oscillation.

plastid: a catchall term for the organelles in plants and algae, including those responsible for photosynthesis.

pleiotropy: a single gene influencing multiple seemingly unrelated traits.

ploidy: the set count of chromosomes in a biological cell. Many prokaryotes are *haploid* (1 set). Most eukaryotes, including most animals, are *diploid* (2 sets), though 30–80% of living plants are *polyploid* (> 2). Polyploidy can occur in animal tissues, such as the human liver.

pluripotency: a stem cell able to differentiate into any cell type.

Pluto: a large ice-encrusted lump on the outskirts of the solar system; once considered a full-fledged planet, but demoted in 2006 to a dwarf planet, owing to its relatively low mass. For consolation, Pluto has 4 moons.

pneumatic chemistry: the quaint term for research into the nature of gases; used from the 17th to early 19th century.

Poincaré group: a group of isometries in a particular (Minkowski) spacetime which corresponds with special relativity. Named after Henri Poincaré.

polarity (chemistry): a molecule that has positive and negative poles; in other words, a molecule with an electric dipole moment. Polar molecules have polar bonds owing to a difference in electronegativity between the bonded atoms. Water is a polar molecule.

polarization (optics): a state of light in which the radiation exhibits distinct properties in different directions.

polaron: a quasiparticle of electron mobility.

polyatomic ion (aka *molecular ion*): an ion with 2 or more atoms covalently bonded which act as a single unit. Historically, a polyatomic icon was referred to as a *radical*.

polycrystal: a substance, typically a solid, comprising many fused crystallites (microscopic crystals).

polymath: a person of encyclopedic learning.

polymer: a macromolecule (large molecule) comprising repeating monomers (molecular units).

polymerization: a process of reacting monomer molecules together to form polymer chains or 3D networked structures.

polymorph: a substance that has multiple potential structures. Polymorphs are typically solids, though helium-4 is a polymorph for its liquid phase.

polynucleotide: a biopolymer of 13 or more nucleotide monomers covalently bonded in a chain. DNA and RNA are polynucleotides.

polyol: an alcohol containing multiple hydroxyl groups.

polypeptide: a short chain of amino acid monomers, linked by peptide bonds.

polyploidy: cells with more than 2 paired (homologous) sets of chromosomes. Polyploidy is common in ferns and angiosperms (flowering plants). Some animals, such as goldfish, salmon, and salamanders, possess polyploidy. In other animals, polyploidy may result from abnormal cell division.

polysaccharide (aka *glycan*): a complex sugar-based macromolecule; a derivative of glucose. Compare *monosaccharide*.

positivism: the philosophical stance that the only authentic knowledge is that which affords empirical verification. Positivism rejects introspection and intuition as knowledge.

positron: the antimatter equivalent of the electron.

post-translational processing: modification of a protein after its translation, including attaching other biosynthetic functional groups or making structural changes.

potassium (K): the element with atomic number 19; an alkali metal with a single valence electron that readily reacts. Potassium is essential to all cells.

potential (electric): see *electric potential*.

potential energy: stored energy that may be released; the energy inherent in an object owing to its position relative to other objects, internal stresses, electric charge, and other factors.

power (physics): the amount of energy transferred via current per unit of time, measured in watts (power = watts / time). Power is a measure of how quickly work is done.

power law: a consistent mathematical relationship between 2 quantities, such as the magnitude of an event as a function of its frequency (e.g. earthquakes or solar flares).

pre-adaptation: a trait which is subsequently adaptively employed in another, distinctive way. Pre-adaptations are a fundamental mechanism of evolvability.

precession: a slow gyration in rotation axis of orbital body.

precipitation: rain, sleet, ice, snow, and fog; also defined as the quality of being precipitate, or hasty.

precipitation (chemistry): formation of a solid within a solution. The solid formed is termed a *precipitate*. The chemical agent that provokes solidity is the *precipitant*.

predeterminism (aka *fatalism*): the idea that events are determined in advance.

presocial: an animal species that lacks 1 of the 3 following characteristics: 1) reproductive division of labor; 2) cooperative care of the young; and 3) overlapping generations. Contrast *eusocial*.

primate: a mammal order, containing prosimians (neither monkey nor ape) and simians (monkeys and apes).

principle: a conceptual construct explaining some countenance of Nature.

principle of least action (aka *principle of stationary action*): a variational principle which can be used to get the equations of motion for a physical system. The principle of least action can derive Newtonian, Lagrangian, Hamiltonian, and general relativity (Einstein–Hilbert) motion equations.

principle of least time: see *Fermat's principle*.

Principia (*Mathematical Principles of Natural Philosophy*) (1687): Isaac Newton's 3-volume work laying the foundation of classical physics; considered one of the most important works in the history of science.

prion: an infectious agent in the form of a misfolded protein.

programmed cell death: cell death mediated by an intracellular program.

prokaryote: an organism that lacks a cell nucleus or other membrane-bound organelles. Archaea and bacteria are prokaryotes. While prokaryotes are single-celled, most can form stable, aggregate communities, such as a biofilm. Compare *eukaryote*.

promote (chemistry): encourage chemical reaction.

promoter (genetics): a region of DNA that facilitates the transcription of a certain gene.

prophase: the 1st stage of mitosis, in which chromatin condenses. See *metaphase, anaphase, telophase, interphase*.

proprioception: the sense of relative position of body parts and effort involved in their movement.

protease (aka *peptidase*, proteinase): an enzyme that abets proteolysis.

proteasome: a protein complex within all eukaryotes and archaea, and in some bacteria. In eukaryotes, proteasomes are in the nucleus and the cytoplasm. The primary work of a proteasome is breaking down unneeded or damaged proteins via *proteolysis*. Enzymes that carry out proteolysis are *proteases*. Proteasomes are part of a major mechanism by which cells regulate the concentration of proteins and recycle portions of misfolded proteins.

protein: a single, long, linear polymer chain of amino acids that typically takes a folded structure; a complex organic macromolecule by which living bodies are intelligently built. See *enzyme*.

protein synthesis: the multiple-stage process of protein production based upon a genetic template.

proteolysis: protein catabolism by hydrolysis of peptide bonds.

proteome: (the idea of) the entire set of proteins expressed by a cell's or organism's genome.

prothallus: the gametophyte stage in the life of a fern or other pteridophyte (a vascular plant that does not produce seeds).

protist: a catchall kingdom of eukaryotic organisms, including algae and amoeba. Most protists are unicellular, though many practice pluricellularity.

protium (chemistry): the most abundant form of hydrogen, comprising a nucleus of a single proton (no neutron). Contrast *deuterium*.

protocell: a cellularly-contained set of chemical reactions with evolutionary potential.

proton: a positively charged hadron that is a constituent in every atomic nucleus. The simplest hydrogen atom comprises a proton nucleus with a single electron orbiting about it.

proton–proton chain reaction: a fusion reaction by which stars convert hydrogen to helium. The proton–proton chain reaction dominates fusion in stars the size of the Sun or smaller. See *carbon–nitrogen–oxygen (CNO) cycle*.

proton flux: the passage of protons through a cell membrane.

proton transfer: movement of a proton from one atom to another.

protozoan (plural: *protozoa*): a single-celled, typically microscopic heterotroph. Protozoa live in aqueous environments and soil. They occupy a range of trophic levels. Protozoa are called animal-like protists because they subsist on other organisms.

Proxima Centauri (aka *Alpha Centauri C*): a small, low-mass, red dwarf star 4.244 light-years from the Sun, in the constellation Centaurus.

Pseudomonas aeruginosa: a common rod-shaped bacterium resident in soil and water. *P. aeruginosa* can cause disease in plants and animals, including humans. *P. aeruginosa* is antibiotic resistant.

pteridophyte: a vascular plant that reproduces and disperses via spores, producing neither flowers nor seeds.

pulsar (portmanteau of *pulsating star*): a magnetized, rotating neutron star that emits electromagnetic radiation.

pulvinus: a joint-like thickening of plant cells at the base of a leaf that facilitates growth-independent movement.

pumice: a solidified, frothy, lava rock.

purine: a chemical class of organic compounds, notably including nucleobases adenine (A) and guanine (G).

Pusey-Barrett-Rudolph theorem: that wave/particle duality is actuality, not merely a mathematical construct.

pyrimidine: a chemical class of organic compounds, notably including nucleobases cytosine (C), thymine (T) and uracil (U).

pyrite (FeS_2): iron sulfide; nicknamed "fool's gold."

pyridoxal ($C_8H_9NO_3$): 1 of 3 natural forms of vitamin B_6.

pyroelectricity: a property of certain crystals to be naturally electrically polarized and thereby have large electric fields.

pyroptosis: an inflammatory form of programmed cell death.

Pythagorean theorem: though previously known by the Babylonians and Indians, a geometry theorem credited to Pythagoras: for any right-angled triangle, $c2 = a2 + b2$.

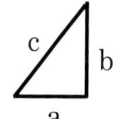

Q

quadratic: an equation involving terms of the 2nd degree at most.

quantum (physics) (plural: *quanta*): an infinitesimal chunk of ripple in a localized energy field that appears particulate (via *quantization*).

quantum chromodynamics (*QCD*): a theory of the nuclear strong force applying to fermions, characterizing the interactions between quarks and gluons which comprise hadrons.

quantum degeneracy pressure: extreme pressure at quantum scale, pushing fermions to the closest possible quarters. According to the Pauli exclusion principle, 2 fermions cannot occupy the same space simultaneously. In a quantum system an energy level is *degenerate* if it corresponds to multiple measurable states.

quantum effect: a physical 4D effect reflecting HD dynamics. Entanglement is a quantum effect.

quantum electrodynamics (*QED*): a relativistic quantum field theory of electrodynamics.

quantum field theory (*QFT*) (aka *quantum theory*, *quantum mechanics*): a theoretical framework explaining subatomic interactions from a particle perspective.

quantum fluctuation: an energy change at a spacetime point arising from the *uncertainty principle*.

quantum foam: the characterization of an energetic ground state as a froth of virtual particles continually perturbed by ghost fields.

quantum mechanics: see *quantum field theory*.

quantum gravity: a quest to explain gravity at the quantum level.

quantum information theory: the idea that a quantum system is a repository of information.

quantum spin liquid: a liquid state of magnetism achieved by quantum entanglement. Compare *ferromagnetism*, *antiferromagnetism*.

quantum tunneling: a particle overcoming its 4D classical confines to move itself through an HD wormhole. The practical size limit of transistors is set by quantum tunneling, as electrons could bypass the carved path in a too-small transistor.

quark: a subatomic particle that serves as the combinational seed for protons, neutrons, and hadrons.

quark star: a star comprising strange quark matter, evolved from an aged neutron star.

quartz: a crystal in a framework of silicon-oxygen (SiO_4) tetrahedra, where each tetrahedron shares an oxygen atom, effectively rendering SiO_2. Quartz is abundant in Earth's continental crust.

quasar: a cosmic energy source caused by the spin-off of a black hole.

quasiparticle: an emergent approximation of fermionic behavior. Localized subatomic energies which mimic bosons are termed *collective excitations*.

quintessence (physics): a hypothetical form of dark energy that is dynamic, unlike the alternately proposed *cosmological constant*.

R

radar: an object-detection system employing radio waves.

radiant energy: the energy of electromagnetic and gravitational radiation.

radiation (physics): a process of traveling electromagnetic waves; also used for a similar sojourn of subatomic particles via *atomic decay* or *beta decay*.

radiative zone: the middle of 3 layers in a star's interior, where core-produced energy is primarily transported by radiative diffusion and thermal conduction, rather than by convection.

radical (chemistry): a reactive atom, molecule, or ion owing to an unpaired valence electron. See *polyatomic*.

radio wave: a long wavelength electromagnetic radiation, ranging between 1 millimeter to 100 kilometers.

radioactivity: a subatomic process of losing energy. See *beta decay*.

radiology: imaging of organic substances using electromagnetic radiation.

radium (*Ra*): the element with atomic number 88; a highly radioactive luminescent metal that glows a faint blue.

radon (*Rn*): the element with atomic number 86; a radioactive, colorless, tasteless, odorless noble gas.

rancidity (aka *rancidification*): chemical decomposition of lipids. Rancidification has 3 pathways: hydrolytic, oxidative, and microbial. Hydrolytic rancidity happens when water peels fatty acid chains off the glycerol backbone in triglycerides. Oxidative rancidity transpires by free radicals on the loose (unbounded oxygen running rampant)–double-bonded unsaturated fats are cleaved, releasing volatile aldehydes and ketones. Microbial rancidity comes with the little ones employing their enzymes to fractionalize fat.

Randall–Sundrum model: a braneworld model that construes a universe of 5 dimensions using warped geometry, with the force of gravity (via gravitons) emanating from the 5th dimension, which is not compact (and so, invisible for being beyond Planck-unit measure); instead, merely phase-shifted from 3D space. There are 2 Randall–Sundrum models: RS1 and RS2. RS1 has 2 branes, while RS2 has 1 brane.

random (adjective): the idea that a system lacks order.

reactive oxygen species (*ROS*): chemically reactive molecules containing oxygen.

reality: that which necessarily is, phenomenal or noumenonal. Contrast *actuality*.

receptor (cytology): a cell signal receiver.

recessive (trait): a genetic trait masked by a *dominant* trait. Recessive traits are part of the phenotype only with homozygous alleles that are recessive.

recoding (genetics): interpretive reading of genetic code by an organism.

recombination (genetics): mixing traits during meiosis that introduces diversity in offspring.

red blood cell (aka *erythrocyte*): the most common type of vertebrate blood cell, employed to deliver oxygen to the tissues via blood flow through the circulatory system.

red giant: a luminous giant star of relatively low mass.

redox: a change during a reaction specific to loss or gain of electrons, with *reduction* a gain and *oxidation* a loss.

redshift: reflected light from a distant object, where the light has a longer wavelength. The longest human-visible wavelength of light is seen as reddish, hence the term *redshift*.

reductant: a chemical species that donates an electron to another species.

reduction (chemistry): a gain of electrons or a decrease in oxidation state to an atom or molecule; typically, reaction with hydrogen. Contrast *oxidation*.

reduction potential: the tendency of a chemical species to acquire electrons, and thereby be reduced.

reductionism: the absurd idea that a complex dynamic phenomenon can be understood by analyzing and ascertaining its constituent elements. Reductionism requires that the something can never be more than the sum of its parts. Reductionists explain biological processes in the same way that chemists and physicists interpret inanimate matter. In adhering to empirical cause-and-effect, reductionism is a tool of matterism. See *synergy*. Contrast *holism*.

reference frame (aka frame of reference): an abstract coordinate system that encompasses location, orientation, and measurement.

reflection (physics): a change in direction for an energy wavefront between 2 different media so that the wavefront returns into the medium from which it originated. Contrast *refraction.*

refraction: energy wave deflection due to passing from one medium into another, each medium having a distinct velocity. Contrast *reflection.*

refractive index (aka *index of refraction*): a dimensionless number indicating the speed of light through a specific material. For instance, the refractive index of water is 1.333: light slows 1/3rd while traversing water (rather than vacuum). The refractive index determines by how much a light path is bent (refracted) when entering a certain material.

reionization: an epoch where the universe's atomic matter reverted into ionized plasma.

relative permittivity (historically, *dielectric constant*): the relative resistance of a material to an electric field.

relativity (physics): the idea that there is an inertial reference frame. See *general relativity, special relativity.*

relativity of simultaneity: the idea that simultaneity is not absolute, instead depending upon an observer's frame of reference. Different observers in relative motion to one another may legitimately disagree as to whether 2 events occurred simultaneously or one before the other.

religion: a belief system encompassing the nature of the universe and life, commonly belied by facts. Religions are frequently faith-based and typically dogmatic. Religions usually involve supernatural agents (gods). Compare *natural philosophy.* Contrast *science.*

renormalization: a mathematical technique to eradicate infinities from physics equations. As infinity is infinitely unwelcome, the erasure of renormalization is liberally applied as needed. Renormalization was initially viewed with suspicion, considered a provisional procedure, but eventually embraced as an acceptable adjunct. The use of renormalization illustrates the travesty of the physical models used in modern physics, which often provide elegant approximations at the expense of ignoring issues that infinities imply.

reprogramming (epigenetics): erasure and remodeling of epigenetic marks. Reprogramming is common during animal early development. Methylation is one reprogramming technique.

resistance (physics, chemistry): a measure of a material's opposition to the flow of electric current; alternately, a measure of the force required to make a current flow through a material; measured in ohms.

resistor (chemistry): a material that resists to a measurable degree passage of electric charges. Contrast *conductor*.

resonance (physics): a periodic synchrony.

respiration (cellular): the metabolic processes and reactions that convert nutrients into ATP, with waste products released.

retrotransposon: (aka *transposon via RNA intermediates*): a genetic element that can amplify itself in a genome. Retrotransposons are considered a subclass of *transposons*.

retrovirus: a family of single-stranded RNA-enveloped viruses that replicate in a host cell via *reverse transcription*.

reverse transcriptase: a DNA enzyme that transcribes single-stranded RNA into single-stranded DNA.

reverse transcription: the process of creating a single-stranded DNA from an RNA template using *reverse transcriptase*.

reversion evolution (aka *reverse evolution, re-evolution, de-evolution, devolution, backward evolution*): evolutionary descent with an unmanifest ancestral trait reactivated (*atavism*).

rheid: a nominal solid at a temperature below melting point, deformed by viscous flow.

rheology: the geological science of matter flow.

rhizosphere: soil managed by plant roots via secretions. By contrast, *bulk soil* is outside the rhizosphere.

ribonucleoprotein (RNP): an RNA-binding protein and associated RNA. RNPs work as regulators in RNA metabolism, DNA replication, and gene expression.

ribose ($C_5H_{10}O_5$): a simple sugar (monosaccharide), finding equilibrium in 5 forms.

ribosome: the cellular factory for synthesizing proteins from peptide pieces.

ribozyme: an RNA-based enzyme.

RNA (*ribonucleic acid* ($C_5H_{10}O_5$; H–(C=O)–(CHOH)$_4$–H)): a macromolecule comprising a long chain of nucleotides. RNA and DNA differ by their sugar (ribose versus deoxyribose (a ribose lacking an oxygen atom)). RNA and DNA also differ by 1 nucleobase: whereas RNA uses uracil (U), DNA employs thymine (T). See *DNA*.

RNA interference (*RNAi*): an epigenetic regulator of gene expression. RNAi limits gene expression. See *miRNA*.

RNA polymerase: an enzyme that unwinds a specific strand of DNA.

RNA splicing: editing precursor messenger RNA (mRNA) after transcription but before translation.

RNA world (abiogenesis): the hypothesis that life began with RNA-based replication.

S

saccharide: sugar (in any form); a sweet-tasting, water-soluble carbohydrate based on 1 ring of 4–5 carbon atoms and 1 oxygen atom.

Sagittarius (aka *Carina–Sagittarius Arm*): a minor spiral arm of the Milky Way. *Sagittarius* is also the constellation in which the core of the Milky Way lies (on its westernmost part).

salamander: an amphibian, typically characterized by a lizard-like appearance, with a short nose, slender body, and long tail. Salamanders have been around for 164 million years.

Salmonella: a genus of rod-shaped bacteria with 2 species, one of which is found in endothermic animals (and the environment), the other in ectothermic animals.

saturated fat: a fat molecule with only single bonds between carbon atoms. Contrast *unsaturated fat*.

Saturn: the 6th planet from the Sun; the 2nd largest in the solar system, behind Jupiter. Saturn is the least dense planet. Saturn is known for its lovely ring system, comprising 9 continuous main rings and 3 discontiguous arcs. The rings consist mostly of ice particles, with bits of rocky debris and dust. Saturn captured 62 satellites.

scalar: a quantity representable as a point on a scale.

scalar field (astrophysics): a hypothetical field independent of the spacetime reference frame. The Higgs field, a spin-zero quantum field, is a hypothesized scalar field. A scalar field is a mathematical construct for which no evidence exists of being actualized in Nature (including the Higgs field).

scale (mathematics, statistics, physics, economics): a relative size or dimensionality.

scale invariance: a feature in physical or mathematical systems of consistency regardless of the scale of the objects or energies in the system.

scar: an area of fibrous tissue replacing normal tissue, typically skin, after injury. A scar occurs as a natural part of wound repair.

Schrödinger's equation: an equation describing how the quantum state of a physical system changes through time.

science: the study of Nature from a strictly empirical standpoint. William Whewell coined the term *scientist* in 1840. See *scientific method*. Contrast *natural philosophy*.

scientific method: a set of techniques for investigating phenomena and acquiring knowledge, ostensibly involving careful observation before guessing what is going on, which is known as forming a *theory*. Guessing prior to intensive observation is making a *hypothesis*.

sea: a large body of saltwater partly or wholly surrounded by land, not as deep as any ocean.

seagrass: a marine angiosperm that resembles grass, of ~60 species. Seagrasses descended from terrestrial grasses 75–100 million years ago.

second law of thermodynamics: see *2nd law of thermodynamics*.

sedimentary (rock): a rock formed by cumulative material deposit. Compare *igneous* and *metamorphic*. See *basement*.

seismic wave: an energy wave traveling through the Earth.

self-energy: the contribution of energy or effective mass a subatomic particle makes in HD interactions. Both fermions and bosons possess self-energy.

self-organized criticality: a property of dynamic systems where a critical threshold (tipping point) exists that, when passed, sets off a substantial reaction.

semiconductor: a material with an electrical conductivity between that of a *conductor* and an *insulator*. Compare *resistor*.

senescence: biological aging; the process of accumulative dysfunction in cells, disrupting metabolism, resulting in deterioration and death. Senescence applies to an organism, organs, and individual cells.

serpentinization: oxidation and hydrolysis of low-silica rocks via heat and water.

sex chromosome: the chromosomes–termed X and Y–employed in sexual reproduction.

sexual reproduction: biological reproduction from 2 haploid cells. Contrast *asexual reproduction*.

shadow partner: see *sparticle*.

Shapley Supercluster (aka *Shapley Concentration*): the largest nearby concentration of galaxies, 650 million light-years away, in the constellation Centaurus. Named after its discoverer, Harlow Shapley. The Shapley Supercluster acts as a gravitationally attractive force to the Milky Way.

shared subjectivity: the principle that shared subjective perceptions creates an illusion of objectivity via showtivity.

shell (physics, chemistry): an electron orbital layer.

shell layering (physics, chemistry): layering of electron shells.

showtivity: the seeming objectivity of Nature via a shared experiential platform provided by Ĉonsciousness and coherence as an ordering principle for the perception of Nature.

SI: the International System of Units; the world standard for measurement since 1960, supplanting the metric system. SI is an abbreviation derived from the French (Le Système International d'unités).

side chain (often designated as R): a defining component of an amino acid, specific to the amino acid to which it belongs.

sigma bond: a covalent bond of electron valence shell sharing.

silica: silicon dioxide (SiO_2). Silicate minerals make up 90% of Earth's crust.

silicon (*Si*): the element with atomic number 14; a hard, brittle, crystalline solid with a blue-gray metallic luster.

single bond: a chemical (covalent) bond of sharing 1 pair of electrons. Compare *double bond*, *triple bond*.

singlet: a diradical with zero spin. Almost all everyday molecules are singlet. Molecular oxygen (O_2) is an exception, existing in a triplet state. Contrast *doublet* and *triplet*.

sine wave (aka *sinusoidal wave*): a mathematical waveform with a smooth periodic oscillation measured by the distance between adjacent peaks or troughs (wavelength).

situs solitus: the anatomical position of organs.

slime mold: a protist that reproduces via zoospores.

Snell's law (aka *Snell-Descartes law*, *law of refraction*): a formula describing the relation between angles of incidence and refraction for waves passing through a boundary of distinct isotropic media.

snow line (planetary): the astronomical line of a star system beyond which ice is deposited on planets.

social Darwinism: a term given to various societal theories that emerged in England and the US in the 1870s, which applied the Darwinian notion of "survival of the fittest" sociologically and politically. The term itself was coined in 1944 as a pejorative by those with a more peaceable mindset.

soil: the surface layer of Earth's crust. Soil is the product of weathering rock, decomposed organic matter, and the cumulative activities of the biotic community. Soil layers are termed *horizons*. A cross-section of soil horizons is a *soil profile*. Soils differ among ecosystems.

Soils are classified as young, mature, or old. A young soil accumulates organic matter, hence continues to develop a profile. Mature soil holds its own, and so has a static profile. Old soil loses material; nutrients are leached away. Old soil's horizon diminishes.

solar (astronomy): relating to the Sun.

solar flare: a sudden flash of light from the Sun. Compare *coronal mass ejection*.

solar maximum: the period of greatest solar activity in the Sun's solar cycle.

solar system: the matter that swirls around the Sun, the formation of which began with the collapse of a giant molecular cloud 4.6 BYA. The largest bodies orbiting the Sun are planets.

solar wind: the constant, fluxing flow of particulate released from the Sun's atmosphere.

solid: a substance with structural rigidity. Crystals and glasses are solids. Contrast *fluid*.

solid-state physics: the study of solids, particularly how solids at the macro scale result from their atomic-scale properties.

soliton (aka *solitary wave*): a self-reinforcing solitary wave that maintains its shape as it travels through a medium at a constant speed.

soluble: capable of being dissolved or liquified.

Solvay Conference: a series of conferences for physicists, held in Brussels. The first was held in 1911. The most famous was the 5th conference, in October 1927, where the newly formulated quantum theory was discussed.

solvent: a substance that dissolves another substance, resulting in a solution.

soma (*somatic cell*): a cell forming the body of a multicellular eukaryote. Contrast *germline*.

sound (physics): an audible, mechanical vibration that propagates as a wave of pressure through a medium.

space: a boundless, non-Euclidean extent as filler for celestial bodies, which are invariably in motion.

spacetime: a treatment of space and time via unified dimensionality.

sparticle (aka *shadow partner* or *superpartner*): a shadow partner particle under *unbroken supersymmetry*. Sparticles are hypothetical.

special relativity: a physical theory of measurement proposed by Albert Einstein in 1905 and since validated empirically: that the speed of light provides an inertial frame of reference. Special relativity has numerous consequences beyond uniform motion being relative, including *relativity of simultaneity*, *time dilation*, and *length contraction*. See *relativity*, *general relativity*.

speciation: the process of species formation.

species (biology): a physically or genetically distinct population of organisms.

species (chemistry): chemically identical molecular entities with distinct interaction characteristics, typified by different ionization or lack thereof.

specific heat capacity: heat capacity per unit mass.

spectral line: a discontiguity in an otherwise uniform and continuous electromagnetic spectrum, caused by emission or absorption of light in a narrow frequency range.

spectrum (plural: *spectra*, *spectrums*): an array of distinguished components of a wave or emission. Discriminative characteristics of a spectrum include wavelength, energy, or mass.

speed: the distance something travels every unit of time. Compare *velocity*.

sperm: a male reproductive cell. Compare *egg*.

spicule (solar physics): a dynamic energy jet in the chromosphere of the Sun.

spin (quantum physics): the mathematically hypothesized internal rotation of a subatomic particle; a form of intrinsic angular momentum. Each particle type has specific spin. In quantum physics' Standard Model, only the Higgs boson is presumed without spin.

spirit plane: the dimensions where extra-dimensional (ED) life resides.

spirochete: a phylum of double-membraned bacteria.

spontaneous emission: the process where an atom or molecule transitions from an excited state to one with a lower energy, emitting a photon as an indication.

spontaneous symmetry breaking (*SSB*): a mathematical concept where the manifestation of a symmetrical system shows a tangible result, which breaks symmetry merely by actualizing. The system may remain symmetrical (hidden symmetry), but its outputs never are, as symmetry has to be broken for any manifestation.

"spooky action at a distance": Albert Einstein's dismissive term for *entanglement*.

spore: a desiccated microbe in hibernation, able to remain dormant and survive adverse conditions, such as cold, heat and radiation. Spores are produced via *sporulation*.

sporophyte: the diploid, spore-producing phase of plants and algae that undergo alternation of generations. Compare *gametophyte*.

spreading ridge: a mid-ocean ridge with a growing rift along its spine, formed by 2 tectonic plates; an underwater divergent plate boundary.

standard cosmological model: See *ΛCDM*.

Standard Model (quantum physics): a quantum field theory focused on theorized fundamental subatomic quanta and their interactions. The Standard Model is known to be incomplete.

star: a massive, luminous sphere of plasma held together by gravity.

Stark shift (aka *Stark effect*): the effect from an external static electrical field shifting and splitting the spectral lines of atoms and molecules. Named after Johannes Stark, who discovered the effect in 1913. The Stark effect is the electrical analogue of the magnetic Zeeman effect.

starquake: an irregularity in the energetic pulse of a pulsar.

stationary critical state: a state that is stable in a system characterized by *self-organized criticality*, but on the edge of a critical point to instability.

statistical mechanics: the study of the average behaviors in a mechanical system where the system is uncertain; a branch of theoretical physics using probability theory. Modeling irreversible processes driven by imbalances is *non-equilibrium statistical mechanics*. Such processes include chemical reactions, thermodynamics, and particle flows.

stem cell: an undifferentiated cell which can differentiate into a specialized cell. Stem cells can divide via mitosis to produce more stem cells. Stem cells are the basis for multicellular organism growth, with differentiation into somatic cells that form tissues with specialized functions. In mature organisms, stem cells serve to maintain and repair tissue in their vicinity. See *germline cell*.

sterile neutrino: a neutrino not interacting with the weak force.

steroid: an organic compound characterized by 4 joined cycloalkane rings with 17 carbon atoms. Eukaryotic cells manufacture steroids for various functions.

sterol (aka *steroid alcohol*): a subgroup of steroids, naturally occurring in the cell membranes of fungi, plants, and animals.

stochastic: probabilistic; appearing random (though nothing is).

stopping power (physics): the retarding force acting upon charged particles from interaction with matter, resulting in loss of energy.

stratum corneum: the outermost layer of the skin (epidermis), comprising (in humans) 15–20 layers of flattened dead cells (corneocytes).

Streptococcus: a genus of spherical bacteria. *Streptococcus pyogenes*, a usually pathogenic bacterium found on human skin, was the basis for the Cas9 enzyme used in gene editing.

stress (biology): a negative influence on well-being. Stress may be received psychologically or physically, but its effect is holistic.

string (physics): a 1-dimensional subatomic particle under string theory.

string theory: a theoretical attempt to reconcile quantum field theory with general relativity, characterizing quanta by their vibrational quality.

strong force: as described by quantum chromodynamics, the force binding quarks and antiquarks to make hadrons, as well the nuclear force gripping protons and neutrons together in atomic nuclei. Compare *weak force*.

subduction: the process of a tectonic plate moving under another; a convergent tectonic boundary.

subduction plate: a tectonic plate undergoing subduction.

subjectivity: the idea that manifestation is necessarily an experience of individual consciousness. Contrast *objectivity*.

sublimation (chemistry): the transition of a substance directly from solid to gas without entering an intermediate liquid phase. Sublimation is an endothermic phase transition occurring at pressures and temperatures below a substance's triple point. The inverse process of sublimation is *deposition*.

substrate (chemistry): a molecule used as a foundation for building a more complex molecule.

sulfur (*S*): the element with atomic number 16; an abundant, multivalent non-metal. Sulfur can react as either a reductant or oxidant. As an organic compound (organosulfur), sulfur is widely employed in biological processes, playing a key role in many enzymes. Sulfur is a component in all proteins.

Sun: the star at the center of the solar system, with a diameter of 1,392,000 km.

sunspot: a temporary phenomenon on the photosphere of the Sun that visibly appears as a dark spot.

supercluster (cosmology): a group of smaller galaxy clusters; the largest known cosmic structure.

superconductivity: zero electrical resistance, resulting from electrons overcoming their mutual repulsion and pairing up, creating a coherent, frictionless flow.

supercontinent: a landmass comprising multiple continental cores. Supercontinents in Earth's history include: Vaalbara (3.1–2.8 BYA), Kenorland (2.7–2.5 BYA), Nuna (1.9–1.5 BYA), Rodina (1.1 BYA–750 MYA), and Pangaea (300–200 MYA).

supercritical: a substance at a temperature and pressure above its *critical point*: the point at which no phase boundaries exist.

superfluid: a matter phase of flowing without friction, via zero viscosity and zero entropy. Helium-4 becomes a superfluid at cooler than 2.17 Kelvin.

supergene: the idea that a group of genes are inherited as an integral unit because of close genetic linkage. While specific to neighboring genes on a chromosome, the concept of supergene also encapsulates the idea of genetic heredity for related traits.

superinsulator: a medium that absolutely resists electrical conductivity. Contrast *superconductivity*.

superluminal: faster than light speed.

supernova (plural: *supernovae* or *supernovas*): a large star in its final death throe, which manifests as a massive explosion of energy and matter.

superposition (quantum physics): a fundamental principle of quantum mechanics that a physical system has all its potentialities (all theoretically possible states) until perceived (measured). Superposition is the assumption that existence itself is emergent.

supersolid: a spatially ordered material with superfluid properties.

superstring theory: a theory integrating fermions into string theory, with supersymmetry tagging along.

supersymmetry (*SUSY*): a unifying field hypothesis for fermions and bosons, bringing together all quantum particles as components of a single master superfield. SUSY lacks essential evidentiary foundation, as requisite partner particles have not been found.

surface tension: a property of the surface of a substance that allows it to resist an external force. Surface tension in a crystal arises from stretching interatomic bonds, whereas liquid surface tension is more about the extra atoms introduced when spreading out in increased surface area.

symbiont: an organism that lives symbiotically with a host.

symbiosis: 2 dissimilar organisms living together, typically in a mutually beneficial association (mutualism).

symmetry: a theoretical situation for a mathematical object, where performing an operation on the object does not alter it. A circle has rotational symmetry, in that a circle is unchanged by rotation. Physicists often see symmetry in their physical models.

sympatric speciation: speciation of a subpopulation when not separated from the population, as contrasted to *allopatric speciation*.

symplectic (mathematics): woven together. Symplectic variables are interdependent.

synergy: an interaction of elements which, in combination, produces a total effect greater than the sum of individual contributions. Contrast *reductionism*.

synthesize: to form (a material or abstraction) by combining parts or elements. Contrast *analyze*.

system: an assemblage of interdependent or interacting constituent concepts that form a whole.

systems biology: modeling of complex biologist systems, focused on interactions.

T

T-cell: an adaptive immune system cell in mammals.

tachyon: a hypothetical particle with imaginary mass that always travels faster than light.

tantalum (Ta): the element with atomic number 73; a hard, blue-gray, lustrous transition metal that is extremely corrosion-resistant.

tectonics: processes related to the movement and deformation of Earth's crust, notably the roving of tectonic plates.

tectonic plate: a sizable chunk of Earth's crust, capable of movement.

teleology (evolutionary biology): the theory that adaptation is goal oriented.

teleology (philosophy): the doctrine that final causes (ends or purposes) exist.

telomerase (aka *terminal transferase*): a ribonucleoprotein responsible for telomere maintenance. An enzymatic subunit–*telomerase reverse transcriptase*–endeavors to refurbish a telomere after cell division.

telomere: a protective region of repetitive nucleotide sequences at each end of a chromosome copy.

telophase: the stage during cell division where 2 daughter nuclei form. The outcome of telophase, after cytokinesis, is 2 daughter cells. See *interphase, prophase,* metaphase, *anaphase.*

tensor: a geometric object describing linear relations between other geometric entities (vectors, scalars, tensors). A tensor is a geometric entity entangled with other tensors. See *tensor network.*

tensor network: a network of tensors.

tetrahedron: a polyhedron with 4 faces.

tetraquark: a hadron with 4 quarks, particularly 2 quarks and 2 antiquarks. Tetraquarks are not accounted for in quantum physics' standard quark model. Hadrons, such as baryons, are made of 3 quarks.

tetraterpenoid: a molecule with a skeleton of 40 carbon atoms.

theory: fact-based explanation about the relations between concepts. See *physical theory*.

> The truth of a theory can never be proven, for one never knows if future experience will contradict its conclusions. ~ Albert Einstein

Theory of Everything: the holy-grail physics theory that explains all (known) phenomena.

thermalization (physics): the process of a system reaching thermal equilibrium via an equipartition of energy that maximizes the system's entropy.

thermoacidophile: an organism that prefers a habitat with temperatures of 70–80 °C and a pH of 2–3; a combination of acidophile and thermophile.

thermodynamic system: a spatial region considered a self-contained system, characterized by certain characteristics, including temperature, pressure, entropy, and internal energy.

thermodynamics: the branch of physics concerned with the dynamics of heat and temperature and their relation to energy and work.

thermophile: an organism that can survive a 60 °C or even hotter habitat.

thymine (*T*) ($C_5H_6N_2O_2$): a DNA nucleobase. Thymine is complementary to *adenine*. In RNA, thymine is replaced by *uracil*.

tidal friction: an effect of tidal forces between an orbiting natural satellite (e.g., the Moon), and the primary planet that it orbits (e.g., Earth).

tidal heating (aka *tidal working*): orbital and rotational energy dissipated as heat in planetary bodies through the tidal friction process.

tidal locking (aka *gravitational locking, captured rotation*): a gravitational process whereby one astronomical body always faces another, such as the Moon always facing Earth. Tidal lock is the eventual outcome of tidal friction.

time: the idea that there is a temporal vector comprising past, present, and future.

time dilation: that concept that time itself is relative to the relative motion of an observer.

titanium (Ti): the element with the autonomic number 22; a silvery, lustrous metal.

tonne: a metric ton (1.102 US (short) tons).

topology: the mathematical study of space. Topology is not constrained to 3D except that the human mind is ill-equipped to envision 4D (or higher) spatial dimensions.

torsional angle: the angle between 2 planes.

trachea: channels in an animal respiratory system.

tracheophyte: a vascular plant.

trait (biology): an organitypic feature of form and/or function; from an evolutionary perspective, a distinct variant of phenotype, mentotype, or envirotype.

transcendence: the state of consciousness where the body is in repose but receptive to stimuli, while the mind is quiet.

transcription (genetics): the process of producing an RNA copy from a DNA sequence. Transcription is an early, major stage of DNA expression.

transcription factor: a protein that controls the flow of genetic information during transcription.

transcription unit: an RNA copy of a DNA sequence which encodes at least 1 gene. If a transcribed gene encodes a protein, the transcription unit is messenger RNA. Otherwise, the transcription unit may encode various other products: a noncoding RNA gene (such as microRNA), ribosomal RNA, a component used in protein assembly, or a ribozyme.

transfer RNA (tRNA): an adapter for bridging the 4-letter genetic code in messenger RNA with the 20-letter code of amino acids; used for protein synthesis.

transistor: a semiconductor device used to switch or amplify electronic signals and electrical power.

translation (genetics): a later stage of gene expression as part of protein biosynthesis, after transcription. Translation transpires in a ribosome.

translocation (genetics): untoward relocation of DNA sequences.

transmember protein: a protein which can travel through a cell membrane.

transmutation (chemistry): change of one element into another.

transporter: a protein within a cell membrane that shuttles material in and/or out of a cell.

transposable element: a transposon or retrotransposon.

transposon: a DNA sequence which can change its position within a genome (typically by placing a copy elsewhere).

transvection (epigenetics): gene activation or repression resulting from allele interactions on homologous chromosomes.

Treponema pallidum: a spirochaete bacterium with subspecies that cause treponemal diseases, including syphilis, bejel, pinta, and yaws.

triglyceride: a fat common in organisms; technically, an ester derived from glycerol and 3 fatty acids.

triple-a process: the process of forming carbon-12 from 3 helium-4 atoms, owing to nuclear clustering.

triple bond: a chemical (covalent) bond of sharing 3 pairs of electrons. Compare *single bond* and *double bond*.

triple point: the temperature and pressure at which a specific molecular structure coexists in the 3 phases: gas, liquid, and solid.

triplet (chemistry): a diradical with a spin = 1. O_2 at room temperature exists in a triplet state. Contrast *singlet* and *doublet*.

Triton: Neptune's largest moon; the 7th largest moon in the solar system.

trivalent: an element with a chemical valence of 3.

troposphere: the atmospheric layer of life: the lowest portion of Earth's atmosphere. The troposphere extends from the Earth's surface 7–20 km up, depending upon location and season.

TRPA (an acronym for: *transient receptor potential ankyrin*): a protein which acts as a cell stress sensor and pain initiator in animals.

tubulin: one of several members of a family of globular proteins, comprising 5 subfamilies. The most common tubulins—α-tubulin and β-tubulin—make up *microtubules*. Tubulins are instrumental in deriving cell organization and organ placement in organisms.

U

ubit (universal quantum bit): an essential adjunct to the real-vector-space quantum model proposed by William Wootters.

ultrarelativistic: very close to the speed of light.

ultraviolet: the 10–400 nm band of the electromagnetic spectrum, shorter than visible light but longer than X-rays.

ultraviolet radiation (UVR): electromagnetic radiation at a wavelength between 10–400 nanometers.

unbroken supersymmetry: a variant of supersymmetry wherein each fermion flavor has a boson shadow and vice versa.

uncertainty principle: the principle that subatomic quanta are inherently probabilistic in their activity: a measurement may yield only an approximation of either a quantum's position or its momentum, but not both simultaneously; an intrinsic property of Nature, not a measurement incapacity; proposed by Werner Heisenberg in 1926 and controversial ever since.

undulipodium (plural: undulipodia): a filamentous, motile, extracellular projection from a eukaryotic cell. See *cilium, flagellum*.

univalent (aka *monovalent*) (chemistry): have a valence of 1.

universal common ancestor: the idea that life arose from a single life form.

universal law (physics): a proven axiom about a relationship between matter and energy.

universe (aka *cosmos*): a presumed self-contained repository of energy – a characterization for which there is no evidence, and which quantum theory disclaims. This universe has ~4 trillion galaxies – half are light (with visible stars), half dark.

unsaturated fat: a molecule of fat with 1 or more double bonds between carbon atoms. A fat molecule with only 1 double bond is *monounsaturated*. Molecules of fat with more than 1 double bond are *polyunsaturated*. Contrast *saturated fat*.

uracil (U) ($C_4H_4N_2O_2$): a nucleobase of RNA. Uracil is complementary to *adenine*. In DNA, uracil is replaced by *thymine*.

Uranus: the 7th planet from the Sun, and the lightest.

urea ($CO(NH_2)_2$ aka *carbamide*): a colorless, odorless, highly soluble, organic solid, crucial for animals to metabolize nitrogen-containing substances.

V

Vaalbara (3.1–2.8 BYA): the 1st known supercontinent.

vacuole: the organelle in cells responsible for autophagy.

vacuum: the idea of empty space. Vacuum has been shown *not* to exist at the quantum level. See *vacuum energy*.

vacuum energy: the underlying energy of 4D empty space. That vacuum has expressed energy shows that *vacuum* is a misnomer. Vacuum energy is the ground state from which 4D virtual particles arise. Vacuum energy is an HD phenomenon.

vacuum polarization (quantum electrodynamics): the process in which a background electromagnetic field produces virtual electron–positron pairs that alter the distribution of charges and currents that generated the original electromagnetic field.

valence (chemistry): the number of electrons involved in forming covalent bonds with an element. The International Union of Pure and Applied Chemistry defines valence as: "the maximum number of univalent atoms that may combine with the atom." Nitrogen has 5 electrons in its outer shell, and so has a valence of 3 (to complete a stable outer shell of 8): hence nitrogen is nominally trivalent.

valence shell: the outermost shell of an atom.

Van Allen belts: radiation belts emanating from Earth's magnetic field, in the inner region of Earth's magnetosphere.

van der Waals interaction: the net sum of attractive or repulsive forces between atoms other than those owing to covalent bonds, electrostatic interaction between ions, or with neutral atoms. The van der Waals interaction is between 2 dipoles; either instantaneously induced (London dispersion force), permanent dipoles (Keesom force), or a permanent dipole and an induced one (Debye force). Relative to covalent and ionic bonding, the attractive power of the van der Waals interaction is subtle; caused by correlations in the fluctuating polarizations of nearby particles. The van der Waals force is an HD interaction: a consequence of quantum dynamics in rapidly fluctuating polarizations among proximate particles. Named after Johannes van der Waals for his work characterizing the behavior of gases, and their condensation to a liquid phase. van der Waals interaction was discovered by Fritz London in 1930.

variational principle: a scientific principle using small changes (variations) to mathematically model. The calculus of variations is used to find minima and maxima of *functionals*: mapping sets of functions to real numbers.

vascular: a life form with vessels to carry fluids; commonly used to identify land plants which are vascular (aka tracheophytes).

vector: a quantity of both magnitude and direction.

vegetative reproduction (aka *vegetative propagation*, vegetative cloning, *vegetative multiplication*): any one of several ways that plants asexually propagate without spores or seeds. Herbaceous and woody perennial plants often practice vegetative propagation.

velocity: speed in a certain direction.

Venus: the 2nd planet from the Sun. Venus has the densest atmosphere of all terrestrial planets in the solar system, comprising mostly carbon dioxide, with an atmospheric pressure 92 times that of Earth.

vernalization: the need for an angiosperm to have a prolonged cold period (winter) before being able to flower.

vertebrate: an animal with a backbone and spinal column. Contrast *invertebrate*.

vertical gene transfer: hereditary genetic transmission from one cell generation to the next. Contrast *horizontal gene transfer*.

vesicle: a membrane-encased bubble within a cell.

vibration: a periodic oscillation about an equilibrium.

vibronic: related to changes in energy levels associated with the vibrational motion of molecules.

virial: the kinetic energy inherent in a system with gravitational bodies.

virion: a virus particle.

virtual particle: a hypothesized HD quantum that significantly affects the properties of 4D quanta. Virtual particles supposedly pop in and out of 4D as a manifestation of vacuum energy: a phase shift in appearance between 4D and ED. See *ghost field*.

virus: an obligate parasite that infects cells of all types of organisms; a domain of life, alongside archaea and bacteria.

vis viva (from the Latin for *living force*): the concept of kinetic energy as proposed by Gottfried Leibniz 1676–1689.

viscous dissipation: heat spreading through a viscous substance.

viscoelasticity: the property of materials which exhibit both viscosity and elasticity. When stressed, viscous materials resist strain and shear flow. Elastic materials bounce back to their original state when unstressed.

viscosity: the resistance of a fluid to flowing.

visual morphology: identification by appearance; used to classify galaxies.

vitamin: an organic compound needed by an organism as a vital nutrient, albeit in minute amounts.

vitamin E: a group of 8 fat-soluble compounds, found in plant oils and the leaves of green vegetables. Vitamin E is an *antioxidant*: stopping ROS production when fat undergoes oxidation.

volcano: a rupture in Earth's surface that affords the flow of hot magma, gases, and ash to escape from below into the atmosphere. Volcanoes are commonly caused by divergent tectonic plates pulling apart.

voltage: a measure of the force that is pushing a current through a material, measured in volts.

W

W boson: an electrically charged massive subatomic particle; carrier of a form of the weak force; sibling of the *Z boson*.

wasp: a flying insect of well over 100,000 species, found on every continent except polar regions. Most wasps are parasites or parasitoids as larvae, feeding on nectar only as adults. Many wasps are predatory, feeding their larvae other insects (often paralyzed). Wasp sociality varies by species, from solitary to social.

water (H_2O): the elixir of life; an odd polar molecule like no other.

water cycle (aka *hydrological cycle*): the cycling of water in the biosphere.

water flea (aka *Cladocera*): a small crustacean of 620 known species, though many more exist. Water fleas are ubiquitous in inland aquatic habitats, but rare in oceans.

wave (physics): a mathematical characterization of a field. Contrast *particle*.

wave/particle duality: the notion that an object simultaneously possesses the properties of a wave and a particle.

wavefront (physics): the locus of a propagating energy wave.

wavelength: the spatial period of a sine wave; commonly used as a statistical measure of the energy of a waveform, which is mathematically the product of a wave's frequency and amplitude.

weak force: the bosonic nuclear force that transforms matter from one variety of into another and causes matter to decay; hypothetically transmitted by the W or Z boson. Contrast *strong force*.

weight: the force, measured in newtons, that gravitation exerts upon an object, equal to the mass (m) of the object times the local acceleration of gravity (g): $W = m \times g$. In a region of constant gravitational acceleration, weight is commonly taken as a measure of mass; hence the easy confusion between the two.

wet (chemistry): competitive interphase bonding.

wettability (chemistry): how wet something can be; the ability of a liquid to maintain contact with a solid surface, as an outcome of the intensity of intermolecular interactions.

Weyl fermion: a massless, but charged, fermion; named after Hermann Weyl. Compare *Dirac fermion, Majorana fermion*.

white dwarf: a high-density star, burned to the nub: a ball of mostly carbon and oxygen.

work (physics): energy in transit; the result of an energetic force applied to matter.

wormhole: a shortcut in spacetime, allowing entanglement.

X

X inactivation (aka *lyonization*): the process in which 1 of 2 copies of the X chromosome in female mammals is inactivated.

X chromosome: one of the sex-determining chromosomes in mammals and some other organisms. The other sex-determining chromosome is termed *Y*.

X-ray: electromagnetic radiation at a wavelength of 0.01–10 nm.

xerophile: an organism that lives in an extremely dry habitat.

Y

YA: years ago.

yarn (genetics): a related group of genes and the regulatory elements necessary for gene activity in a chromosome.

yeast: a eukaryotic microorganism classified in the fungus kingdom. There are ~1,500 known species of yeast. Yeast are famous for brewing beer and making bread rise.

yin-yang: the dynamic balance of order (coherence) in Nature; an essential concept in Chinese philosophy, dating to the 14th century BCE or even earlier.

yttrium (*Y*): the element with the atomic number 39; a slivery metal.

Z

Z boson: an electrically neutral massive subatomic particle; carrier of a form of the weak force; sibling of the *W boson*.

Zeeman effect: the effect from an external static magnetic field splitting the spectral lines of atoms and molecules. Named after Pieter Zeeman, who discovered the effect in 1896.

Zeno effect (aka *Turing paradox*): a static quantum state created by continuous observation.

zero-point energy: the lowest energy a particle can have when confined to a finite region of space.

zeroth law of thermodynamics: see *0th law of thermodynamics*.

zoonosis: an infectious animal disease that can be transmitted to humans. Most human diseases originated in other animals but only diseases that routinely involve transmission between other animals to humans are considered zoonosis. When humans infect other animals, the term used is *reverse zoonosis* or *anthroponosis*.

zygote: a cell formed by the union of 2 gametes (male & female).

Zyklon B: the trade name of a cyanide-based pesticide invented by Fritz Haber in the early 1920s, comprising hydrogen cyanide (HCN) (prussic acid), and an adsorbent, such as diatomaceous earth.

❧ People ❦

Abraham, Max (1875–1922): German physicist who hypothesized in 1902 that the electron was a perfect sphere, with its charge evenly distributed around its surface.

Abrahamsen, Hilde: Norwegian cytologist.

Abzug, Bella (1920–1998): American politician and civil rights advocate.

Adams, Douglas (1952–2001): English writer and wry humorist, best known for *The Hitchhiker's Guide to the Galaxy*.

Adelson, David L.: Australian geneticist.

Adivarahan, Srivathsan: Indian geneticist.

Aepinus, Franz Maria Ulrich Theodor Hoch (1724–1802): German physicist who discovered pyroelectricity (1756) and published the 1st mathematical theory of electricity and magnetism (1759). Aepinus studied medicine and was also interested in astronomy.

al-Haytham, Ibn (965–1040): Iraqi scientist, mathematician, astronomer, and philosopher.

Alatalo, Katherine: American astronomer.

Albrecht, Andreas J.: American theoretical physicist and cosmologist.

Alegría-Torres, Jorge Alejandro: Mexican environmental toxicologist.

Aleksandrova, Antoniya: theoretical physicist.

Amann-Winkel, Katrin: Austrian chemist.

Amemiya, Chris: American biologist, interested in the evolution of vertebrates.

Anaximander of Miletus (610–546 BCE): Turkish Greek philosopher, astronomer, geographer, mathematician, and proponent of science.

Anaxagoras of Clazomenae (~510–428 BCE): Turkish-born Greek philosopher and cosmologist who proposed panspermia.

Anderson, Carl D. (1905–1991): American physicist who discovered the positron and the muon.

Anderson, Don L. (1933–): American geophysicist.

Anderson, Peter W. (1923–): American theoretical physicist, interested in particle physics, localization, emergence, symmetry breaking, and superconductivity.

Andrulis, Erik D.: American microbiologist who works on gyre theory.

Antonucci, Robert: American astrophysicist.

Apkarian, Ara: Indian American physical chemist.

Archimedes (287–212 BCE): Greek mathematician, physicist, engineer, inventor, and astronomer; considered one of the leading scientists in antiquity, and one of the greatest mathematicians of all time.

Aristarchus of Samos (310–230 BCE): Greek astronomer and mathematician who first speculated that the Earth orbited the Sun. His astronomical ideas were rejected in favor of Ptolemy and Aristotle, who touted a geocentric model with Earth as the center of the universe.

Aristotle (384–322 BCE): Greek philosopher and polymath. Prolific Aristotle had views on a wide range of subjects, and was considered authoritative for centuries, sometimes stymying further investigation that might have gone against his belief.

Armstrong, Neal (1930–): American astronaut; first to set foot on the Moon.

Aschenauer, Elke-Caroline: American nuclear physicist.

Ashtekar, Abhay (1949–): Indian theoretical physicist who was a founder of loop quantum gravity.

Asimov, Isaac (born *Isaak Ozimov*) (1920–1992): Russian-born American writer and biochemist, known for his works of science fiction and popular science.

Auber, Daniel François Esprit (1782–1871): French composer.

Augustine of Hippo (354–430): prolific Latin theologian. Augustine's writings were very influential in the development of European Christian thought.

Ausländer, Simon: Swiss biologist.

Avogadro, Amedeo (1776–1856): Italian physicist, mathematician, and ecclesiastical lawyer who contributed to molecular theory. Avogadro's work was ignored for almost a century.

Ayala, Francisco J. (1934–): Spanish American biologist.

Babu, Kaladi S.: Indian physicist, interested in the fundamental constituents of matter.

Bacon, Francis (1561–1626): English philosopher, scientist, and jurist. Bacon has been called the father of empiricism.

Bahcall, Neta A.: American astrophysicist, interested in the large-scale structure of the universe.

Bainer, Russell O.: American biologist, interested in mechanical biology, gene regulation, and computational biology.

Baker, Daniel: American astrophysicist.

Bakker, Huib J.: Dutch chemical physicist, immersed in water.

Balasubramanian, Shankar (1966–): Indian chemist, interested in nucleic acids.

Balazs, Lajos: Hungarian astronomer.

Balents, Leon: American physicist.

Ball, Philip: English physicist, chemist, and science writer.

Barge, Laurie: American geochemist.

Barr, Murray (1908–1995): Canadian physician.

Barrangou, Rodolphe: French American geneticist, molecular biologist, and food scientist.

Barrett, Jonathan: English particle physicist, known for the Pusey-Barrett-Rudolph theorem.

Barry, Dave (1947–): American writer and humorist.

Bastian, Nate: English science writer.

Bateson, William (1861–1926): English evolutionary biologist who coined the term *genetics* based upon a Mendelian conception of heredity.

Baum, Buzz: American cytologist.

Baum, David A.: American evolutionary biologist and botanist.

Baum, Lauris M.: American astrophysicist.

Becher, Johann Joachim (1635–1682): German alchemist and physician; concocter of the *phlogiston theory*.

Becquerel, Henri (1852–1908): French physicist who accidentally discovered radioactivity.

Beichler, James E.: American theoretical physicist and cosmologist, interested in a theory of everything.

Bejan, Adrian: Romanian American mechanical engineer who conceived *constructal law*.

Bekenstein, Jacob (1947–): Israeli theoretical physicist who has contributed to understanding black hole thermodynamics. Bekenstein is an adherent of physics as information theory.

Beliveau, Brian: American geneticist, interested in DNA organization.

Bell, John Stewart (1928–1990): Irish physicist who developed *Bell's theorem*, which posits nonlocality (e.g. entanglement).

Belshaw, Robert: English zoologist, interested in the evolution of viruses and selfish genetic elements.

Benea-Chelmus, Ileana-Cristina: Swiss physicist.

Benner, Steven: American chemist who hypothesizes that life on Earth came from Mars. Brenner was first to design a gene, and the first to artificially augment the DNA alphabet.

Benveniste, Jacques (1935–2004): French immunologist, interested in allergies and homeopathy.

Bergliaffa, S.E. Perez: Brazilian astrophysicist.

Bernauer, Jan C.: German physicist.

Berra, Yogi (1925–2015): American baseball player, coach, and manager, remembered for his dry wit, pithy paradoxical statements, and malapropisms.

Bettini, Alessandro: Italian particle physicist.

Bizzarro, Martin: Dutch astronomer and chemist.

Black, Joesph (1728–1799): Scottish chemist and physician, best known for the rediscovery of "fixed air" (carbon dioxide), the concept of latent heat, and the discovery of bicarbonates.

Bleicken, Stephanie: German cytologist.

Blencowe, Miles P.: English physicist, interested in quantum mechanics and condensed matter.

Bloch, Felix (1905–1983): Swiss physicist.

Bohm, David J. (1917–1992): American theoretical physicist.

Bohr, Niels (1885–1962): Danish physicist who contributed to atomic theory and quantum mechanics.

Bojowald, Martin (1973–): German physicist, working in cosmology and loop quantum gravity.

Bolton, Adam S.: American astrophysicist.

Bolton, Charles Thomas (Tom) (1943–): American astronomer who was the first to show evidence of a black hole.

Boltzmann, Ludwig (1844–1906): Austrian physicist who made significant contributions to mechanics and thermodynamics. Boltzmann advocated atomic theory when it was still quite controversial.

Bolyai, János (aka *Johann*) (1802–1860): Hungarian mathematician, one of the founders of non-Euclidian geometry.

Born, Max (1882–1970): German-English physicist and mathematician, instrumental in developing quantum mechanics. Born also contributed to optics and solid-state physics.

Bose, Satyendra Nath (1894–1974): Indian mathematician and physicist who worked on electromagnetic radiation and statistical mechanics.

Boulding, Kenneth E. (1910–1993): English economist and sociologist who had wide-ranging beliefs about economic behavior as part of a larger systemic web.

Boyle, Robert (1627–1691): Anglo Irish chemist, physicist, natural philosopher, and inventor.

Braakman, Rogier: Dutch chemical physicist.

Bradford, Charles M. (Brad): American astrophysicist.

Bragg, William Henry (1862–1942): English physicist, chemist, mathematician, and active sportsman who discovered the elemental dynamics of ionizing radiation in 1903. Bragg's science legacy is unique, in sharing a Nobel Prize in Physics with his son, fellow physicist William Lawrence Bragg, in 1915, for analysis of crystalline structures using X-rays.

Bragg, William Lawrence (1890–1971): Australian-born English physicist and chemist, known for *Bragg's law*, on the diffraction of X-rays by crystals. Bragg was instrumental in the discovery of the structure of DNA, providing support to Francis Crick and James Watson, who worked under his aegis.

Brahe, Tycho (1546–1601): Danish nobleman and astronomer. Brahe refuted the Aristotelian belief in a static celestial realm. Brahe was the last major astronomer to work without a telescope. Skeptical of Copernican heliocentricity, Brahe worked out a system where the rest of the cosmos whirled about the Earth, which he thought a "lazy" body too bulky to move.

Brand, Hennig (1630–1692 or 1710): German merchant and alchemist who discovered phosphorous in 1669.

Brandengerger, Robert (1956–): Canadian theoretical cosmologist and physicist

Brenner, Sydney (1927–): South African biologist who studied genetics.

Brody, Dorje: English mathematician.

Bronowski, Jacob (1908–1974): Polish mathematician, biologist, science historian, playwright, poet, and inventor; best known for his 1973 book and BBC TV documentary series *The Ascent of Man*, which traced the development of human society.

Brønsted, Johannes Nicolaus (1879–1947): Danish physical chemist who introduced the protonic theory of acid-base reactions in 1923 (as did Thomas Lowry).

Brout, Robert (1928–2011): Belgian theoretical physicist, interested in particle physics.

Brown, Harrison (1911–1986): American chemist, known for his work in geological aging by counting lead isotopes in igneous rocks.

Brown, Margaret Wise (1910–1952): American author of children's books, best known for *Goodnight Moon* and *Runaway Bunny*.

Brown, Robert (1773–1858): Scottish botanist who made contributions to botany by peering through a microscope. Credited with discovering *Brownian motion*, noted 2,000 years earlier (by Lucretius).

Browne, Thomas (1605–1682): English author of diverse works in medicine, science, religion, and more esoteric subject matter.

Bruce, David (1855–1931): Scottish pathologist and microbiologist who investigated brucellosis (then called *Malta fever*) and trypanosomes, the parasitic protozoa behind sleeping sickness.

Brukner, Caslav: Austrian theoretical physicist.

Bruno, Giordano (1548–1600): Italian Dominican friar, philosopher, mathematician, and astronomer who proposed that the Sun was just a star, and that the cosmos was populated by other worlds with intelligent life. For his far-fetched speculations, the Catholic Church convicted Bruno of heresy and burned him at the stake.

Buchert, Thomas: German cosmologist

Buchhave, Lars A.: Dutch astrophysicist.

Bulkley, L. Duncan: American physician and cancer researcher.

Butlerov, Alexander (1828–1886): Russian chemist who was one of the principal theorists of chemical structure (1857–1961), the first to put double bonds into chemical formulas, and discoverer of hexamine (1859), formaldehyde (1859) and the formose reaction (1961).

Caetano-Anollés, Gustavo: Argentinian biochemist and geneticist, interested in evolutionary and comparative genomics.

Caffau, Elisabetta: French cosmologist.

Cairns-Smith, Graham (1931–): Scottish organic chemist and molecular biologist who originated the hypothesis that life may have originally replicated via mineral scaffolding. Cairns-Smith explored the evolution and nature of consciousness in *Evolving the Mind* (1996).

Caldecott, Keith W.: English biochemist, interested in DNA repair.

Canton, John: English chemist.

Capra, Fritjof (1939–): Austrian-born American physicist and systems theorist.

Carlin, George (1937–2008): American comedian.

Carnot, Nicolas Léonard Sadi (1796–1832): French military engineer who developed a half-baked theory of heat engines (the *Carnot cycle*), anticipating the 2nd law of thermodynamics. Carnot was fascinated with steam engines.

Carr, Lincoln D.: American physicist.

Carroll, Sean M. (1966–): American theoretical cosmologist, interested in dark energy and general relativity.

Carson, Rachel (1907–1964): American marine biologist, famous for *Silent Spring* (1962), which chronicled the environmental devastation caused by synthetic pesticides, especially DDT. American chemical companies were incensed by the book.

Carvunis, Anne-Ruxandra: French geneticist.

Casimir, Hendrik (1909–2000): Dutch physicist, best known for his work on superconductors.

Cavendish, Henry (1731–1810): English pneumatic chemist who discovered "factitious air," later termed *hydrogen*.

Celsius, Anders (1701–1744): Swedish astronomer, physicist, and mathematician who in 1742 proposed an inverse of the Celsius temperature scale, which bears his name.

Chadwick, James (1891–1974): English physicist who discovered the neutron, which had been predicted Ettore Majorana.

Chambers, Scott A.: American chemist.

Chandra, Fiona A.: Indonesian biologist.

Chandrasekhar, Subramanyan (1910–1995): Indian astrophysicist.

Charlesworth, Brian (1945–): English evolutionary biologist, interested in population genetics.

Charlesworth, Deborah (1943–): English evolutionary biologist, interested the genetic evolution.

Cheng Zhu: Chinese microbiologist.

Cherenkov, Pavel A. (1904–1990): Russian physicist who discovered Cherenkov radiation.

Cherry, Colin (1914–1979): English cognitive scientist.

Chess, Barry: American molecular biologist.

Cho, Adrian: American physicist and science writer.

Chourrout, Daniel: French molecular biologist.

Christin, Jean-Pierre (1683–1755): French physicist, mathematician, astronomer, and musician. Christin is remembered for his thermometer and suggestion about the Celsius scale.

Clairaut, Alexis (1713–1765): French mathematician, astronomer, and geophysicist; a child prodigy who became dissolute in later life. His friend Charles Bossut remarked:

> He was focused with dining and with evenings, coupled with a lively taste for women, and seeking to make his pleasures into his day-to-day work; he lost rest, health, and finally life at the age of 52.

Clapeyron, Benoît Paul Émile (1799–1864): French engineer and physicist; one of the founders of thermodynamics; known for writing *Driving Force of the Heat* (1834), which only sounds like pulp fiction.

Clarke, Andrew: English ecologist.

Clausius, Rudolf (1822–1888): German mathematical physicist who formulated the 2nd law of thermodynamics (1850) and introduced the concept of entropy (1865).

Clayton, David F.: English psychologist.

Clifford, William Kingdon (1845–1879): brilliant English mathematician and philosopher who anticipated the most important developments in 20th century physics, including relativity and quantum field theory.

Close, Frank (1945–): English particle physicist. Close wrote in his book *Lucifer's Legacy: The Meaning of Asymmetry* (2000):

> Fundamental physical science involves observing how the universe functions and trying to find regularities that can be encoded into laws. To test if these are right, we do experiments. We hope that the experiments won't always work out, because it is when our ideas fail that we extend our experience. The art of research is to ask the right questions and discover where your understanding breaks down.

Coelho, Ricardo Lopes: Portuguese physicist.

Colding, Frederik: Danish physicist.

Collings, Peter J.: American physicist.

Conklin, Bruce R.: American geneticist.

Conradt, Barbara: German cytologist.

Conselice, Christopher J.: English astrophysicist.

Cooper, Leon (1930–): American physicist who contributed to understanding superconductivity.

Copernicus, Nicolaus (1473–1543): Prussian astronomer who developed a comprehensive heliocentric cosmology, displacing the Earth from the center of the universe. Copernicus's work was published posthumously, as he worried about the scorn that his crazy idea would provoke.

Corn, Jacob: American geneticist and cytologist.

Costa, Fabio: quantum physicist.

Cotler, Jordan: American theoretical physicist.

Coulomb, Charles-Augustin de (1736–1806): French physicist, best known for elucidating the attraction and repulsion of the electrostatic force. Coulomb also worked on friction.

Cox, Robert M.: American evolutionary biologist.

Cowen, Ron: American science writer.

Crick, Francis (1916–2004): English molecular biologist, known as the co-discoverer of the structure of DNA in 1953, with James Watson.

Crisp, Alastair: English biochemist.

Cronin, Leroy (1973–): English chemist, interested in the origin of life.

da Vinci, Leonardo (1452–1519): Italian painter, draftsman, sculptor, architect, musician, inventor, scientist, mathematician, engineer, geologist, cartographer, anatomist, botanist, and writer. Best known for a small portrait of a drab woman with a half-smile (Mona Lisa).

Dalton, John (1766–1844): English chemist, meteorologist, and physicist, known for his work on atomic theory and color blindness.

Darvish, Behnam: American astrophysicist.

Darwin, Charles (1809–1882): English naturalist who developed a theory of biological evolution.

Davies, Paul C.W. (1946–): English theoretical physicist, cosmologist, and astrobiologist, working on finding extraterrestrial life. Davies courted controversy by noting that the faith of scientists is in the immutability of physical laws; a faith with roots in Christian theology. Davies called the claim that science is "free of faith": "bogus."

Davis, J.C. Séamus: Scottish Irish American physicist interested in superconductivity.

Davis, Tamara M.: Australian astrophysicist and ultimate Frisbee player.

De, Sandip: Indian epigeneticist.

de Broglie, Louis (1892–1987): French physicist who developed the *pilot wave theory*.

de Chancourtois, Alexandre-Emile Béguyer (1820–1886): French mineralogist and geologist who first arranged the chemical elements by atomic weight and noticed their periodicity.

de Coulomb, Charles-Augustin: see *Coulomb*.

de Fermat, Pierre: see *Fermat* (as the man is commonly called).

de La Rochefoucauld, François (1613–1680): French author.

de Leeuw, Nora H.: English chemist.

de Vries, Hugo (1848–1935): Dutch botanist and one of the first geneticists. de Vries coined the term *mutation*.

Dehnel, August (1903–1962): Polish zoologist, known for discovering *Dehnel phenomenon*.

Dekker, Job: American geneticist.

Deming, W. Edwards (1900–1993): American statistician, best known for his contribution to industrial production quality control, for which he is considered a hero in Japan.

Democritus (460–370 BCE): Greek rationalist philosopher who formulated an atomic theory of existence.

Dennett, Daniel C. (1942–): American philosopher and cognitive scientist.

Descartes, René (1596–1650): French mathematician and philosopher who believed in dualism.

Devreotes, Peter N.: American cytologist.

DeYoung, Tyce: American astrophysicist.

Dickman, Mark J.: English biochemist, interested in bioanalytics.

Dillin, Andrew: American cytologist.

Diderot, Denis (1713–1784): French philosopher, writer, and art critic; a prominent figure in the Enlightenment.

Dirac, Paul (1902–1984): brilliant English theoretical physicist who contributed to the early development of quantum mechanics and quantum electrodynamics. Dirac was a precise and taciturn man. Raised Catholic, Dirac once remarked, "religion is a jumble of false assertions, with no basis in reality."

Döbereiner, Johann Wolfgang (1780–1849): German chemist whose grouping of chemical elements foreshadowed the periodic law.

Dobos, László: Hungarian astrophysicist.

Dobzhansky, Theodosius (1900–1975): Ukrainian geneticist and evolutionary biologist.

Dombeck, Mark: American physician, interested in autism.

Dongfeng Gao: Chinese physicist, interested in quantum cosmology theory.

Dongshan He: Chinese physicist, interested in quantum cosmology theory.

Doppler, Christian (1803–1853): Austrian mathematician and physicist who proposed the Doppler effect in 1842.

Downie, Evangeline J.: Scottish nuclear physicist.

Doyle, Arthur Conan (1859–1930): Irish-Scots novelist and physician, best known for the crime fiction tales of detective Sherlock Holmes.

du Châtelet, Émilie (1706–1749): French physicist, natural philosopher, mathematician.

Duffy, Ken R.: Irish immunologist.

Dufourc, Erick J.: French biochemist.

Duret, Laurent: French geneticist.

Dyer, Adrian: Australian vision scientist.

Dylan, Bob (1941–): American songwriter and musician.

Dyson, Freeman (1923–): English-born American physicist, cosmologist, and mathematician.

Eccleston, Alex: English cytologist.

Ecker, Ullrich K.H.: Australian psychologist.

Eddington, Arthur Stanley (1882–1944): English physicist, mathematician, astronomer, and philosopher of science.

Eisenhower, Dwight D. (1890–1969): American army general; Supreme Commander of the Allied Forces in Europe during World War 2; 34th US President (Republican) (1953–1961).

Einstein, Albert (1879–1955): German theoretical physicist, best known for his theories of relativity.

> Positivism states that what cannot be observed does not exist. I am not a positivist. ~ Albert Einstein

Eiseley, Loren (1907–1977): American anthropologist.

Eisenberg, Eli: Israeli geneticist.

Eisert, Jens: German physicist.

Elf, Johan: Swedish molecular biologist.

Elgar, Mark A.: Australian zoologist, interested in unusual animal behaviors and use of chemical communication.

Ellis, George F.R. (1939–): South African mathematician, logician, and cosmological theorist.

Emerson, Ralph Waldo (1803–1882): American essayist and poet.

Emons, Anne Mie C.: Dutch botanist and cytologist.

Empedocles (490–430 BCE): eclectic Greek philosopher who originated the cosmogenic theory of the 4 classical elements: earth, water, wind, and fire. Empedocles considered chemical changes similar to emotional relations.

Enríquez, José Antonio: Spanish biochemist and molecular biologist.

Erdmann, Hugo (1862–1910): German chemist who coined *noble gas*.

Engelhardt, Netta: American physicist.

Ertter, Barbara: American botanist.

Euclid of Alexandria (~300 BCE): Greek mathematician, the father of geometry. Euclid wrote the most influential mathematics book of all time: *Elements*, the primary textbook for math, especially geometry, for over 2,000 years, into the early 20th century. Euclidean geometry was extended into higher dimensions via independent work by János Bolyai and Nikolai Lobachevsky.

Euler, Leonhard (1707–1783): Swiss mathematician, logician, engineer, and physicist who introduced much modern mathematical terminology and notation; also known for his work in mechanics, fluid dynamics, optics, astronomy, and music theory; considered one of the greatest mathematicians of all time.

Eulgem, Thomas: American plant cytologist.

Everett III, Hugh (1930–1982): American physicist who first proposed a many-worlds interpretation of quantum physics.

Fahrenheit, Daniel Gabriel (1686–1736): German glassblower, engineer, and physicist who advanced thermometry by inventing the first practical, accurate thermometer (mercury in glass). Fahrenheit devised the namesake temperature scale.

Faraday, Michael (1791–1867): largely self-taught English chemist, physicist, and philosopher who greatly contributed to understanding electromagnetism and electrochemistry. Faraday invented electric motors.

Farley, Francis J.M.: English physicist.

Fathi, Kambiz: Swedish astrophysicist.

Fedoroff, Nina V. (1942–): American biologist and chemist.

Feldman, Gerald: American nuclear physicist.

Fermat, Pierre de (1607–1665): French lawyer and mathematician who contributed discoveries in calculus, analytic geometry, probability, and optics; best known for *Fermat's principle* for light propagation and *Fermat's last theorem*, a number theory.

Fermi, Enrico (1901–1954): Italian-born physicist, best known for his work on developing a nuclear reactor. Fermi was a rare physicist in excelling in both experimental and theoretical work.

Fernald, Russell D.: American biologist.

Ferrell, James E., Jr. (1955–): American systems biologist.

Feynman, Richard (1918–1988): eccentric American theoretical physicist who made contributions to particle physics, including quantum mechanics, electrodynamics, and superfluidity.

Fischer, André: German molecular biologist.

Fischer, Debra: American astronomer.

Flambaum, Victor V.: Australian physicist.

Forterre, Patrick: French molecular biologist.

Frampton, Paul H. (1943–): American astrophysicist.

Franklin, Rosalind (1920–1958): English chemist and X-ray crystallographer who managed the first snapshots of DNA via X-ray diffraction imagery. Franklin significantly forwarded understanding of DNA's intricate structure, providing the foundation of information used by Watson & Crick in their finalizing the structure of DNA.

Frauchiger, Daniela: Swiss theoretical physicist.

Frenk, Carlos: English Mexican computational cosmologist.

Friedan, Daniel H. (1948–): American theoretical physicist who works on string theory and condensed matter theory, focusing on (1+1)-dimensional models.

Fresnel, Augustin (1788–1827): French engineer and physicist whose study of optics led to widespread acceptance of light as a waveform phenomenon, as contrasted to Newton's particle (corpuscular) theory.

Fuchs, Elaine (1950–): American cytologist, interested in mammalian dermatology.

Fuller, Franklin D.: American physicist.

Furey, Terrence S.: American geneticist.

Fussenegger, Martin: Swiss biologist.

Gabrielse, Gerald: American physicist, interested in antimatter and the electron.

Gaiti, Federico: evolutionary and molecular biologist.

Galileo Galilei (1564–1642): Italian physicist, mathematician, astronomer, and philosopher. Galileo was a seminal figure in development of science as a discipline, and a scourge to the Catholic Church for buying into Copernicus' notion of heliocentricity.

Gallio, Marco: American neurobiologist.

García, Pedro David: Spanish physicist.

Garman, Scott C.: American biochemist and molecular biologist.

Gell-Mann, Murray (1929–2019): American particle physicist, linguist, collector of antiquities, and avid bird watcher who developed theories about quarks, neutrinos, and the weak force.

Giacomini, Flaminia: Italian quantum physicist.

Gilbert, Walter (1932–): American physicist, biochemist, and molecular biologist who developed an RNA-world hypothesis of abiogenesis.

Gierasch, Lila: American biochemist.

Gisin, Nicolas: Swiss quantum physicist.

Glashow, Sheldon Lee (1932–): American theoretical physicist who proposed the first grand unified theory in 1973, as an extension to the Standard Model. Glashow is an outspoken skeptic of superstrings owing to its lack of testable predictions.

Globus, Rea: Israeli microbiologist.

Gobley, Theodore Nicolas (1811–1876): French chemist and pharmacist.

Goff, Jon: English physicist.

Goldman, Nir: American chemist with an interest in the origin of life on Earth.

Golgi, Camillo (1843–1926): Italian physician and pathologist, known for his work on the human central nervous system.

Gould, Stephen Jay (1941–2002): American evolutionary biologist, best known for positing *punctuated equilibrium*: evolution being marked by rare bursts of speciations, with long periods of stability.

Gräter, Frauke: German molecular biomechanist, interested in biomaterials and protein evolution, dynamics, and mechanics.

Grazier, Kevin R.: American planetary physicist.

Green, Brian (1963–): American theoretical physicist, mathematician, and string theorist.

Gregoryanz, Eugene: condensed matter physicist.

Gribbin, John R. (1946–): English astrophysicist and science-fiction writer.

Gross, David (1941–): American particle physicist and string theorist.

Grosseteste, Robert (1175–1253): English scholastic philosopher, theologian, and scientist who proposed that the universe began by expanding from a singularity of light. Grosseteste also posited the possibility of a multiverse.

Grotthuss, Theodor (1785–1822): German chemist who first theorized electrolysis in 1806 (the *Grotthuss mechanism*) and formulated in 1817 the *1st law of photochemistry*: that light must be absorbed by a chemical substance for a photochemical reaction to occur.

Grove, William Robert (1811–1896): Welsh jurist, civil servant, and physical scientist who anticipated the theory of *conservation of energy*. Grove invented the fuel cell.

Guiber, Sylvain: French geneticist.

Guth, Alan (1947–): American cosmologist, credited with concocting cosmic inflation.

Haacke, Johann Wilhelm (1855–1912): German zoologist who hypothesized orthogenesis in 1893 and introduced the concept of *genes* as hereditary units, which he called *gemmaria*.

Haber, Fritz (1868–1934): German chemist who invented the *Haber process* of synthetic nitrogen fixation; considered the father of chemical warfare.

Hagen, Gaute: American nuclear physicist.

Hahnemann, Samuel (1755–1843): German physician, known for creating *homeopathy*. Hahnemann had a knack for languages: proficient in English, French, Italian, Greek, Latin, Arabic, Syriac, Chaldaic, and Hebrew.

Hall, Harriet: American physician.

Halpern, Paul: American physicist.

Hamilton, Paul: American astrophysicist.

Hamilton, William Rowan (1805–1865): Irish physicist, astronomer, and mathematician who contributed to classical mechanics, optics, and algebra.

Hamlin, James J.: American physicist.

Han, Tian-Heng (*Harry*): Chinese American physicist.

Hannon, Gregory: American epigeneticist, interested in RNA interference.

Harris, Frederick A.: American physicist and astronomer.

Hartmann, Marie-Andrée: French molecular biologist.

Harvey, David: English astrophysicist.

Hawking, Stephen (1942–2018): English theoretical physicist and cosmologist, interested in general relativity, especially black holes.

Hazen, Robert M. (1948–): American mineralogist and astrobiologist.

Heard, Dwayne: English chemist, interested in photochemistry, atmospheric and interstellar chemistry.

Heaven, Alan: English astronomer.

Heaviside, Oliver (1850–1925): self-taught English electrical engineer, mathematician, and physicist who used complex numbers to study electrical circuits, invented techniques to solve differential equations, formulated vector analysis, and reformulated Maxwell's field equations in terms of energy flux and electromagnetic forces.

Heckman, Timothy M.: American astrophysicist.

Heisenberg, Werner (1901–1976): German theoretical physicist, best known for asserting the *uncertainty principle* of quantum field theory, which states that measurement of subatomic particles is tricky to the point of indeterminate.

Held, Karsten: Austrian solid-state physicist.

Henking, Hermann (1858–1942): German cytologist who discovered the X chromosome.

Heraclitus (535–475 BCE): Turkish Greek energyist philosopher who believed in an ever-changing universe and a force of coherence creating a unity of existence.

Hero of Alexandria (aka *Heron of Alexandria*) (10–70 CE): Greek mathematician and engineer; considered the greatest experimenter of antiquity.

Herschel, William (1738–1822): German-born English astronomer and composer of 24 symphonies. Herschel discovered Uranus and 2 moons of Saturn. He and his sister Caroline compiled the first map of the Milky Way galaxy.

Hevelius, Johannes (1611–1687): German astronomer and civic leader. Hevelius described 10 new constellations, 7 of which are still recognized.

Hewish, Tony: English astronomer.

Hicks, William M. (1850–1934): English mathematician and physicist who proposed negative gravity as an adjunct to a vortex theory of gravity.

Hilbert, David (1862–1943): German mathematician; one of the most influential mathematicians of his time.

Hiley, Basil J. (1935–): Burma-born British quantum physicist.

Hippocrates (460–370 BCE): Greek physician; considered the father of western medicine.

Hobson, Art: American theoretical physicist.

Hoekstra, Hopi E.: American biologist.

Hoffman, Yehuda: Israeli cosmologist.

Holliday, Robin: English biologist.

Holliger, Philipp: English molecular biologist, interested in abiogenesis.

Holst, Gustav (1874–1934): English composer, best known for his orchestral suite *The Planets*. Holst composed numerous works in various musical genres, but none achieved comparable success; a classical one-hit wonder.

Hooper, Dan: American theoretical astrophysicist.

Horava, Petr: Czech string theorist who works on D-brane theory.

Hoscheit, Benjamin: American astrophysicist.

Hossenfelder, Sabine: German astrophysicist, interested in physics beyond the Standard Model, with a special emphasis on the phenomenology of quantum gravity.

Hoyle, Fred (1915–2001): English astronomer, mathematician, and science fiction writer. One of Hoyle's science fiction beliefs was in a steady-state universe. Einstein shared that belief for a time.

Huang, Chuan-Hsiang (*Bear*): Chinese cytologist.

Hubble, Edwin (1889–1953): American astronomer, often incorrectly credited with discovery of other galaxies and galactic Doppler shift (inaptly termed *Hubble's law*). Hubble did devise the *Hubble sequence*: a simple way of classifying galaxies by how they look.

Huber, Patrick: American particle physicist who works on neutrinos.

Huygens, Christiaan (1629–1695): Dutch mathematician, astronomer, physicist, probabilist, and horologist.

Hunziker, Alexander: Hungarian geneticist.

Huppert, Julian L.: English chemist.

Hutsemékers, Damien: Belgian cosmologist.

Hyman, Anthony A. (1962–): English cytologist.

Ibba, Michael: English biochemist.

Ingenhousz, Jan (1730–1799): Dutch physiologist, credited with discovering photosynthesis and cellular respiration. Not credited for discovering Brownian motion.

Iovino, Nicola: Italian geneticist.

Ishi, Hope: American cosmochemist.

Isinger, Marcus: Swedish atomic physicist.

Ishino, Yoshizumi: Japanese molecular biologist who accidently discovered pais in 1987, though he had no idea what it meant.

Itano, Wayne M.: American physicist.

Ivanova, Natalia N.: Russian geneticist.

Jablonka, Eva (1952–): Polish-born Israeli geneticist, interested in epigenetics and evolution.

Jacob, François (1920–2013): French biologist.

Jakšic, Ana Marija: Serbian geneticist.

Jakubczyk, Daniel: Polish physicist interested in physical chemistry.

Jaramillo, Rafael: American solid-state physicist.

Jaynes, E.T. (1922–1998): outspoken American physicist who worked on statistical mechanics.

Jeans, James (1877–1946): English physicist, astronomer, and mathematician, who was interested in quantum theory, radiation, and stellar evolution. In 1928, Jeans was first to concoct a steady-state cosmology, based upon the assumption that matter continually accreted in the cosmos. The hypothesis was disproved by the 1965 discovery of cosmic microwave background radiation.

Jensen, Henrik (1961–): Dutch economist.

Jiggins, C.D.: English zoologist.

Johannsen, Wilhelm (1857–1927): Danish botanist who coined the term *gene* in 1909.

Johnson, Jennifer A.: American astronomer.

Joule, James Prescott (1818–1889): English physicist, mathematician, and brewer who studied the nature of heat, and discovered it as a form of energy (i.e., mechanical work), which led to the conservation of energy law.

Juan, David de: Spanish biologist.

Jun, Suckjoon: Korean Canadian molecular biologist and physicist.

Junjie Li: Chinese physicist.

Justinian I (*Flavius Justinianus*, born *Petrus Sabbatius*) (482–565): Byzantine emperor (527–565), best remembered for his codification of civil laws (*Corpus Juris Civilis*).

Kalantry, Sundeep: American geneticist.

Kaluza, Theodor (1885–1954): German mathematician and physicist who developed a model unifying electromagnetism and gravitation via a 5-dimensional space.

Kant, Immanuel (1724–1804): influential German philosopher and rationalist.

> It always remains a scandal of philosophy and universal human reason that the existence of things outside us should have to be assumed merely on faith, and that if it occurs to anyone to doubt it, we should be unable to answer him with a satisfactory proof. ~ *Critique of Pure Reason* (1781)

Yet Kant rejected positivism, warning of the seduction of perception as truth.

> Up to now it has been assumed that all our cognition must conform to the objects; but let us once try whether we do not get farther with the problems of metaphysics by assuming that the objects must conform to our cognition.

Karlseder, Jan: Austrian cytologist, interested in telomeres.

Karnkowska, Anna: Polish molecular evolutionary biologist.

Kashina, Anna: Russian biochemist and fantasy novelist.

Kassis, Judith A.: American epigeneticist.

Katajisto, Pekka: Finnish cytologist.

Kauffmann, Guinevere: American astrophysicist.

Kelvin, Lord: see *Thomson, William*.

Kepler, Johannes (1571–1630): German mathematician, astronomer, and astrologer, known for his laws of planetary motion.

Kerr, Richard A.: American science writer.

Kim, Seohyun (*Chris*): Korean American chemist.

King, Nicole: American cytologist and molecular biologist.

King, Scott D.: American geophysicist.

Kirchhoff, Gustav (1824–1887): German physicist who contributed to understanding electrical circuits, black-body radiation, and spectroscopy.

Kirschvink, Joseph L. (*Joe*): American geobiologist, interested in magnetism.

Klein, Oskar (1894 –1977): Swedish physicist, credited with originating the notion that extra dimensions exist compacted (smaller than a Planck length). This insight was an adjunct to work by Theodor Kaluza; hence the *Kaluza–Klein theory*, which became the fountainhead of follow-on HD theories.

Koç, Ibrahim: Turkish geneticist, interested in the origin of life.

Kolata, Gina Bari: American science journalist.

Kolesnikov, Alexander I.: Russian nuclear physicist.

Kollmeier, Juna A.: American astronomer.

Kolodrubetz, Michael: American physicist.

Koonin, Eugene (1956–): Russian biologist who works in evolutionary and computational biology.

Koroidov, Sergey: Swedish biochemist.

Kovac, John M.: American cosmologist interested in the cosmic microwave background.

Krishnamurthy, Ramanarayanan: Indian organic chemist, interested in the origin of life.

Krog, Jens: Danish physicist.

Kubarych, Kevin: Canadian chemist.

Kubas, Daniel (1974–): German astronomer.

Kuhn, Jeff: American astronomer.

Kühne, Wilhelm (1837–1900): German physiologist, known for coining the term *enzyme*.

Lagrange, Joseph-Louis (1736–1813): Italian French mathematician and astronomer.

Lamb, Marion J. (1939–): English evolutionary biologist.

Lamb, Willis (1913–2008): American physicist who contributed to understanding the magnetic moment of the electron.

Landau, Lev (1908–1968): Azerbaijanian quantum physicist who made important contributions to many areas of theoretical physics.

Lane, Nick: English biochemist, interested in evolutionary biology and the origin of life.

Langacker, Paul G. (1946–): American particle physicist, interested in unified field theories.

Langer, Fabian: German physicist.

Lao Tzu (aka *Laozi, Lao-Tsu, Lao-Tze*) (6th or 5th century BCE): Chinese scholar and philosopher; inadvertent founder of

Daoism, which teaches reverence of Nature, the value of patience, and a path to judicious existence. A legendary figure, when and even whether Lao Tzu lived is speculative. His name is an honorary title.

Laplace, Pierre-Simon (1749–1827): French mathematician and astronomer who made important contributions to mathematical astronomy, physics, and statistics.

Lassell, William (1799–1880): English beer brewer and astronomer who discovered Triton, Neptune's largest moon, in 1846, and 2 moons of Uranus in 1851. Lassell started the tradition of naming all Uranus' moons after characters in the works of William Shakespeare and Alexander Pope. In his honor, craters on the Moon and Mars have been named after Lassell, and a ring of Neptune as well.

Lavoisier, Antoine Laurent (1743–1794): French nobleman and first-rate scientist. Lavoisier is considered the father of modern chemistry, in part by demonstrating the value of methodology in experimentation. He put together the first extensive list of elements; named *oxygen* and *hydrogen*; established that sulfur was an element, not a compound as previously supposed; introduced the concept of chemical species. Lavoisier contributed to biology by explaining oxygen's role in plant and animal respiration, and the nature of metabolism. Lavoisier helped develop the metric system.

Lederberg, Joshua (1925–2008): American molecular biologist who discovered horizontal gene transfer among bacteria.

Lederman, Leon M. (1922–): American experimental physicist.

Lee, Dung-Hai: Taiwanese American physicist, interested in strongly correlated many-particle systems.

Lee, Jeannie T.: Chinese American geneticist and molecular biologist.

Lehnmann, Ruth: American cytologist and molecular biologist, interested in germ cells and embryogenesis.

Leiber, Fritz Jr. (1910–1992): American writer of fantasy, science fiction, and horror.

Leibniz, Gottfried von (1646–1716): German mathematician and philosopher.

Lemaître, Georges (1894–1966): Belgian Roman Catholic priest and astrophysicist. Lemaître conceived the Big Bang origin of the universe and discovered Hubble's law.

Lemeshko, Mikhail: Russian physicist, studying atomic and molecular interactions and ultracold quantum gases.

Leucippus (early 5th century BCE): Greek rationalist philosopher who developed a theory of atomism.

Levene, Phoebus Aaron Theodore (1869–1940): Russian American biochemist who first identified the components of DNA and RNA and coined the term *nucleotide.*

Levy, Emmanuel D.: French structural biologist.

Lewis, Gilbert N. (1875–1946): American physical chemist, known for his discovery of the covalent bond, and his concept of electron pairs. His valence bond theory shaped current theories of chemical bonding. Lewis also contributed to thermodynamics, photochemistry, isotope separation, and an electronic theory of acid-base reaction.

Libeskind, Noam I.: German cosmologist.

Licausi, Francesco: Italian botanist, interested in plant physiology.

Lisi, Antony Garrett (1968–): American theoretical physicist.

Liske, Jochen: German astronomer.

Litvinyuk, Igor V.: Russian physicist, interested in physical chemistry.

Lloyd, Karen G.: American microbiologist.

Lobachevsky, Nikolai I. (1802–1860): Russian mathematician, early developer of non-Euclidean geometry, known primarily for his work in hyperbolic geometry.

Loeb, Avi: Israeli American astrophysicist.

Lollar, Barbara Sherwood: Canadian biochemist, interested in deep crustal fluids and life therein.

London, Fritz (1900–1954): German American physicist who made fundamental contributions in understanding chemical bonding and intermolecular forces (London dispersion forces).

Long, Hannah K.: English geneticist, interested in epigenetics.

Longfellow, Henry Wadsworth (1807–1882): American poet.

Lorentz, Hendrik (1853–1928): Dutch physicist who derived the transformation equations which Einstein's special relativity theory was based upon.

Lorenz, Edward (1917–2008): American meteorologist who coined the term *butterfly effect.*

Lorenz, Ralph D.: American astrophysicist, interested in planets and moons.

Lowell, Percival (1855–1916): American astronomer who fueled speculation about life on Mars. Percival wrote extensively of the "non-natural features" on the planet's surface. From peering through his telescope at Mars for 15 years, Percival convinced himself that the planet sustained an advanced alien civilization.

Lowry, Thomas Martin (1874–1936): English physical chemist who developed a protonic theory of acid-base reactions in 1923 (as did Johannes Brønsted).

Lucretius (99–55 BCE): Roman philosopher and poet who noted Brownian motion 2 millennium before Robert Brown got his named pinned to jiggling bits (Brownian motion).

Lynch, Vincent J.: American evolutionary biologist.

Mach, Ernst (1838–1916): Austrian physicist and philosopher.

MacKay, Harvey (1932–): American businessman and best-selling business writer, including such succinct titles as *Swim with the Sharks Without Being Eaten Alive* and *Beware the Naked Man Who Offers You His Shirt.*

Madesh, Muniswamy: Indian biochemist.

Maeder, André: Swiss theoretical astrophysicist.

Mahler, Gustav (1792–1856): Austrian composer whose exquisite symphonies often exhibit a temporal fractal quality.

Majorana, Ettore (1906–?): gifted Italian physicist who first predicted the neutron and Majorana fermion.

Maldacena, Juan (1968–): Argentinian theoretical physicist who works on the holographic principle.

Mallamace, Francesco: Italian physicist fascinated by water.

Malyutin, Sergey Vasilyevich (1859–1937): Russian painter, architect, and stage designer, who designed the first Russian Matryoshka doll.

Mandelbrot, Benoît B. (1924–2010): Polish-born French American mathematician, known for his work in fractal geometry.

Mann, Adam: American science writer, interested in cosmology.

Manning, Gerard: American geneticist who studies pseudoenzymes.

Martin, Steve (1945–): American comedian and banjo player, known for being "a wild and crazy guy."

Martinez, Todd J. (1968–): American chemist.

Martins, Zita: English geologist, interested in abiogenesis.

Mattick, John S.: Australian molecular biologist.

Mathur, Harsh: Indian theoretical physicist, studying condensed matter theory, particularly superconductivity, with interest in theoretical particle astrophysics and cosmology.

Maupertuis, Pierre Louis (1698–1759): French mathematician and philosopher who worked in classical mechanics, heredity, and natural ecology. Maupertuis made the first known suggestion that all life had a common ancestor.

Maxwell, James Clerk (1831–1879): Scottish physicist, most famous for formulating classical electromagnetic theory in 1865. Maxwell is widely considered the 19th century physicist most influential on 20th century physics. In 1861, Maxwell invented the first durable color photograph.

Mayer, Julius Robert von (1814–1878): German physician, chemist and physicist who was one of the founders of thermodynamics.

McClintock, Barbara (1902–1992): American cytogeneticist and botanist.

McMahon, Sean: British cosmologist.

McNamara, Paul W. (1973–): Scottish astrophysicist.

Meissner, Fritz Walther (1882–1974): German technical physicist who contributed to superconductivity. Working with Robert Ochsenfeld, Meissner discovered the *Meissner effect* in 1933.

Mendel, Gregor (1822–1884): Austrian monk and botanist, interested in heredity.

Mendeleyev, Dmitry (aka *Dmitri Mendeleev*) (1834–1907): Russian chemist who created the modern table of periodic elements.

Menninger, Charles (1862–1952): American physician.

Meurer, Gerhardt R.: American astrophysicist.

Meyer, S.N.: Swiss molecular biologist.

Michell, John (1724–1793): English natural philosopher and geologist, best known for his work on gravity.

Michelson, Albert (1852–1931): American physicist who worked on measuring the speed of light; best known for the failed *Michelson-Morley experiment*, which sought the presumed "aether wind," which does not exist.

Miescher, Friedrich (1844–1895): Swiss physician and biologist who first identified nucleic acid.

Milanković, Milutin (1879–1958): Serbian geophysicist who suggested long-term climatic changes based upon Earth's cosmological movements, known as *Milankovitch cycles*. Milanković was also a mathematician, astronomer, climatologist, civil engineer, and popularizer of science.

Mill, John Stuart (1806–1873): English philosopher, political economist, and civil servant; proponent of individual liberty, in opposition to unlimited state control; adherent to *utilitarianism*, an ethical precept of right action by maximizing overall "happiness"; contributed to the *scientific method* via the premise of falsification.

Miller, Arthur (1915–2005): American playwright.

Miller, Stanley L. (1930–2007): American chemist who made landmark experiments in prebiotic chemistry aimed at understanding the chemical origin of life.

Millman, Dan (1946–): American athletic coach turned self-help book author and lecturer.

Minkowski, Hermann (1864–1909): Lithuanian mathematician who created and developed the geometry of numbers. By 1907, Minkowski realized that Einstein's 1905 special theory of relativity could best be understood in a 4-dimensional spacetime, where space and time are integrated. Einstein was a former student of Minkowski, of whom Minkowski thought at the time would never amount to anything.

Misner, Charles W. (1932–): American physicist, interested in general relativity and cosmology.

Misra, Baidyanath (1937–): Indian physicist and mathematician.

Mitchell, Joni (1943–): Canadian singer, songwriter, and musician.

Molaro, Antoine: French geneticist, interested in embryogenesis and heredity.

Montell, Denise J.: American biochemist and cytologist.

Moore, Stanford (1913–1982): American biochemist who worked on ribonuclease and protein sequencing.

Moore, Thomas: American physicist.

Morley, Edward (1838–1923): American scientist, whose claim to fame is the failed *Michelson-Morley experiment*, which sought the presumed "aether wind," which does not exist.

Morr, Dirk K.: German American physicist.

Moskalenko, Andrey S.: Russian quantum physicist.

Motlagh, Hesam N.: American molecular biophysicist and financial economist.

Mpemba, Erasto (1950–): Tanzanian scientist who, as a hasty childhood ice cream maker, serendipitously rediscovered that hot water freezes faster than cold water. Mpemba had the good fortune to have the oddity named after him (the *Mpemba effect*). Others throughout history, including Aristotle, Francis Bacon, and René Descartes, had noted this phenomenon.

Murayama, Hitoshi: Japanese physicist who works on supersymmetry.

Nagaoka, Hantaro (1865–1950): Japanese physicist who contributed to atomic theory and radio wave communication.

Nair, Gautham: American geneticist.

Natland, James H.: American marine geologist and geophysics.

Nemenman, Ilya: Russian American theoretical biophysicist, interested in neuroscience, biological communication, learning and evolutionary adaptation.

Nernst, Walther (1864–1941): German physicist and chemist who is best known for developing the 3rd law of thermodynamics.

Newcomb, Simon (1835–1909): self-taught Canadian astronomer, mathematician, economist, and author (science books, and 1 science fiction novel). Newcomb spoke French, German, Italian, and Swedish.

Newton, Isaac (1642–1727): English physicist, astronomer, alchemist, mathematician, natural philosopher, and theologian; widely considered to be one of the most influential scientists. Classical mechanics are typically termed *Newtonian physics*.

Nicolis, Alberto: Italian physicist, interested in theoretical high-energy physics.

Nietzsche, Friedrich (1844–1900): German philosopher that embraced existentialism and nihilism. Existentialism embraces individual experience as the proper path to understanding. Nihilism posits that life is subjectively valuated.

Noffke, Nora: American geomicrobiologist.

Nora, Elphege: American geneticist.

Nordenfelt, Pontus: Swedish engineer.

Norman, Eric B.: American nuclear physicist.

Novello, Mario: Brazilian astrophysicist.

Ochsenfeld, Robert (1901–1993): German physicist who worked with Fritz Meissner on superconductivity, co-discoverer of the *Meissner effect*.

O'Donoghue, John: English astronomer.

Ohm, Georg Simon (1789–1854): German physicist and mathematician, interested in electrochemical cells.

Ohno, Susumu (1928–2000): Japanese American geneticist and evolutionary biologist who helped popularize the wrong-headed notion that most human DNA was useless ("junk").

O'Neill, Craig: Australian geophysicist.

Onnes, Heike Kamerlingh (1853–1926): Dutch physicist who contributed to refrigeration; first to liquefy helium; discovered superconductivity in 1911.

Oort, Jan (1900–1992): Dutch astronomer who was a pioneer in radio astronomy. The Oort cloud of comets in the deep solar system bears his name.

Oparin, Alexander (1894–1980): Russian biochemist, best known for his book: *The Origin of Life* (1936).

Orgel, Leslie E. (1927–2007): English chemist, interested in abiogenesis.

Osborn, Raymond: American physicist, interested in electron systems.

Padmanabhan, Thanu (1957–): Indian theoretical physicist; adherent to the holographic principle.

Pal, Sourav: Indian chemist.

Pandey, Akhilesh: Indian molecular biologist, pathologist, oncologist, and geneticist.

Papenbrock, Thomas: American nuclear physicist.

Parkinson, David: Australian cosmologist.

Parkinson, Gareth S. (1981–): English physicist, interested in the atomic-scale processes underlying metal-oxide surface chemistry.

Pascal, Blaise (1623 –1662): French mathematician, physicist, inventor, and Christian philosopher; a child prodigy.

Pascal, Robert: French molecular biologist.

Pasteur, Louis (1822–1895): French chemist and microbiologist, remembered for his discoveries regarding the causes and preventions of infectious diseases. Pasteur's experiments supported the *germ theory of disease*: that pathogenic microorganisms cause many diseases.

Patterson, Clair Cameron (1922–1995): American geochemist who developed a geological dating method based upon radioactive decay of uranium; the first to accurately age Earth at 4.55 billion years, in 1948.

Pauli, Wolfgang (1900–1958): sharp-tongued and sharp-witted Austrian theoretical physicist; a quantum physics pioneer.

> The best that most of us can hope to achieve in physics is simply to misunderstand at a deeper level. ~ Wolfgang Pauli

Pauling, Linus (1901–1994): American chemist, peace activist, and admirer of vitamin C.

Pearson, Karl (1857–1936): English mathematician, credited with establishing mathematical statistics. Pearson contributed to biometrics and meteorology. Pearson was a proponent of social Darwinism and favored eugenics.

Penzias, Arno A. (1933–): American astrophysicist who co-discovered cosmic background radiation with Robert Wilson in 1964.

Perrin, Jean (1870–1942): French physicist who worked on atomic theory and the nature of matter. Perrin explained cathode rays as negatively charged corpuscles, and solar energy as thermonuclear hydrogen reactions.

Peruzzo, Alberto: Italian physicist, interested in quantum information.

Petit, Jean-Pierre (1937–): French astrophysicist, interested in fluid mechanics (particularly magnetohydrodynamics), plasma physics, kinetic theory of gases, and topology.

Pettersson, Lars G.M.: Swedish physiochemist.

Planck, Max (1858–1947): German physicist who founded quantum field theory, then rejected it out of philosophic revulsion, owing to the indeterminate nature of wave/particle duality (Heisenberg's uncertainty principle). Planck philosophically preferred determinism.

Plutarch (46–120): Greek essayist and biographer.

Pohl, Randolf: German physicist.

Poincaré, Henri (1854–1912): brilliant French mathematician, theoretical physicist, engineer, and philosopher of science.

Polanyi, Michael (1891–1976): Hungarian–English polymath who made contributions in philosophy, chemistry, and economics. Polanyi argued that positivism gives an incomplete account of the knowledge of reality.

Polchinski, Joseph (1942–): American string theorist working on D-brane theory and the idea of wormholes.

Polimann, Frank: German physicist, interested in quasiparticles and solid-state physics.

Polkovnikov, Anatoli: Russian American physicist.

Popper, Karl (1902–1994): Austrian British philosopher, interested in the philosophy of science. Popper rejected the classical inductivist view on the scientific method (attributed to Francis Bacon) in favor of inductive (empirical) falsifiability. See *falsifiability*, *inductivism*.

Porter, Cole (1891–1964): American songwriter, particularly fond of musical theater.

Powner, Matthew (1982–): English organic chemist, interested in abiogenesis.

Pratchett, Terry (1948–2015): English fantasy novelist.

Priestley, Joseph (1733–1804): English political theorist, Unitarian minister, and theologian who is often credited with discovering oxygen. Best known for his advocacy of *utilitarianism*: an ethical precept that right action maximizes holistic happiness.

Prost, Jacques (1946–): French physicist.

Prugh, Laura R.: American biologist.

Pruitt, Jonathan N.: American ecologist, interested in species variation.

Ptashne, Mark (1940–): American molecular biologist and violinist; first to demonstrate specific binding between protein and DNA.

Ptolemy, Claudius (90–168): Egyptian mathematician, astronomer, geographer, and astrologer. His lasting fame owed to 3 treatises: 1 on astronomy, 1 on geography, and 1 on astrology, which was based upon Aristotelian natural philosophy.

Pulleyblank, David E.: Canadian geneticist.

Pusey, Matthew F.: English particle physicist, known for the Pusey-Barrett-Rudolph theorem.

Pythagoras (570–495 BCE): Ionian Greek mathematician and philosopher, best known for the Pythagorean theorem, which was previously known by the Babylonians and Indians.

Qing-yu Cai: Chinese physicist, interested in quantum cosmology theory.

Qingdi Wang: Chinese physicist.

Quayle, Dan (1947–): American politician; 44th US Vice President under George H.W. Bush; former congressman and senator from Indiana. Quayle was widely ridiculed for his frequent gaffes.

Qimron, Udi: Israeli microbiologist.

Raj, Arjun: Indian American molecular biologist, cytologist, and bioengineer.

Ralph, Timothy C.: Australian quantum physicist.

Ramond, Pierre (1943–): French physicist working on superstring theory.

Randall, Lisa (1962–): American theoretical physicist who works on string theory; best known for the Randall–Sundrum braneworld models (developed with Raman Sundrum).

Ranjan, Sukrit: Indian astrophysicist and astronomer, interested in the evolution of rocky planets and the origin of life on Earth.

Rankine, William J.M. (1820–1872): Scottish mechanical engineer who made contributions to civil engineering, physics, and mathematics.

Raspail, François-Vincent (1794–1878): French chemist, naturalist, physiologist, and socialist politician; one of the founders of cytology.

Rees, Martin J.: English astrophysicist.

Renner, Renato: Swiss theoretical physicist.

Reuveni, Shlomi: Israeli systems biologist.

Rey, Felix: French virologist.

Rich, Alexander (1924–2015): American biologist and biophysicist who first suggested the RNA-world hypothesis in 1962.

Riess, Adam G. (1969–): American astrophysicist.

Robertson, Brant: American astronomer.

Robinson, Gene E.: American systems biologist and genomist.

Rodejohann, Werner: German physicist.

Rømer, Ole (1644–1710): Danish astronomer who made the first quantitative measurements of the speed of light.

Romero, Jacquiline: Australian quantum physicist, interested in photons.

Röntgen, Wilhelm (1845–1923): German physicist who first accidentally generated X-rays.

Rosen, Nathan (1909–1995): American Israeli physicist.

Roth, V. Louise: American evolutionary biologist.

Rothbart, Scott B.: American epigeneticist.

Rothmann, Christoph (CA 1555–1605): German mathematician and astronomer who was a convinced follower of Copernican heliocentricity. This early stargazer fell into oblivion compared to his contemporaries, notably correspondent Tycho Brahe, who was skeptical of Earth moving about.

Rovelli, Carlo: Italian theoretical physicist.

Rudolph, Terry: English particle physicist, known for the Pusey-Barrett-Rudolph theorem.

Rumsfeld, Donald (1932–): American politician, bureaucrat, and businessman; US defense secretary (2001–2006). Once popular for his candor, admiration wore thin as the wars he helped conduct in Iraq and Afghanistan slogged on to no positive outcome.

Rupprecht, Jean-Francois: French biophysicist.

Rushworth, Stuart: English cytologist, interested in immunology and hematology.

Russell, John Scott (1808–1882): Scottish engineer who discovered solitons. Russell was a naval architect and shipbuilder.

Russell, Michael J.: American geochemist.

Russell, Peter: English physicist.

Rutherford, Daniel (1749–1819): Scottish physician, chemist, and botanist; known for isolating atmospheric nitrogen in 1772 without really appreciating what he was doing.

Rutherford, Ernest (1871–1937): English physicist and chemist, known as the father of nuclear physics.

Sabatini, David: American biologist.

Sahai, Erik: English cell pathologist.

Sahl, Ibn (940–1000): Persian physicist and mathematician.

Sakai, Hideaki: Japanese physicist.

Salam, Abdus (1926–1996): Pakistani theoretical physicist who worked on the unification of electromagnetic and weak forces (electroweak unification).

Sana, Hughes: Dutch astronomer.

Sanbonmatsu, Karissa: American epigeneticist.

Sandage, Allan (1926–2010): American astronomer.

Sandford, Scott: American astrophysicist.

Sarkar, Subir: Indian theoretical physicist.

Sassone-Corsi, Paolo: Italian biochemist.

Scheck, Marcus: Scottish nuclear physicist.

Scheele, Carl Wilhelm (1742–1786): Swedish pharmaceutical chemist; called "hard-luck Scheele" for making numerous unaccredited chemical discoveries, including oxygen, hydrogen, chlorine, barium, tungsten, and molybdenum. Scheele was a tad slow to publish.

Schleich, Wolfgang P. (1957–): German theoretical physicist, interested in the foundations of quantum mechanics.

Schon, Eric A.: American cytologist, biochemist, and neurologist, interested in mitochondria.

Schrenk, Matthew O.: American geomicrobiologist, interested in subsurface ecosystems.

Schrödinger, Erwin (1887–1961): Austrian physicist and theoretical biologist who was one of the fathers of quantum field theory, and later disowned it. Best known for *Schrödinger's equation,* regarding the dynamics of quantum systems.

Schuetz, Robert: Swiss microbiologist.

Schuster, Arthur (1851–1934): German-born British physicist who worked on electrochemistry, optics, spectroscopy, and X-radiography.

Schuur, Edward A.G.: American ecologist.

Schwarz, Dominik J.: astrophysicist.

Schwarz, Melvin (1932–2006): American physicist.

Schwarzschild, Karl (1873–1916): German physicist, best known for deriving the first exact solution to the Einstein field equations of general relativity. Einstein was only able to produce an approximate solution.

Sciama, Dennis W. (1926–1999): English physicist, interested in cosmology.

Seiberg, Nathan (1956–): Israeli theoretical physicist who works on string theory.

Sethupathy, Praveen: Indian geneticist.

Severinov, Konstantin: Russian geneticist, interested in the mechanics of gene expression in bacteria, and development of new antibiotics.

Shabala, Stanislav: Australian astrophysicist, best known for his work on black holes.

Shalm, Lynden K.: American physicist.

Shannon, Claude E. (1916–2001): American mathematician, electrical engineer, and cryptographer, interested in implementing symbolic logic via machinery. Shannon founded circuit design theory (1937) and information theory (1948).

Shapiro, James A.: American molecular biologist and bacterial genetics maven.

Shapley, Harlow (1885–1972): American astronomer who discovered the nearby galactic supercluster, now named the *Shapley Supercluster*.

Shaw, George Bernard (1856–1950): Irish playwright, angered by exploitation of the working class; an ardent socialist.

Shockley, William Jr. (1910–1989): American physicist who invented semiconductor transistors.

Silva, Isabela A.: Brazilian physicist.

Simcoe, Robert A.: American astrophysicist.

Simons, Kai (1938–): Finnish biochemist.

Simplicius of Cilicia (490–560): Greek philosopher; one of the last Neoplatonists who wrote extensively on Aristotle; a pagan persecuted by Justinian in the early 6th century.

Skenderis, Kostas: Dutch theoretical physicist and mathematician.

Skyttner, Lars: Swedish systems theorist.

Slipher, Vesto (1875–1969): American astronomer, discoverer of galaxies beyond the Milky Way.

Sluse, Dominique: Belgian astrophysicist.

Smith, D. Eric: American chemical physicist.

Smith, John Maynard (1920–2004): English theoretical evolutionary biologist and geneticist. Smith applied game theory to evolution and studied the evolution of sex and the nature of communication.

Smolin, Lee: Canadian theoretical physicist.

Snellius, Willebrord (known in the English-speaking world as *Snell*) (1580–1626): Dutch astronomer and mathematician.

Soddy, Frederick (1877–1956): English radiochemist and monetary economist who contributed to understanding radioactivity. In understanding the inherent folly of finance and the limits to growth, Soddy anticipated ecological economics.

Söll, Dieter: German biochemist.

Sorek, Rotem: Israeli molecular geneticist, interested in bacteriophage strategies and epigenetic regulation in microbes.

Spatz, Joachim: German cytologist.

Stairs, Shaun: English organic chemist, interested in abiogenesis.

Stark, Johannes (1874–1957): German physicist and enthusiastic Nazi who agitated against the "Jewish physics" of Albert Einstein and Werner Heisenberg (who was not Jewish).

Starkman, Glenn D.: astrophysicist.

Stefan, Jožef (1835–1893): Austrian physicist and mathematician, best known for stating in 1879 the physical power law that the total radiation from a black body is proportional to the 4th power of its temperature.

Stefani, Frank: German physicist.

Steinberg, Peter: American physicist.

Steinhardt, Paul J.: American theoretical physicist. Steinhardt helped develop the notion of cosmic inflation, but later rejected it, instead embracing cyclic cosmology.

Stenger, Victor J. (1935–2014): American particle physicist, atomist philosopher, and godless heathen who advocated science and reason.

Stenmark, Harald: Norwegian cytologist.

Stergachis, Andrew B.: American geneticist.

Stevenson, Adlai (II) (1900–1965): thoughtful and eloquent American liberal politician (Democrat).

Stewart, Balfour (1828–1887): Scottish physicist, interested in solar dynamics.

Stradler, Lewis J. (1896–1954): American geneticist, interested in the mutagenic effects of radiation.

Su-Yang Xu: Chinese physicist.

Sudarshan, E.C. George (1931–): Indian theoretical physicist, interested in various quantum phenomena.

Sundrum, Raman: American theoretical particle physicist, known for the Randall–Sundrum braneworld models (developed with Lisa Randall).

Suntzeff, Nicholas B. (1952–): American astronomer.

Susskind, Leonard (1940–): American theoretical physicist; a pioneer in string theory who also works in quantum field theory, quantum statistical mechanics, and quantum cosmology.

Sutherland, John D.: English biochemist.

Sweatt, J. David: American biochemist.

Swedenborg, Emanuel (1688–1772): Swedish scientist, theologian, and Christian mystic who developed the *nebular hypothesis*: that the solar system formed by gyral matter accretion.

't Hooft, Gerard (1946–): Dutch theoretical physicist, interested in quantum gravity, black holes, gauge theory, holistic dimensionality, and the holographic principle.

Tait, Peter (1831–1901): Scottish mathematical physicist, best known for his work on knot theory.

Talaro, Kathleen Park: American molecular biologist.

Tamm, Igor (1895–1971): Russian physicist who conceptualized phonons in 1932. Tamm helped discover Cherenkov radiation.

Tanaka, Hajime: Japanese physicist.

Tanurdžić, Miloš: geneticist and molecular biologist, interested in plant development and epigenetics.

Taroni, Andrea: English physicist.

Tegmark, Max (1967–): Swedish American cosmologist.

Temple, William (1881–1944): English Anglican clergyman who favored socialism.

Theobald, Douglas: American biochemist who believes in universal common ancestry.

Thomas, Lewis (1913–1993): American physician and writer.

Thompson, Benjamin (aka *Count Rumford*) (1753–1814): American-born English physicist, inventor, and military man who helped shape the modern understanding of thermodynamics.

Thompson, M.J.: English zoologist.

Thomson, Joseph John (J.J.) (1856–1940): English physicist, credited with discovering electrons and isotopes.

Thomson, William (1837–1907) (better known as *Lord Kelvin*): English mathematical physicist and engineer, best known for suggesting that there is an absolute lower limit to temperature; hence the Kelvin temperature scale.

Thurber, Andrew R.: American oceanographer.

Ting Wang: Chinese geneticist.

Tolkien, J.R.R. (1892–1973): English writer, poet, philologist, and university professor; best known for his fantasy novels *The Hobbit* and *Lord of the Rings*.

Tolman, Richard C. (1881–1948): American mathematical physicist and physical chemist, interested in statistical mechanics.

Tomkins, Gordon M. (1926–1975): American biochemist.

Torbert, Roy B.: American astrophysicist.

Toschi, Alessandro: Italian physicist.

Trakhtenbrot, Benny: Israeli astronomer, interested in the evolution of black holes.

Trefil, James (1938–): American physicist and science writer. Trefil argued that human intelligence is special in his book *Are We Unique?* (1997).

Tryon, Edward P. (1940–): American physicist.

Tsuchiya, Tokuji: Japanese plant cytologist.

Tully, R. Brent (1943–): Canadian astronomer.

Turing, Alan (1912–1954): influential English logician, mathematician, computer scientist, cryptanalyst, and theoretical biologist; influential in the conceptualization of computer science; persecuted by the British government for homosexuality to the point of suicide (torture which Queen Elizabeth called "appalling" in 2009).

Turner, Michael S. (1949–): American theoretical cosmologist and physicist. Turner coined the term *dark energy*.

Tūsī, Nasīr al-Dīn (1201–1274): Persian polymath who proposed a hierarchical theory of evolution.

Twain, Mark (pen name of *Samuel Langhorne Clemens*) (1835–1910): talented American author prized for his satire and wit. Best known for his novels *The Adventures of Tom Sawyer* (1876), and its sequel, *Adventures of Huckleberry Finn* (1885).

Unruh, William G. (Bill) (1945–): Canadian physicist, interested in gravity. Among other eccentricities, Unruh thinks that quantum nonlocality is actually local, as quantum bits need not subscribe to Bell's theorem.

Valencia, Alfonso: Spanish biologist.

Valéry, Paul (1871–1945): French poet, essayist, and philosopher. While best known as a poet, he published fewer than 100 poems and none of them drew much attention.

Van Allen, James A. (1914–2006): American astrophysicist who discovered the radiation belt surrounding Earth in 1958.

van der Marel, Dirk: Dutch physicist.

van der Waals, Johannes Diderik (1837–1923): Dutch theoretical physicist and thermodynamicist, famous for his work modeling gases and liquids.

van der Wal, Casper H. (1971–): Dutch quantum physicist.

van Dokkum, Pieter: Dutch astronomer.

van Helmont, Jan Baptist (1579–1644): Belgian chemist, physiologist, and physician, best remembered for his advocacy of spontaneous generation, which turned out to be rot. van Helmont is considered the father of pneumatic chemistry and credited with introducing the term *gas* (from the Greek word for *chaos*) into scientific nomenclature.

Van Raamsdonk, Mark: Canadian theoretical physicist, working on a unified field theory involving wormhole entanglement.

Vedral, Vlatko: Serbian-born British physicist, working on theories of entanglement and quantum information theory.

Veneziano, Gabriele (1942–): Italian string theorist.

Venuti, Lorenzo Campos: Italian theoretical physicist.

Verlinde, Erik P.: Dutch theoretical physicist.

Verresen, Ruben: German quantum physicist.

Vilenkin, Alexander: Ukrainian American physicist, interested in cosmology.

Villarreal, Luis P.: American virologist, biochemist, and molecular biologist, interested in the role of viruses in evolution.

Vogel, Steven (1940–2015): American zoologist and biomechanist.

von Helmholtz, Hermann (1821–1894): German physician, physicist, and philosopher; known for his contributions in understanding vision and thermodynamics. von Helmholtz's philosophy of science considered the relation between laws of Nature and perception.

von Neumann, John (1903–1957): Hungarian American mathematician and polymath.

Vultaggio, Janelle: American microbiologist, interested in microbiomes and microbial relationships.

Wächtershäuser, Günter (1938–): German organic chemist and patent lawyer who developed the *iron-sulfur origin of life theory*: that life originated in seafloor hydrothermal vents, nestled in pyrite.

Waddington, Conrad H. (1905–1975): English geneticist, developmental biologist, paleontologist, embryologist, and philosopher. Waddington laid the foundation for systems biology.

Wallberg, Andreas: Swedish geneticist.

Walker, Sarah Imari: American astrobiologist.

Ward, Peter: American marine biologist and paleontologist.

Watanabe, Haruki (1986–): Japanese physicist who works on spontaneous symmetry breaking and Nambu-Goldstone bosons.

Watson, James D. (1928–): American molecular biologist, known as the 1953 co-discoverer of the structure of DNA, with Francis Crick.

Weaver, Valerie M.: American biologist and biochemist, interested in oncology (tumors).

Weber, Jesse N.: American biologist.

Weber, Michael: French geneticist.

Wegner, Gary A. (1944–): American astronomer; one of the discoverers of *The Great Attractor*: a massive gravity anomaly in deep space.

Weiguo Yin: Chinese quantum physicist.

Weimer, Hendrik: German quantum physicist.

Weinberg, Marc S. (Marco): South African geneticist.

Weinberg, Steven (1933–): American theoretical physicist who contributed to electroweak theory.

Weitzman, Jonathan B.: French geneticist interested in epigenetics.

Wells, H.G. (1866–1946): prolific English author, best known for his science fiction works, including *The Time Machine*, *The Invisible Man*, and *The War of the Worlds*. Wells was a socialist, and a proponent of world government. Wells advocated improving the human breeding stock (eugenics).

Wesson, Paul S. (1949–2015): English theoretical physicist and astrophysicist.

Weyl, Hermann (1885–1955): German mathematician and theoretical physicist; one of the first to conceive of combining electromagnetism with general relativity.

Wheeler, John A. (1911–2008): American theoretical physicist who worked on the principles behind nuclear fission. Wheeler collaborated with Albert Einstein on a relativity-based unified field theory which came to naught. Wheeler later bought into the idea that information is fundamental to physics. Wheeler coined the terms *black hole, wormhole,* and *quantum foam.*

Whewell, William (1794–1866): English polymath, scientist, science historian, economist, philosopher, theologian, and Anglican priest. Whewell's legacy was wordsmithing: he coined the terms *scientist, physicist, linguistics, catastrophism,* and *uniformism,* among others. To Michael Faraday, Whewell suggested: *ion, dielectric, anode,* and *cathode.* Whewell coined the term *consilience* to characterize the unification of knowledge between different branches of learning.

Whitesides, George M. (1939–): American chemist.

Wienert, Beeke: Australian geneticist and molecular biologist.

Wiens, John J.: American ecologist.

Wigner, Eugene P. (1902–1995): Hungarian American theoretical physicist and mathematician.

Wilczek, Frank (1951–): American theoretical physicist.

Wilde, Oscar (1854–1900): Irish writer and poet.

Wiles, Andrew (1953–): English mathematician, best known for proving Fermat's last theorem in 1994.

Wilkins, Adam S.: English biologist.

Wilkinson, Miles F.: American obstetrician and gynecologist.

William of Ockham (~1287–1347): English Franciscan friar, theologian, and scholastic philosopher; one of the major figures in medieval thought.

Willbanks, Amber: American geneticist and cytologist.

Williams, Loren D.: American biochemist.

Wilson, Robert W. (1936–): American astrophysicist who co-discovered cosmic background radiation with Arno Penzias in 1964.

Wiltshire, David L.: New Zealander astrophysicist.

Winkel, Brenda S.J.: American biochemist and geneticist.

Witten, Ed (1951–): American theoretical physicist who developed M-theory.

Woese, Carl (1928–2012): American microbiologist and physicist who declared in 1977 archaea a new domain of life (distinct from bacteria).

Wöhler, Friedrich (1800–1882): German chemist who initiated modern organic chemistry with his synthesis of urea. Wöhler was also the first to isolate several chemical elements, including aluminum, beryllium, silicon, titanium, and yttrium.

Woit, Peter: American mathematician and theoretical physicist.

Wolf, Christian: Australian astronomer.

Wolff, Suzanne: American cytologist.

Wootters, William K.: American theoretical physicist, interested in quantum entanglement; one of the founders of quantum information theory.

Wright, Addison V.: American molecular biologist.

Wright, Jason T.: American astronomer.

Wright, Peter E.: American molecular biologist.

Wright, Steven (1955–): American comedian.

Xianrui Cheng: Chinese cytologist.

Young, Ross D.: Australian physicist.

Zahn, Laura M.: American geneticist.

Zamore, Phillip: American biochemical geneticist.

Zanardi, Paolo: Italian theoretical physicist, interested in quantum entanglement and quantum information theory.

Zeeman, Pieter (1865–1943): Dutch physicist who discovered the *Zeeman effect* in 1896.

Zenklusen, Daniel: Canadian geneticist.

Zeno of Elea (495–430 BCE): Greek philosopher and mathematician whom Aristotle credited with inventing dialectic. Zeno is best known for his paradoxes, which contributed to the development of logical and mathematical rigor and were insoluble until precise notions of continuity and infinity developed. None of Zeno's writings are extant intact. The main sources on Zeno are Aristotle and Simplicius of Cilicia.

✎ Index ✎